CHEMICAL STRUCTURES OF THE AMINO ACIDS

Neutral, Nonpolar

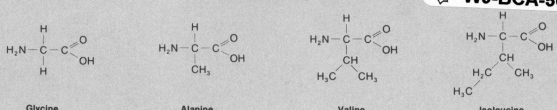

Glycine   Alanine   Valine   Isoleucine

Leucine   Phenylalanine   Proline   Methionine

Neutral, Polar

Serine   Threonine   Asparagine   Glutamine

Tyrosine   Tryptophan   Cysteine

Acidic

Aspartic acid   Glutamic acid

Basic

Lysine   Arginine   Histidine

Microbial Genetics

The Jones and Bartlett Series in Biology

Microbial Genetics

Second Edition

Stanley R. Maloy
Department of Microbiology
University of Illinois, Urbana

John E. Cronan, Jr.
Departments of Microbiology and Biochemistry
University of Illinois, Urbana

David Freifelder
University of California, San Diego

Jones and Bartlett Publishers
Boston   London

We dedicate this book to our late friend and colleague William D. Nunn

Editorial, Sales, and Customer Service Offices
Jones and Bartlett Publishers
One Exeter Plaza
Boston, MA 02116
1-800-832-0034
617-859-3900

Jones and Bartlett Publishers International
P.O. Box 1498
London W6 7RS
England

Library of Congress Cataloging-in-Publication Data

Maloy, Stanley R.
    Microbial genetics / Stanley R. Maloy, John E. Cronan, David
Freifelder. -- 2nd ed.
        p.   cm.
    Includes bibliographical references and index.
    ISBN 0-86720-248-3
    1. Microbial genetics.   I. Cronan, John E.   II. Freifelder,
David.   III. Title.
QH434.F74   1994
576'.139--dc20   94-7472
                                                                         CIP

Acquisitions Editors: David E. Phanco, Arthur C. Bartlett
Production Editor: Mary Cervantes
Manufacturing Buyer: Dana Cerrito
Editorial Production Service: Ruttle, Shaw & Wetherill, Inc.
Typesetting: Weimer Graphics, Inc.
Cover Design: Marshall Henrichs
Printing and Binding: Courier Corporation
Cover Printing: John P. Pow Company

Cover: The figure on the cover represents a physical map of the *Escherichia coli* chromosome. The outer circle shows regions of the chromosome where the DNA sequence has been determined (compiled in 1992). The eight inner circles show the positions of recognition sites of eight restriction endonucleases: *Bam*HI, *Hin*dIII, *Eco*RI, *Eco*RV, *Bgl*I, *Kpn*I, *Pst*I, and *Pvu*II. This figure was kindly provided by Kenneth E. Rudd from the National Center for Biotechnology Information, National Library of Medicine, Bethesda, MD. The physical map is described in detail in: K. Rudd. 1993. Maps, genes, sequences, and computers: an *Escherichia coli* case study. ASM News 59, 335.

Printed in the United States of America
98 97 96 95 94   10 9 8 7 6 5 4 3 2 1

Brief Contents

Contents

PART 2
MOLECULAR ASPECTS OF GENE EXPRESSION

Chapter 6
Gene Expression   101

Chapter 7
Regulation of Gene Expression   121

PART 3
MAINTENANCE OF GENETIC INFORMATION

Chapter 8
DNA Replication   145

Chapter 9
DNA Damage and Repair   161

Chapter 16
Lytic Growth of Phage λ   335

Chapter 17
Lysogeny   351

Preface

Although many general genetics and molecular biology textbooks are available, few recent textbooks focus on microbial genetics. Furthermore, although many of the general genetics and molecular biology textbooks have sections on microbial genetics, they do not cover a wide variety of specialized aspects of microbial genetics. The first edition of David Freifelder's *Microbial Genetics* sought to fill this niche.

Since the first edition was published, there have been numerous advances in microbial genetics. The development of new techniques such as polymerase chain reaction and electroporation made it possible to do molecular genetics in a variety of interesting and important bacteria that were previously genetically intractable. Genetic approaches have been essential for dissecting important microbiological problems, such as the mechanisms of virulence of bacterial pathogens and the emergence of antibiotic-resistant pathogens. Genetic analysis of Archaea (previously called Archaebacteria) has contributed to our understanding of evolution of the major groups of living organisms and the diversity of life. Each of these advances required a solid knowledge of classic microbial genetics and modern molecular biology. Because David Freifelder is now deceased, we undertook the task of rewriting and updating the second edition of *Microbial Genetics*. (It might be argued that the book would be more appropriately titled *Bacterial and Phage Genetics* because it does not deal with eukaryotic microbes. We tried to include Archaea when relevant, however, and we hope these sections will be expanded in the future. Thus, because the content is not strictly limited to bacteria, we retained the original title.) Some sections are extensively rewritten, and other sections are only slightly changed, but we retained the basic format, organization, and style of the first edition.

We were fortunate to have scientific mentors who are not only outstanding bacterial geneticists, but also have a strong passion for genetics. Dan Wulff (SUNY at Albany), John Roth (University of Utah), and Tom Silhavy (Princeton University) left a profound influence on our style of thinking about genetics.

We greatly appreciate the suggestions of several reviewers, especially Jeffrey Gardner (University of Illinois, Urbana), Michelle Igo (University of California, Davis), and Ken Noll (University of Connecticut, Storrs). In addition, we appreciate the help of a number of graduate students and undergraduates at the University of Illinois who read the manuscript and checked the problems: Scott Allen, Karen Colletta, Matt Lawes, Min-Ken Liao, Aileen Rubio, and Paula Ostrovsky de Spicer. Art Bartlett (Jones and Bartlett Publishers) gave us the encouragement and gentle prodding needed to keep us from getting terminally sidetracked. Finally, we would like to thank our families for their support and patience with our constant preoccupation with science.

Stanley Maloy
John Cronan, Jr.

Preface to the First Edition

Experiments in microbial genetics carried out in the 1940s and 1950s gave birth to what ultimately became known as molecular biology. The elucidation of the structure of DNA by James Watson and Francis Crick, neither of whom were geneticists, had enormous impact on all of genetics and biology. Many physicists, fascinated by the new discoveries and attracted by the quantitative aspects of microbial genetics, joined classic microbiologists, geneticists, and biochemists, and what has been called the Golden Age of Biology began. During this period, genetic analysis took on new importance in that it provided the insights for developing detailed theories of gene action. Genetic results obtained with bacteria and bacteriophages led to hypotheses about gene organization, chromosome structure, regulation of gene action, genetic recombination, and many other phenomena. Although genetic results could not rigorously prove a molecular mechanism, they placed limits on hypotheses and were enormously suggestive about mechanisms. In fact, hypotheses based solely on genetic results so often proved to be correct that it became commonplace among molecular biologists to assume the validity of a mechanism before biochemical proof was obtained; certainly mechanisms that were inconsistent with genetic results were not thought worthy of consideration. A great deal of the early work in microbial genetics was compiled in an excellent and influential text—*The Genetics of Bacteria and Viruses* by William Hayes—the most recent edition of which was published in 1968. Since that time, few texts on microbial genetics have appeared, the most recent one being more than 5 years old. *Microbial Genetics* represents my attempt to bring this still-useful subject up to date.

The application of microbial genetics led to the accumulation of a huge body of knowledge and a continually greater understanding of the nature of the gene. In the early 1970s, microbial genetics itself underwent a revolution, with the development of the recombinant DNA technology. This collection of remarkable but straightforward techniques, usually called genetic engineering, allows manipulation and exchange of fragments of DNA outside of the cell and reintroduction of recombinant DNA into a new cell. In this way, novel organisms can be created, with characteristics drawn from distant species and genera. Even human genes can be transferred to a bacterium, and genes of microorganisms can be placed in animal cells. In fact, the glitter of this new technology has led to the conversion of hundreds of research laboratories to gene-cloning factories and to the development of a new industry, known as bioengineering. This new technology is the topic of the later chapters in this book.

Microbial Genetics is the result of a good deal of prodding by many colleagues who urged me to write such a book. Although I am a "modern" molecular biologist (though I have never cloned a gene), I did acquire my research experience during the first Golden Age and worked intensively in various aspects of microbial genetics. Furthermore, since I have had the experience of writing several books, I was told that the task was clearly mine. Thus, several months ago I started to re-read ancient books and papers. This brought back many pleasant memories of lectures and meetings that were thoroughly exciting because in retrospect we seem to have known, comparatively speaking, very little about the details of genetic processes. These recollections, as well as thoughts about many former colleagues, made the process of book-writing more of a pleasure than a task.

Microbial Genetics begins with five review chapters. These include basic material about classic genetics, DNA and protein structure, and the basic biology of bacteria and bacteriophages. Students who have had a course in biochemistry can surely omit Chapters 2 and 3, and those having studied classic genetics can undoubtedly skip Chapters 1, 4, and 5. Because these chapters are meant as a review, the material is presented fairly succinctly. Furthermore, only topics needed in later discussions have been included.

Every chapter in this book ends with a substantial set of questions and problems. Most of the beginning questions simply test the memory of the student; they are designed to ensure that the reader has learned the definitions and understood the most elementary concepts. Later questions are more difficult. Complete answers to all questions and problems, with explanations, are given in the Answers section at the back of the book.

David Freifelder

PART 1

ESSENTIALS OF GENETICS
AND MICROBIOLOGY

Essentials of Genetics

A unifying concept of modern biology is that species not only persist over long periods of time through heredity, but also change by evolution. Heredity is the process by which all living things produce offspring like themselves. This capacity for self-reproduction involves the transmission of information that specifies a particular pattern of growth and organization from parent to offspring. **Genetics** is the science of heredity; it is concerned with the physical and chemical properties of the hereditary material, how this material is transmitted from one generation to the next, and how the information it contains is expressed during the development of an individual.

Because species continue to exist through many generations, the hereditary material must be inherently stable. Within most species, however, immense hereditary diversity is present; for example, although most humans have the same overall appearance, even persons as closely related as sisters or brothers usually look quite different. Hereditary variability is an essential tool for evolution, which is the cumulative change in the genetic characteristics of populations of organisms through time. Evolution is the most important generalization in biology; it is the process that enables us to understand the origin of the enormous number of species of living things and the relation between them. Evolution provides a logical basis for understanding how the complex organisms existing today were derived through continuous lines of descent, with modification, from the first primitive organisms.

Genetic analysis provides information about the functions of genes and their products in vivo. Genetic analysis has been a driving force in the development of molecular biology. Because genetic analysis is indirect, it cannot prove molecular mechanisms. Genetic analysis, however, has been the major source of intuition about molecular mechanisms and has elucidated phenomena that might otherwise be ignored. Furthermore, genetic approaches are often the only way to determine whether a specific mechanism functions in vivo.

To discuss genetics, you need to know the vocabulary for describing the genetic properties of an organism. A uniform nomenclature is used for bacteria but not for all organisms.

Generally all strains of a species are compared with an arbitrarily chosen **wild-type** strain. A **mutation** is any change in the DNA sequence compared with the wild-type strain. A strain with a mutation is called a **mutant**. The mutation may or may not affect some observable property or **phenotype** compared with the wild-type or parent strain. Thus, a cell that can synthesize the amino acid proline is denoted Pro$^+$; if it cannot do so, it is denoted Pro$^-$. Note that the phenotype is indicated by a three-letter symbol, the first letter is capitalized, and it is not italicized. **Genotype** refers to the genetic composition of an organism. A Pro$^-$ cell certainly must have some defective gene that keeps it from synthesizing proline; this would be denoted *pro$^-$* (or often simply, *pro* without the minus sign). Note that the genotype is indicated by three lowercase letters in italics or underlined. Several genes may be required to synthesize proline. Different genes are indicated by adding a fourth italicized letter: *proA*, *proB*. A Pro$^+$ cell must have one functional copy of every gene required for synthesis of proline. Thus, its genotype must be *proA$^+$, proB$^+$, proC$^+$*, but the genotype is normally summarized by writing *pro$^+$*, unless it is important for some reason to state the genotype for each gene.

Typical bacteria have only a single set of genes, hence they are **haploid**. Eukaryotic organisms (whose cells contain a nucleus enclosed in a membrane) usually have two sets of genes, hence they are **diploid**. A diploid cell might have one normal copy of a particular gene and one defective copy. In this case, the genotype includes both copies, and the gene symbols are separated by a diagonal line: *pro$^+$/proA$^-$*. A cell could also have two normal copies or two defective copies of a gene.

Some genes are responsible for resistance and sensitivity to certain extracellular products, such as antibiotics. The genotypes of an ampicillin-resistant cell and an ampicillin-sensitive cell are written *amp^r* and *amp^s*; the corresponding phenotypes are written Amp^r and Amp^s.

In summary, the following conventions are used for bacteria. A different convention is used for bacteriophages, with many one-letter and two-letter symbols.

1. Phenotypes are indicated by three roman letters with the first letter capitalized and the superscript + or – to denote presence or absence of the designated character or the superscript "s" for sensitivity or "r" for resistance.
2. Genotypes are indicated by three lowercase letters and with a fourth capitalized letter to indicate specific genes. All components of the genotype are italicized.

Often many different mutations are isolated in a single gene, so it becomes necessary to identify different mutations (or **alleles**). Mutations are assigned numbers in the order in which they have been isolated. Thus, the Pro$^-$ mutations numbered 58 and 79 are written *pro–58* and *pro–79*. Once the gene affected by the mutation is identified, the letter indicating the gene is added. For example, if mutation *pro–58* was located in the *proA* gene, it would be written as *proA58*.

It is often useful to describe the protein products of a particular gene. There is a convention for gene products. For example, the protein product of the *dnaA* gene would be called simply DnaA. However, it might also be called the *dnaA* product or the DnaA protein. Each of these designations is convenient and is often seen in scientific literature.

Numerous abbreviations are commonly used for various genes and substances. These are listed in Table 1-1.

The term **genome** refers to the set of all genes in a cell or virus. Often genome is used to describe the chromosome for bacteria or phages that contain a single chromosome. The functional form of a gene is sometimes called **wild-**

Table 1-1 Some genotype abbreviations used in this book

Phenotype	Genotype symbol[1]	Phenotype	Genotype symbol
Auxotrophic mutants		**Other mutants**	
Amino acid biosynthesis		Utilization of carbon sources	
Alanine	ala	Acetate	ace
Arginine	arg	Arabinose	ara
Aromatic	aro	Fatty acids	fad
Asparagine	asn	Galactose	gal
Aspartic acid	asp	Glucose	glu
Cysteine	cys	Lactose	lac
Glutamic acid	glu	Maltose	mal
Glutamine	gln	Proline	put
Glycine	gly	Sorbitol	srl
Histidine	his	Antibiotic resistance	
Isoleucine	ile	Ampicillin	amp
Leucine	leu	Chloramphenicol	cam
Lysine	lys	Kanamycin	kan
Methionine	met	Penicillin	pen
Phenylalanine	phe	Streptomycin	str
Proline	pro	Tetracycline	tet
Serine	ser	DNA and RNA functions	
Threonine	thr	DNA polymerase	pol
Tryptophan	trp	DNA gyrase	gyr
Tyrosine	tyr	Mutator	mut
Valine	val	Recombination	rec
Nucleotide biosynthesis[2]		Repair of UV damage	uvr
Purine	pur	RNA polymerase	rpo
Pyrimidine	pyr	Suppressors	sup
Adenine	ade		
Cytosine	cyt		
Guanine	gua		
Thymine	thy		
Uracil	ura		
Vitamin biosynthesis			
Biotin	bio		
Lipoic acid	lip		
Nicotinamide	nad		
Thiamine	thi		

[1]When describing a protein or a phenotype rather than a genotype, the same three-letter symbol is used, but it is capitalized and not italicized.

[2]When describing base sequences, a purine is usually abbreviated Pu and a pyrimidine is usually abbreviated Py. Particular bases are designated A (adenine), G (guanine), T (thymine), U (uracil), and C (cytosine).

type genotype because presumably this is the form found in nature. Sometimes, however, functional genes are not present in the wild-type organism. For example, *Escherichia coli* and *Salmonella typhimurium* are closely related enteric bacteria, yet the commonly used wild-type strain of *E. coli* (designated *E. coli* K-12) carries genes for metabolism of lactose (lac^+), and these genes are missing (lac^-) in the wild-type *S. typhimurium* (designated *S. typhimurium* LT2).

The term **allele** is used to indicate that there are alternative forms of a gene; sometimes the + and − forms are called the wild-type allele and the mutant allele. Each different mutation in a gene is a different allele of the gene. An organism in which the members of a pair of alleles are different is **heterozygous**, and one in which the two alleles are alike (either both wild-type or both with the same mutation) is **homozygous**. In a heterozygote, the expression of one of the genes usually predominates; that allele is said to be **dominant**, and the other allele is **recessive**. For example, if pro^+/pro^- heterozygote has Pro$^+$ phenotype, the pro^+ allele is dominant.

MUTANTS AND MUTATIONS

The terms mutant and mutation are commonly confused. Strictly speaking, a **mutant** is an organism whose genotype (or, more precisely, its DNA base sequence) differs from the wild-type. Mutations are the raw material of evolution, providing the changes on which selection can act. For the microbial geneticist, mutations are tools; for example, **genetic markers** can be used to identify a particular chromosome or a specific region of a chromosome. Furthermore, study of the changes induced by mutations gives a great deal of information about the role of the normal gene product in the cell.

The process of generating mutations is called **mutagenesis**. In nature and in the laboratory, mutations sometimes arise spontaneously as a result of rare errors during DNA replication or repair. This is called **spontaneous mutagenesis**. Mutagenesis can also be induced by the addition of chemicals called **mutagens** or by exposure to radiation, both of which result in chemical alterations of the DNA. Mutagenesis and mutations are discussed in detail in Chapter 10.

Types of Mutants

Mutants can be classified in several ways. One classification is based on the conditions in which the mutant phenotype is expressed. A **nonconditional** mutant displays the mutant phenotype under all conditions; for example, a mutant bacterium that requires proline for growth in all culture media and at all temperatures. Nonconditional mutants that completely eliminate activity of a gene product are called null mutants, and mutants that retain some residual activity are called leaky mutants.

In contrast, a **conditional mutant** does not always show the mutant phenotype; its behavior depends either on environmental conditions or on the presence of other mutations. An example of a conditional mutant is a **temperature-sensitive (Ts) mutant**, which has the wild-type phenotype at one temperature (for example, at 30°C) and has a mutant phenotype at another temperature (for example, at 42°C); often an intermediate phenotype is observed between these temperatures. Note that the gene itself is not altered at 42°C; rather the product of the gene is inactive at 42°C. Another common type of conditional mutant are **suppressor-sensitive mutants**; these exhibit the mutant phenotype in some bacterial strains but not in others. The difference is due to the presence of other gene products made by **suppressor** genes. These suppressor-gene products either substitute for the defect in the mutant or enable an altered gene to produce a functional gene product. Thus, the phenotype of a suppressor-sensitive mutant "depends on the genetic background": that is, whether or not the mutant strain also contains a suppressor. Suppressor-sensitive mutations are designated by adding Am (for "amber" or UAG mutations), Oc (for "ochre" or UAA mutations), or Op (for "Opal" or UGA mutations) in parentheses. The symbols are capitalized because they represent a phenotype; if the mutation has a number, this is also added. Thus, if mutation 35 in the *pro* gene is a suppressor-sensitive Am mutation, it would be written *pro*–35(Am); if the mutation were temperature-sensitive, it would be written *pro*–35(Ts). Suppressors are discussed in detail in Chapter 10.

Another way of classifying mutants is based on the number of changes that have occurred in the genetic material (that is, the number of DNA base pairs that have changed). If only one base pair has been changed, the mutation is called a **point mutation**. If two changes are present, the mutation is called a **double mutation**. Some mutations remove all or part of a gene; in that case, the mutation is called a **deletion**. Occasionally other material replaces the deleted gene or gene segment. Such a mutation is called a **deletion-substitution**. If genetic material is added without removal of any other material, it is called an **insertion mutation**. Some of these types of mutations are shown schematically in Figure 1-1.

Isolation and Characterizations of Mutants

Microbial genetics depends on the analysis of mutant bacterial or phage strains that affect the process of interest (for example, proline biosynthesis). Much of this book is concerned with genetic analysis, but without mutants, no analysis is possible. Because mutations are very rare, mutants must be isolated from large populations of cells. Thus, some means is needed to identify the rare mutants from the population of wild-type cells. Detection of mutants requires a selection, or screen.

1. *Genetic selections*. A selection is a condition that allows specific mutants to grow but not the parental cells. Genetic selections are very powerful because they allow isolation of rare mutations from a population of cells. For example, antibiotic-resistant mutants can be selected by simply plating a large number of bacteria on solid medium containing the antibiotic—only resistant mutants can form colonies.

2. *Screens*. If the mutation is relatively common and there is no direct selection for mutants, it is often necessary to resort to screening for mutants on media on which both the mutant and the parental cells grow, but the phenotype of the mutant can be distinguished from the phenotype of the parental cells. For example, sugar-utilization mutants can be identified on indicator plates (described in Chapters 4 and 7). Both the sugar-utilization mutant and the parental cells can grow on a color-indicator plate lacking the sugar because other carbon sources are present, but the two types of cells can be distinguished by the color of the colonies. In

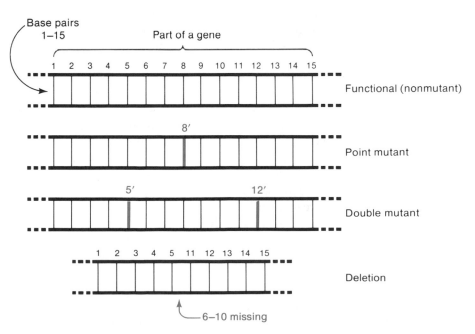

Figure 1-1. Schematic diagram of a normal DNA molecule and various kinds of mutants. In the point mutant, base pair 8' replaces the normal base pair 8. The double mutant has two base pairs (5' and 1') not present in the nonmutant organism. In the deletion, five base pairs (6 to 10), present in the nonmutant, are absent.

contrast, to isolate a leucine auxotroph (Leu⁻), you cannot simply look for growth on media lacking leucine because the desired mutant will not grow. Auxotrophs can be identified by replica-plating. The cells are plated on nutrient agar such that there are several hundred colonies on a plate. This master plate is then replicated onto two minimal agar plates, one containing leucine and the other lacking leucine. Any Leu⁻ auxotrophs will form colonies on the plate with leucine but not on the plate without leucine. The mutants can then be isolated from the reference plate with leucine. Replica-plating can also be used to screen for conditional mutants. For example, to isolate temperature-sensitive (Ts) mutants, the colonies can be initially grown at 30°C, and then the colonies can be replica-plated onto two plates with one incubated at 30°C and the other at 42°C. Ts mutants will form a colony on the 30°C plate but not on the 42°C plate.

The difference between a selection and a screen is very important. For example, compare the selection for a streptomycin-resistant (Str^r) mutant with a screen for a Leu⁻ mutant. Isolation of a Str^r colony is a positive selection because only Str^r cells can grow on a plate containing streptomycin. Thus, as many as 10^{10} cells can be put on a plate, allowing a mutant as rare as $1/10^{10}$ to be detected. Isolating a Leu⁻ mutant, however, involves simply screening for the rare mutant: No more than a few hundred colonies can be examined on a single plate. Thus, thousands of plates would be needed to find a single mutant if the fraction of mutant was 10^{-5} (a tedious, expensive job). This problem can be minimized by two techniques: (1) The starting culture can be treated with a mutagen to increase the mutation frequency; (2) an enrichment procedure can be used to increase the proportion of mutants in the population. In certain situations, mutagenesis can increase the fraction of mutants to 10^{-3}, allowing screening for the desired mutant on a reasonable number of replica plates. Often random chemical or physical mutagenesis, however, does not increase the mutant fraction to more than one mutant per 10^5 cells. Isolating such rare mutants may require the use of an enrichment procedure (discussed in Chapter 10).

Most phage mutants are detected in two ways. "Plaque-morphology" mutants are found because they look different from wild-type plaques. For example, some mutants make plaques that are turbid, or clear, or have a halo, or are a combination of these. Conditional mutants—that is, mutants that are either temperature sensitive or host dependent (forming plaques on one host but not on another)—are often found by replica-plating. With a temperature-sensitive mutant, plaques are formed at 30°C and then transferred to two plates already seeded with bacteria and incubated at either 30°C or 42°C. A temperature-sensitive mutant forms a plaque on the 30°C plate but not on the 42°C plate. Similarly, a host-dependent mutant (a type of conditional mutant) can be detected by plating on a permissive host and replica-plating onto either a permissive or a nonpermissive host. Such host-dependent mutations, called suppressor-sensitive mutations, are described in Chapter 10.

Revertants and Reversion

A mutant organism sometimes regains its wild-type characteristics. This occurs by means of an additional mutation in the mutant, which restores function. The process of regaining the original phenotype is called **reversion**, and an organism that has reverted is called a **revertant**.

Reversion can result from spontaneous mutagenesis and, similar to the formation of all spontaneous mutations, is a nearly random process. Unless there is some selective pressure for revertant phenotype, the frequency of reversion is rare, so revertants must be isolated from the population by appropriate selection or screening methods. Of course, as with other mutations, it is possible to stimulate reversion by mutagenesis.

The **reversion frequency**—that is, the fraction of cells in a population of mutants that after many generations of growth have regained the original phenotype—is a useful criterion for identifying the nature of a mutation. The reversion frequency can distinguish a single point mutant from a double mutant or a deletion mutant in a gene: Single point mutants (which can be repaired by a single base change in the DNA) revert at a much higher frequency than double mutants (which require two changes and thus revert very rarely) or deletion mutants (which cannot directly revert). The effect of mutagens on the reversion frequency can also be used to distinguish different types of mutations: Reversion of frameshift mutations is stimulated by frameshift mutagens, and so on. The reversion of known bacterial mutations has also been extensively used to detect potential human carcinogens. The uses of such reversion analysis are discussed in more detail in Chapter 10.

Uses of Mutations

Mutations are extremely valuable for understanding molecular biology, physiology, and genetic regulation. A few examples of the uses of mutations are described briefly here.

1. *Mutations can elucidate functions.* For example, before the early 1950s, it was not known whether nutrients were taken up by bacteria by passive diffusion through the cell membrane or by specific transport systems. Wild-type *E. coli* can grow on low concentrations of lactose. Some mutants (called *lacY*) that were unable to grow on lactose were unable to take up radiolabeled lactose, indicating that they were defective for lactose transport. The properties of the *lacY* mutants indicated that lactose transport is mediated by the *lacY* gene product. Temperature-sensitive mutant organisms are especially useful in defining essential functions. For example, Ts mutants of *E. coli* have been isolated that fail to synthesize DNA. The mutations fall into at least 10 distinct classes, suggesting that there may be at least 10 different proteins required for DNA synthesis.

2. *Mutations can introduce biochemical blocks that aid in the elucidation of metabolic pathways.* For example, the metabolism of the sugar galactose requires the activity of three distinct genes, called *galK, galT,* and *galE*, identified by the isolation of three classes of Gal$^-$ mutants. If radioactive galactose ([^{14}C]Gal) is added to a culture of Gal$^+$ cells, the radioactivity accumulates in the products of galactose metabolism. At very early times after addition of [^{14}C]Gal, three related compounds are detectable: [^{14}C]galactose-1-phosphate (Gal-1-P), [^{14}C]uridine diphosphogalactose (UDP-Gal), and [^{14}C]uridine diphosphoglucose (UDP-Glu). Mutations in different genes block distinct steps of the metabolic pathway. If the mutant cell possesses a *galK$^-$* mutation, the [^{14}C]Gal label is found only in galactose. Thus, the *galK* gene product is responsible for the first metabolic step. A *galT$^-$* mutant accumulates Gal-1-P. Thus, the first step in the reaction sequence is found to be the conversion of galactose to Gal-1-P by the *galK* gene product (the enzyme galactokinase). If a *galE$^-$* mutant is used, some Gal-1-P is found, but the principal radiochemical is UDP-Gal. Thus, the biochemical pathway must be:

$$\text{Gal} \xrightarrow{\textit{galK} \text{ product}} \text{Gal-1-P} \xrightarrow{\textit{galT} \text{ product}} \text{UDP-Gal} \xrightarrow{\textit{galE} \text{ product}} X$$

Note that the identity of X cannot be determined from these genetic experiments.

3. *Mutants enable one to learn about metabolic regulation.* Many mutants have been isolated in which there is an alteration in either the amount of a particular protein that is synthesized or the way the amount synthesized responds to external signals. Such mutations define regulatory systems. For example, the enzymes corresponding to the *lacZ, lacY,* and *lacA* genes are normally made at low levels, unless lactose is added to the growth medium. Mutants have been isolated, however, in which these enzymes are always present, whether or not lactose is also present. This indicates that some gene is responsible for turning the system of enzyme production on and off, and this regulatory gene must somehow respond to the presence or absence of lactose.

4. *Mutations allow an in vitro biochemical activity to be correlated with an in vivo function.* The *E. coli* enzyme, DNA polymerase I (Pol I), was purified and studied in great detail. Purified Pol I is capable of synthesizing DNA in vitro, so it seemed likely that this enzyme was also responsible for DNA synthesis in vivo. An *E. coli* mutant ($polA^-$), however, was isolated in which the activity of Pol I was reduced 50-fold without any detectable effect on either the growth rate of the cells or the ability to synthesize DNA in vivo. This observation strongly suggested that Pol I could not be the only enzyme that synthesizes DNA in the cell. Biochemical analysis of cell extracts from the $polA^-$ mutant showed the existence of two other enzymes, DNA polymerase II (Pol II) and DNA polymerase III (Pol III), which could also synthesize DNA. Later a mutant with a temperature-sensitive mutation in a gene called *dnaE* (later called *polC*) was found, which was unable to synthesize DNA at 42°C, although synthesis was normal at 30°C. The three enzymes—Pol I, II, III—were isolated from cell extracts of the *dnaE*(Ts) mutant, and each enzyme was assayed. Although Pol I and Pol II were active at both 30°C and 42°C, Pol III was active at 30°C but not at 42°C; therefore, Pol III was identified as the enzyme responsible for DNA synthesis in vivo.

5. *Mutations can identify the site of action of external agents.* The antibiotic rifampicin prevents synthesis of RNA. When first discovered, it was not known whether rifampicin acted by preventing synthesis of precursor molecules; by binding to DNA and thereby preventing the DNA from being translated into RNA; or by binding to RNA polymerase, the enzyme responsible for synthesizing RNA. Mutants were isolated that were resistant to rifampicin (Rif^r). In most of these mutants, the RNA polymerase was slightly altered. The isolation of Rif^r mutants that alter RNA polymerase indicated that the antibiotic acts by binding to RNA polymerase.

6. *Mutations can indicate relations between apparently unrelated systems.* Bacteriophage λ normally infects *E. coli*, but it is unable to adsorb to certain *E. coli* mutants that are unable to metabolize maltodextrans (*lamB*$^-$). λ adsorbs normally to mutants unable to metabolize other sugars, and other phage adsorb normally to the *lamB* mutants; hence these results indicate that the *lamB* gene product is specifically required for the adsorption of λ.

7. *Mutations can indicate that two proteins interact.* How this occurs is best shown by a hypothetical example. Suppose that mutants carrying mutations in either of the two genes *a* and *b*, which are responsible for syn-

thesizing the proteins A and B, fail to carry out a particular process. Clearly the products of both genes are necessary for the process to occur. A and B may act consecutively in the process, or they may interact to form a single functional unit consisting of both proteins. Evidence that the two proteins interact may be obtained by reversion studies. When revertants of an a^- mutant are sought, it is sometimes found (by additional genetic analysis) that reversion is a result of a second mutation in gene b. Likewise, reversion of some b^- mutants may be due to mutations in gene a. One explanation for these results is that the A and B proteins interact. For example, proteins A and B interact to form the functional heterodimer (Fig. 1-2). Some mutations in either A or B may affect a protein-protein binding site. A compensating alteration in the other protein could allow the interaction to occur again. For example, some mutations in a phage λ DNA-replication gene (P) are compensated for by a nonlethal mutation in an *E. coli* DNA-replication gene ($dnaB$), and the corresponding gene products, P protein and DnaB protein, bind together to form a protein active in the replication of λ DNA.

GENETIC ANALYSIS OF MUTANTS

Two of the most important tools for analysis of mutations are genetic recombination and complementation. These two methods are often confused, but the actual methods and the conclusions they reveal are quite different. Recombination allows you to make new combinations of genes and to determine the positions of genes on a chromosome with respect to one another. Complementation allows you to determine the number of genes responsible for a particular phenotype

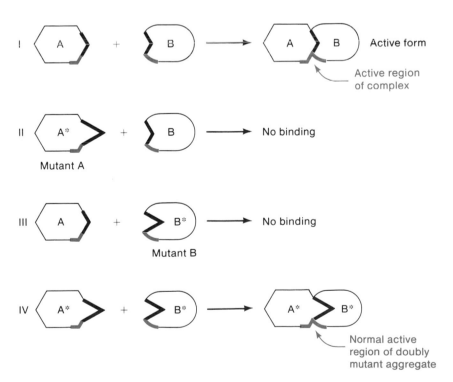

Figure 1-2. Schematic diagram showing how two separately inactive mutants can interact to make a functioning protein complex. Sites of interaction of proteins A and B are shown by heavy lines. Components of the active site of the A–B complex are shown in red. Only complexes I and IV have active binding sites.

and to distinguish regulatory genes from regulatory sites. Genetic recombination and complementation are described. Numerous examples of the use of these techniques are found throughout the book.

Genetic Recombination

Genetic recombination is the process of physically exchanging two genetic loci, initially on two different DNA molecules, onto a single DNA molecule. The molecular mechanisms are complex and are not yet fully understood (see Chapter 14). For many genetic experiments, however, recombination can be visualized quite simply: Two DNA molecules align with one another, then a cut is made in both DNA molecules at random but matching points, and then the four fragments are joined together to form two new combinations of genes (Fig. 1-3). (In fact, most genetics was done long before any understanding of the molecular mechanism of recombination.) For example, two parental chromosomes with the genotypes a^+b^- and a^-b^+ are cut and are then joined to form two recombinant chromosomes, whose genotypes are a^+b^+ (wild-type) and a^-b^- (double mutant).

Genetic recombination can be easily observed when a bacterium is simultaneously infected with two mutant phage. If the parental phage have the genotypes a^+b^- and a^-b^+, among the hundred or so progeny phage released when the infected bacterium lyses, there will be a few a^+b^+ and a^-b^- recombinant phage (Fig. 1-3). The ratio:

$$\frac{\text{Number of recombinant phage}}{\text{Number of total phage}}$$

is called the **recombination frequency**. More generally, the recombination frequency is defined as:

$$\frac{\text{Number of recombinant types}}{\text{Number of parental + recombinant types}}$$

Genetic Mapping

The distance along the chromosome between two genes determines the recombination frequency. The relative position of genes is inferred from their **linkage**. Linkage refers to the frequency that two genes segregate with one another. Thus, two genes or markers are considered linked if they are close to one another on the chromosome. (In classic Mendelian genetics, linkage refers to the genes on a single chromosome, so genes are considered linked if they segregate from one another only by recombination. Most prokaryotes have a single linkage group because most prokaryotes contain a single chromosome.) Except when two genes

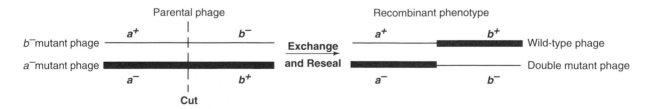

Figure 1-3.   A simplified cartoon showing genetic exchange between two phages resulting in recombinants.

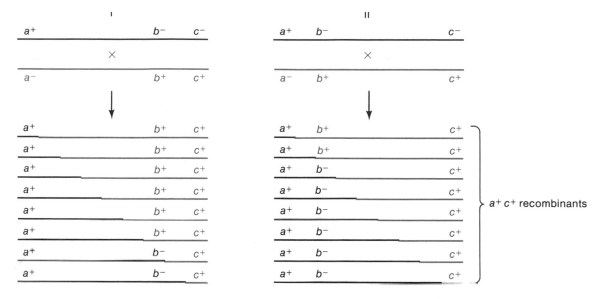

Figure 1-5. A three-factor cross. Eight possible $a^+ c^+$ progeny arising from equally spaced exchange points are shown for each arrangement. In I, 6 of 8 $a^+ b^- c^+$ are $a^+ b^+ c^+$; in II, 6 of 8 $a^+ c^+$ are $a^+ b^- c^+$.

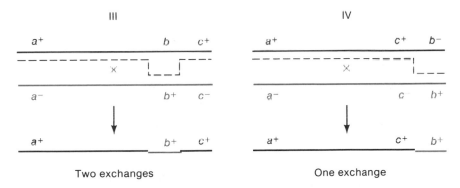

Figure 1-6. Determination of gene order by a three-factor cross. The frequency of appearance of $a^+ b^+ c^+$ is much higher for order IV because only one exchange is required.

order III and a^+ for order IV. In the second method, the frequency of $a^+b^+c^+$ recombinants is measured. These recombinants arise by two exchanges in case III but only one exchange in case IV. The frequency for a double exchange is the product of the frequencies of each simple exchange. Thus, for case III, the recombination frequency is $(0.07)(0.002) = 0.00014 = 0.014\%$, but for case IV, in which only a single exchange is needed, the frequency is 0.2% (the same as that for forming b^+c^+ recombinants). Hence one three-factor cross yields the gene order; in the next section, it is shown that the cross also generates a quantitative genetic map for the three markers. A real example of genetic mapping using three-factor crosses is shown in Box 1-2.

Multiple Exchanges and the Recombination Frequency for Distant Markers

When two genes are quite distant in a chromosome, more than one exchange may often occur in a single pairing event. This phenomenon complicates the interpretation of recombination data, primarily because if two exchanges occur

BOX 1-2: GENETIC MAPPING BY THREE-FACTOR CROSSES

Example: To confirm the gene order determined from two-factor crosses (see Box 1-1), the following three-factor crosses were done.

Donor	*fadL*	*purF*$^+$	*aroC*$^+$	*dsdA*$^+$
Recipient	*fadL*$^+$	*purF*	*aroC*	*dsdA*

Selected phenotype	Recombinants	Number obtained
a. PurF$^+$	*fadL*$^+$ *aroC*$^+$	183
	fadL aroC$^+$	129
	fadL$^+$ *aroC*	238
	fadL aroC	20
b. AroC$^+$	*fadL*$^+$ *purF*$^+$	305
	fadL purF$^+$	246
	fadL$^+$ *purF*	328
	fadL purF	266
c. AroC$^+$	*fadL*$^+$ *dsdA*$^+$	5
	fadL dsdA$^+$	21
	fadL$^+$ dsdA	39
	fadL dsdA	24

From these data, determine the order of the *fadL, purF, aroC*, and *dsdA* genes.

Answer:

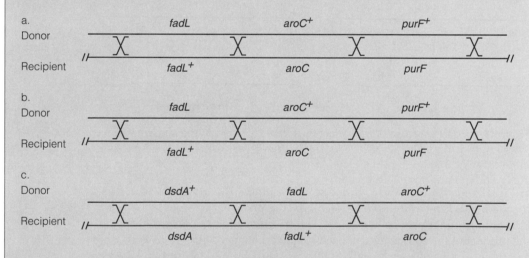

a. Select *purF*$^+$
Rarest class of recombinants will require four crossovers. Rarest class of recombinants was *aroC fadL*, thus *aroC* is the middle gene.
b. Select *aroC*$^+$
No rare class of recombinants were obtained, thus the selected marker (*aroC*) is the middle gene.
c. Select *aroC*$^+$
Rarest class of recombinants was *dsdA*$^+$ *fadL*$^+$, thus *fadL* is the middle gene.

between two markers, recombination is not observed. If three exchanges occur, recombinants form (Fig. 1-7). In general, recombination is not observed in a two-factor cross if the number of exchanges is an even number. Because of the failure to detect all exchanges, an observed recombination value is usually a

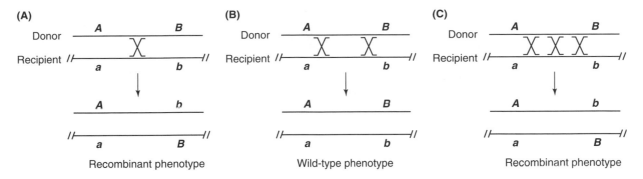

Figure 1-7. If there is an odd number of crossovers between two genes, recombinant combinations of the two genes can be observed (panels A and C). However, if there is an even number of crossovers between the two genes, recombinant combinations of the two genes are not observed (panel B).

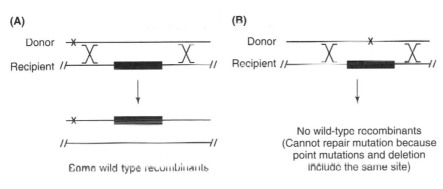

Figure 1-8. Deletion mapping. (A) The mutation in the donor lies outside of the region deleted in the recipient, allowing formation of wild-type recombinants. (B) The mutation in the donor lies within the same region as that deleted in the recipient, so no wild-type recombinants can be formed.

slight underestimate of the true exchange frequency and hence of the map distance between two genes.

Deletion Mapping

Deletion mutations remove a portion of the genetic material. The extent of a deletion can be determined by recombination experiments. The principle is straightforward: In a cross between a deletion mutant and a second mutant carrying a point mutation, wild-type progeny cannot be formed if the deletion spans the region of the map that includes the point mutation. This principle is illustrated in Figure 1-8. Often the end points of deletions are indicated by a box or brackets, but it is important to remember that the DNA at both ends of a deletion is contiguous. When a mutant carrying a deletion that removes genes c and d is crossed with mutants that have a defective allele of the a, b, or e genes, wild-type progeny can be formed because the deletion does not span these genes. When crossed with the c^- or d^- mutants, however, wild-type progeny cannot be formed because the deletion does not have any genetic material corresponding to mutant alleles. Thus, in contrast to two-factor or three-factor crosses, in which the genetic map is inferred from the recombination frequencies, deletion mapping involves a simple "yes" or "no" answer; if any recombinants are obtained, a functional copy of the deleted gene must be present.

Collections of deletions in which various amounts of genetic material are missing are valuable for genetic mapping because they reduce the number of crosses required to map a new mutation, and the results are unambiguous. For example, an uncharacterized mutation can be mapped by crossing it with a set of deletions having different, known boundaries. The lack of wild-type recombinants in a cross with a particular deletion (I) indicates that the mutation is in the

BOX 1-3: DELETION MAPPING

Example: A donor strain with an uncharacterized *his⁻* mutation was mapped against a set of nested *his⁻* deletion mutations. The extent of the deletion mutations (indicated by the thick bars) and the results are shown below: Where does the *his⁻* mutation in the donor strain map?

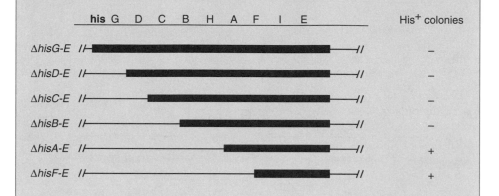

Answer: The *hisA-E* deletion can be repaired by the point mutation, therefore the mutation is not in the *hisAFI* or *E* genes. Deletions that remove *hisB* or *hisH* cannot be repaired, indicating that the mutation is in either *hisB* or *hisH*.

region spanned by the deletion. Furthermore, if wild-type recombinants are produced in crosses with a slightly smaller deletion (II), the mutation must be located between the boundaries of deletions I and II. An example of deletion mapping is shown in Box 1-3.

Complementation

As discussed earlier, a particular phenotype is frequently the result of the activity of many genes. To understand any genetic system, it is essential to know the number of genes and regulatory elements that constitute the system. Because multiple genes that affect the same function may map very close to each other, it is not possible to determine if two mutations are in the same or different genes simply from the recombination frequency. Instead the genetic test used to determine this number is called **complementation**.

Complementation is best explained by example. The test requires that two copies of the genes are present in the same cell. In bacteria, this can be done by constructing a **partial diploid or merodiploid**—that is, a cell containing one complete set of genes and a duplicated copy of part of the genome. (How cells of this type are constructed is described in Chapter 14). A partial diploid is described by writing the genotype of each set of genes on either side of a diagonal line. As an example of this, $a^+b^-c^-d^-e^+ \ldots z^+/b^+c^+d^+$ indicates that a second copy of genes b, c, and d is present in a cell whose single chromosome contains all of the genes $a, b, c, \ldots, z$. Usually only the duplicated genes are indicated, so this partial diploid would be designated $b^+c^+d^+/b^-c^-d^-$.

Consider the three genes required for *E. coli* to grow on galactose, *galK*, *galT*, and *galE*. The genes encode the enzymes galactokinase, galactose transacetylase, and galactose epimerase, which act sequentially to metabolize galactose. If there is a mutation in any of these genes, the mutant cannot grow on galactose (Gal$^-$). The partial diploid $galK^- \, galT^+ \, galE^+ \, / \, galK^+ \, galT^- \, galE^+$ will grow on galactose, however, because the cell contains at least one copy of each gene that produces functional GalK, GalT, and GalE proteins; GalT will be made from one copy of the *gal* genes and GalK from the second copy of the *gal* genes. The *galK* and *galT* mutations in the diploid are said to complement one another because the phenotype of the partial diploid containing them is *Gal$^+$*. In contrast, in a partial diploid $galK1 \, galT^+ \, galE^+ / galK2 \, galT^+ \, galE^+$ (where $galK1$ and $galK2$ are two different mutations in the *galK* gene), no functional GalK protein will be made because both copies of the *galK* gene are mutated. The $galK1$ and $galK2$ mutations do not complement one another because the phenotype of the partial diploid containing these mutations is Gal$^-$.

When a new *gal$^-$* mutation (*gal-101*) is isolated that prevents growth on galactose, initially the gene in which the mutation has occurred is not known. The mutant gene may be identified by a complementation test. The following partial diploids are constructed: *gal-101 / galK$^-$ galT$^+$ galE$^+$*, *gal-101 / galK$^+$ galT$^-$ galE$^+$*, and *gal-101 / galK$^+$ galT$^+$ galE$^-$*. If the *gal-101 / galK$^-$ galT$^+$ galE$^+$* and *gal-101 / galK$^+$ galT$^-$ galE$^+$* diploids are Gal$^+$, the mutation is not in the *galK* or *galT* genes. If the *gal-101 / galK$^+$ galT$^+$ galE$^-$* diploid is Gal$^-$, the *gal-101* mutation is most likely in the *galE* gene (Fig. 1-9). If all three diploids were Gal$^+$, the *gal-101* mutation is not in the *galK*, *galT*, or *galE* genes. Furthermore, if the *gal-101* mutation is not in any of these genes, yet the *gal-101* mutant is unable to metabolize galactose, this would indicate that galactose metabolism requires at least four genes ("at least" because additional genes that affect galactose metabolism might still be discovered).

A few additional examples of complementation analysis are shown in Box 1-4. These examples are straightforward, but sometimes complementation analysis is more complex. The basic rules for complementation analysis are:

1. If two mutations complement, they are in different genes. (There is one exception to this rule—rare examples of intragenic complementation occur in proteins with multiple, identical subunits. Intragenic complementation is described in more detail in the next section.)
2. If two mutations do not complement, they are not necessarily in the same gene. There are three possible reasons why two mutations may not complement each other:
   a. The mutations may be in the same gene.
   b. One of the mutations may affect expression of the other gene.
   c. One of the mutations may make a gene product that inhibits the other gene product.

Two practical points should be considered before doing complementation analysis: (1) Complementation analysis is usually done after mutations have been grouped by genetic mapping experiments. This greatly simplifies the number of partial diploids that must be constructed because genes that map far from each other clearly lie in different genes. (2) Complementation analysis should be done

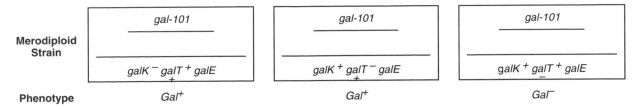

Figure 1-9.   Complementation tests showing three different *gal* merodiploids. The Gal$^+$ phenotype requires a functional   product from each of the *gal* genes: *galK*, *galT*, and *galE*.

BOX 1-4: COMPLEMENTATION ANALYSIS

Example 1: *Salmonella typhimurium* makes proline from glutamate by the pathway shown below.

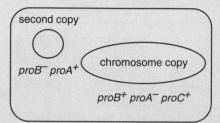

Will the following merodiploid strains be proline auxotrophs?

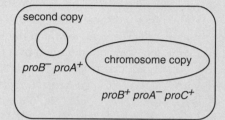

Answer: At least one copy of each functional gene is present so the cell can make proline. Genes on the plasmid **complement** the mutation on the chromosome. Therefore, the phenotype will be Pro⁺: the merodiploid will not be a proline auxotroph.

Answer: No functional copy of the *proB* gene is present so the cell cannot make proline. Genes on the plasmid **do not** complement the mutation on the chromosome. Therefore, the phenotype will be Pro⁻: the merodiploid will be a proline auxotroph.

Example 2: A new proline auxotrophic mutant (*pro-53*) was isolated. By constructing partial diploid strains, a complementation table was constructed as shown below. One copy of the genes is shown on the top of the table and the other copy of the genes is shown down one side of the table. Only the mutant genes are shown in the table. "−" indicates failure to complement and "+" indicates complementation. (Note that a *proA⁻* mutation cannot complement another *proA⁻* mutation, a *proB⁻* mutation cannot complement another *proB⁻* mutation, and a *proC⁻* mutation cannot complement another *proC⁻* mutation). What gene does the *pro-53* mutation affect?

	proB⁻	proA⁻	proC⁻
proB⁻	−	+	+
proA⁻	+	−	+
proC⁻	+	+	−
pro-53	+	−	+

Answer: The new mutation affects the *proA* gene.

in partial diploids with an equal number of copies of each gene. If complementation analysis is done with one of the genes present in excess to the other gene (for example, with genes cloned on multicopy plasmids, as described in Chapter 11), artifacts can occur, which may be misleading.

Several examples demonstrate how complementation analysis can be used to identify mutations in different genes. First, we consider a set of mutations (1 through 8), in which rules 1 and 2a account for all of the data. After studying

Table 1-2 An example of complementation results. The interpretation of these results is described in the text.

Mutation Number										
	1	2	3	4	5	6	7	8	9	10
Mutation Number										
1	−									
2	+	−								
3	+	+	−							
4	+	−	+	−						
5	−	+	+	+	−					
6	+	+	−	+	+	−				
7	+	+	−	+	+	−	−			
8	+	−	+	−	+	+	+	−		
9	−	−	−	−	−	−	−			
10	±	−	±	−	±	±	±	−	−	

A "+" indicates complementation, a "−" indicates no complementation, and a "±" indicates weak complementation. Note that the results on the diagonal of the table should always be − because these results represent a complementation test between two copies of the same mutation. For example, mutation 3 will complement mutation 2, but mutation 2 will not complement another copy of mutation 2 and mutation 3 will not complement another copy of mutation 3. The data shown for mutations 9 and 10 represent unusual complementation results as described in the text.

this simple example, we consider a mutation that requires that rule 2b be invoked and then a second mutation that can be explained by rule 2c. The data are presented in Table 1-2. To analyze the first example, use rows 1 through 8 of the table. Each entry designates a single pairing of the mutation numbers shown at the top and side of the table. The results along the diagonal indicate that none of the mutations can complement a second copy of the same mutation. A − entry indicates that the pair does not complement; a + entry indicates complementation. In particular, notice any − entries that can be aligned down one column or across one row. These are noncomplementing mutations within a single gene. For example, in column 1, mutations 1 and 5 fail to complement, indicating that both mutations affect the same gene. The noncomplementing mutations identify three genes: The first contains mutations 1 and 5; the second contains mutations 2, 4, and 8; and the third contains mutations 3, 6, and 7. Mutations in each of these genes can complement mutations in each of the other two genes. We can give each complementing gene an identifying letter for convenience:

Group	Mutation
A	1,5
B	2,4,8
C	3,6,7

The simplest explanation for the data is that the phenotype being studied consists of at least three genes.

Now consider mutation 9, a case in which rule 2b would be applied:

	1	2	3	4	5	6	7	8	9
9	−	−	−	−	−	−	−	−	−

Alone these data might suggest that all of the mutations are in the same gene, but the data from mutations 1 through 8 indicate that this is not the case. For example, mutation 9 may be deletion mutation that removed all three genes, or a ***cis*-dominant mutation**, that is, a mutation that prevents expression of related genes residing on the same chromosome in which the mutation is located. For instance, a *cis*-dominant mutation might be in a site that signals the start of synthesis of all three gene products.

This interpretation of mutation 9 has not been done realistically because mutations 1 through 8 were classified first, and then 9 was introduced. How would all nine mutations have been classified if they were examined simultaneously? Once again, – entries would be observed and clustered. Mutation 9, however, would immediately be anomalous because it would appear in more than one complementation group—in fact, in all groups. Whenever a mutation appears in more than one group, the need for rule 2b should be suspected, and that mutation should temporarily be ignored in making the initial classification.

Now consider mutation 10, the rare case in which rule 2c would be applied. Because mutation 10 fails to complement mutations 2, 4, and 8, it might appear to be an additional mutation in the B gene. In contrast to mutations in the A or C genes, however, mutation 10 weakly complements mutations 1, 3, 5, 6, and 7. Thus, the simplest interpretation is that mutation 10 is in the B gene, but that the mutant gene product also inhibits the activity of a good copy of B. Thus, in each cell in which one would expect B to be fully active, B activity is weak because the total activity of the inhibited B is not adequate for a normal + response. In this case, mutation 10 is said to be negative dominant.

Intragenic Complementation

Some proteins are composed of multimers of identical polypeptide chains. When two copies of the mutant gene are provided, the two different mutant polypeptides can aggregate to form a functional or partly functional protein. This phenomenon, called **intragenic complementation**, occurs only with certain combinations of mutations, and therefore it is quite rare. One of the first examples of intragenic complementation involved the enzyme alkaline phosphatase. To be enzymatically active, alkaline phosphatase must form dimers with two identical subunits. Garen and Garen isolated many mutations that lacked alkaline phosphatase activity but still made the protein based on immunological tests— the protein was detected as cross-reacting material (CRM) that was bound by anti–alkaline phosphatase antibody. When certain combinations of these CRM$^+$ mutant genes were provided as merodiploids in a single cell, alkaline phosphatase activity was restored. Furthermore, intragenic complementation occurred when the mutant proteins were purified separately and mixed in vitro.

THE NEED FOR ISOGENIC STRAINS FOR GENETIC ANALYSIS

Mutations have great value in elucidating both genetic and biochemical properties of living systems. To infer genetic or biochemical properties from analysis of mutations, however, it is essential that the mutant bacterium differs from the parent bacterium by only one mutation. If several mutations were present, you would not know which one was responsible for an observed phenotypic change. Two organisms that differ by only a single mutation are said to be **isogenic**, or to have the same genetic background.

A variety of procedures are used to ensure that strains are isogenic. When mutations have been introduced by mutagenesis, the reversion frequency can be measured to ascertain that the mutation is a single point mutation. This procedure is fairly reliable but may not indicate the presence of other mutations that do not directly affect the phenotype. To avoid this problem, the small region of the chromosome carrying the mutation obtained in one strain is introduced into a desired strain by recombination. Strain construction is discussed further in Chapters 18 and 19.

KEY TERMS

allele	linkage
cis-dominant	merodiploid
complementation	mutagen
conditional mutant	mutagenesis
deletion mutation	mutant
diploid	mutation
dominant	negative dominant
genetic mapping	null mutant
genetic marker	phenotype
genetic screen	point mutation
genetic selection	recessive
genetics	recombination
genome	recombination frequency
genotype	reversion
haploid	reversion frequency
heterozygous	revertant
homozygous	spontaneous mutagenesis
insertion mutation	suppressor
intragenic complementation	temperature-sensitive mutant
isogenic	wild-type

QUESTIONS AND PROBLEMS

1. What is the difference between genotype and phenotype?

2. What is the difference between a null mutation and a conditional mutation?

3. What are two types of conditional mutations?

4. What is the difference between a genetic selection and a genetic screen?

5. What is the difference between complementation and recombination?

6. The frequency of producing mutant *arb* is 2×10^{-6} per generation and mutant *snd* is 8×10^{-5}. What is the frequency of producing an *arb snd* double mutant in a single event?

7. Rank the following mutations in order of reversion frequency: deletion, point mutation, double mutation.

8. A mutation prevents the synthesis of substance Z and results in the accumulation of large amounts of substance R, which is normally present in only very small amounts. What can probably be said about the relation between R and Z and about the gene product inactivated by the mutation?

9. Two phages with genotypes x^+y^- and x^-y^+ are crossed. What are the genotypes of the possible recombinant types? Are any of them wild-type?

10. The recombination frequencies between three genes are: *a-b* 2.6%, *b-d* 1.4%, and *a-d* 1.2%. What is the gene order?

11. Mutants that fail to synthesize a substance X have been found in four complementation groups, none of which are *cis*-acting. How many proteins are required to synthesize X?

12. In a haploid organism mutation *x* eliminates production of a pigment; another mutation *y* does the same. An x^-/y^- merodiploid cell makes pigment. Do *x* and *y* complement? How many genes are required for pigment formation?

13. The following recombination frequencies occur between the indicated markers: $a \times c$ 2%; $b \times c$ 13%; $b \times d$ 4%; $a \times b$ 15%; $c \times d$ 17%; $a \times d$ 19%.

    a. What is the gene order?

    b. In the cross $aBd \times AbD$, what is the frequency of getting ABD progeny?

14. In a cross between two phage having genotypes EFG and efg, 1000 progeny were analyzed. The number of phage having each of the eight possible genotypes were as follows: efg 396; EFG 406; eFg 23; efG 1; EfG 25; Efg 75; eFG 73; EFg 1. Construct a map showing the positions of the genetic markers.

15. Four genes, $kyuA$, $kyuB$, $kyuC$, and $kyuQ$, are known to be required to synthesize substance Q, and each biochemical reaction can be detected. The reaction sequence is P → B → A → Q in which the product of a gene $kyuX$ is needed to synthesize substance X. Addition of ^{14}C-P yields ^{14}C-Q.

    a. A mutant is found for which addition of ^{14}C-P yields ^{14}C-A but no ^{14}C-Q. In what gene is this mutation?

    b. Another mutant is found for which there is no conversion of ^{14}C-P to any other substance. Furthermore, addition of ^{14}C-A fails to yield ^{14}C-Q. What kind of mutant would have this phenotype?

16. The complementation data shown in the following table are observed. The numbers refer to particular mutations. The symbols + and – indicate that the two mutations do and do not complement, respectively. How many genes are represented? Assign the mutations to the genes.

	Mutants						
	1	2	3	4	5	6	7
1	–	+	+	+	+	+	–
2		–	+	–	+	+	+
3			–	+	+	–	+
4				–	+	+	+
5					–	+	+
6						–	+
7							–

17. A new $putP$ mutation was mapped against a set of $putP$ deletion mutations. The regions removed by the deletion mutations are indicated by open boxes below the $putP$ gene, and the results showing whether or not recombinants were obtained are shown to the right of each deletion. Based on these results, where does the new $putP$ mutation map?

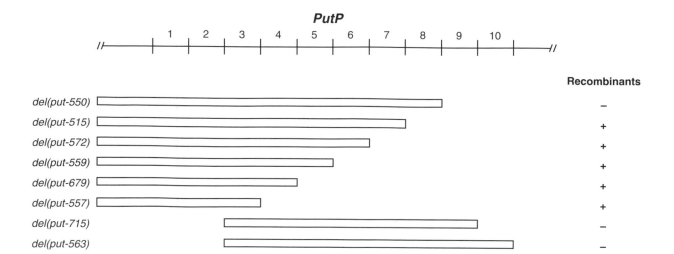

18. To map a new mutation that affects proline transport, Liao and Maloy did two-factor crosses with linked *livA* and *tet* genes. The results are shown below:

Donor	Recipient	Selected phenotype	Recombinants	Number obtained
proZ tet[r]	*proZ*[+] *tet*[s]	Tet[r]	*proZ*	78
			proZ[+]	120
livA tet[r]	*livA*[+] *tet*[s]	Tet[r]	*livA*	70
			livA[+]	128
proZ livA[+]	*proZ livA*	LivA[+]	*proZ*	68
			proZ[+]	78

From these data, draw a linkage map of the *proZ, livA,* and *tet* genes.

19. To confirm the gene order determined from two-factor crosses, the following three-factor cross was done to map *proZ:*

Donor	Tet[r]	proZ	livA[+]
Recipient	Tet[s]	proZ[+]	livA

Selected phenotype	Recombinants	Number obtained
Tet[r]	*proZ livA*[+]	10
	proZ livA	68
	proZ[+] *livA*[+]	2
	proZ[+] *livA*	118

Based on these data, determine the order of the *tet, proZ,* and *livA* genes.

20. The nine genes in the *his* operon are required for biosynthesis of histidine. Will the following merodiploid strain be a histidine auxotroph?

$$hisG^+\ D^-\ B^+\ C^+\ H^+\ A^-\ F^+\ I^+\ /\ hisA^+\ F^-\ I^+$$

21. Hughes and Roth isolated a mutation (*nadD*) in a gene required for NAD biosynthesis in *Salmonella*. Hfr mapping indicated that the gene mapped at 14 min near the *lip* gene. They did two-factor crosses with phage P22 to determine the linkage map of *nadD* with other genes that mapped in this region. From the following two-factor cross data construct a linkage map of the *nadD, lip,* and *leuS* genes.

Donor	Recipient	Selected phenotype	Recombinants	Number obtained
nadD[+] *lip*	*nadD lip*[+]	NadD[+]	*lip*	47
			lip[+]	53
nadD lip[+]	*nadD*[+] *lip*	Lip[+]	*nadD*	57
			nadD[+]	43
nadD[+] *leuS*	*nadD leuS*[+]	NadD[+]	*leuS*	77
			leuS[+]	23
lip[+] *leuS*	*lip leuS*[+]	Lip[+]	*leuS*	37
			leuS[+]	63

22. Hughes and Roth also did three-factor crosses to confirm the order of the *nadD* gene relative to the adjacent genes. Do these data agree with the linkage map constructed from the two-factor crosses?

Donor strain	*nadD*[+]	*lip*[+]	*leuS*
Recipient strain	*nadD*	*lip*	*leuS*[+]

Selected phenotype	Recombinants	Number obtained
Lip⁺	nadD⁺ leuS	94
	nadD⁺ leuS⁺	32
	nadD leuS	2
	nadD leuS⁺	170
NadD⁺	lip⁺ leuS	90
	lip⁺ leuS⁺	15
	lip leuS	78
	lip leuS⁺	25

23. Beck and Maloy isolated mutants defective for the glyoxylate shunt (*aceBAK*) in *Salmonella*. The mutations mapped near the *metA* gene at 90 min on the *Salmonella* chromosome. They did two-factor crosses to determine the linkage of *ace* to nearby genes. Draw the linkage map from the results of the following two-factor crosses.

Donor	Recipient	Selected marker	Recombinants	Number obtained
ace metA⁺	ace⁺ metA	metA⁺	ace	1631
			ace⁺	569
ace⁺ metA	ace metA⁺	ace⁺	metA	2207
			metA⁺	303
ace purD⁺	ace⁺ purD	purD⁺	ace	183
			ace⁺	620
ace⁺ purD	ace purD⁺	ace⁺	purD	217
			purD⁺	283
metA purD⁺	metA⁺ purD	purD⁺	metA	124
			metA⁺	76
metA⁺ purD	metA purD⁺	metA⁺	purD	190
			purD⁺	110
thiA purD⁺	thiA⁺ purD	purD⁺	thiA	31
			thiA⁺	169
thiA metA⁺	thiA⁺ metA	metA⁺	thiA	22
			thiA⁺	446
thiA aceA⁺	thiaA⁺ aceA	aceA⁺	thiA	0
			thiA⁺	315

24. Beck and Maloy also did three-factor crosses to determine the order of the *ace* gene with the nearby genes. Determine the gene order from the following three-factor crosses.

a. Donor strain   *metA⁺ ace⁺ purD⁺*
     Recipient strain   *metA⁺ ace⁺ purD*

Selected phenotype	Recombinants	Number obtained
Ace⁻	metA purD⁺	139
	metA purD	54
	metA⁺ purD⁺	2
	metA⁺ purD	80
PurD⁺	metA ace	101
	metA ace⁺	36
	metA⁺ ace	9
	metA⁺ ace⁺	71

b. Donor strain   *metA⁺ ace purD*
     Recipient strain   *metA ace⁺ purD⁺*

Selected phenotype	Recombinants	Number obtained
Ace⁻	metA purD	1
	metA purD⁺	16
	metA⁺ purD	59
	metA⁺ purD⁺	84
MetA⁺	ace purD	26
	ace purD⁺	39
	ace⁺ purD	10
	ace⁺ purD⁺	13

25. What is the phenotype of a *leu*(Ts)/*leu*⁺ merodiploid? Does the phenotype depend on the temperature?

REFERENCES

Bachmann, B. 1990. Linkage map of *Escherichia coli* K-12, Edition 8. *Microbiol. Rev.* 54: 130.

Demerec, M., E. Adelberg, A. Clark, and P. Hartman. 1966. A proposal for a uniform nomenclature in bacterial genetics. *Genetics 54.* 61.

Fincham, J. R. S. 1966. *Genetic Complementation.* Benjamin-Cummings, Menlo Park, CA.

*Fogiel, M. 1985. *The Genetics Problem Solver.* Research and Education Association, NJ.

*Gonick, L., and M. Wheelis, 1991. *The Cartoon Guide to Genetics*, updated edition. Harper Perennial, New York.

*Hayes, W. 1968. *The Genetics of Bacteria and Their Viruses*. John Wiley and Sons, New York.

*King, R., and W. Stansfield. 1985. *A Dictionary of Genetics*, Third Edition. Oxford University Press, New York.

Sanderson, K., and J. Roth. 1988. Linkage map of *Salmonella typhimurium*, Edition VII. *Microbiol. Rev.* 52: 485.

*Smith-Keary, P. 1975. *Genetic Structure and Function*. John Wiley and Sons, New York.

*Snyder, L. A., D. Freifelder, and D. L. Hartl. 1985. *General Genetics*. Jones and Bartlett, Boston.

*Stahl, F. W. 1987. Genetic recombination. *Scientific Am.* 52: 90.

*Stent, G., and R. Calendar. 1978. *Molecular Genetics: An Introductory Narrative*. W. H. Freeman, San Francisco.

*Resources for additional information.

2

Nucleic Acids

All hereditary information resides in nucleic acids. In most organisms, genes are segments of DNA molecules, but in a few phages and many animal and plant viruses, RNA is the genetic material. A great deal of what we know about the nature of genes and gene expression has been obtained from knowledge of the structure of DNA or RNA, from experiments in which DNA is used to carry genetic information from one organism to another, or from studies in which DNA or RNA is altered. Hence, a review of the properties of nucleic acids is provided in this chapter.

COMPONENTS OF NUCLEIC ACIDS

DNA and RNA are polynucleotides—that is, polymers of nucleotides. Each **nucleotide** has three components (Fig. 2-1).

1. A *cyclic five-carbon sugar*. This is ribose in ribonucleic acid (RNA), and deoxyribose in deoxyribonucleic acid (DNA). The structures of ribose and 2'-deoxyribose differ only in the absence of a 2'-OH group in deoxyribose. The bulky 2-OH group in RNA limits the range of secondary structures the RNA can form and makes RNA more susceptible to chemical and enzymatic degradation than DNA.
2. A *purine or pyrimidine base* attached to the 1'-carbon atom of the sugar by an *N*-glycosidic bond. The bases are the purines adenine (A) and guanine (G) and the pyrimidines cytosine (C), thymine (T), and uracil (U) (Fig. 2-2). DNA and RNA both contain A, G, and C; however, T is usually found only in DNA, and U is usually found only in RNA.
3. A *phosphate* attached to the 5'-carbon of the sugar by a phosphoester linkage. This phosphate is responsible for the strong negative charge of both nucleotides and nucleic acids. A base linked to a sugar is called a **nucleoside**; thus, *a nucleotide is a nucleoside phosphate*.

The nucleotides in nucleic acids are covalently linked by a second phosphoester bond that joins the 5'-phosphate of one nucleotide and the 3'-OH group of the adjacent nucleotide (Fig. 2-3). This bond between the phosphate and the 3'- and 5'-carbon atoms is called a **phosphodiester bond**. The result of successive linkage of nucleotides in a polynucleotide is an alternating sugar-phosphate backbone having one 3'-OH terminus and one 5'-phosphate terminus.

A typical RNA molecule is the single-stranded polyribonucleotide. In contrast, except in unusual cases, DNA contains two polydeoxynucleotide strands wrapped around one another to form a double-stranded helix.

THE DOUBLE HELIX

X-ray diffraction studies in the early 1950s indicated that the common form of DNA, called B-DNA, is an extended chain with a highly ordered structure: DNA is helical with the nucleotide bases stacked with their planes separated by 34 nm. Chargaff did chemical analysis of the molar content of the bases (generally called the **base composition**) in DNA molecules isolated from many organisms—in each case, [A] = [T] and [G] = [C] (where the brackets denote mole fraction). James Watson and Francis Crick combined the physical and chemical data and determined that the two strands are coiled about one another to form a double-stranded helix (Fig. 2-4). The sugar-phosphate backbones follow a helical path at the outer edge of the molecule, and the bases are in a helical array in the central

Figure 2-1. A typical nucleotide showing the three major components, the difference between DNA and RNA, and the distinction between a nucleoside and a nucleotide.

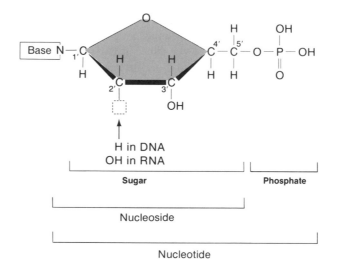

Figure 2-2. The bases found in nucleic acids. The weakly charged groups are shown in red. Adenine and guanine are derivatives of purines. Cytosine, thymine, and uracil are derivatives of pyrimidine.

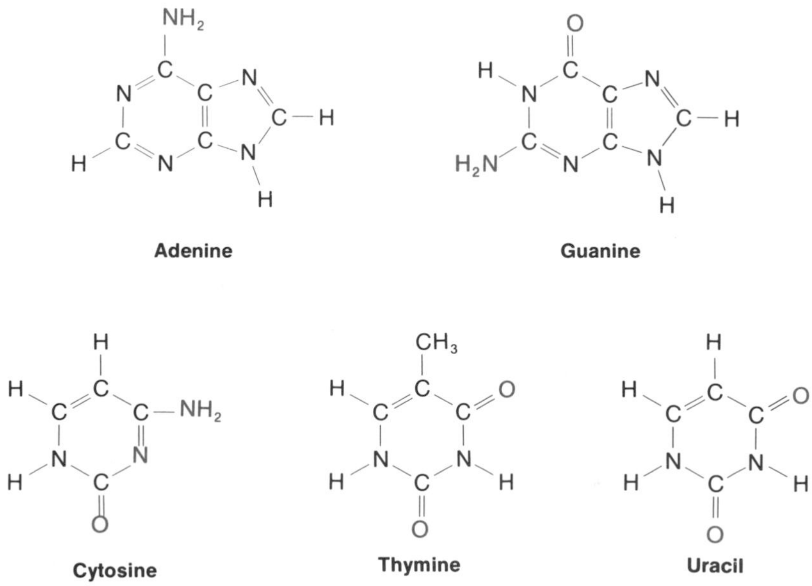

core. The bases of one strand are hydrogen bonded to those of the other strand to form the purine-pyrimidine base pairs AT and GC (Fig. 2-5). Because each pair contains one two-ringed purine (A or G) and one single-ringed pyrimidine (T or C), the length of each pair is about the same, and the helix can fit into a smooth cylinder.

The two bases in each base pair lie in the same plane, and the plane of each pair is perpendicular to the helix axis. The base pairs are rotated with respect to each other to produce 10 pairs per helical turn. The diameter of the double helix is 200 nm, and the molecular weight per unit length of the helix is approximately

Base   Base

5′-P terminus   **Phosphodiester group**   **3′-OH terminus**

Figure 2-3. The structure of a dinucleotide. The vertical arrows show the bonds in the phosphodiester group about which there is free rotation. The horizontal arrows indicate the *N*-glycosidic bond about which the base can freely rotate. A polynucleotide would consist of many nucleotides linked together by phosphodiester bonds.

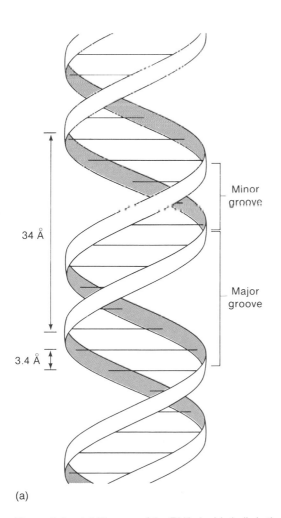

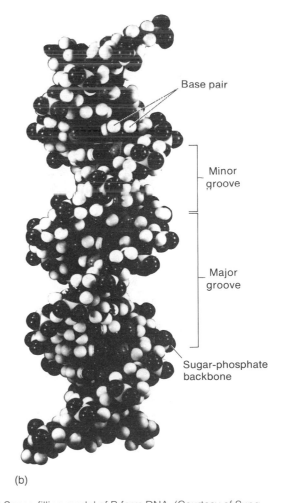

(a)   (b)

Figure 2-4. (a) Diagram of the DNA double helix in the common B form.

(b) Space-filling model of B form DNA. (Courtesy of Sung-Hou Kim.)

2×10^6 per micrometer. Because the molecular weight of a typical bacterial DNA molecule is about 2×10^9, the DNA molecule is very long: about 1 mm long.

DNA is a right-handed helix. This means that each strand appears to follow a clockwise path moving away from a viewer looking down the helical axis. The DNA helix has two external helical grooves, a deep wide one (the **major groove**) and a shallow narrow one (the **minor groove**), as shown in Figure 2-4. Most specific DNA-binding proteins bind in the major groove.

The two polynucleotide strands of the DNA double helix are antiparallel—that is, the 3'-OH terminus of one strand is adjacent to the 5'-P terminus of the other (Fig. 2-6). Thus, in a linear double helix, there is one 3'-OH and one 5'-P terminus at each end of the helix. We see in Chapter 8 that the antiparallel helix poses a constraint on the mechanism of DNA replication.

By convention, the sequence of bases of a single chain is usually written with the 5'-P terminus at the left; for example, ATC denotes the trinucleotide P-5'-ATC-3'-OH, which may also be written pApTpC.

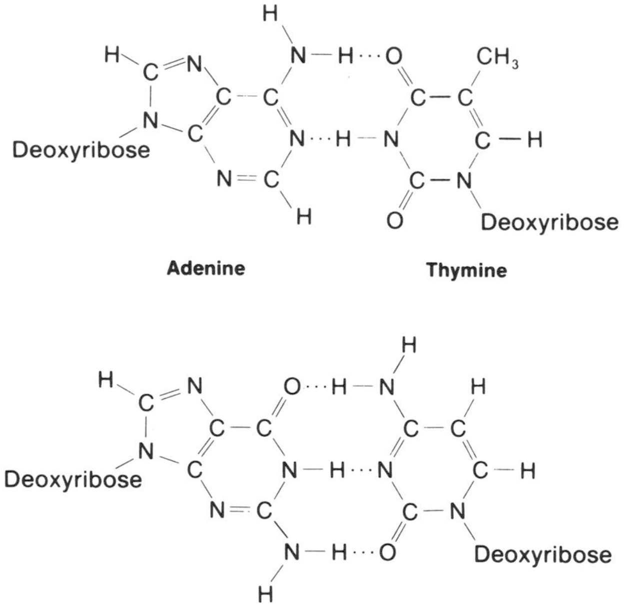

Figure 2-5. The two common base pairs of DNA.

Adenine   Thymine

Guanine   Cytosine

Figure 2-6. A stylized drawing of a segment of a DNA duplex showing the antiparallel orientation of the complementary chains. The arrows indicate the 5' → 3' direction of each strand. The phosphates (P) join the 3' carbon of one deoxyribose (horizontal line) to the 5' carbon of the adjacent deoxyribose.

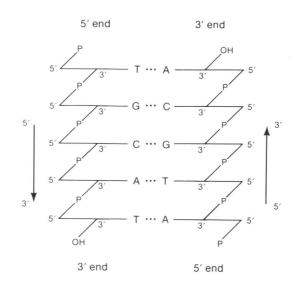

DENATURATION AND MELTING CURVES

The attractive forces that produce the three-dimensional structure of molecules, such as the hydrogen bonding between DNA base pairs, are fairly weak and easily disrupted by heat. When the hydrogen bonds are disrupted, the DNA is said to be **denatured**; when the hydrogen bonds are intact, as in double-stranded DNA in nature, the DNA is said to be **native**. The transition from the native to the denatured state is called **denaturation**. When native double-stranded DNA is heated, the hydrogen bonds between the strands are broken, and the two strands separate; thus, *denatured DNA is single-stranded*.

A great deal of information about structure and stabilizing interactions in DNA has been obtained by studying denaturation. This is typically done by measuring some property of the molecule that changes as denaturation proceeds. For example, DNA absorbs ultraviolet light with a wavelength of 260 nm. At this wavelength, single-stranded DNA absorbs more strongly than double-stranded DNA. In early experiments, denaturation was accomplished by heating a DNA solution, so a graph of a varying property as a function of temperature is called a **melting curve**. A convenient measure of the light absorption is the absorbance (A) of a solution:

$$A = -\log_{10}\left[\frac{\text{Intensity of light transmitted by a solution 1 cm in diameter}}{\text{Intensity of incident light}}\right]$$

An example of a melting curve is shown in Figure 2-7. The state of a DNA molecule in different regions of the melting curve is also shown. Before the rise in A begins, the molecule is fully double-stranded. In the rise region, base pairs in various segments of the molecule are broken; the number of broken base pairs increases with temperature. In the initial part of the upper plateau, a few base pairs remain to hold the two strands together, until a critical temperature is reached at which the last base pair breaks, and the strands separate completely.

A convenient parameter to characterize a melting transition is the temperature at which the rise in A_{260} is half complete. This temperature is called the melting temperature and is designated T_m. AT pairs are held together by two hydrogen bonds, and GC pairs are held together by three hydrogen bonds, so a higher temperature is required to disrupt GC pairs than AT pairs. For this reason, the value of T_m is related to the base composition of the DNA. In solutions

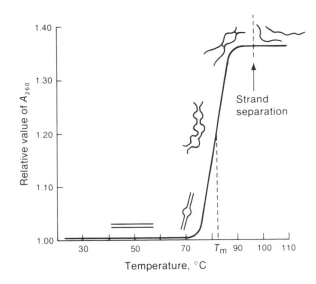

Figure 2-7. A melting curve of DNA showing T_m and possible molecular conformations for various degrees of melting.

standardized with respect to salt concentration and pH, the T_m can be used to measure the base composition. At high pH, the charge of several of the groups engaged in hydrogen bonding is changed, preventing the base from forming hydrogen bonds. At a pH greater than 11.3, all the hydrogen bonds are disrupted, and DNA is completely denatured. At high temperatures and neutral pH, the phosphodiester bonds can be broken. The phosphodiester bonds are quite resistant to alkaline hydrolysis, so treatment at high pH is the method of choice for denaturing DNA without breaking covalent bonds.

When heated DNA is quickly returned to room temperature or when alkali-denatured DNA is restored to neutral pH, the single strands remain separate. If the salt concentration of the solvent is low, the strong negative charge of the phosphate groups keeps the strands extended and single-stranded throughout. If the salt concentration is high enough to neutralize the negative charge, however, the single strands fold back on themselves, forming compact molecules with intrastrand base-pairing. Because lengthy complementary sequences are rare, however, the paired segments rarely contain more than 10 nucleotides.

RENATURATION

A solution of denatured DNA can be treated in such a way that native DNA reforms. The process is called **renaturation** or **reannealing**, and the reformed DNA is called renatured DNA. Renaturation has proved to be a valuable tool in microbial genetics because it can be used to demonstrate genetic relatedness between different organisms, to detect particular species of RNA, to determine whether certain sequences occur more than once in the DNA of a particular organism, and to locate specific base sequences in a DNA molecule. Two requirements must be met for renaturation to occur.

1. The salt concentration must be high enough that electrostatic repulsion between the phosphates in the two strands is eliminated. Usually 0.15 to 0.50 molar NaCl is used.
2. The temperature must be high enough to disrupt the random, intrastrand hydrogen bonds described in the previous section. The temperature cannot be too high, however, or stable interstrand base-pairing will not occur. The optimal temperature for renaturation is 20 to 25°C below the T_m.

Renaturation is a slow process compared with denaturation. The rate-limiting step is not the actual rewinding of the helix (which occurs rapidly—within seconds) but the precise collision between complementary strands such that base pairs are formed at the correct positions. Because this is a result only of random motion, it is a concentration-dependent process; at concentrations normally encountered in the laboratory, renaturation may take several hours. Renaturation can be detected by the decrease in absorbance of a DNA solution.

Filter Hybridization

Thin filters (membrane filters) made of nitrocellulose are commercially available. These filters bind single-stranded DNA tightly but do not bind either double-stranded DNA or RNA. They provide a useful method for measuring hybridization, as shown in Figure 2-8. A sample of denatured DNA is filtered. The single strands bind tightly to the filter along the sugar-phosphate backbone, but the bases remain free. The filter is then placed in a vial with a solution containing a reagent that prevents additional binding of single-stranded DNA to the filter and a small amount of radioactive denatured DNA. After a period of renaturation, the filter is washed. Radioactivity is found on the filter only if renaturation has occurred.

This filter-binding assay can be used to determine whether two organisms have common base sequences. For example, if excess *Escherichia coli* DNA is on the filter and a small amount of denatured ^{14}C-labeled DNA isolated from *Salmonella typhimurium* is added, the fraction of the applied ^{14}C that is retained on the filter is proportional to the fraction of *S. typhimurium* DNA that has a sequence with sufficient complementarity with that of *E. coli* to hybridize under the annealing conditions used. The value of this fraction indicates the degree of sequence similarity. This type of experiment has confirmed the basic expectations of evolutionary theory—that is, taxonomically related organisms have common sequences, and the sequence similarity reflects the relatedness determined by other criteria.

Another important use of filter-binding assays is the detection of RNA encoded by a specific region of DNA. This is called **DNA-RNA hybridization**. In this procedure, a filter to which single strands of DNA have been bound, as previously, is placed in a solution containing radioactive RNA. After renaturation, the filter is washed, and hybridization is detected by the quantitating radioactive RNA on the filter using a liquid scintillation counter.

Filter hybridization can also be used to assess the sequence similarity between two DNA segments. This is done by varying the pH, salt concentration, and renaturation temperature. Certain conditions (termed stringent) allow annealing only when complementarity is nearly perfect. Less stringent conditions permit annealing of DNA segments binding with less complementarity.

DNA Heteroduplexes

Renaturation has been combined with electron microscopy in a procedure that allows the localization of common, distinct, and deleted sequences in DNA. This procedure is called **heteroduplex mapping**. Consider the DNA molecules #1 and #2 shown in Figure 2-9a. These molecules differ in sequence only in one region. If a mixture of the two molecules is denatured and then renatured, in addition to parental molecules, hybrid molecules having unpaired single strands are produced, as shown in Figure 2-9b. Figure 2-10 shows an actual electron micrograph of a heteroduplex. Measurement of the lengths of the single-stranded and double-stranded regions yields the end points of the regions of nonhomology. If the sequences by which the molecules differ have the same or nearly the same number of nucleotides, the two single strands of the bubble have the same length.

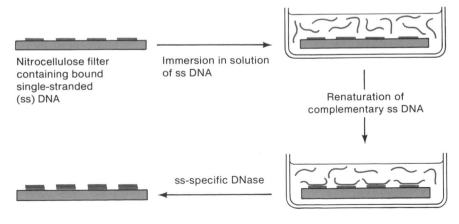

Nitrocellulose filter
containing bound
single-stranded
(ss) DNA

Immersion in solution
of ss DNA

Renaturation of
complementary ss DNA

ss-specific DNase

Figure 2-8. DNA hybridization on nitrocellulose filters containing bound, single-stranded DNA. In the final step, the filter is treated with a single strand–specific DNase, an enzyme that degrades single-stranded but not double-stranded DNA.

Consider now molecule #3, shown in Figure 2-9a. In this molecule, region A is deleted. If a hybrid is made between this molecule and molecule #2, the result is a molecule with a single loop, as shown in Figure 2-9b.

Heteroduplex mapping is also possible between a double-stranded DNA molecule and an RNA molecule that is complementary to part of the DNA sequence. Denaturation of the DNA and reannealing with the RNA produces a molecule with a bubble, called an **R-loop**. One branch of the bubble is a DNA-RNA hybrid, and the other is single-stranded DNA. This technique can be used to map DNA sequences from which particular RNA molecules are copied.

CIRCULAR AND SUPERHELICAL DNA

The intact chromosomes of most prokaryotes and plasmids are circular. A circular molecule may be a **covalently closed circle**, which consists of two unbroken complementary single strands, or it may be a **nicked circle**, which has one or more interruptions (**nicks**) in one or both strands, as shown in Figure 2-11. With few exceptions, covalently closed circles are twisted, as shown in Figure 2-12. Such a circle is said to be a **supercoiled**. What is a supercoiled DNA molecule? The first point to be understood is that the word "super" in supercoil does not mean that the DNA double helix is overwound, but that there is another form of coiling **superimposed** on that found in linear DNA.

It is worth reviewing what is meant by a right-handed or positive coil. Helical coiling is positive if, when looking down the helical axis, the coil follows a clockwise path and moves away from the viewer. If the path is counterclockwise, the coil is left-handed, or negative. Recall that DNA is a right-handed helix.

The two ends of a linear DNA helix can be brought together and joined in such a way that each strand is continuous. If, in so doing, one of the ends is rotated 360 degrees with respect to the other to produce some unwinding of the double helix and then the ends are joined, the resulting covalent circle, if the hydrogen bonds reform, twists in the opposite sense (here, opposite to the unwinding direction) to form a twisted circle, to relieve strain. Such a molecule looks like a figure 8 (that is, have one crossover point or **node**.) If it is instead twisted 720 degrees before joining, the resulting superhelical molecule has two nodes. The reason for the twisting is the following. In the case of a 720-degree unwinding of the helix, 20

Figure 2-9. (a) Three DNA molecules to be heteroduplexed. Sequences A and A' of molecules 1 and 2 differ. Neither sequence is present in the deletion molecule 3. The dashed lines indicate reference points. (b) Heteroduplexes resulting from renaturing molecules 1 and 2 or 2 and 3.

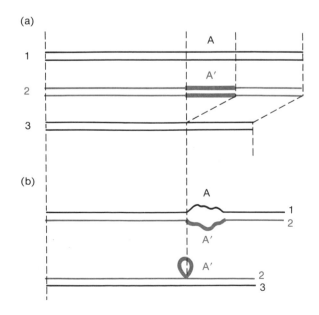

base pairs must be broken (because the helix has been unwound two turns, and the linear molecule has 10 base pairs per turn of the helix). To maintain a right-handed (positive) helical structure with 10 base pairs per turn, the DNA deforms in such a way that the underwinding is rewound and compensated for by negative (left-handed) twisting of the circle. Similarly, the initial rotation might instead be in the direction of overwinding, in which case the joined circle twists in the opposite sense, forming a positive superhelix. Both underwound and overwound molecules are supercoils because the twisting is the superimposed coiling. Generally naturally occurring superhelical DNA molecules are initially underwound and hence form negative superhelices. (The DNA from an Archae phage is the one possible exception to this rule). Furthermore, the degree of twisting is about the same for all molecules; one negative twist is produced per 200 base pairs, or 0.05 twists per turn of the helix. In bacteria, the underwinding of superhelical DNA is not a result of unwinding before joining but is introduced into preexisting circles by an enzyme called **DNA gyrase**, which is described in Chapter 8, when DNA replication is examined.

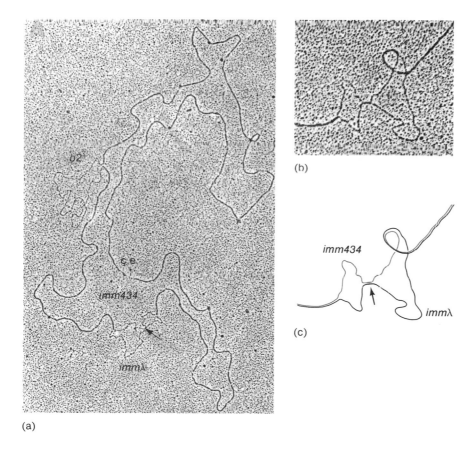

(a)

(b)

(c)

Figure 2-10. An electron micrograph of a heteroduplex between λ *imm*λ *b2* DNA, which carries the *b2* deletion, and λ *imm434 b2⁺*, in which the *imm*λ segment is replaced by the shorter, nonhomologous *imm434* segment. (a) Two bubbles of nonhomology are seen. The identity of each single-stranded segment is indicated. The arrow is explained in part (c) (b) An enlargement of the *imm434-imm*λ segment. (c) An interpretive drawing of panel (b). The arrow indicates a region of homology between the *imm434* and *imm*λ segments. The same region is indicated by the arrow in panel (a). (Courtesy of Barbara Westmoreland and W. Szybalski.)

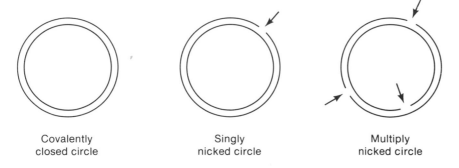

Covalently closed circle

Singly nicked circle

Multiply nicked circle

Figure 2-11. A covalently closed circle and two kinds of nicked circles. Arrows indicate the nicks. A nicked circle is also called an open circle.

Single-Stranded Regions in Superhelices

We have just pointed out that the strain of underwinding can be accommodated by negative supercoiling. Three other arrangements that could counteract the strain of underwinding are possible: (1) The number of base pairs per turn of the helix could change. This does not happen, however, for thermodynamic reasons. (2) All of the underwinding could be taken up by having one or more large single-stranded bubbles (Fig. 2-13). (3) The underwinding could be taken up in part by superhelicity and in part by bubbles. This is what is actually observed—DNA molecules are dynamic structures that constantly undergo transient unwinding. If a circular molecule were made that was not underwound, transient breakage and remaking of hydrogen bonds (**breathing**) would introduce compensating transient negative twists. If the DNA is initially superhelical, the degree of supercoiling fluctuates as breathing occurs: The strain produced by the underwinding is relieved in a superhelix both by the superhelicity and by an increase in the number and size of the bubbles and the duration of their existence. Thus, in a supercoil, the fraction of the molecule that is single-stranded at any moment is greater than in a nicked circle.

Figure 2-12. Nicked circular and supercoiled DNA of phage PM2. (Courtesy of K. G. Murti.)

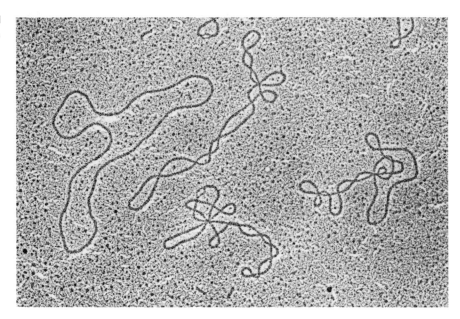

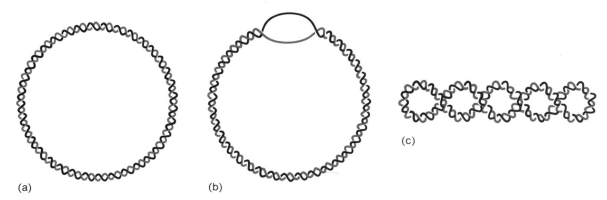

(a)   (b)   (c)

Figure 2-13. Different states of a covalent circle. (a) A non-supercoiled covalent circle having 36 turns of the helix. (b) An underwound covalent circle having only 32 turns of the helix. (c) The molecule in part (b) but with four superhelical turns to eliminate the underwinding. In solution, (b) and (c) would be in equilibrium; the equilibrium would shift toward (b) with increasing temperature.

Experimental Detection of Covalently Closed Circles

In the life cycles of many organisms, the DNA molecules cycle through the various circular forms that have just been described. Two techniques are commonly used to distinguish these forms.

1. *Sedimentation at denaturing alkaline pH.* Above pH 11.3, a linear DNA molecule in a salt concentration of 0.3 M denatures to yield two single strands, each of which sediments about 30% faster than native DNA. The two strands of a covalently closed circle, however, cannot separate (because there are no free ends to allow unwinding), so the molecules collapse in a tight tangle that sediments about three times faster than native DNA. If the circle has a single nick, one linear molecule and one single-stranded circle result (Fig. 2-14a), and the circle sediments 14% more rapidly than the linear molecule. Figure 2-14b shows a sedimentation pattern for a mixture composed of linear molecules, nicked circles, and covalently closed circles in an alkaline solution.

2. *Equilibrium centrifugation in CsCl-containing ethidium bromide.* **Ethidium bromide** (Fig. 2-15) binds tightly to DNA and, in so doing, decreases the density of the DNA by approximately 0.15 g/cm^3. It binds by intercalating between the DNA base pairs, thereby causing the DNA molecule to unwind as more ethidium bromide is bound. Because a covalently closed DNA molecule has no free ends as it unwinds, the entire molecule twists in the opposite direction; the degree of twisting increases as more molecules intercalate. Ultimately the DNA molecule is unable to twist any more, so no more ethidium bromide molecules can be bound. On the contrary, a linear DNA molecule or a nicked circle does not have the topological constraint of reverse twisting and can therefore bind more of the ethidium bromide molecules. Because the density of the DNA and ethidium bromide complex decreases as more ethidium bromide is bound and because more ethidium bromide can be bound to a linear molecule

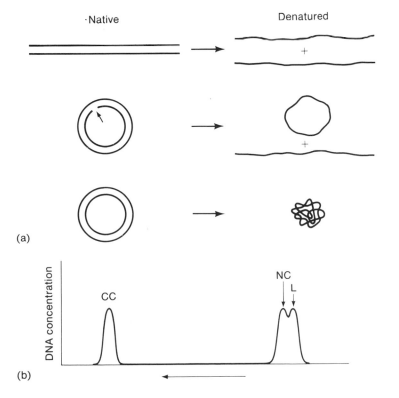

(a)

(b)

Figure 2-14. (a) Products of the denaturation of different forms of DNA. (b) Separation of covalently closed circles (CC), nicked circles (NC), and linear (L) molecules by sedimentation in alkali. The horizontal axis represents the length of a centrifuge tube. Sedimentation is from right to left.

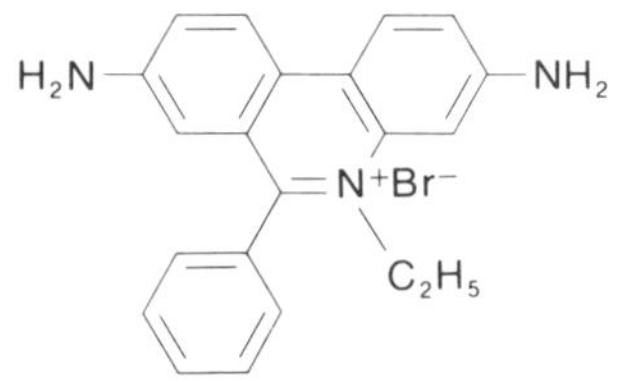

Figure 2-15. Chemical structure of ethidium bromide.

or an open circle than to a covalent circle, the covalent circle has a higher density at saturating concentrations of ethidium bromide. Therefore, covalent circles can be separated from the other forms in an equilibrium density gradient, as shown in Figure 2-16. Equilibrium centrifugation is reviewed in a later section of this chapter.

3. Supercoiling can also be detected by electron microscopy (see Fig. 2-12) or by agarose gel electrophoresis as described later in this chapter.

STRUCTURAL CONSEQUENCES OF SPECIAL BASE SEQUENCES

Certain base sequences impart unique structures to nucleic acids. Several types of repeated sequences commonly occur in regulatory regions and to sites of enzymatic activity and may in some cases impart special properties to either double-stranded or single-stranded nucleic acids. Other sequences in which purines and pyrimidines alternate can cause DNA to form a left-handed helical region.

One special type of sequence is a palindrome. A **palindrome** is a sequence of the general form:

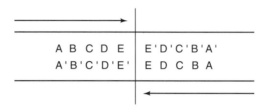

where A and A', B and B', and so forth are complementary bases able to pair. The dashed vertical line is an axis of symmetry: The double-stranded segment to the right of the axis can be superimposed on the one to the left by a 180-degree rotation in the plane of the page. Other terms used to describe palindromes are

Figure 2-16. Effect of ethidium bromide on the density of DNA in a CsCl solution. A mixture of equal parts of nicked circles (NC) and covalently closed circles (CC) is centrifuged in CsCl containing various concentrations of ethidium bromide. The density of the DNA molecules decreases until, at saturation, the two components separate. The covalent circles bind less ethidium bromide and therefore sediments at a higher density.

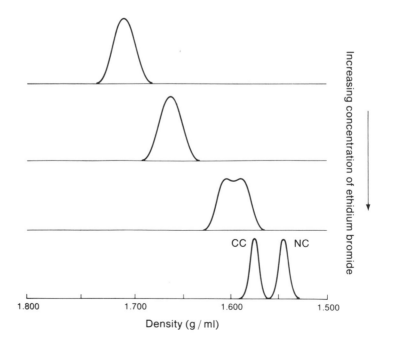

inverted repeat and region of **dyad symmetry**. Palindromes range in length up to about 50 base pairs. The two inverted sequences may be separated by a spacer, for example;

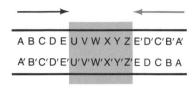

A B C D E U V W X Y Z E'D'C'B'A'
A'B'C'D'E'U'V'W'X'Y'Z'E D C B A

in which case only the term inverted repeat is used. Because DNA breathes, molecules containing palindromes and inverted repeats can in theory assume alternative structures (Fig. 2-17). Once complementary strands have separated, intramolecular base pairing can cause a double-stranded branch to form between adjacent complementary sequences. This structure is referred to as a **cruciform**. Cruciforms are not as stable as the double stranded DNA because they have less hydrogen bonding. They can be produced in the laboratory but are not usually stable in double-stranded DNA in vivo. Both palindromes and interrupted inverted repeats, however, have significant effects on the structure of RNA and single-stranded DNA found in certain phages. Both single-stranded DNA and RNA can readily form intrastrand hydrogen bonding between adjacent or nearby complementary sequences. Thus, a palindrome produces an intrastrand, double-stranded segment called a **hairpin** (Fig. 2-18a), and an interrupted inverted repeat produces a structure consisting of a double-stranded segment with a terminal single-stranded loop, known as a **stem-and-loop** (Fig. 2-18b). We see in Chapters 6 and 7 that stem-and-loop structures are important in the regulation of RNA synthesis.

A sequence may also be repeated in the same orientation with or without a spacer:

A B C D E   U V W X Y Z   A'B'C'D'E'
A'B'C'D'E'   U'V'W'X'Y'Z'   A B C D E

Such sequences are called **direct repeats**. They do not provide alternative structures to double-stranded DNA and do not form secondary structures in single-stranded DNA or RNA.

STRUCTURE OF RNA

A typical cell contains about 10 times as much RNA as DNA. With the exception of some RNA phages and a few eukaryotic viruses, RNA usually exists as a single-stranded polynucleotide. In bacteria, there are three major types of RNA—ribosomal RNA (of which there are three types in bacteria), transfer RNA (of which there are about 50 different types), and messenger RNA (of which there are almost as many different types as there are genes). All of these molecules superficially resemble single-stranded DNA in that single-stranded regions are

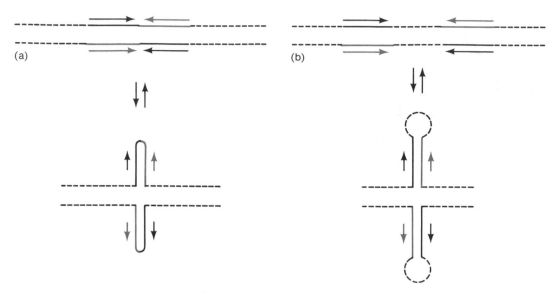

Figure 2-17. Possible alternative forms of a DNA molecule containing two inverted repeats that are (a) adjacent or (b) separated by a spacer. The horizontal arrows denote orientation of the sequences.

interspersed with intramolecular, double-stranded regions. Typically between one-half and two-thirds of the bases in RNA are paired. In single-stranded DNA, the pairing is random with short regions containing six or fewer base pairs. Because the base pairing in single-stranded DNA is between short complementary sequences that occur by chance in any stretch of DNA, if a sample of identical DNA molecules is denatured and intramolecular hydrogen bonds are allowed to form, the base-pairing pattern may differ from one molecule to the next. In contrast, in RNA, the double-stranded regions may contain up to 50 base pairs, and each molecule has a specific base-pairing pattern. The structures of the different classes of RNA molecules are discussed in Chapter 6.

NUCLEASES

Both DNA and RNA can be hydrolyzed to free nucleotides either chemically or enzymatically. For example, at pH 1, the phosphodiester bonds and the *N*-glyosidic bond between the base and the sugar of DNA and RNA are broken, releasing the free bases.

Nucleases are enzymes that cleave nucleic acids. Most nucleases show chemical specificity and are either a deoxyribonuclease (DNase) or a ribonucle-

Figure 2-18. (a) Hairpin and (b) stem-and-loop structures that can form from two types of palindromes in single-stranded nucleic acids.

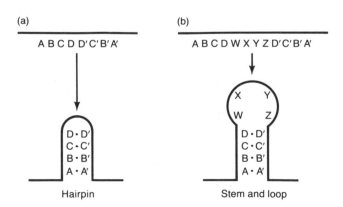

Gel Electrophoresis

Most biological macromolecules carry an electrical charge and thus can move in an electric field. For example, if the terminals of a battery are connected to the opposite ends of a horizontal tube containing a solution of positively charged protein molecules, the molecules migrate from the positive end of the tube to the negative end. The direction of migration obviously depends on the sign of the charge, but the rate of movement depends on the magnitude of the charge and, as in sedimentation, on the shape of the molecule (that is, its frictional resistance to movement). The mass of the molecule plays no direct role in the rate of migration (in contrast with sedimentation) and influences the rate only indirectly when the surface area of the molecule, which affects its frictional resistance, is a function of its mass.

The most common type of electrophoresis used in molecular genetics is zonal electrophoresis through a gel, or **gel electrophoresis**. This procedure can be performed such that the rate of movement depends only on the molecular weight of the molecule. Two experimental arrangements for gel electrophoresis of DNA is shown in Figure 2-20. A thin slab of an agarose or polyacrylamide gel is prepared containing small slots ("wells") into which samples are placed. An electric field is applied, and the DNA molecules begin to penetrate and move through the gel. DNA is negatively charged because of the phosphate groups on the DNA backbone. Because DNA is negatively charged, it will move toward the positive pole in an electric field. A gel is a complex network of cross-linked molecules, so the migrating macromolecules must squeeze through the narrow maze. Thus, when the DNA moves through a gel matrix the speed of its migration will depend on its size (smaller fragments run faster than larger fragments) and on its tertiary structure (supercoiled circular, relaxed or open circular [i.e., not supercoiled], or linear). Supercoiled DNA is very compact and thus migrates through an agarose gel the fastest. For linear DNA molecules, the rate of migration increases as the molecular weight, M, decreases. The distance moved, D, depends roughly logarithmically on M, obeying an equation of the form $D = a - b \log M$, in which a and b are empirically determined constants that depend on the buffer, the gel concentration, and the temperature. Figure 2-21 shows the result of gel electrophoresis of a collection of DNA molecules.

Pulsed Field Gel Electrophoresis

Standard gel electrophoresis can separate DNA molecules that are less than about 20 kb, but DNA molecules that are much larger cannot be separated because they do not efficiently enter the gel. Separation of large DNA molecules in an electric field requires **pulsed field gel electrophoresis** (PFGE). DNA molecules as large as 200 to 3000 kb can be separated by PFGE. During PFGE, the orientation of the electric field changes periodically during electrophoresis. The theoretical basis of PFGE is not yet known, but it is believed that the change in the electric field allows large DNA molecules to become reoriented so they can begin to "snake" through the agarose pores.

DNA molecules this large are sensitive to breakage by shearing. Shearing of the DNA is avoided by embedding the cells in agarose before lysis. While embedded in a small agarose block, the cells are gently lysed to release the DNA. The DNA in the agarose block can then be digested with restriction enzymes. By using restriction enzymes that cut the DNA rarely (about once every 10^3 to 10^4 kb), it is possible to construct restriction maps of entire chromosomes using PFGE. Furthermore, it is possible to map cloned genes physically on the restriction map by DNA-DNA hybridization. This approach has been recently used to construct physical maps of the chromosome from many different organisms.

ISOLATION OF NUCLEIC ACIDS

Isolation of DNA is an essential step in many experiments. Although the methods are straightforward, the particular procedure must be tailored to the organism from which the DNA is to be obtained because the structure and composition of organisms vary. The differences between the techniques follow from the

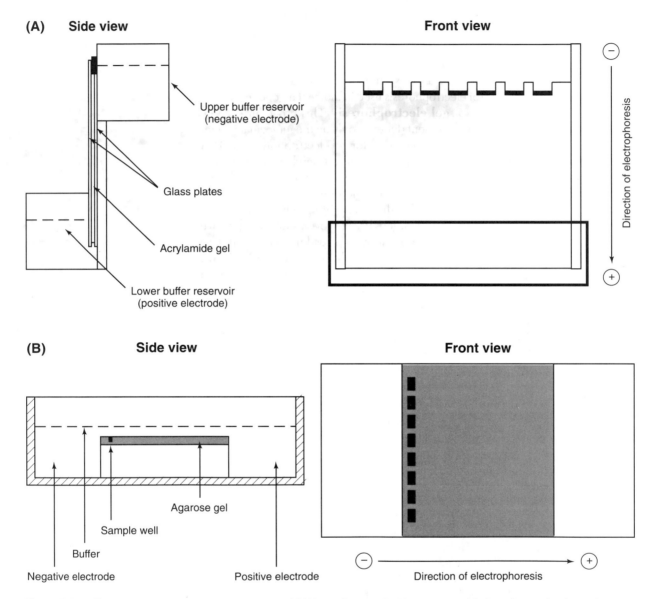

Figure 2-20. Two types of set-ups for electrophoreses of DNA. Polyacrylamide gels are typically used to separate small DNA fragments and agarose gels are used to separate larger DNA fragments. In both cases, after electrophoresis, the gel is removed and soaked in a solution of ethidium bromide. The ethidium bromide is incorporated into the DNA, causing the DNA to fluoresce when visualized under ultraviolet light. The region of migration below each well is called a lane. (a) Vertical gel set-up for polyacrylamide gel electrophoresis. The polyacrylamide is formed between two glass plates separated by thin spacers along each side and a "comb" along the top that forms the sample wells. After the polyacrylamide has solidified, the spacers and comb are removed and the gel and the glass plates are clamped into the gel apparatus. Buffer is poured into the top and bottom chambers and the set-up is attached to a power supply with the negative electrode at the top and the positive electrode at the bottom. (b) Horizontal set-up for agarose gel electrophoresis. Typically the agarose gel is submerged beneath the buffer.

organization of the DNA; the fraction of the total dry weight that is DNA (which varies from about 10% in bacteria to about 50% in phages); and the amount of contaminating polysaccharides, nucleic acids, and nucleases. The common feature in all procedures is that a cell or virus is first broken open, and then DNA is separated from such other components as protein, RNA, lipid, and carbohydrates. The basic procedures are different for phage and bacteria.

1. *Phage DNA.* The DNA can be easily purified from phages that simply contain DNA in a protein coat. An aqueous suspension of phage is vigorously mixed with phenol, a hydrophobic agent that denatures proteins. The phenol denatures the proteins in the phage head, releasing the DNA. Most of the denatured protein either enters the phenol layer or precipitates at the phenol-water interface. The DNA remains in the aqueous layer. The aqueous layer is removed, and ethanol is added to precipitate the DNA. The DNA can then be collected and redissolved in an appropriate solution.

2. *Bacteria.* To isolate chromosomal DNA from bacteria, the cells are first gently lysed by adding proteinase, an enzyme that degrades proteins, and sodium dodecyl sulfate (SDS), an ionic detergent that disrupts the lipid bilayer of the cytoplasmic membrane. The lysate is then extracted with phenolchloroform which removes any remaining proteins. The DNA remains in the aqueous phase. The DNA can be concentrated by precipitation with ethanol. The DNA precipitates as long fibers which can be collected by winding the fibers around a glass rod. After the DNA is collected, it is usually dissolved in a buffer (Tris HCl) containing ethylenediamine tetraacetic acid (EDTA). The EDTA chelates divalent cations, which are required by DNA degrading enzymes (DNases), and thus protects the purified DNA from degradation by any contaminating nucleases. Any contaminating RNA can be removed by treatment with RNase, an enzyme that specifically degrades RNA.

3. *Plasmid DNA.* It is often necessary to purify plasmid DNA from bacteria. Plasmids are extrachromosomal DNA molecules (see Chapter 11). The DNA from most plasmids is a double-stranded, covalently closed circular, supercoiled molecule. Purifying plasmids from bacteria requires separation of the plasmid DNA from chromosomal DNA and RNA. The "alkaline lysis" procedure is one of the most commonly used methods for purifying plasmids from gram negative bacteria such as *Escherichia coli*. First, the cells are treated with EDTA which chelates divalent cations. Because divalent cations are required to stabilize the outer membrane, this disrupts the integrity of the outer membrane. Sometimes lysozyme is added to degrade the cell wall peptidoglycan, but for *E. coli* this step is not required. The cells are then lysed with a solution of SDS and sodium hydroxide. After the cells are lysed, the solution must be mixed gently to avoid shearing the chromosomal DNA into small fragments which might copurify with the plasmid DNA. SDS disrupts the cytoplasmic membrane and also denatures much of the cell protein. The sodium hydroxide raises the pH which denatures double-stranded DNA. Ammonium acetate is then added to neutralize the pH, allowing the denatured single-stranded DNA to reanneal. However, the rate of reassociation is proportional to the length of the DNA. Because the plasmid DNA is small and intertwined, it quickly reanneals reforming the properly paired, double-stranded DNA. In contrast, the chromosomal DNA anneals much more slowly and forms large aggregates of improperly paired, partially double-stranded DNA. When the mixture is centrifuged, the reannealed plasmid DNA remains in solution, but the large aggregates of denatured chromosomal DNA, RNA, and protein are pelleted. The plasmid DNA is then precipitated from the supernatent with ethanol.

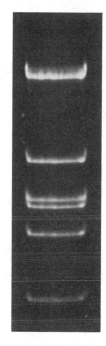

Figure 2-21. Agarose gel electrophoresis of *E. coli* phage λ DNA digested with the *Eco*RI restriction endonuclease. *Eco*R1 cuts λ into six discrete fragments. The direction of electrophoresis is from top to bottom. The DNA is made visible by the fluorescence of bound ethidium bromide. (Courtesy of Arthur Landy and Wilma Ross.)

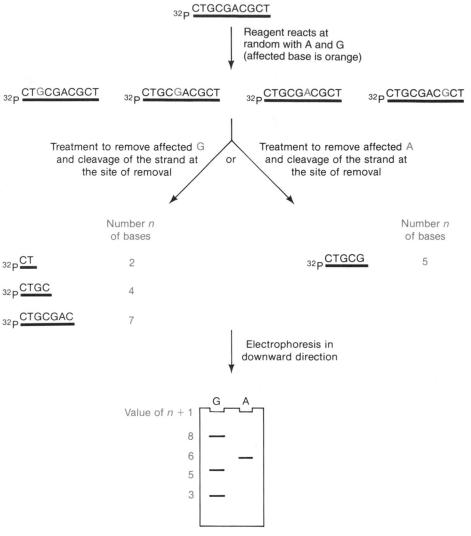

Figure 2-22.   Determination of the positions of G and A in a DNA fragment containing ten bases. The value of $n + 1$ for all four bases would be determined by noting positions in all four bands in a gel containing the A, C, G, and C + T samples.

Isolation of RNA is somewhat different than isolation of DNA owing to differences in the properties of RNA and DNA and because contaminating ribonucleases can rapidly destroy the RNA. Thus, cells must be quickly broken under conditions that inhibit ribonucleases.

Many experiments require that the complementary strands of a particular DNA molecule be separately purified. This can be accomplished in several ways, described in the references.

DETERMINING THE BASE SEQUENCE OF DNA

A great deal of information about the structure and function of genes has come from direct determination of the base sequence of DNA. Two procedures are commonly used: the chemical cleavage method and the dideoxy method.

The chemical cleavage technique developed by Maxam and Gilbert is shown in Figure 2-22. The procedure begins by cleavage of a DNA segment of interest into a set of overlapping fragments of several hundred base pairs each. The DNA is radiolabeled at the 5' ends, then the two complementary strands are separated. Next four samples of a solution containing only one of the two complementary strands are subjected to four distinct chemical treatments that cause phosphodiester cleavage at the 5' side of a particular nucleotide. Reaction conditions are chosen so only one cleavage occurs in each strand. The number of molecules in the sample is so large that all potential cleavage sites are cut in some molecules in the sample, resulting in a set of fragments whose lengths are determined by the cleavage sites. The fragments are then separated by gel electrophoresis, on polyacrylamide gels that can resolve DNA fragments differing in length by one nucleotide. The individual bands in the gel are identified by autoradiography: The fragments radiolabeled on the 5' end expose silver grains in a photographic film, producing an image of the positions of the fragments on the gel.

An example of a Maxam-Gilbert sequencing gel is shown in Figure 2-23. The result of this protocol is that the existence of a band corresponding to, say, 16 nucleotides among the set of fragments obtained by the treatment that cleaves next to guanine, for example, means that the 17th base from the 5' end is guanine. Actually the four chemical treatments do not by themselves identify each base. Instead they identify purines (A or G), G, pyrimidines (C or T), and C. G is identified by the presence of an intense band in the G column and a weak band in the G + A column, A is represented by a band only in the G + A column, T is identified by the presence of a band in the C+T column, and C is found by noting a band in both the C+T and the C columns. The base sequence is read directly from the gel; the bottom of the gel represents the 5' terminus. In practice, both of the complementary strands are sequenced, so the two sequences determined serve as a check on one another.

A second method has replaced the Maxam-Gilbert method in most laboratories because it is much easier and quicker. In the dideoxy method developed by Sanger, the fragments are made by synthesis by a DNA polymerase. Dideoxy sequence analysis is based on the random incorporation of analogs of the deoxynucleoside triphosphates (dNTPs) into a growing DNA chain by DNA polymerase. The dideoxynucleoside triphosphate (ddNTP) analogs lack the 3'-OH group on the ribose moiety of the nucleotide:

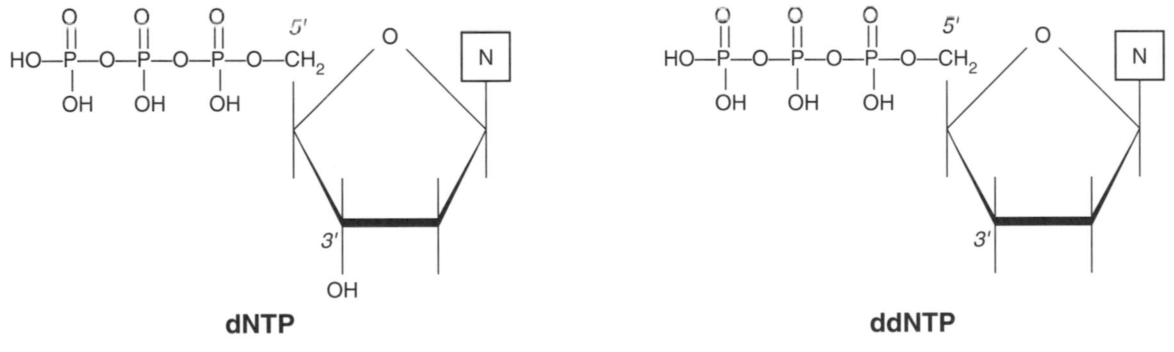

dNTP **ddNTP**

The 3'-OH group is necessary for the formation of the next phosphodiester bond, so incorporation of a ddNTP into a growing DNA chain causes termination of chain elongation. To determine the sequence of a DNA template, four separate reactions are run (Fig. 2-24). Each reaction contains all four dNTPs but only one of the four ddNTPs. For every nucleotide on the template, DNA polymerase inserts the complementary nucleotide during synthesis of the new DNA strand. If a dNTP is inserted, chain elongation continues, but if a ddNTP is inserted, synthesis stops at that position. For example, in the reaction with ddATP, when the enzyme needs to incorporate an adenine nucleotide, it has the choice between

Figure 2-23. DNA sequencing by the chemical cleavage technique. Determination of the positions of G and A in a DNA fragment containing 10 bases. The value of n + 1 for all four bases would be determined by noting positions in all four bands in a gel containing the A, C, G, and C+T samples.

the substrates dATP and ddATP. If it incorporates the ddATP, the reaction stops (chain termination). If it incorporates the dATP, the reaction continues until another adenine nucleotide is needed, then the enzyme again has a choice between the dATP or ddATP. By controlling the ratio of ddATP to dATP in the reaction, incorporation of the ddATP will be random. This results in a nested set of DNA fragments of different lengths, each terminated at a different adenine residue. By determining the length of fragments produced with each of the four ddNTPs, it is possible to deduce the nucleotide sequence of the template DNA (see Fig. 2-24).

Initiation of DNA synthesis requires double-stranded DNA as a primer. The primer is provided by using a small oligonucleotide that hybridizes to the single-stranded DNA adjacent to the region to be sequenced. DNA polymerase begins synthesis from the 3' end of the primer and adds dNTPs in the 5' to 3' direction. Thus, every DNA fragment has the same 5' end but different 3' ends (wherever a ddNTP was inserted). The Klenow fragment of DNA polymerase is used for dideoxy sequencing because it lacks the 5' to 3' exonuclease activity present in the *E. coli* DNA polymerase I holoenzyme. This exonuclease activity would degrade the common 5' end of the DNA fragments, making interpretation of the DNA sequence impossible. (There are continual minor modifications that improve the dideoxy DNA sequencing technology. For example, a variety of DNA polymerases from phage or other bacteria with useful properties can be used instead of the *E. coli* DNA polymerase I.)

By including a radioactive dNTP (for example, α-^{32}P-dATP or α-^{35}S-dATP), the DNA fragments become radioactively labeled during synthesis. After the reactions are stopped, the sequencing fragments are denatured from the template and resolved according to size by polyacrylamide gel electrophoresis. If the DNA fragments are denatured, fragments that differ in length by a single nucleotide can be separated on polyacrylamide gels. The gels contain urea and are run at high voltages (which makes them hot) to keep the DNA denatured. The DNA sequence is determined after autoradiography of the gel by reading the order of the bands in the four lanes from each ddNTP reaction. The bands form a ladder corresponding to the size of the DNA fragments. The first band at the bottom of the gel represents the shortest fragment synthesized from the sequencing primer that terminated with the ddNTP used (usually the smallest readable band is 5 to 10 bp from the end of the primer). The sequence is determined by reading up the four lanes of the autoradiogram in order of the occurrence of the bands on the ladder (see Fig. 2-25).

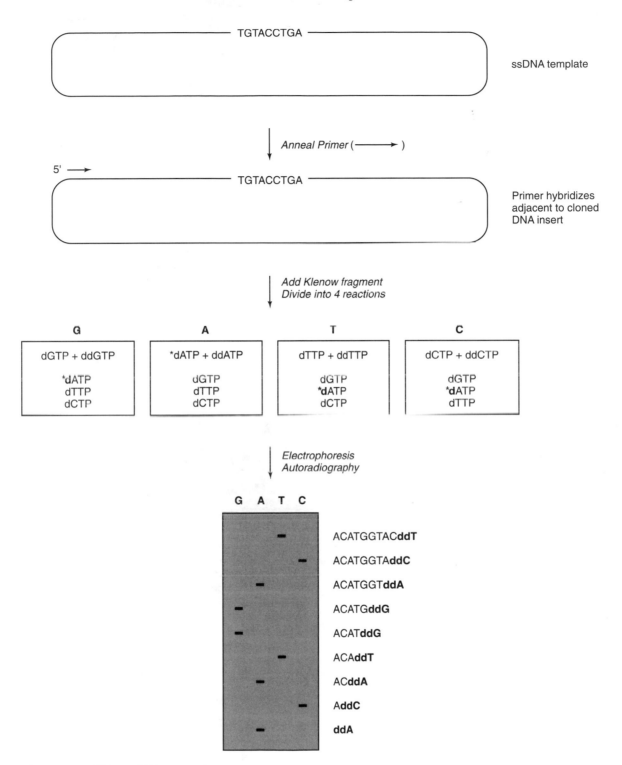

Figure 2-24.   Dideoxy DNA sequencing.

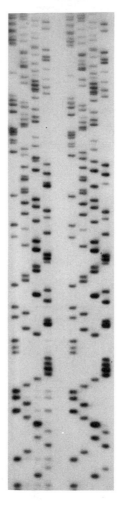

Figure 2-25. An example of a dideoxy sequencing gel. Note that the shortest DNA fragments are located at the bottom of the gel (because the shorter DNA migrates faster through the polyacrylamide gel) and the longer fragments are located at the top of the gel. The dideoxy nucleotide used in each lane is labeled at the top of the lane.

KEY TERMS

A,T,G,C	minor groove
denatured	native
dideoxynucleotides	nuclease
direct repeats	nucleic acid
DNA	nucleotide
dyad symmetry	palindrome
endonuclease	purine
ethidium bromide	pyrimidine
exonuclease	reannealing
heteroduplex	renaturation
hybridization	RNA
inverted repeat	supercoiled
major groove	T_m
melting curve	

QUESTIONS AND PROBLEMS

1. Which nucleic acid base is unique to DNA and to RNA?

2. How do ribose and 2'-deoxyribose differ?

3. What is the name of the bond formed between an N atom in a purine or pyrimidine and a C atom in ribose or deoxyribose?

4. In a nucleic acid, which carbon atoms are connected by a phosphodiester group?

5. What chemical groups are at the end of a polynucleotide?

6. How many phosphate groups are there per base in DNA and in RNA?

7. How many hydrogen bonds are there in an AT and a GC base pair?

8. What chemical groups are at the end of a single polynucleotide strand?

9. How many turns of a double helix are there in a molecule consisting of 45 base pairs?

10. In what sense are the two strands of DNA antiparallel?

11. What is meant by a nuclease, and how do endonucleases and exonucleases differ?

12. Could a single-stranded DNA molecule with base sequence 5'-GATTGCCGGCAATC-3' fold back on itself to form a hairpin?

13. One of the complementary strands of two DNA molecules is given. Which DNA molecule would denature at a lower temperature? Why?
(a) AGTTGCGACCATGATCTG   (b) ATTGGCCCCGAATATCTG

14. ^{15}N-labeled DNA from phage T4 is mixed with T4 DNA of normal density. The solution is then heat denatured and renatured. The resulting DNA is analyzed by centrifugation in a CsCl density gradient. How many bands will be observed, and what will their relative proportions be?

15. Consider a long linear DNA molecule, one end of which is rotated four times with respect to the other end, in the unwinding direction. The two ends are then joined.
   a. If the molecule is to remain in the underwound state, how many base pairs will be broken?
   b. If the molecule is allowed to form a supercoil, how many nodes will be present?

16. The sedimentation velocity properties of a supercoiled DNA molecule are being studied as a function of the concentration of added ethidium bromide. It is found that s decreases, reaches a minimum, and then increases. Explain. (*Hint*: Recall that s is a function of both molecular weight and shape.)

17. In gel electrophoresis, what feature of the method causes DNA molecules to move at a rate that is dependent on their molecular weight? Do larger molecules move more slowly or faster than smaller molecules?

18. A DNA fragment containing 17 base pairs is sequenced by the Maxam-Gilbert procedure. The data are shown here; panels 1 and 2 correspond to the two complementary strands. Note that the 5'-terminal base does not appear in either lane because it would have to be identified by a sugar-phosphate lacking a base; such molecules do not migrate with the nucleotides. What are the complete sequences of the two complementary strands, including the 5'-terminal bases?

19. What are the DNA sequences shown at left for a wild-type gene (lanes 1–4) and mutant gene (lanes 5–8)?

REFERENCES

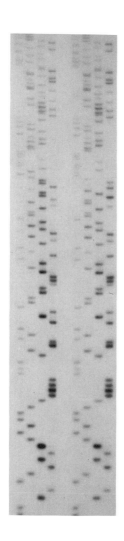

*Adams, R., J. Knowler, and D. Leader. 1990. *The Biochemistry of the Nucleic Acids*, Eleventh Edition. Chapman and Hall, London.

*Bauer, W. R., et al. 1980. Supercoiled DNA. *Scientific Am.* July, p. 118.

Bickle, T., and D. Kruger. 1993. Biology of DNA restriction. *Microbiol. Rev.* 57: 293.

*Blackburn, G., and M. Gait. 1990. *Nucleic Acids in Chemistry and Biology*. IRL Press.

*Cech, T. R. 1986. RNA as an enzyme. *Scientific Am.* November, p. 64.

*Cooper, T. G. 1977. *The Tools of Biochemistry*. Wiley, New York.

*Darnell, J. E., Jr. 1985. RNA. *Scientific Am.* October, p. 68.

Echols, H., and M. Goodman. 1991. Fidelity mechanisms in DNA replication. *Ann. Rev. Biochem.* 60: 477–511.

*Felsenfeld, G. 1985. DNA. *Scientific Am.* October, p. 58.

Forterre, P., F. Charbonnier, E. Marguet, F. Harper, and G. Henckes. 1991. Chromosome structure and DNA topology in extremely thermophilic archaebacteria. *Biochem. Soc. Symp.* 58: 99.

Freifelder, D. (ed.). 1978. *The DNA Molecule: Structure and Properties*. W. H. Freeman and Co., New York.

*Freifelder, D. 1993. *Molecular Biology*. Third edition. Jones and Bartlett Publishers, Inc., Boston.

Freifelder, D. 1982. *Physical Biochemistry*. W. H. Freeman and Co., New York.

Horgan, J. 1991. In the beginning . . . . *Scientific Am.* February, p. 116.

*Krawiec, S., and M. Riley. 1990. Organization of the bacterial chromosome. *Microbiol. Rev.* 54: 502.

*Kornberg, A., and T. Baker. 1992. *DNA Replication*, Second edition. W. H. Freeman and Co., New York.

Linn, S. M., and R. J. Roberts (eds.). 1982. *Nucleases.* Cold Spring Harbor Laboratory, NY.

Maxam, A. M., and W. Gilbert. 1977. A new method for sequencing DNA. *Proc. Natl. Acad. Sci. USA* 74: 560.

Meselson, M., F. W. Stahl, and J. Vinograd. 1957. Equilibrium sedimentation of macromolecules in density gradients. *Proc. Natl. Acad. Sci. USA* 43: 581.

*Mullis, K. B. 1990. The unusual origin of the polymerase chain reaction. *Scientific Am.* April, p. 56.

*Radman, M., R. Wagner. 1988. The high fidelity of DNA duplication. *Scientific Am.* August, p. 40.

*Reeve, J. 1992. Molecular biology of methanogens. *Ann. Rev. Microbiol.* 46: 165.

*Ross, J. 1989. The turnover of messenger RNA. *Scientific Am.* April, p. 48.

Sanger, F., S. Nicklen, and A. R. Coulson. 1977. DNA sequencing with chain-terminating inhibitors. *Proc. Natl. Acad. Sci. USA* 74: 5463.

Trifonov, E. 1991. DNA in profile. *Trends Biochem. Sci. 16*: 467–470.

Vinograd, J., R. Radloff, and W. Bauer. 1967. A dye-buoyant density method for the detection and isolation of closed circular duplex DNA. *Proc. Natl. Acad. Sci. USA* 57: 1514.

*Wang, J. 1982. DNA topoisomerases. *Scientific Am.* July, p. 94.

Watson, J. D., and F. H. C. Crick. 1953. Molecular structure of nucleic acid. A structure for deoxyribose nucleic acid. *Nature 171*: 737.

*Weinberg, R. A. 1985. The molecules of life. *Scientific Am.* October, p. 48.

*Resources for additional information.

Proteins

Proteins are polymers of amino acids. Each type of protein molecule has a unique three dimensional structure determined principally by the amino acid sequence. This results in the enormous diversity of protein structures that catalyze the thousands of different processes required by a cell.

Because many mutations have subtle effects on protein structure, a brief review of protein structure is useful. The detailed structure of proteins, however, is beyond the scope of this book, so for further information, the reader should consult the references at the end of this chapter.

CHEMICAL STRUCTURE OF A POLYPEPTIDE CHAIN

A typical protein molecule consists of one or more polypeptide chains. The building blocks of a polypeptide are 20 different amino acids. Most of these amino acids have the basic structure shown in Figure 3-1: a carbon atom (the α carbon) to which is attached an amino group, a carboxyl group, and a **side chain or R-group** (Fig. 3-2). (Proline is the one exception.) The side chains are of several types (for example, acidic, basic, hydrophobic, sulfhydryl-containing), and the chemical and physical properties of each amino acid are determined mainly by these side chains. In a polypeptide chain, amino acids are covalently joined via the amino group of one and the carboxyl group of the adjacent one, forming a **peptide bond** (Figure 3-3a). Thus, a polypeptide is a polymer of amino acids in which α-carbon atoms and peptide groups alternate to form a linear backbone with the side chains projecting from each α-carbon atom (Fig. 3-4), without participating in the backbone structure. The term "linear" refers to the fact that the backbone is not branched. As is seen in the

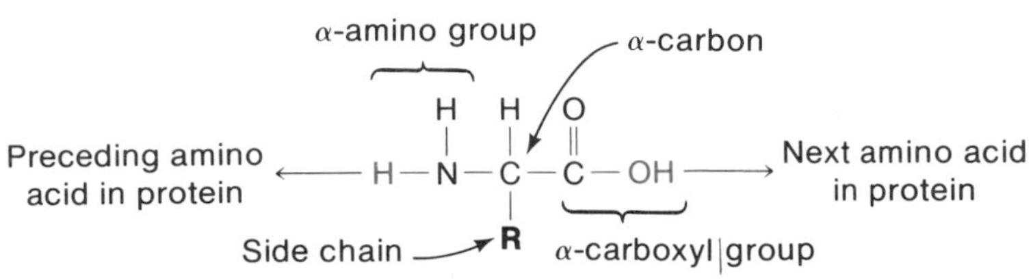

Figure 3-1. Basic structure of an α-amino acid. The NH_2 and COOH groups are used to connect amino acids to one another. The red OH of one amino acid and the red H of the next amino acid are removed when two amino acids are linked together (see Fig. 3-3).

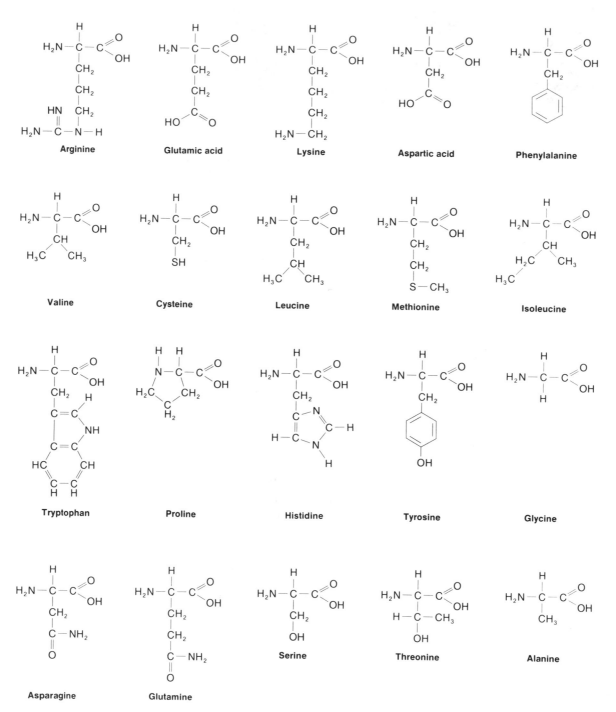

Figure 3-2. Chemical structures of the amino acids.

following sections, however, a polypeptide chain is highly folded and can assume a variety of three-dimensional shapes.

A typical polypeptide chain contains about 300 to 700 amino acids (with molecular weights of 30,000 to 70,000) although both smaller and larger polypeptides are known. Polypeptides containing more than 1000 amino acids are rare. Many proteins contain several polypeptide chains, a phenomenon that is discussed later. The largest proteins of this type have molecular weights of about 500,000.

PHYSICAL STRUCTURE OF A POLYPEPTIDE CHAIN

A fully extended polypeptide chain, if it were to exist, would have the conformation shown in Figure 3-5. (The chain is not perfectly straight because the C—N and C—C bonds in which the α-carbon atom participates are not co-linear.) Such an extended zigzagged molecule could not exist without stabilizing interactions to maintain the extension. In fact, a single polypeptide is never completely extended but is folded in a complex way, as described in the next section.

Folding of a Polypeptide Chain

Several rules govern the folding of polypeptide chains.

1. The peptide bond has a partial double-bond character and hence is constrained to be planar. Free rotation occurs only between the α-carbon atom and the peptide group. Thus, the polypeptide chain is flexible but is not as flexible as would be the case if there were free rotation about all of the bonds.
2. The side chains of the amino acids cannot overlap. Thus, the path of the backbone can never be truly random because certain orientations are forbidden.
3. Two charged groups having the same sign will not be very near one another. Thus, similar charges tend to cause extension of the chain.
4. Amino acids with polar side chains tend to be on the surface of the protein in contact with water.
5. Amino acids with nonpolar side chains tend to be internal. Very hydrophobic side chains tend to cluster.
6. Hydrogen bonds tend to form between the carbonyl oxygen of one peptide bond and the hydrogen attached to a nitrogen atom in another peptide bond. This hydrogen bonding gives rise to two fundamental polypeptide structures called the α-helix and the β-structure, which are described in the next section.

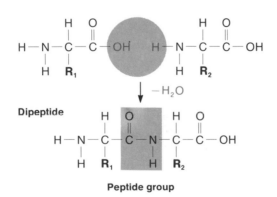

Figure 3-3. Formation of a dipeptide from two amino acids by elimination of water (shaded circle) to form a peptide group (shaded rectangle).

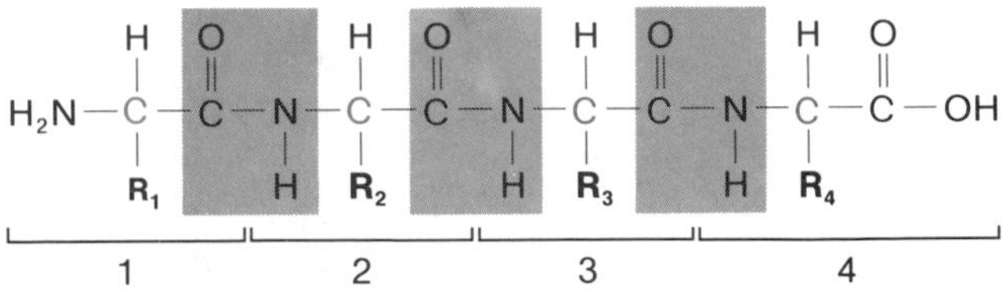

Figure 3-4. A tetrapeptide showing the alternation of α-carbon atoms (red) and peptide groups (shaded). The four amino acids are numbered below.

Figure 3-5. The conformation of a hypothetical, fully extended polypeptide chain. The length of each amino acid residue is 36.1 nm; the repeat distance is 72.3 nm. The α-carbon atoms are shown in red. Side chains are denoted by R.

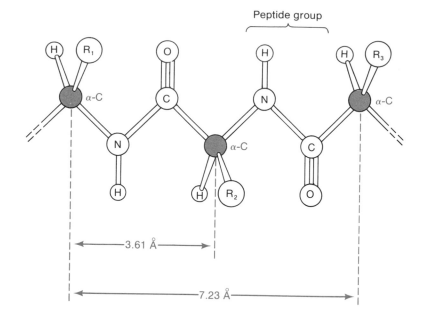

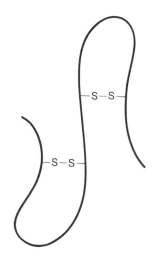

Figure 3-6. A polypeptide chain in which four cysteines are engaged in two disulfide bonds.

7. The sulfhydryl group (SH) of the amino acid cysteine (see Fig. 3-2) can react with a SH group of a second cysteine to form a covalent S—S (disulfide) bond. Such bonds pose powerful constraints on the structure of a protein (Fig. 3-6). Although S—S bonds are common in secreted proteins, the reducing environment in the cytoplasm of bacteria prevents S—S bonds in cytoplasmic proteins.

Because the structure of a protein is determined by its amino acid sequence, **the structure of a protein may be dramatically changed by a single amino acid substitution** (for example, substituting a polar amino acid for a nonpolar one). On the other hand, a single amino acid substitution may have little or no effect on the structure of a protein (for example, substitution of one nonpolar amino acid for another nonpolar one). This notion is encountered again in Chapter 10 when mutations are considered.

The three-dimensional shape of a polypeptide chain is a result of a balance between each of the rules just described and can be very complex. In examining many polypeptide chains, however, it has become apparent that certain geometrically regular arrays of the chain are found repeatedly in different polypeptide chains and in different regions of the same chain. These arrays result from hydrogen bonding between different peptide groups, as described in the next section.

Proper protein folding is essential because the properties of most proteins are determined by the three-dimensional folded structure. For example, the catalytic sites of enzymes are formed by bringing together distant regions of a polypeptide chain. Disruption of the precise pattern of folding invariably results in loss of biological activity of proteins.

Hydrogen-Bonded Conformations: The α-Helix and the β-Structure

In the absence of any interactions between different parts of a polypeptide chain, free rotation of each bond except for the peptide bond would occur continually, and the chain would assume a large number of changing conformations collectively called a **random coil**. Interactions within the polypeptide chain, however, do occur; for example, hydrogen bonds easily from between the H of the peptide N—H group and the O of the carbonyl of another peptide unit. In the

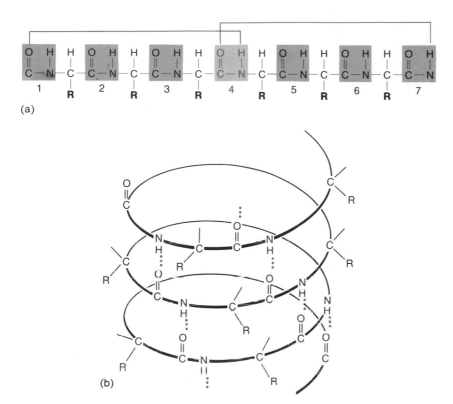

(a)

(b)

Figure 3-7. Properties of an α-helix. (a) The two hydrogen bonds in which peptide group 4 (red) is engaged. The peptide groups are numbered below the chain. (b) An α-helix drawn in three dimensions, showing how the hydrogen bonds stabilize the structure. The red dots represent the hydrogen bonds. The hydrogen atoms that are not in hydrogen bonds are omitted for the sake of clarity.

absence of all side-chain interactions, the most stable hydrogen-bonded structure of this sort is the **α-helix**. In the α helix, the polypeptide chain follows a helical path that is stabilized by hydrogen bonding between peptide groups. Each peptide group is hydrogen bonded to two other peptide groups, one three units ahead and three units behind the chain direction (Fig. 3-7). The helix has a repeat distance of 5.4 nm, has a diameter of 2.3 nm, and contains 3.6 amino acids per turn. Thus, it is a much tighter helix than the DNA helix. The α-helix is the preferred form of a polypeptide chain because, in this structure, all monomers are in an identical orientation, and each one forms the same hydrogen bonds as any other monomer. Thus, polyglycine, which lacks side chains (the R-group of glycine is a hydrogen atom) and hence cannot participate in any interactions other than those just described, forms an α-helix. Study of polylysine indicates how the composition of the medium can affect protein structure. Lysine, which also has an amino group in its side chain, is charged in a certain pH range and uncharged otherwise. When uncharged, it forms an α-helix. If the pH is altered and the side chain becomes charged, however, the repulsion caused by the similar charges destroys the helical structure, and the molecule becomes highly extended.

If the amino acid composition of a real protein is such that the helical structure is extended a great distance along the polypeptide backbone, the protein will be somewhat rigid and fibrous (not all rigid, fibrous proteins are α helical, though). This structure is common in many structural proteins, such as the α-keratin in hair.

Another common hydrogen-bonded conformation is the **β-structure**. In this form, the molecule is almost completely extended (repeat distance = 7 nm), and hydrogen bonds form between peptide groups of polypeptide segments lying adjacent and parallel with one another (Fig. 3-8a). The side chains lie alternately above and below the main chain.

Two segments of a polypeptide chain (or two chains) can form two types of β-structure, which depend on the relative orientations of the segments. If both segments are aligned in the same orientation (for example, both in the N-terminal to C-terminal direction or both in the C-terminal to N-terminal direction), the β-structure is **parallel**. If one segment is in the N-terminal to C-terminal orientation and the other is in the C-terminal to N-terminal orientation, the β-structure is **antiparallel**. Figure 3-8b shows how both parallel and antiparallel β-structures can occur within a single polypeptide chain.

When many polypeptides interact in the way just described, a pleated structure results called the β-**pleated sheet** (Fig. 3-8c). These sheets can be stacked and held together in rather large arrays by Van der Waals forces and are often found in fibrous structures such as silk.

Fibrous Versus Globular Proteins

Few proteins are pure α-helix or β-structure; usually regions having each structure are found within a protein. Because these conformations are rigid, a protein in which most of the chain has one of these forms is usually long and thin and is called a **fibrous protein**.

The fibrous proteins are typically responsible for the structure of cells, tissues, and organisms. Some examples of structural proteins are collagen (the protein of tendon, cartilage, and bone) and elastin (a skin protein). Some of the fibrous proteins are not soluble in water—examples are the proteins of hair and silk.

In contrast, the **globular proteins** have α-helices and β-structures that are short and interspersed with randomly coiled regions held together by numerous intrastrand interactions that create a compact quasispherical structure.

The catalytic and regulatory functions of cells are performed by globular proteins that have a well-defined but deformable structure (for example, the catalytic proteins or enzymes and the regulatory proteins are the largest group). Large segments of the polypeptide backbone of a typical globular protein are α-helical. The molecule, however, is extensively bent and folded. Usually the stiffer α-helical segments alternate with flexible, randomly coiled regions, which permit bending

Figure 3-8. β-structures. (a) Two regions of nearly extended chains are hydrogen bonded (red dots) in an antiparallel array (arrows). The side chains (R) are alternately up and down. (b) Antiparallel and parallel β-structures in a single molecule. (c) A large number of adjacent chains forming a β-pleated sheet.

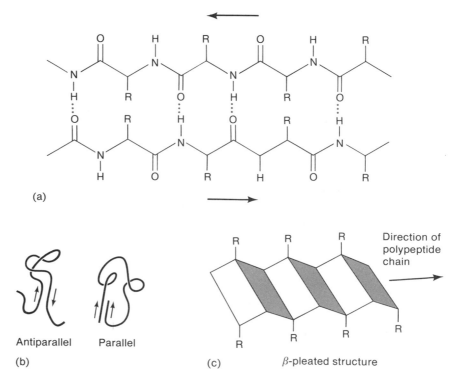

(a)

Antiparallel   Parallel

(b)

Direction of polypeptide chain

(c)   β-pleated structure

of the chain without excessive mechanical strain. Numerous segments of the chain, which might be quite distant along the backbone, form short parallel and antiparallel β-structures; these also are responsible in part for the folding of the backbone. The α-helix and β-structures are called the **secondary structure** of the protein. The extensive folding of the backbone is called the **tertiary structure** of the protein. The amino acid sequence is called the **primary structure**.

Secondary structure results from hydrogen bonding between peptide groups, whereas tertiary structure is formed from α-helices and β-structures and several different side-chain interactions.

The tertiary structure of polypeptides is mainly determined by five types of interactions.

1. Hydrophobic clustering between the hydrocarbon side chains of phenylalanine, leucine, isoleucine, and valine. This is the most important stabilizing feature.
2. Ionic bonds between oppositely charged groups in the side chains.
3. Hydrogen bonds between the hydroxyl group in tyrosine and a carboxyl group of aspartic or glutamic acids.
4. Van der Waals forces, which produce specific interactions between clusters of amino acids that may or may not be nearby in the polypeptide chain.
5. Metal-ion coordination complexes between amino, hydroxyl, and carboxyl groups; ring nitrogens; and pairs of SH groups.

Figure 3-9 shows a schematic diagram of a hypothetical protein (in two dimensions) in which several of these interactions determine the structure. This

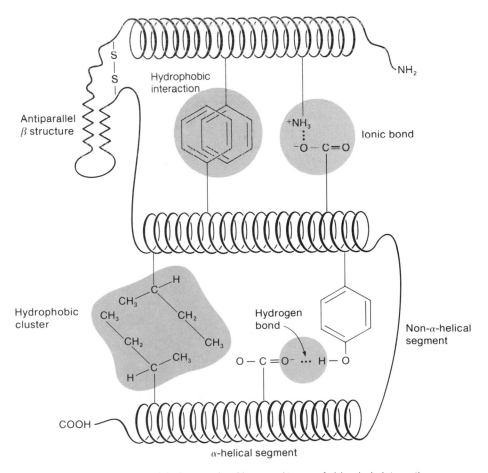

Figure 3-9.   A hypothetical globular protein with several types of side-chain interactions.

figure indicates the role of different features of a protein molecule in determining the overall conformation of the molecule.

1. Most interactions except for those between similar charges bring distant amino acids together.
2. Hydrogen bonds sometimes bring distant amino acids together, but usually a single hydrogen bond makes a more subtle change in position.
3. Both β-structures and α-helices make portions of the polypeptide backbone stiff and linear.

Figure 3-10 is an idealized drawing of the three-dimensional structure of the enzyme carbonic anhydrase, showing a β-structure, a hydrophobic cluster, and three amino acids in a coordination bond with a metal ion.

Although the first three-dimensional structures of proteins were solved over thirty years ago, solving protein structures was a slow and laborious process. However, recent technical advances in protein production, data collection, and data handling have greatly accelerated the rate at which new protein structures become known. At present the structures of over 300 proteins are known and this large data base indicates that the β-structures and α-helices of proteins are arranged in simple combinations having specific geometric arrangements. These specific geometric arrangements (which are called **motifs**) are found in numerous proteins. Thus, a protein structure can now be viewed as being composed of a set of motifs. Some motifs are designed for specific functions such as interaction with nonprotein molecules such as coenzymes or nucleic acids, whereas other motifs function only as parts of larger assemblies. Folding of some specific motifs requires special sequences of amino acids (primary structures), and thus the appearance of one of these special amino acid sequences in a protein can provide a useful clue to the function of the protein.

PROTEINS WITH SUBUNITS

A polypeptide chain usually folds such that nonpolar side chains are internal—that is, out of contact with water. It is rarely possible, however, for a polypeptide chain to fold in such a way that all nonpolar groups are internal. Thus, nonpolar amino acids on the surface often form clusters to minimize contact with water. A polypeptide having a large hydrophobic patch can further reduce contact with

Figure 3-10. An idealized drawing of the tertiary structure of human carbonic anhydrase. Shown are the peptide chain and the three histidines (orange) that coordinate to a zinc ion at the active site. Individual β-sheet strands are drawn as arrows from the amino to carboxyl ends. Note the twist of the β-sheets. A hydrophobic cluster is shaded in orange. (After Kannan, K. K., et al. *Cold Spring Harbor Symp. Quant. Biol. 36*:221, 1971.)

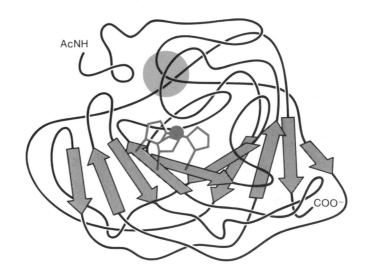

water by pairing with a hydrophobic patch on another polypeptide. Similarly, if a molecule has several distantly located hydrophobic patches, a structure consisting of several polypeptides in contact effectively minimizes contact with water. The protein is then said to consist of identical **subunits**; this is in fact a common phenomenon, with two, three, four, and six subunits occurring most frequently. (A multisubunit protein may also contain unlike subunits. This is discussed shortly.) Such multisubunit systems may have distinct advantages because the activity of multisubunit proteins can be efficiently and rapidly switched on and off. How this occurs (allosteric effects) is described in references given at the end of the chapter.

The multisubunit proteins described here consist of several identical subunits. Different polypeptide chains can also aggregate and form proteins made up of nonidentical subunits. For example, RNA polymerase, which catalyzes synthesis of RNA, has five subunits of which four are different. The subunit structure of a protein is occasionally called **quaternary structure**.

ENZYMES

Enzymes are biological catalysts that stimulate chemical reactions. Their catalytic power exceeds all man-made catalysts. A typical enzyme accelerates a reaction 10^8-fold to 10^{10}-fold, although some enzymes increase the reaction rate by a factor of 10^{15}. Enzymes are also highly specific in that each catalyzes only a single reaction or set of closely related reactions. Furthermore, only a small number of substrates, often only one, can participate in a single enzyme-catalyzed reaction. Because nearly every biological reaction is catalyzed by an enzyme, a large number of distinct enzyme molecules are required.

Many genes encode enzymes. Hence the ability to measure enzyme activity is an important part of microbial genetics because enzyme activity is often used as a reliable measure of gene expression. There are numerous ways to study enzyme reactions. Two common procedures are optical assays and radioactivity assays. In an optical assay of an enzyme, some component in the reaction mixture is detected by the ability of that component to absorb light of a particular wavelength. An important, commonly used optical assay is that for the enzyme β-galactosidase, which we encounter frequently in this chapter.

The enzyme β-galactosidase catalyzes cleavage of the disaccharide lactose into the monosaccharides galactose and glucose.

Lactose   **Galactose**   **Glucose**

The bond broken is a β-galactoside linkage, as indicated by the arrow. To detect this cleavage, the synthetic substrate *o*-nitrophenyl galactoside (ONPG) is used. A β-galactoside linkage is also present in ONPG, so β-galactosidase can hydrolyze ONPG. The value of this substrate is that it is colorless but when cleaved yields galactose and *o*-nitrophenol, which is intensely yellow.

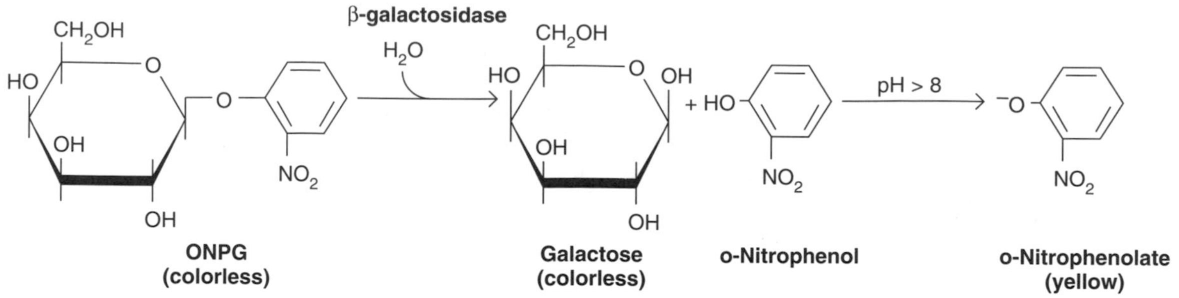

ONPG (colorless) → β-galactosidase, H₂O → Galactose (colorless) + o-Nitrophenol → pH > 8 → o-Nitrophenolate (yellow)

Thus, the activity of the enzyme can easily be followed by assaying the concentration of o-nitrophenol produced at a wavelength of 420 nm.

In a radioactive assay, a substrate that is radioactive is added to a reaction mixture, and either the appearance of a radioactive product or the loss of the radioactive substrate is measured. There are many ways of doing such assays. One common radioactive assay used to measure polymerization of proteins or nucleic acids is based on the fact that all proteins and nucleic acids are insoluble in 1 M trichloracetic acid (TCA), whereas amino acids and nucleotides are TCA soluble. For example, to measure protein synthesis, a radioactive (for example, ^{14}C) amino acid can be added to a reaction mixture containing the other 19 nonradioactive amino acids and the appropriate enzymes and factors. After a period of time, TCA is added, and the mixture is filtered and washed with TCA. The ^{14}C-labeled amino acid is soluble and passes through the filter, but if ^{14}C-protein has been made, it will be precipitated and ^{14}C will be retained on the filter. The filter-bound radioactivity can be counted to determine the extent of reaction.

KEY TERMS

α-helix
antiparallel
β-structure
parallel
peptide bond

protein
   primary structure
   secondary structure
   tertiary structure
   quaternary structure
R-group
side chain

QUESTIONS AND PROBLEMS

1. Which of the following sets of three amino acids are probably clustered within a protein? (1) Asn, Gly, Lys; (2) Met, Asp, His; (3) Phe, Val, Ile; (4) Tyr, Ser, Lys; (5) Ala, Arg, Pro.

2. How would you expect the ratio of polar to nonpolar amino acids to change, on the average, with increasing size of a typical protein? (Think about the relative change in surface area and volume.)

3. Glycine is often found in regions of a polypeptide that are tightly folded. Why might this be the case?

4. What would you guess to be the environment of a glutamine that is internal? What is the environment of an internal lysine?

REFERENCES

Anfinsen, C. G. 1973. Principles that govern the folding of protein molecules. *Science 181*: 223.

*Branden, C., and J. Tooze. 1991. *Introduction to Protein Structure*. Garland Publishing, Inc., .

*Dickerson, R. E., and I. Geis. 1980. *The Structure and Action of Proteins*. Harper & Row, New York.

*Doolittle, R. F. 1985. Proteins. *Scientific Am*. October, p. 88.

Fersht, A. 1985. *Enzyme Structure and Mechanism*, Second edition. W. H. Freeman and Co., New York.

*Freifelder, D. 1982. *Physical Biochemistry*. W. H. Freeman and Co., New York.

Gething, M.-J., and J. Sambrook. 1992. Protein folding in the cell. *Nature 355*: 33.

*Richards, F. 1991. The protein folding problem. *Scientific Am*. January, p. 54.

*Schultz, G. E., and R. H. Schirmer. 1979. *Principles of Protein Structure*. Springer-Verlag, New York.

*Stryer, L. 1981. *Biochemistry*. W. H. Freeman and Co., New York.

*Resources for additional information.

Bacteria

Bacteria are prokaryotes. Their genome is not enclosed in a nuclear membrane, and it usually consists of a single circular chromosome. The physical organization of bacteria is much simpler than eukaryotes. Bacteria have many features that make them ideal for studying fundamental biological processes. The two bacteria that have provided the most insight into molecular genetics are *Escherichia coli* and *Salmonella typhimurium*, although other bacteria have also played important roles. Features of these bacteria that make them useful model systems are their simple growth requirements and short generation time. This chapter discusses how to grow these bacteria, how to determine the number of bacteria, and some of the essential properties of bacteria.

GROWTH OF BACTERIA

Bacteria can be grown in a variety of media and conditions. Careful control of growth conditions, such as media composition, pH, and aeration, is necessary to obtain bacterial populations with reproducible properties.

Growth Media

Bacteria can be grown in a **liquid growth medium** or on a solid surface. A population growing in a liquid medium is called a bacterial **culture**. A culture is initiated by placing a small amount of bacteria—an **inoculum**—into sterile medium in a flask or tube. If the liquid is a complex extract of biological material, it is called rich medium or a **broth**. An example is tryptone broth, which is the milk protein casein hydrolyzed by the digestive enzyme trypsin to yield a mixture of amino acids and small peptides. Another common broth is prepared from an extract of beef. If the growth medium contains no organic compounds other than a carbon source (such as a sugar), it is called a **minimal medium**. A typical minimal medium contains the ions, Na^+, K^+, Mg^{++}, Ca^{++}, Fe^{++}, NH_4^+, Cl^-, phosphate buffered to neutral pH, and SO_4^{-2}, and a source of carbon, such as glucose or glycerol. Other metal ions are also required but in such small quantities that they are usually provided as contaminants in the other salts used to prepare the medium. The best source of carbon for most bacteria is glucose; they grow more rapidly in a minimal medium with glucose than with any other single carbon source.

If a bacterium can grow in a minimal medium—that is, if it can synthesize all necessary organic substances, such as amino acids, vitamins, and lipids from simple precursors—the bacterium is called a **prototroph**. If any organic substances other than a carbon source must be added for growth to occur, the bacterium is called an **auxotroph**. For example, if the amino acid proline is required in the growth medium, the bacterium is a proline auxotroph; the genetic symbol for a bacterium with this phenotype is Pro$^-$. A bacterium that does not require proline would be designated Pro$^+$.

Bacteria can also be grown on solid surfaces. Agar, obtained from a type of seaweed, has several properties that make it the ideal solidifying agent. It dissolves when heated to boiling (100°C) but does not resolidify until cooled to about 45°C. Thus, the media can be cooled somewhat before pouring plates so it is not necessary to add labile reagents directly to the hot media. Furthermore, few bacterial species make enzymes that degrade agar (which would liquefy the media). When agar is added to a rich broth, the corresponding solid growth medium is called **rich agar**. Solid media are typically prepared by pouring a hot molten agar solution into a **petri dish** and allowing the medium to cool and harden. In lab jargon, a petri dish containing a solid medium is called a **plate**, and the act of depositing bacteria on the agar surface is called **plating**.

Some Parameters of a Bacterial Culture

When bacteria are inoculated in a liquid medium, they slowly start to grow and divide. After an initial period of slow growth called the **lag phase**, they begin a period of rapid growth in which they divide at a fixed rate called the **doubling time**. The number of cells per milliliter, the **cell density**, doubles repeatedly, giving rise to a logarithmic increase in cell number; this stage of growth of the - bacterial culture is called the **exponential** or **log phase**. For *E. coli* and *S. typhimurium* growing in rich medium, exponential growth continues until a cell density of about 10^9 cells/ml is reached. Then the growth rate decreases due to depletion of nutrients including oxygen and accumulation of waste products. (The maximum cell density varies in different growth media for some mutants and for other bacterial species.) Ultimately, at a cell density of about 10^9 cells/ml, no further increase in population density is possible and the cell number becomes constant; this stage is called the **stationary phase**. A typical growth curve for a bacterial culture is shown in Figure 4-1.

The doubling time of a culture varies with temperature and the composition

Figure 4-1. A typical growth curve for a bacterial culture. Note that they y axis is logarithmic, so the curve is a straight line in log phase.

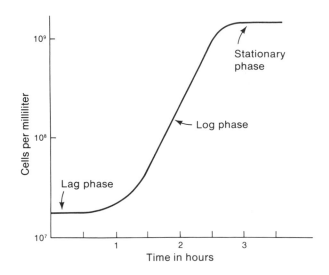

of the growth media depending on the bacterial species. For both *E. coli* and *S. typhimurium*, whose natural host is mammals, the optimal temperature is about 37°C. Maximum growth rates are invariably achieved in rich media because the bacteria do not need to use energy to synthesize a large number of essential organic compounds. When grown in rich broth with good aeration at 37°C, *E. coli* and *S. typhimurium* have a doubling time of about 20 to 25 minutes. Because cells grow exponentially, the culture multiplies quickly. For instance, one bacterium with a doubling time of 20 minutes passes through 30 generations in 11 hours, generating 10^9 cells. In theory, under optimal conditions, a single bacterium could multiply in less than 2 days to form a number of bacteria equal in mass to that of the earth.

In minimal media, the rate-limiting substance in the growth medium is usually the carbon source. With glucose, the optimal carbon source, the doubling time of *E. coli* and *S. typhimurium* at 37°C is about 45 minutes. With less efficiently used sugars, the doubling time can take up to 60 minutes, and with carbon compounds that are not sugars, several hours can be required for a culture to double in cell density. Most physiological and genetic studies of bacteria utilize exponentially growing cultures since the cells have very active metabolism and a defined chemical composition and are readily reproducible. However, stationary phase more closely approximates the physiological state of bacteria in nature where bacteria are generally in environments that do not allow rapid growth. Survival in stationary phase requires the expression of specific gene products as the cells stop growing. An extreme example is the stationary phase cells of *Bacilli* which undergo a patterned program of gene expression that results in the formation of spores, a morphologically different form of these bacteria that is extremely resistant to environmental stresses.

Chemostat Cultures

To maintain continual growth, a culture must be repeatedly diluted. If a stationary-phase culture is diluted, it enters lag phase. If a log-phase culture is diluted into fresh medium, however, with the same composition and prewarmed to the same temperature at which the cells have been growing, the culture remains in log phase. Cultures can be maintained in continuous growth by repeated dilution during log phase. A technique for maintaining a culture at a constant cell density uses a modified culture vessel called a **chemostat**. With a chemostat, fresh medium drips into the culture vessel at a carefully controlled, constant rate. Each drop causes a drop of the culture to overflow through a siphon. A nutrient required by the bacterium is provided in the medium at a concentration sufficiently low that cell growth is limited by it. Furthermore, the rate of flow of the fresh medium is adjusted to be so low that the cell population can fully use all of the nutrient; that is, the nutrient is present at a concentration that is growth-limiting. In such a system, the flow rate of the fresh medium determines the growth rate of the culture. Because the mass of the bacteria remains constant in a chemostat, growth of the culture is linear, not exponential. Hence the time required to add a volume of liquid equal to the capacity of the chemostat is the doubling time; that is, if liquid flows into the vessel at a rate of 0.2 ml/minute and the capacity of the chemostat is 25 ml, the doubling time will be 25/0.2 = 125 minutes. Chemostat cultures are useful in measuring mutation rates, as described in Chapter 10.

COUNTING BACTERIA

For many experiments, it is important to know the number of cells per unit volume, the **cell density**. To determine the cell density requires counting the number of bacteria in a known volume. Three methods are commonly used: The number of bacteria can be directly determined by measuring colony formation or counting the number of cells in a known volume with a microscope or indirectly determined by measuring optical absorbance. Electronic counters, which measure both cell number and cell size, are available, but they are sufficiently expensive that they are not widely used to count bacteria.

Only viable cells form colonies on agar medium. Because most bacteria are not very motile on a solid surface, as the bacteria divide, the progeny bacteria remain adjacent to the original bacterium. As the number of progeny increases to about 10^6 cells, a visible cluster of bacteria appears. This cluster is a population of cells called a **bacterial colony** (Fig. 4-2). Because each viable cell can form a single colony, by counting the number of colonies formed when a known volume of culture is plated you can determine the number of bacteria in a culture. For instance, if 100 cells are plated, 100 colonies will be visible the next day. Usually a culture has such a high cell density, however, that it must be diluted before plating, and the dilution factor must be taken into account when calculating the cell density. The initial cell density can be calculated as follows:

$$\frac{(\text{Number of colonies formed})}{(\text{ml plated}) \times (\text{dilution before plating})} = \frac{\text{Number of viable cells}}{\text{ml of undiluted culture}}$$

For example, of 0.1 ml of a 10^6-fold dilution of a bacterial culture is plated and 200 colonies appear, the cell density in the original culture is $(200 \text{ cells}/0.1 \text{ ml})(10^6)$ = 2×10^9 cells/ml. The total number of cells can also be directly determined by counting the cells in a known small volume of the culture under a microscope. This method, however, counts both viable and dead cells, and it is tedious.

Often it is necessary to determine the cell density while a culture is growing. Counting colonies is not adequate to monitor growth of a culture because the colonies take many hours to appear. The most common method is to measure the optical absorbance in a spectrophotometer or colorimeter. If a bacterial culture is placed in the path of a narrow beam of light, the intensity of the transmitted light is less than that of the incident light. If the wavelength of the incident light is not absorbed by any of the intracellular molecules, the decreased transmission is caused entirely by **scattering** of the light by the bacteria. The light scattering is proportional to the optical density (OD) or absorbance, defined as:

$$A = -\log_{10} \left[\frac{\text{Intensity of light transmitted by a solution 1 cm thick}}{\text{Intensity of incident light}} \right]$$

As long as the cell density is low, the OD is proportional to the dry weight per milliliter of cells and hence to the cell density. All that is needed to use this method is a standard curve relating absorbance and cell density. Such a curve is obtained

Figure 4-2. A petri dish with bacterial colonies that have formed on agar. (Courtesy of Gordon Edlin.)

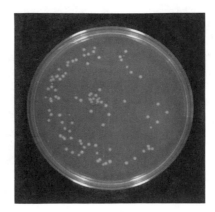

by removing samples from a growing culture, measuring the optical absorbance, and determining the cell density by colony formation. The cell concentration of a culture can then be calculated by comparing the absorbance to this standard curve. However, the curve will vary considerably among spectrophotometers and colorimeters since these instruments are designed to measure light absorbance rather than light scattering. The variability in response to bacterial suspensions is due to differing instrumental geometries. If the light detector is large or close to the tube holding the bacterial culture, some or most of the light scattered by the bacteria will contact the detector thus decreasing the response compared to an instrument with a more distant or smaller detector. Therefore, a standard curve is required for each instrument. In addition to the number of cells, the optical density is affected by the size of the cells. Cell size depends on the particular growth medium and on the bacterial species and strain; thus, a standard curve is required for each growth condition and each strain.

PREPARATION OF A PURE CULTURE

Genetic experiments must be done with a pure culture, that is, a population of bacteria obtained from a single cell. It is essential to isolate a pure culture of bacteria whenever you isolate new mutants or strains. Often a colony containing the desired mutant initially contains a mixture of the mutant and the parental cells. This problem can be eliminated by resuspending a colony in a buffer, diluting it, and replating to obtain single colonies. Each resulting colony will be an individual clone from single cells of the original colony and will no longer contain a mixed population. Instead strains are usually purified by **streaking**. A sterile wire is first touched to a colony. The wire, which carries several million cells, is dragged across the surface of an agar plate ("streaked"). Cells are transferred to the agar, but the number decreases continually as the wire is dragged. After streaking for a few centimeters, the wire is sterilized by heating in a flame. It is allowed to cool and touched to the last part of the streak and then streaked back and forth across the plate. The number of cells in the streak decreases as streaking proceeds until individual cells are well separated. The plate is then incubated until growth occurs. Because bacterial geneticists often restreak large numbers of colonies, it is common to use sterile toothpicks instead of a wire to eliminate the time required to sterilize the wire in a flame. A typical streak plate is shown in Figure 4-3. Once the colonies grow, a single colony is taken from the streak plate to inoculate a culture.

Figure 4-3.   A streak plate.

IDENTIFICATION OF NUTRITIONAL REQUIREMENTS

Auxotrophic mutants can be identified by testing for growth on rich and minimal agar. Most auxotrophic mutants can grow on rich agar because it contains most of the amino acids, nucleotides, and vitamins required by bacteria. If colonies also grow on minimal agar, the bacterium is a prototroph; if no colonies grow on the minimal agar, it is an auxotroph that requires some substance that is absent in the minimal agar. Minimal plates are then prepared with various supplements. If the bacterium is a leucine auxotroph, the addition of leucine alone enables a colony to form. An example of this approach is shown in Table 4-1. If both leucine and isoleucine must be added, the bacterium is auxotrophic for both of these substances. In Chapter 10, where mutant selection is discussed, procedures for identifying auxotrophic mutants among a population of prototrophs are described.

A variety of **color-indicator plates** are available for determining whether a bacterium can use a particular substance as a carbon source. For example, **MacConkey agar** contains dyes, and the color of the dyes is sensitive to pH. The medium also contains a sugar, such as lactose, as a carbon source and a complete mixture of amino acids. Both Lac$^+$ and Lac$^-$ cells are able to form colonies on this medium. A Lac$^+$ cell ferments the lactose, producing a local decrease in pH that causes the dyes to stain the colony red. A Lac$^-$ cell cannot use the lactose and instead uses the amino acids as carbon sources. One of the

BOX 4-1. IDENTIFYING AUXOTROPHIC REQUIREMENTS

Auxotrophic mutants can be identified by simply screening for growth on many different minimal medium plates which each contains a different nutritional supplement. However, this approach is tedious and expensive. Instead the auxotrophics can be screened for growth on a relatively small number of pools of nutritional supplements. A pool is simply a mixture of several different components. The set of eleven pools contains the common auxotrophic supplements that account for most of the major biosynthetic pathways in *E. coli* and *S. typhimurium*. The composition of these eleven pools is shown in the table below. The supplements in pools 1 to 5 are listed in the vertical columns and the supplements in pools 6 to 11 are listed in the horizontal rows.

	1	2	3	4	5
6	adenine	guanine	cysteine	methionine	thiamine
7	histidine	leucine	isoleucine	lysine	valine
8	phenylalanine	tyrosine	tryptophan	threonine	proline	p-aminobenzoic acid, dihydroxybenzoic acid
9	glutamine	asparagine	uracil	aspartic acid	arginine
10	thymine	serine	glutamate	diaminopimelic acid	glycine
11	pyridoxine, nicotinic acid, biotin, *panthothenate*, alanine

Each amino acid and nucleotide is present in two pools of the eleven. A mutant requiring one amino acid or nucleotide will grow only on the two pools that contain the required supplement. For example, proline is present in pool 5 and pool 8, so a Pro$^-$ auxotroph will grow only on the plates that contain these two pools of supplements. Some auxotrophs require two supplements. Such a mutant will grow only on the pool that contains both required supplements. For example, *pyrA* mutants require both uracil and arginine, so a *pyrA* mutant will grow only on plates that contain the pool 9 supplements. Several vitamins are present only in pool 11, so an auxotroph requiring one of these vitamins will grow only on plates containing the pool 11 supplements. Thus, by simply testing for growth on minimal medium plates with each of these eleven pools many different types of auxotrophs can be identified.

The concept of using pools greatly simplifies initial genetic analysis. A variety of types of pools can be used in many other situations as well.

Table 4-1   Plating data enabling the determination of the nutritional requirements of a bacterium

Medium supplement	Growth	Conclusion
1. His, Leu, Thy	−	Needs some nutrient
2. Leu, Ala, His	+	Thy not needed
3. Ala, Thy, His	+	Ala needed (cf. plate 1)
		Leu not needed
4. Leu, Ala, Thy	−	His needed (cf. plates 1 and 3)

Note: Abbreviations of names of amino acids are given in Table 1-1, note 1.

products of amino acid metabolism is ammonia, which increases the local pH, decolorizes the dyes, and causes the colonies to be white.

A medium on which all bacteria form colonies is called a **nonselective medium**. Wild-type and mutant bacteria may or may not be distinguishable on a nonselective medium. For example, although both the wild-type and mutant may grow on a color-indicator medium, the wild-type bacteria may form different colored colonies than mutants unable to use the carbon source. If the medium allows growth of only one type of cell (for example, only the wild-type cells or a specific type of mutant cells), it is said to be **selective**. For example, a medium containing streptomycin is selective for streptomycin-resistant (Str^r) mutants, preventing growth of streptomycin-sensitive (Str^s) cells; minimal medium containing lactose as the sole carbon source is selective for Lac^+ cells because Lac^- mutants cannot use lactose as a carbon source. Selective media are valuable for isolating rare mutants or recombinants from a population of cells.

PHYSICAL ORGANIZATION OF A BACTERIUM

The general features of a typical bacterial cell are shown in Figure 4-4. The bacterial cell does not have its genetic material enclosed in a nucleus and hence is a prokaryote. Bacteria are enclosed in a rigid, multilayered cell envelope that gives the cell a defined shape—spherical, rod-shaped, and so forth (Fig. 4-5). The organization of the chromosome and the cell envelope are unique in bacteria, and these unique characteristics have many special roles in bacterial genetics.

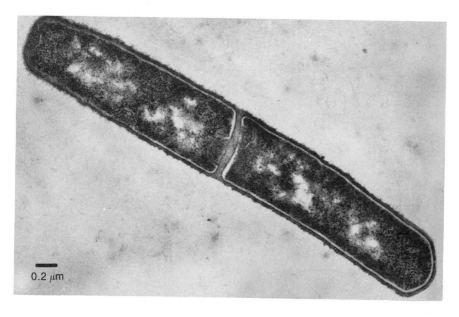

0.2 μm

Figure 4-4. An electron micrograph of a dividing bacterium. The layers of the cell wall can be seen. The light areas inside the cell are the DNA; note that the DNA is distributed throughout the cell. The fine dark particles are ribosomes, the units on which proteins are synthesized. (Courtesy of A. Benichou-Ryter.)

Bacterial Chromosome

The chromosome of *E. coli* and *S. typhimurium* and of most other bacteria is a single supercoiled, double-stranded circular DNA molecule. The chromosome of *E. coli* is about 4.7×10^6 bp, which, if stretched out, would be about 1 mm of B form DNA, although the length of the bacterium is only about 1 to 2 µm (Fig. 4-6). Thus, the bacterial DNA must be condensed about 1000-fold to fit into the cell.

When the DNA is isolated by a technique that avoids both DNA breakage and protein denaturation, the DNA is organized in a highly compact structure called a **nucleoid**. This structure contains a single DNA molecule, protein, and RNA. An electron micrograph of the *E. coli* nucleoid is shown in Figure 4-7. Two

Figure 4-5. Two forms of bacteria. (a) Cocci (spheres that sometimes form chains). (b) Bacilli (rods). (From Shih, G., Kessel, R. *Living Images: Biological Microstructures Revealed by Scanning Electron Microscopy.* Jones & Bartlett, 1982.)

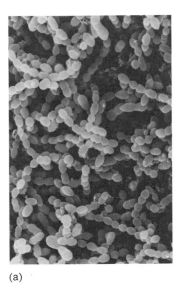
(a)

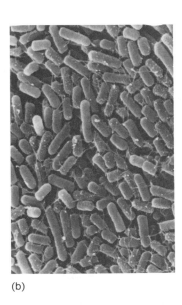
(b)

Figure 4-6. (a) Schematic diagram showing the relative sizes of *E. coli* and its DNA molecule, drawn to the same scale except for the width of the DNA molecule, which is enlarged approximately 10^6 times. (b) The localization of DNA in *E. coli*. Bacteria were exposed to a fluorescent dye that binds to DNA and then observed by fluorescence microscopy. The mode of sample preparation causes the DNA to condense slightly; in a living cell, the DNA occupies about twice as much space. (Courtesy of Todd Steck and Karl Drlica.)

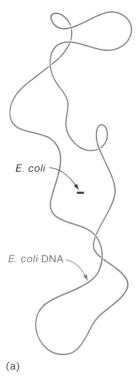

E. coli

E. coli DNA

(a)

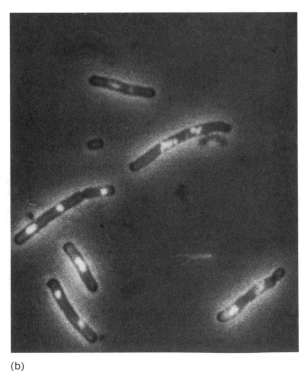
(b)

features of the structure should be noted: (1) The DNA is arranged in a series of loops, and (2) each loop is supercoiled. The nucleoid seems to be held together by a dense region containing proteins and RNA.

Introduction of a single-strand break into supercoiled DNA by a DNase causes an abrupt transition to a relaxed (nonsupercoiled) form because the nick allows free rotation about the opposing sugar-phosphate bond (see Chapter 2). With each nick, however, only one of the supercoiled loops is relaxed: The nucleoid requires about 50 nicks before it becomes an open circle. This indicates that each loop is independently supercoiled. The structure of the *E. coli* chromosome that has been deduced from these data is shown in Figure 4-8. The ends of each supercoiled loop are believed to be held by proteins in a way that allows the individual supercoiled loops to be isolated from one another.

The degree of supercoiling of the DNA is carefully controlled. The enzyme DNA gyrase, which plays an important role in DNA replication (see Chapter 8), is responsible for the supercoiling. DNA gyrase introduces negative superhelical twists to covalently closed circular DNA. The drug coumermycin inhibits *E. coli* DNA gyrase. If coumermycin is added to a culture of *E. coli*, the chromosome quickly loses its supercoiling. The activity of DNA gyrase is opposed by another

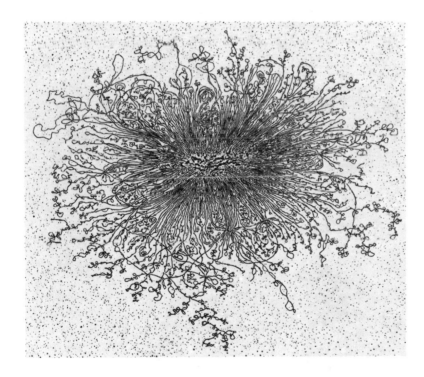

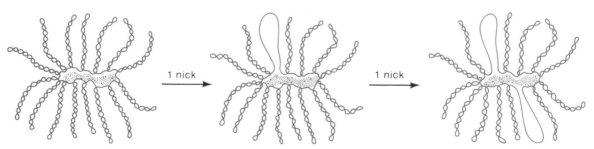

Figure 4-8. A schematic drawing of the highly folded, supercoiled *E. coli* chromosome, showing only 15 of the 40 to 50 loops attached to proteins (stippled region) of unknown organization and the opening of a loop by a single-strand break (nick).

enzyme (topoisomerase I) that removes supercoils; in vitro purified topoisomerase I unwinds supercoiled DNA. Mutants lacking topoisomerase I activity (*topA*) have been found. Initially nucleoids isolated from these mutants have increased supercoiling, about 32% greater than normal. These mutants, however, quickly acquire secondary mutations in the genes encoding DNA gyrase; these secondary mutations reduce gyrase activity slightly, restoring the degree of supercoiling to normal. Studies have shown that expression of many genes is sensitive to supercoiling, explaining the strict homostatic regulation of supercoiling in the cell.

Bacterial Cell Walls

Most bacteria can be divided into two groups: **gram-positive** and **gram-negative**. Operationally gram-positive and gram-negative bacteria are distinguished by their ability to retain a crystal violet-iodine stain when treated with alcohol. Bacteria that retain the stain are gram-positive; those that do not are gram-negative. The cell walls of gram-negative bacteria, such as *E. coli* and *S. typhimurium*, are much more complicated than that of gram-positive bacteria. Figure 4-9a shows electron micrographs of a section of gram-negative and gram-positive bacteria. A cartoon of the gram-negative cell wall is shown in Figure 4-10. In contrast to gram-positive bacteria, gram-negative bacteria are surrounded by two membranes, the inner and outer membrane, separated by an aqueous region known as the **periplasmic space**. A relatively thin **peptidoglycan layer** is located between the inner membrane and outer membrane. Gram-positive bacteria lack the outer membrane, the peptidoglycan layer is thicker, and there is no periplasmic space (Figures 4-9b and 4-11).

The peptidoglycan layer, which can be seen most clearly as the thick outer layer in Figure 4-9b, is an unusual substance. It is a polymer consisting of both sugar and peptide units. Individual polysaccharide chains are cross-linked by a pentaglycine peptide to form the large sheetlike structure (Figure 4-12). The remarkable feature of peptidoglycan is that the sheet ultimately closes on itself to

Figure 4-9. Two types of bacterial cell walls. (a) An electron micrograph of a thin section of *E. coli* (a gram-negative bacterium) showing the multiple layers. (Courtesy of Jack Pangborn.) (b) An electron micrograph of a thin section of a gram-positive bacterium, *Staphylococcus stapholyticus*. The thick outer layer is the peptidoglycan. (Courtesy of Harriet Smith.) (c) An electron micrograph of the cell wall of a gram-positive bacterium prepared in a way that shows the cell wall structure more clearly than in (b). (d) An electron micrograph of the cell wall of a gram-negative bacterium prepared in a way that shows the three layers more clearly than in (a). The segments in (c) and (d) are enlarged about three times more than in (a) and (b).

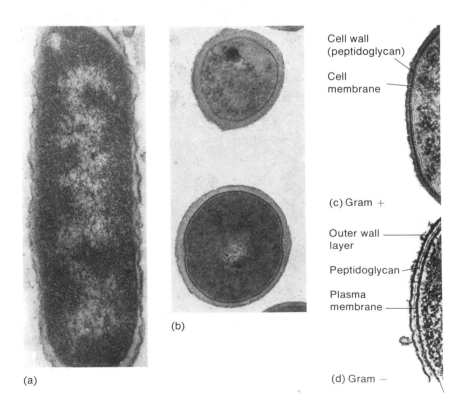

(a)

(b)

Cell wall
(peptidoglycan)

Cell
membrane

(c) Gram +

Outer wall
layer

Peptidoglycan

Plasma
membrane

(d) Gram −

form one enormous saclike macromolecule, which encloses the entire inner membrane and cytoplasm of the bacterium. The variations in the cross-linking pattern of the peptidoglycan determine the shape of each bacterial species.

Peptidoglycan is the site of attack of the enzyme lysozyme. Because peptidoglycan is the only rigid component of the cell wall, a lysozyme-treated bacterium assumes a spherical shape (whether the bacterium was initially spherical, rod-shaped, or any other shape); a cell treated in this way is called a **spheroplast** if some cell wall material remains and a **protoplast** if the cell wall is completely stripped off. The antibiotic penicillin interferes with the synthesis of peptidoglycan. Thus, if a bacterium is allowed to grow in the presence of penicillin, it enlarges without growth of the strong peptidoglycan layer; without the rigid peptidoglycan layer surrounding the membrane, the volume of the cell swells until the cell bursts. Thus, penicillin kills only growing cells.

The inner or cytoplasmic membrane provides the major osmotic barrier and determines which molecules can enter and leave the cytoplasm. Transport of most molecules across the cytoplasmic membrane requires specific transport systems. Some molecules cannot enter the cell because of lack of the appropriate

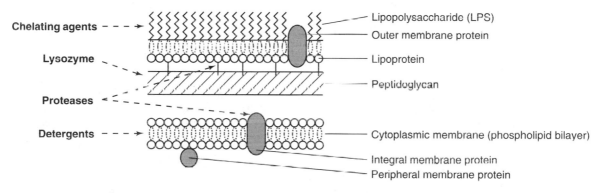

Figure 4-10. A schematic cartoon showing the cell envelope gram negative bacteria. The sites affected by chelating agents, proteases, lysozyme, and detergents are indicated by dashed arrows.

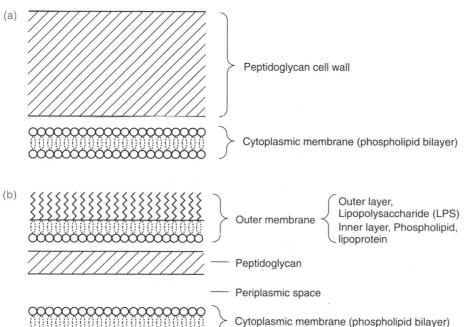

Figure 4-11. A schematic cartoon showing the basic differences between the cell walls of gram positive (a) and gram negative (b) bacteria.

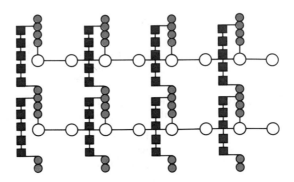

Figure 4-12. Schematic diagram of the peptidylglycan. The sugars are the large open circles, the tetrapeptides are small red circles, and the pentaglycine bridges are the five linked squares. The cell wall is a single, enormous macromolecular sac because of contiguous cross-linking.

transport system. For example, most phosphorylated organic molecules (such as nucleotides) cannot enter cells. Treatment of a cell with lipid solvents (such as toluene) disrupts the membrane lipids and thus also the permeability barrier. Thus, cells treated with toluene are **permeabilized**. They cannot grow and divide but are able to carry out many biochemical reactions. For example, permeabilized cells have been used to study DNA synthesis because, in contrast to untreated cells, they are able to take up deoxynucleoside triphosphates, the immediate precursor of DNA. Permeabilizing cells is often also a convenient way to assay certain enzymes. For example, synthesis of the enzyme β-galactosidase is extensively used to study the regulation of gene expression (see Chapter 7). It is often assayed by permeabilizing cells before adding a chromogenic substrate.

The outer membrane of gram-negative bacteria is a complex structure. The outer leaflet of the phospholipid-protein bilayer is coated with lipopolysaccharide. The lipopolysaccharide (LPS) has several protective functions. It gives the cell a hydrophilic surface, which protects the cell from hydrophobic agents (such as bile salts and detergents). The outer membrane contains proteins (called porins) that form pores, which allow entry of small hydrophilic molecules but exclude large molecules (such as many antibiotics) from the periplasm.

Between the outer membrane and the inner membrane is an aqueous region called the **periplasmic space**. Many proteins are localized in the periplasmic space. These proteins include nucleases and proteases, which break down large impermeable nutrients so they can be transported across the cytoplasmic membrane. Another major class of proteins in the periplasmic space binds specific ions, sugars, and amino acids and facilitates their transport across the inner membrane.

METABOLIC REGULATION IN BACTERIA

Bacteria rarely synthesize macromolecules that are not needed. For example, the enzymes required for synthesis of the amino acid tryptophan are not formed if tryptophan is present in the growth medium; however, when the tryptophan in the medium is used up, the enzymes are rapidly made. The systems responsible for use of various energy sources are also efficiently regulated. A well-studied example (discussed in detail in Chapter 7) is the metabolism of the sugar lactose as an alternate carbon source to glucose.

Glucose is metabolized by a series of chemical conversions in which the molecule is progressively broken down. Glucose is a primary carbon source in the sense that other sugars must be converted either to glucose or to one of the products of glucose degradation to be metabolized. Thus, most bacteria use glucose more efficiently than other sugars, such as lactose. The first step in the metabolism of lactose is its hydrolysis into the two sugars, glucose and galactose. The enzyme needed to catalyze this cleavage, however, is not present in cells in significant quantities unless lactose is in the growth medium. If lactose is provided to a cell, synthesis of the necessary enzyme is turned on and glucose produced from lactose is broken down and used as a source of energy. The galactose that is also produced is converted to an intermediate that enters the glucose-degradation pathway. Furthermore, if a cell is supplied with both glucose and lactose in the growth medium, there is no reason for the cell to synthesize the lactose-cleaving enzyme, so synthesis of the enzyme is turned off until all the glucose is used.

The control of both tryptophan synthesis and lactose degradation are two examples of **genetic regulation**. Genetic regulation is discussed in more detail in Chapter 7.

KEY TERMS

auxotroph	minimal medium
broth	nucleoid
chemostat	peptidoglycan
chromosome	periplasmic space
colony	prototroph
exponential phase	rich medium
gram negative	selective
gram positive	stationary phase
lag phase	

QUESTIONS AND PROBLEMS

1. If the doubling time of a bacterial culture is 20 minutes, by what factor does the cell density increase in 2 hours?

2. 0.1 ml of a bacterial culture is diluted into 9.9 ml of buffer; 0.1 ml of this dilution is again diluted in 9.9 ml of fresh buffer. Plating 0.1 ml from the second dilution tube yields an average of 72 colonies per plate on four plates. What is the cell density of the culture?

3. A microscopic count of a culture that has been in the refrigerator for 2 weeks shows that 453 cells are present, on the average, in each 0.001 ml of a 1000-fold dilution. Plating gives a cell density of 1.2×10^7 per milliliter. What information does the discrepancy in the two values give you?

4. A standard curve relating optical absorbance and cell density (determined by plating) has the following points: Absorbance values of 0.2, 0.4, 0.6, and 0.8 correspond to 7.5×10^7, 1.5×10^8, 3×10^8, and 6×10^8 cells per milliliter. A cell sample taken from a growing culture has an absorbance of 0.7. What is its cell density?

5. A bacterial strain can grow on agar supplemented with arginine (Arg), tryptophan (Trp), and leucine (Leu). It fails to grow on agar containing (A) only Arg and Trp or (B) only Leu and Trp. It will grow if (C) only Arg and Leu are present. What is the genotype of the bacterium with respect to these three amino acids?

6. What is the doubling time of a culture growing in a 75-ml chemostat if the flow rate for the growth medium is 0.3 minute?

7. In an effort to grow a Leu$^-$ mutant, you prepare minimal medium containing leucine. The cells grow to a maximum cell density of 2×10^8 per milliliter. In contrast, Leu$^+$ cells in the same medium reach a cell density of 2×10^9 per milliliter. How might you explain this observation?

8. An A^- auxotrophic bacterium does not grow in minimal medium but does grow in minimal medium containing substance A. It also grows, however, in minimal medium containing substance B. What information does this give you about the metabolic pathway for synthesis of B?

9. What evidence supports the notion that the loops of a bacterial chromosome are in some way isolated from one another?

10. A collection of new mutants were isolated which could grow well on rich medium plates but could not grow on minimal medium plates. The auxotrophic mutations were identified by checking for growth on minimal plates with the eleven pools of supplements shown in Box 4-1. Given the results shown below, what is the most likely auxotrophic mutation in each of the mutants?

Mutant	Growth on pools:										
	1	2	3	4	5	6	7	8	9	10	11
aux-2001	–	–	–	–	–	–	–	–	–	–	+
aux-2002	+	–	–	–	–	–	+	–	–	–	–
aux-2003	–	–	–	–	–	+	–	–	–	–	–

REFERENCES

*Bretscher, M. S. 1985. The molecules of the cell membrane. *Scientific Am.* October, p. 100.

Cooper, S., and C. E. Helmstetter. 1968. Chromosome replication and the division cycle of *E. coli Br. J. Mol. Biol. 31*, 519.

Drlica, C., and M. Riley. 1990. *The Bacterial Chromosome.* American Society for Microbiology, Washington, DC.

Dykhuizen, D., and D. Hartl. 1983. Selection in chemostats. *Microbiol. Rev. 47:* 150.

*Glass, R. E. 1982. *Gene Function:* E. coli *and its Heritable Elements.* University of California Press, Berkeley.

Helmstetter, C. E., and S. Cooper. 1968. DNA synthesis during the division cycle of rapidly growing *E. coli Br. J. Mol. Biol. 31:* 507.

*Neidhardt, F., J. Ingraham, and M. Schaechter. 1990. *Physiology of the Bacterial Cell.* Sinauer Associates, Inc., MA.

Novick, A., and L. Szilard. 1950. Experiments with the chemostat on spontaneous mutations in bacteria. *Proc. Natl. Acad. Sci. USA 36:* 708.

*Schmid, M. 1988. Structure and function of the bacterial chromosome. *Trends Biochem. Sci. 13:* 131.

*Shapiro, J. A. 1988. Bacteria as multicellular organisms. *Scientific Am.* June, p. 82.

Sinden, R. R., and D. E. Pettijohn. 1981. Chromosomes in living *E. coli* cells are segregated into domains of supercoiling. *Proc. Natl. Acad. Sci. 78:* 224.

*Stanier, R. Y., J. Ingraham, M. Whellis, and P. Painter. 1986. *The Microbial World*, Fifth edition. Prentice-Hall, Englewood Cliffs, NJ.

*Stent, G. S., and R. Calendar. 1978. *Molecular Genetics.* W. H. Freeman and Co., New York.

Suwanto, A., and S. Kaplan. 1989. Physical and genetic mapping of the *Rhodobacter sphaeroides* 2.41 genome: Presence of two unique circular chromosomes. *J. Bacteriol. 171:* 5850.

*Unwin, N., R. Henderson. 1984. The structure of proteins in biological membranes. *Scientific Am. 56:* 78.

Vaara, M. 1992. Agents that increase the permeability of the outer membrane. *Microbiol. Rev. 56:* 395.

*Resources for additional information.

Phage Biology

Phage (or bacteriophage) are viruses that grow in bacterial cells. Phages have played an important role in the development of molecular genetics. At present, a few phages are the most completely understood of any organisms. Because phages are less complex than bacteria and higher cells, they have been extremely useful in the study of replication, transcription, and regulation. This chapter presents a general description of phage biology, then specific uses of phages and their genetics are described in subsequent chapters. A comment on terminology may be useful before beginning: The plural word **phages** refers to different types of phage, whereas in common usage the word **phage** can be both singular and plural, referring in the plural sense to particles of the same type of phage. Thus, P1 and P22 are both phages, but a test tube might contain either 1 P22 phage or 100 P22 phage.

GENERAL PROPERTIES OF PHAGES

A bacteriophage is an obligate bacterial parasite. By itself, a phage can persist, but it cannot replicate except within a bacterial cell. Most phages possess genes encoding a variety of proteins. All known phages, however, use the protein-synthesizing system, amino acids, and energy-generating systems of the host cell, and hence a phage can multiply only in a metabolizing bacterium. Each phage must perform some minimal functions for continued survival:

1. Protection of its nucleic acid from environmental chemicals that could alter the molecule (for example, break the molecule or cause a mutation).
2. Delivery of its nucleic acid to the inside of a bacterium.
3. Conversion of an infected bacterium into a phage-producing system, which yields a large number of progeny phage.
4. Release of progeny phage from an infected bacterium.

These functions are carried out in a variety of ways by different phages. All phages have certain features in common, but differences in detail show the many ways in which specific biological functions can be accomplished. An important observation that has been made is of the degree to which an individual phage particle uses parts of the machinery of the cell. Some phage have fewer than 10 genes and depend almost entirely on cellular functions, whereas others have 30 to 100 genes and depend more on proteins encoded by their own genetic material. A few of the largest phage particles have so many of their own genes that, for certain functions such as DNA replication, they need no

81

host genes. Surprisingly in a few cases, phage genes duplicate genes present in the host.

STRUCTURE OF PHAGES

Different types of phage differ in their physical structures, and often certain features of their life cycles are correlated with their structure. There are three basic phage structures: icosahedral head with no tail, icosahedral head with a tail, and filamentous. Usually the phage particle consists of a single nucleic acid molecule—which may be single-stranded or double-stranded, linear or circular DNA, or single-stranded, linear RNA—and one or more proteins. (One known exception is phage Ø6, which contains three linear, double-stranded RNA molecules, whose base sequences differ from one another.) The proteins form a shell, called either the **coat** or the **capsid**, around the nucleic acid; the nucleic acid is thereby protected from nucleases and harmful substances. Figure 5-1 shows electron micrographs of the three basic structures; the components of a tailed phage are shown in Figure 5-2.

1. In both icosahedral tailless and tailed phages, the nucleic acid is contained in a hollow region formed by the capsid and is highly compact. In a filamentous phage, the nucleic acid is embedded in the capsid and is present in an extended helical form.
2. The tail is a complex multicomponent structure that often has tail fibers.
3. In icosahedral phages, the length of the DNA molecule is much greater than any dimension of the head.

Figure 5-1. The three major morphological classes of phages. (a) Icosahedral, tailless: φX174. (b) Icosahedral, tailed: T4. (c) Filamentous: M13. (Courtesy of Robley Williams.)

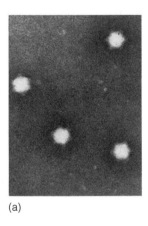

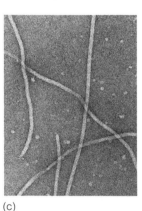

(a)   (b)   (c)

Figure 5-2. Diagrams of the three basic phage structures. The tailed phages do not always have a collar and can have from 0 to 6 tail fibers, the number depending on the phage type. The nucleic acid is shown in red.

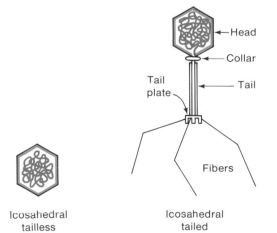

Head
Collar
Tail plate
Tail
Fibers

Icosahedral tailless

Icosahedral tailed

Filamentous

There are many variations on the basic structure of the tailed phages. For example, the length and width of the head may be the same, or the length may be greater than the width. The tail may be very short (barely visible in electron micrographs) or up to four times the length of the head, and it may be flexible or rigid. A complex baseplate may also be present on the tail; when present, it typically has from one to six tail fibers.

STAGES IN THE PHAGE LIFE CYCLE

Phage life cycles fit into two distinct categories: **lytic** and **lysogenic** cycles. A phage in the lytic cycle converts an infected cell into a phage factory, and many phage progeny are produced. A phage capable only of lytic growth is called **virulent**. The lysogenic cycle, which has been observed only with phages containing double-stranded DNA, is one in which no progeny particles are produced; the phage DNA usually becomes part of the bacterial chromosome. A phage capable of such a life cycle is called **temperate**. Most temperate phages also undergo a lytic cycle in certain circumstances. In this section, only the lytic cycle is outlined. The lysogenic cycle is described later. There are many variations in the details of the life cycles of different virulent phages. The typical lytic cycle of phages containing double-stranded DNA is described here (Fig. 5-3).

1. *Adsorption of the phage to specific receptors on the bacterial surface* (Fig. 5-4). Many different types of phage receptors exist. Typically phage receptors are proteins or carbohydrates on the surface of the bacteria that normally serve purposes other than phage adsorption.
2. *Passage of the DNA from the phage through the bacterial cell wall*. Some phage have long tails that may directly inject their DNA into the cytoplasm by a hypodermic syringe–like mechanism (Fig. 5-5). In this process, the nucleic acid could be directly transferred into the cell without exposure to the medium surrounding the recipient cell. Little is known about how the DNA is transferred by other phage types. Some phage with short tails seem to transfer the DNA to the periplasm. With tailless phages, the nucleic acid is transiently susceptible to nuclease attack, so it is thought that the phage coat may break open and release its nucleic acid first onto the cell wall before entering the cell. Some unknown mechanism must then transport the DNA across the cell membrane into the cytoplasm.
3. *Conversion of the infected bacterium into a phage-producing cell*. After phage infection, bacteria often lose the ability either to replicate or to transcribe their own DNA. This shutdown of host DNA or RNA synthesis is accomplished in many different ways (for example, by degradation of the host DNA) depending on the phage species. The shutdown is less common with phages containing single-stranded DNA or RNA.
4. *Production of phage nucleic acid and proteins*. Often the phage directs the synthesis of a replication system that specifically copies the phage nucleic acid. This programming is accomplished either by synthesis of phage-specific DNA and RNA polymerases or by phage proteins that modify the specificity of bacterial polymerases. In both cases, many bacterial replication proteins are used. Synthesis of phage mRNA from the phage DNA is almost always initiated by the bacterial RNA polymerase; but after the first phage mRNA is made, either the bacterial polymerase is modified to recognize other start points for mRNA synthesis (**promoters**), or a phage-specific RNA polymerase is synthesized. RNA synthesis is regulated, and phage proteins are synthesized sequentially in time as they are needed. Usually there is a fairly distinct difference in the

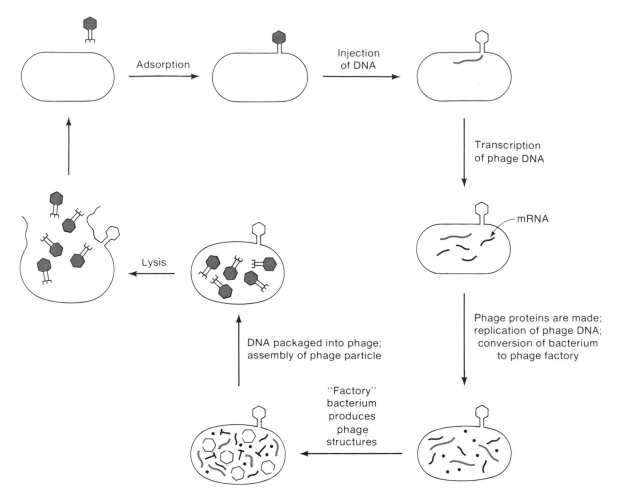

Figure 5-3.   Schematic diagram of the life cycle of a typical phage.

time of synthesis of phage-specified enzymes (made early in the life cycle and called **early proteins**) and the structural proteins of the phage particle (made late in the life cycle). The temporal difference is accomplished by the timing of mRNA synthesis. The early proteins are encoded by **early mRNA**, and the structural proteins are encoded by **late mRNA**. RNA-containing phages and single-stranded DNA phages differ with respect to use of host enzymes. RNA phages must encode their own replication enzymes because bacteria do not contain the enzymes needed to replicate RNA.

5. *Assembly of phage particles* (**morphogenesis**). Two types of proteins are needed for the assembly process: **structural proteins**, which are present in the phage particle, and **catalytic proteins**, which participate in the assembly process but do not become part of the phage particle. A subset of the latter class consists of the **maturation proteins**, which convert intracellular phage DNA into a form appropriate for packaging in the phage particle. With the icosahedral phages assembly occurs in several stages: (a) Aggregation of phage structural proteins to form a phage head and, when needed, to form a phage tail. At this point, the tail is not attached to the head. (b) Condensation of the nucleic acid and entry into a preformed head. (c) Attachment of the tail to a filled head. With filamentous phages, the nucleic acid and the protein form a phage particle in a single step. The mechanism of nucleic acid condensation is not

completely known. Usually 50 to 100 phage particles are produced per cell, the number depending on the particular phage and the physiology of the host cell.

6. *Release of newly synthesized phage.* Late in the infection cycle, most phages synthesize enzymes that lyse the host cell. Two enzymes are typically made: (a) an enzyme that disrupts the cytoplasmic membrane, called "holin," and (b) an enzyme called lysozyme, which degrades the cell wall

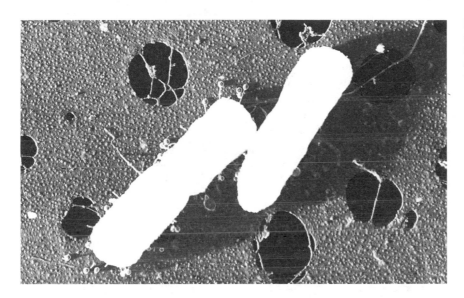

Figure 5-4. An electron micrograph of an *E. coli* cell to which numerous λ phage particles are adsorbed by their long tails. (Courtesy of T. F. Anderson.)

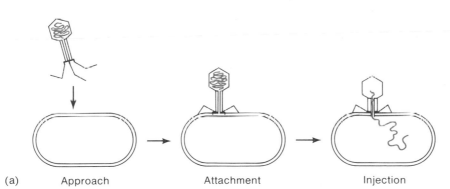

(a)　Approach　　　Attachment　　　Injection

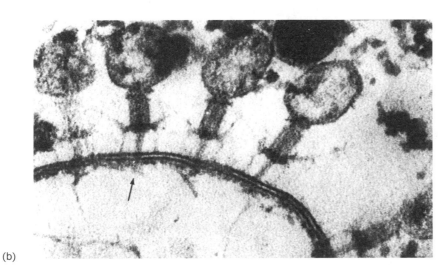

(b)

Figure 5-5. DNA injection by phage T4. (a) After T4 adsorbs to the host cell, the tail sheath contracts and seems to drive a core protein tube through the cell wall like a hypodermic syringe. (b) Electron micrograph of T4 phage adsorbed to the cell wall of *E. coli*, observed in thin section. The tail sheath is contracted, and the core is fixed firmly against the cell wall. The arrow shows a portion of the core projecting through the cell wall. DNA can be seen entering the cell from the two phage at the right. (Courtesy of Lee Simon.)

peptidoglycan. These enzymes disrupt the cell membrane and cell wall, causing the cell to burst (**lysis**), and phage are released to the surrounding medium. The suspension of newly released phage is called a **lysate**. A few filamentous phages release progeny continuously by outfolding of the cell wall; this process is called **extrusion**, and it does not cause major damage to the cell. Cells infected with such filamentous phages can continue to produce virus particles for long periods of time.

COUNTING PHAGE

Phage are easily counted by a technique known as the **plaque assay**. If 10^8 bacteria are spread on a rich agar plate, the 10^8 colonies that result are so close to each other that they appear as a confluent, turbid layer of bacteria called a **lawn**. Alternatively the bacteria can be mixed into a small volume of warm, slightly dilute, liquid agar, which is then poured onto the surface of the solid medium. The liquid, known as **top agar** or **soft agar**, rapidly hardens, providing a smooth surface, with a uniform lawn of bacteria (Fig. 5-6). If a phage is present in the hardened top agar, it can adsorb to one of the bacteria in the agar; the infected bacterium lyses and releases about 100 phage, each of which adsorbs to nearby bacteria. These bacteria in turn release a burst of phage, which then can infect other bacteria in the vicinity. These multiple cycles of infection continue, and, after several hours, the phage will have destroyed all of the bacteria at a single localized area in the agar, giving rise to a clear, transparent circular region in the turbid, confluent layer. This region is called a **plaque**. Because one phage forms one plaque, the number of phage particles added can be calculated from the number of plaques.

The **efficiency of plating (EOP)** is defined as the fraction of phage particles that can form a plaque. The value of EOP is 1 or nearly 1 for many phages but can be less than 1 (0.1 to 0.5) for phages that make very small plaques. When phage lysates are stored, the EOP may decrease; when this occurs, it is usually a result of accumulated chemical damage or denaturation of phage proteins. Storage of phage at 4°C often reduces the loss of phage viability. Addition of cations, glycerol, and proteins also protects phage from damage.

Figure 5-6. (a) Schematic drawing of plaque formation. Bacteria grow and form a translucent lawn. There are no bacteria in the vicinity of the plaque, which remains transparent. (b) Plaques of *E. coli* phage T4. Two types of plaques are present. The smaller plaques are made by wild-type phage; the larger plaques are those of an *rII* mutant. Note the halo around the larger plaques—it is a result of a large amount of lysozyme diffusing outward and lysing uninfected cells. (Courtesy of A. H. Doermann.)

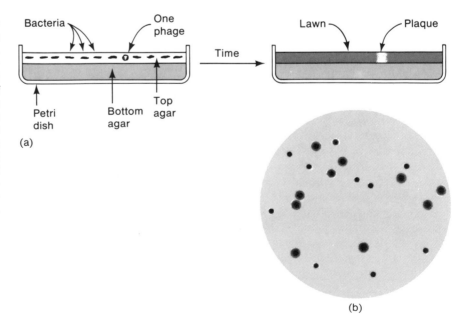

PROPERTIES OF A PHAGE-INFECTED BACTERIAL CULTURE

Rather than infecting a single cell with a phage as described previously, in the laboratory, one usually infects a bacterial culture with a large number of phage particles. Special techniques are needed to analyze the results of these interacting populations.

Number of Participating Phage and Bacteria

The adsorption of phage to a bacterial culture is a random process, and the variation in the distribution of phage among the cells is described by the Poisson distribution:

$$P(n) = \frac{m^n e^{-m}}{n!}$$

where $P(n)$ is the fraction of the bacteria to which n phage have adsorbed when m is the average number of adsorbed phage per bacterium (the **multiplicity of infection or MOI**). That is, the fraction of bacteria infected with $0, 1, 2, 3, \ldots, i$ phage is $P(0), P(1), P(2), P(3), \ldots, P(i)$. Thus, if 3×10^8 phage adsorb to 10^8 bacteria ($m = 3$), the values of $P(0), P(1), P(2), P(3), \ldots$, are 0.05, 0.15, 0.22, ... Because $P(0) = 0.05$, the sum of $P(1) + P(2) + \ldots + P(i)$ must equal $1 - 0.05 = 0.95$. In other words, 95% of the bacteria will be infected by at least one phage.

Note also that the value of $P(0)$ tells the fraction of the phage particles that have adsorbed. For example, in the infection just described, $P(0)$ should equal 0.05 for $m = 3$, if all of the added phage had adsorbed to the bacteria. In a particular experiment using 3×10^8 phage and 10^8 bacteria, however, if one observed that 12% of the bacteria remained uninfected, the value of $P(0)$ would be 0.12; using this value, one can calculate from the Poisson term $P(0) = e^{-m}$ that $m = 2.12$, which is the true value of the MOI. Thus, $2.12/3 = 0.71$, or 71% of the added phage particles actually adsorbed.

It is possible to measure $P(0)$ in a simple way (Fig. 5-7). A known number of cells are used in an infection. After an adsorption period, the bacterial suspension is diluted and plated, and the fraction of the bacteria able to form a colony is measured. Because the lytic phage will kill any infected cells, only uninfected cells can form a colony. Thus, $P(0)$ is the number of colonies formed divided by the total number of cells. As a check, the number of infected cells can be measured in the following way. First, antibodies that inactivate unadsorbed phage are added to the infected culture. (Antibodies can be obtained from the blood of a rabbit that has been injected with a purified suspension of the phage.) Then the phage suspension is plated on a lawn of phage-sensitive cells, where each infected cell, provided that it is plated before lysis, will produce a single plaque. A cell that can form a plaque in this way is called an **infective center**.

The number of phage produced by an infected cell is called the **burst size**. This is an important parameter in many experiments because it is a measure of the efficiency of phage production. It is measured by determining (by counting plaques) the number of phage produced after lysis of the culture and dividing by the number of infective centers.

Production of a Phage Lysate

Lytic phage multiply much more rapidly than bacteria. That is, bacteria double in one generation time, whereas, in one life cycle, the number of phage is increased by a factor equal to the burst size. This is easily seen in the example shown in Table 5-1, in which a single phage whose average burst size is 100 and whose life cycle lasts 25 minutes infects a 1 ml bacterial culture with a 25-minute doubling time. In this calculation, it is assumed that adsorption is always instantaneous and complete. Note that, in four generations, the number of bacteria has increased 14-fold, whereas the number of phage has increased 10^8-fold. At this time, there are approximately eight times as many phage as bacteria; hence all bacteria are infected. Thus, after 125 minutes, the bacteria are gone, and the original phage particle has produced 1.4×10^9 progeny.

By careful choice of media, growth conditions, and time of injection, phage lysates containing high concentrations of phage can be prepared. For example, to grow phage P22, a rich, well-aerated medium is used, and the culture is infected at a cell density of about 10^8 cells/ml with a MOI of 0.1. One-tenth of these cells (10^7 cells/ml) are infected and released phage at a concentration of about 50 phage per cell. By the time the concentration of uninfected bacteria has reached 10^9 cells/ml, the MOI is 5, all bacteria are infected, and after one more phage cycle, the phage concentration is about 5×10^{10} per milliliter.

High concentrations of phage can also be prepared by growth on solid media using an amount of phage such that all bacteria are ultimately infected and lysed. For example, if 10^6 phage are placed in soft agar with about 10^8 bacteria, after about 6 hours the soft agar layer (which is completely clear) will contain about 10^{11} phage. Such a lysate is called a **plate lysate**. A plate lysate can be prepared from a single plaque by the following procedure. A sterile toothpick or wire is stabbed into the center of an isolated plaque and then stirred in liquid soft agar containing bacteria. The soft agar is then poured on a plate, where it hardens. Usually about 10^6 phage are transferred, so confluent lysis results.

Figure 5-7. Scheme for determining the number of uninfected bacteria (A), unadsorbed phage (B), infective centers (C), progeny phage (D), and the burst size (D–C).

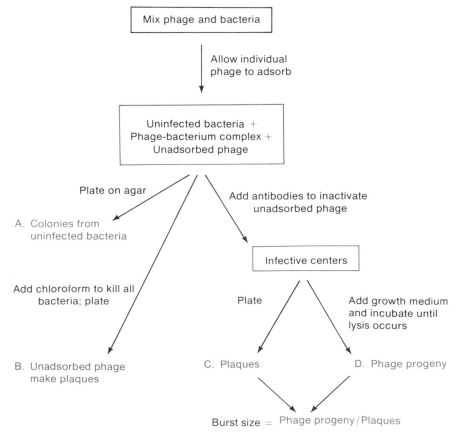

Table 5-1 Calculation of the phage and bacterial concentrations after various numbers of bacterial generations*

		Concentrations	
		Bacteria / ml	
Number of generations	Phage / ml	Approximate	Precise
0	1	10^6	10^6
1	10^2	2×10^6	$2 \times (10^6 - 1)$
2	10^4	4×10^6	$(4 \times 10^6) - (2 \times 10^2) - 4$
3	10^6	7.98×10^6	$(8 \times 10^6) - (2 \times 10^4) - (4 \times 10^2) - 8$
4	10^6	1.4×10^7	$(1.6 \times 10^7) - (2 \times 10^6) - (4 \times 10^4) - (8 \times 10^2) - 16$
5	$1.4 \times 10^9 = 100 \times (1.4 \times 10^7)$	0	0

*Initially a bacterial culture at a concentration of 10^6 cells/ml is infected with 1 phage. The doubling time of the bacteria and the life cycle of the phage are equal.

One-Step Growth Curve

Certain kinetic parameters of the phage infection can be determined by studying an infected culture. A classic experiment is the **one-step growth curve**. In this experiment, a culture is infected with a MOI of about 0.1 (to ensure than no cell is infected with more than one phage). Antibody to the phage is then added to inactivate any unadsorbed phage. The infected cells are diluted about 1000-fold into fresh warm medium (to prevent inactivation of progeny phage by the antibody), and at various times aliquots of the supernatant are taken and plated on a lawn of sensitive bacteria. At first the number of plaques is constant (Fig. 5-8) because plaques are formed only by phage released later by infected unlysed cells. This period is called the latent period. Some time after infection (a length of time is characteristic for each phage), the number of plaques increases. During this short time interval (the rise period), the infected cells are lysing. When all infected cells have lysed, the phage concentration remains constant. The ratio of phage produced to the initial number of infective centers is the burst size, and the number of minutes before the increase in plaque number occurs is the lysis time.

A modification of this experiment (also shown in Fig. 5-8) can be used to determine the kinetics of phage production within the cell. In this procedure, chloroform is added at various times after infection. Chloroform destroys the cell membranes, resulting in **premature lysis** of the infected cells. In this case, the increase in the number of plaques represents intracellular production of phage.

Single-Burst Technique

Studies of phage production in infected cultures yield information that has been averaged over a phage population. Sometimes, however, it is also important to know the events in a single infected cell. For example, the burst size of a particular phage is the average number of phage produced per cell but does not indicate whether the burst size varies from cell to cell. A **single-burst experiment** can provide important information about cell-to-cell variation.

In a single-burst experiment, a bacterial culture is infected, and, after the phage are adsorbed, the infected cells are diluted in growth medium to low concentration—usually about 0.05 infected cells/ml. Then 1 ml aliquots are dispensed into hundreds (or thousands) of test tubes. According to the Poisson distribution, at this concentration, 95.1% of the tubes will not contain infected cells, 4.8% will contain one infected cell, and 0.1% will contain more than one infected cell. Thus, 4.8/4.9, or 98%, of the tubes *containing infected cells* will contain only one infected cell. The single cell in each tube is allowed to lyse, and the contents of each tube are plated in a single petri dish with indicator bacteria, so plaques will form. Thus, the plaques on one plate are formed by the phage progeny of a single infected cell. The number of plaques observed on the various plates yields the distribution of burst sizes, which are found to range from less than 10 to several hundred. The wastefulness of this technique should be noted: To study 100 infected cells requires using about 2000 petri dishes, of which roughly 1900 will contain no plaques. The single-burst technique, however, is useful for studying the result of phage crosses in a single cell.

SPECIFICITY IN PHAGE INFECTION

Several thousand different types of phages have been isolated. The ability of a particular phage to infect a bacterium is almost always limited to a single bacterial species and often to a few strains of that species. For example, phage P22 infects *S. typhimurium* but not *E. coli*. Several factors contribute to this specificity. One of these is the ability to adsorb. Phage P22 cannot adsorb to *E. coli* because *E. coli* lacks the specific receptor for P22 on the cell surface. The phage bacterium interaction, however, is subject to genetic selection: It is possible to isolate bacterial mutants that are resistant to phage and phage mutants with new host specificity. For example, if 10^8 *E. coli* B cells are infected with 10^{10} T6 phage particles and the infected cells are put on an agar surface, about 100 colonies (1 in 10^6) form. Cultures prepared from these colonies invariably consist of mutants that have lost the ability to be infected by T6; in particular, phage no longer adsorb to the mutant cells. (These mutant cells are not produced by the infection but preexist in the culture and are selected by growth with an excess of phage.) These phage-resistant bacterial mutants are called T6 resistant or Tsx^r (Fig. 5-9). If 10^8 Tsx^r cells are used to form a bacterial lawn on agar and 10^8 T6 phage are added, about 10 plaques result. The phage in these plaques carry a mutation in the tail fiber gene and have thereby regained the ability to adsorb to Tsx^r cells. These phage are called *h* mutants (for *host* range). They usually retain the ability to form plaques on Tsx^s bacteria and hence are said to have an "extended host range."

An interesting phenomenon occurs if a bacterial culture is infected with both wild-type (T6h^+) and T6h phages at a MOI such that all bacteria are infected with both phage types. One would expect that all progeny phage would

Figure 5-8. A hypothetical one-step growth curve. Cells at 10^8 cells/ml are infected at MOI = 0.1. Antiserum is added to inactivate unadsorbed phage, and then the culture is diluted 1000-fold. The black curve shows the number of plaques produced per milliliter of culture, without further treatment. The red curve shows the result of premature lysis by lysozyme and chloroform. The burst size is 100.

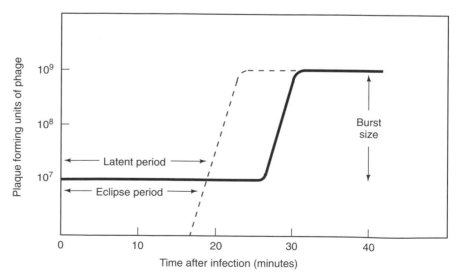

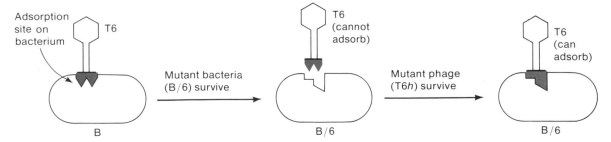

Figure 5-9. The generation of a phage-resistant bacterium, and a phage *h* mutant. Although the tail fiber is modified, for graphic purposes, the tail plate has been modified in the drawing.

form plaques on strain B (which plates both phage types), and half of the progeny (the *h* mutants) would plate on the Tsx^r mutants. Only one-fourth, however, form plaques on the Tsx^r mutants. The reason is that during the phage-assembly process, no mechanism exists to match a DNA molecule with the *h* genotype with an *h*-type tail fiber. Thus, half of the phage with *h* DNA are in h^+ particles and cannot adsorb to the Tsx^r bacteria. Similarly, there are h^+ DNA molecules in *h* particles; they can adsorb to Tsx^r bacteria but produce h^+ progeny, which cannot go through further cycles of infection. This phenomenon is called **phenotypic mixing**. If the initial lysate resulting from the mixed infection is allowed to infect a B culture at a MOI well below 1, no phenotypically mixed progeny result.

HOST RESTRICTION AND MODIFICATION

Even if the phage can infect a different bacterial species, most bacteria have another barrier called the **host restriction and modification** system. This is a phenomenon in which a bacterium of a type X is able to distinguish a phage that has been grown in type X bacterium from one grown in a different type such as Y and is able to prevent the phage grown in Y from carrying out a successful infection. A phage P grown in a bacterium X is denoted P(X). Host modification and restriction are illustrated by the data in Table 5-2. Note that λ(K), which has been grown in *Escherichia coli* strain K, forms plaques at a low efficiency in strain B. Thus, λ(K) is **restricted** by strain B. The phage population in these rare plaques (λ(B)) has been modified by strain B, so the phage grow efficiently in strain B; however, λ(B) now fails to grow in strain K—that is, it is restricted by K. The restriction is due to a specific restriction endonuclease produced by the recipient. For example, *E. coli* B contains an enzyme called a restriction endonuclease (the *Eco*B nuclease), a site-specific nuclease that cuts DNA strands only near a specific base sequence (many restriction enzymes cut within a target sequence, but *Eco*B cuts near the sequence; see Chapter 20). Phage λ(K) contains this sequence; so when its DNA is injected into *E. coli* B, the phage DNA is degraded. *E. coli* B also contains this sequence

Table 5-2 The restriction and modification pattern of *E. coli* phage λ

Bacterial strain	Phage		
	λ(K)	λ(B)	λ(C)
K	1	10^{-4}	10^{-4}
B	10^{-4}	1	10^{-4}
C	1	1	1

Note: Numbers indicate relative plating efficiency.

and would destroy its own DNA unless the sequence were modified. A site-specific methylating enzyme (*Eco*B methylase) methylates an adenine in the sequence, thereby rendering the sequence resistant to the *Eco*B nuclease. When λ(K) infects strain B, a few parental phage-DNA molecules in the large population of infected cells are methylated before they are restricted. Thus, restriction is avoided in these rare phage so a small population of phage having the B modification (λ(B)) is produced. *E. coli* K also contains a restriction enzyme (*Eco*K). It attacks a base sequence that is different from the sequence recognized by *Eco*B. An *Eco*K methylase also protects *E. coli* K from self-destruction, producing the K modification. A λ phage that has always been grown in strain K (λ(K)) is methylated in the *Eco*K-specific sequence and is resistant to *Eco*K nuclease. However, λ(B) has an unmethylated *Eco*K sequence, so λ(B) DNA is usually cleaved when a strain K cell is infected. Occasionally a λ(B) DNA molecule escapes restriction and replicates, and its replicas have a methylated K-specific sequence. Thus, the rare progeny phage that result when λ(B) successfully infect *E. coli* K are λ(K); they now lack the B modification and are restricted when infecting strain B.

Note in Table 5-2 that one grown on strain C (λ(C)) also fails to grow well in strains B and K, but neither λ(B) nor λ(K) is restricted by strain C. The reason for the lack of restriction is that strain C has no restriction nuclease active against any base sequence in λ DNA. The λ(C) phage are restricted by both strains B and K because, of course, strain C does not have the *Eco*B and *Eco*K methylases. Host restriction and modification is widespread in bacteria, probably serving to destroy foreign DNA.

LYSOGENIC CYCLE

Lysogeny is an alternative reproductive pathway to the lytic cycle described earlier. There are two types of lysogenic cycles. The most common lysogenic pathway, for which the *E. coli* phage λ pathway is the paradigm, is described in the following simplified outline (Fig. 5-10).

1. The linear phage DNA molecule is injected into a bacterium.
2. After a brief period of mRNA synthesis, which is needed to synthesize a repressor protein (which inhibits the synthesis of the mRNA species that encode the lytic functions) and a site-specific recombination enzyme, phage mRNA synthesis is turned off by the repressor.
3. Recombination between the phage DNA molecule and the DNA of the bacterium inserts the phage DNA into the bacterial chromosome.
4. The bacterium continues to grow and multiply, and the phage genes replicate as part of the bacterial chromosome.

The second type of lysogenic pathway, which is less common, differs from the preceding one in that there is no DNA-insertion system, and the phage DNA becomes a plasmid (an independently replicating circular DNA molecule) rather

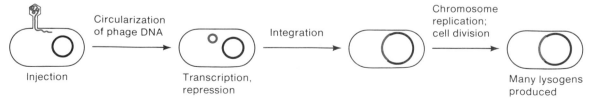

Figure 5-10. The general mode of lysogenization by insertion of phage DNA into a bacterial chromosome.

than a segment of the host chromosome. *E. coli* phage P1 is an example of this type of lysogenic pathway. In this chapter, we mainly consider the λ type pathway.

General Properties of Lysogens

The following terms describe various aspects of lysogeny.

1. A phage capable of entering either a lytic or a lysogenic life cycle is called a **temperate** phage.
2. A bacterium containing a complete set of phage genes is called a **lysogen**.
3. If the phage DNA is contained within the bacterial DNA, the phage DNA is said to be **integrated**. The process by which this state of the DNA is achieved is called **integration** or **insertion**. Phage DNA in plasmid form is nonintegrated. Whether integrated or not, the phage DNA in lysogens are called a **prophage**.

Two important properties of lysogens are the following:

1. Lysogens are resistant to reinfection by a phage of the type that first lysogenized the cell; this resistance to superinfection is called **immunity**.
2. Even after many cell generations, a lysogen can initiate a lytic cycle; in this process, which is called **induction**, the phage genes are excised as a single segment of DNA (Fig. 5-11).

The molecular mechanism for immunity and the circumstances that give rise to induction are discussed in Chapter 17.

More than 90% of the thousands of known phages are temperate. These phages are often unable to produce bursts as large as many highly virulent phages, such as T4 and T7, but compensate by their ability to multiply in environmental conditions that are not suitable for rapid production of progeny. For example, consider a bacterial population that is actively dividing. If a phage can infect one cell and multiply (in a lytic cycle), the number of progeny phage increase rapidly, as shown in Table 5-1. If the bacteria, however, were growing very slowly because limiting nutrients are available in the surrounding medium (a common condition in nature), the infecting phage may not be able to reproduce because phage grow only in bacteria that are actively metabolizing.

When bacteria are starved of nutrients, they degrade their own mRNA and protein before they become dormant. Restoration of nutrients enables the bacteria to grow again. This is not true of a phage-infected cell in which the lytic life cycle has been interrupted: Usually the ability to produce phage is permanently lost, probably because essential phage functions are destroyed by the protein and mRNA degradation. In contrast, the phage can survive in the host if it lysogenizes the bacterium because the phage DNA can become dormant. When growth of the bacterium resumes, the phage genes replicate as part of the

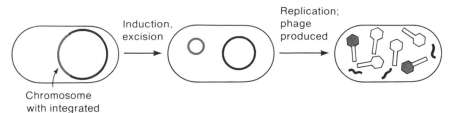

Chromosome
with integrated
prophage (red)

Figure 5-11. An outline of the events in prophage induction. The prophage DNA is in orange. The bacterial DNA is omitted from the third panel for clarity.

chromosome. Even though production of progeny phage is delayed in the lysogen, it can resume on induction of the lytic cycle.

What happens during an infection of an actively growing bacterial population in which phage are multiplying rapidly? When the number of phage exceeds the number of bacteria, the phage cannot multiply further because there are no more sensitive bacteria. It is possible that years could pass before these phage particles might encounter another sensitive host bacterium, and during this time various deleterious agents might damage the phage particles. Until a host cell appears, the phage particles have no chance to increase in number. If lysogenization could occur at a high MOI, however, the phage genes could be maintained indefinitely because the lysogen would grow whenever nutrients were available. Indeed, the two conditions that stimulate a lysogenic response of a temperate phage are depletion of nutrients in the growth medium and a high MOI. In contrast to temperate phages, phages that have only a lytic lifestyle frequently possess exceedingly stable head and tail structures.

Prophage Insertion of *E. coli* Phage λ

Phage λ DNA integrates into a specific site in the *E. coli* chromosome, between the *gal* and *bio* (biotin) genes. The insertion site is called the λ attachment site and designated *att*λ or *attL*. Most other temperate phages also integrate almost exclusively at a single site. Integration is a result of recombination between an attachment site in the phage DNA and one in the bacterial DNA (see Chapter 17). The attachment sites have a common base sequence, designated *O*, in which the exchange occurs, and are flanked by sequences that are specific to the bacterium or the phage. The bacterial and phage attachment sites are written BOB and POP, respectively. The insertion process is shown schematically in Figure 5-12. The essence of the mechanism is circularization of λ DNA followed by physical breakage and rejoining of phage and host DNA—precisely in the two *O* regions. The exchange is catalyzed by a phage enzyme **integrase**. Because the attachment sites in the phage and bacteria are not the same, the recombination forms **prophage attachment sites** that are hybrids of the bacterial and phage sites, BOP and POB. When the prophage is induced, the excision reaction is due to recombination between these sites; another phage enzyme, **excisionase**, is required as a result

Figure 5-12. The mechanism of prophage integration and excision of phage λ. The phage attachment site has been denoted POP' in accord with subsequent findings. The bacterial attachment site is BOB'. The prophage is flanked by two new attachment sites denoted BOP' and POB'.

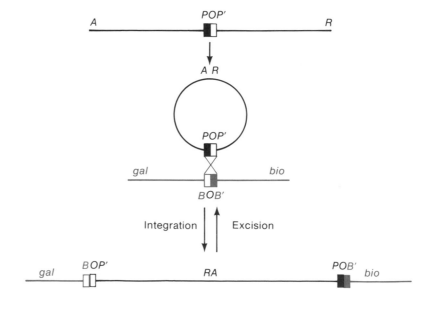

of recombination to catalyze excision. Both insertion and excision are examples of site-specific recombination.

Nonintegrative Lysogeny

Most temperate phages form lysogens in the way described for λ: A prophage is inserted at a unique site in the host chromosome. In contrast, lysogeny with *E. coli* phage P1 is markedly different because the prophage is not inserted into the chromosome. After infection, P1 DNA circularizes and, similar to λ, is repressed. In the lsyogenic mode, it remains as a free supercoiled plasmid DNA molecule, roughly one or two per cell. Once per bacterial life cycle the P1 DNA replicates, and this replication is coupled to chromosomal replication (the coupling is controlled by a phage gene). When the bacterium divides, each daughter cell receives one copy of the P1 plasmid. The mechanism of prophage maintenance is not as foolproof in phage P1 as in temperate phages that insert their phage DNA into a chromosome; for example, in each round of cell division, about 1 cell per 1000 fails to receive a copy of the P1 plasmid. It is not known whether this is due to occasional failure in replication or to imperfect segregation of plasmids into the daughter cells.

Plaques of Temperate Phages

When a virulent phage forms a plaque on a lawn of growing bacteria, the plaque is clear because all bacteria in the center of the plaque are killed and lysed. Temperate phages, such as λ, however, form a plaque with a **turbid center** (Fig. 5-13). The turbidity is caused by the growth of phage-immune lysogenic cells in the plaque. When plating phage to obtain plaques, phage and bacteria are usually mixed in soft agar in a ratio of about 1 phage per 10^7 bacteria. The bacteria grow rapidly, and the MOI is low, so the lytic cycle ensues. After several lytic cycles, the local MOI becomes high, and a few cells are lysogenized; because availability of nutrients does not yet limit development of phage or cells, most cells are lysed. When the nutrients in the agar are depleted, the uninfected cells stop growing, and the plaque stops increasing in size. Because there has been less bacterial growth within the plaque, however, nutrients are still present there. Therefore the lysogenic cells, which are immune to subsequent infection by λ, continue to grow, forming a turbid center in the plaque.

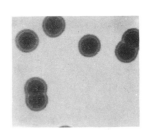
(a)

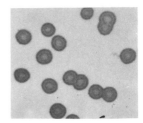

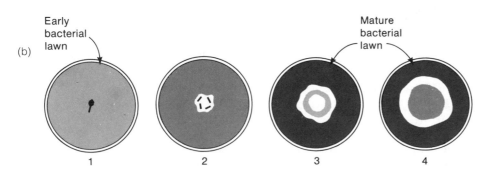
(b)

Figure 5-13.　(a) Clear (*cl⁻*) and turbid (*cl⁺*) plaques of phage λ. (Courtesy of A. D. Kaiser.) (b) Diagram showing the development of a turbid plaque of a temperate phage. (1) A single phage infects a bacterial lawn; (2) a small clear plaque (usually invisible) contains a few lysogens (shown as rods); (3) the clear region enlarges, but lysogens grow within the plaque; (4) clear region reaches maximum size, and lysogens stop growing as nutrient is exhausted.

ADDENDUM

This chapter described the basic features of the life cycles of temperate and virulent phages. Methods for counting phage by plating and various aspects of a phage-infected bacterium have been described, but little or no genetics has been discussed. We return to phages, and particularly phage genetics, later in the book. In Chapter 15, we see the elementary aspects of phage genetics, examining genetic recombination and mapping. Surprising features of certain genetic maps and their molecular explanations are given. Chapter 15 goes on to emphasize *E. coli* phage T4, which is the grandfather of phage genetics. Chapters 16 and 17 stress phage λ, the best-understood phage. Chapter 16 examines the genetics and physiology of its lytic cycle, and Chapter 17 considers λ lysogeny. In Chapter 17, the role of genetic analysis in elucidating the phenomenon of lysogeny is emphasized, and physical experiments that confirm many conclusions originally derived from genetic analysis are presented briefly. In Chapter 18, the mechanism by which phages can transfer bacterial genes between different bacteria (transduction) is examined. These chapters have been relegated to a later point in the book because a good understanding of phage phenomena requires more knowledge of molecular biology and bacterial genetics than has yet been presented.

KEY TERMS

adsorption	multiplicity of infection (MOI)
burst size	phage
early and late mRNA	plaque
efficiency of plating (EOP)	Poisson distribution
immunity	prophage
induction	prophage attachment sites
infective center	restriction and modification
integrase	single-burst experiment
lawn	temperate
lysogenic/lysogen	top or soft agar
lytic/lysis/lysate	virulent
morphogenesis	

QUESTIONS AND PROBLEMS

1. How many plaques can be formed by a single phage particle?

2. A phage adsorbs to a bacterium in a liquid growth medium. Before lysis occurs, the infected cell is added to a large number of bacteria, and a lawn is allowed to form on a solid medium. How many plaques will result?

3. If 10^6 phage are mixed with 10^6 bacteria and all phage adsorb, what fraction of the bacteria will not have a phage?

4. What is meant by the term phage-host specificity, and what is the most frequent cause of this specificity?

5. A particular bacterial mutant cannot use lactose as a carbon source. If a phage adsorbs to such a bacterium and the infected cell is put in a growth medium in which lactose is the sole carbon source, can progeny phage be produced?

6. Why do phage plaques not enlarge indefinitely?

7. A sample of a wild-type virulent phage grown on strain A of *E. coli* plates on strain X with an efficiency of 10^{-4}. What is the most likely explanation for this low efficiency of plating?

8. Roughly speaking, what is a typical burst size of a phage whose nucleic acid is double-stranded DNA?

9. Because infection of a bacterium by a phage is usually lethal to the bacterium, why have bacteria not evolved to lose their phage receptors?

10. Bacteria are allowed to grow on an agar surface until a confluent turbid layer appears. Then 10^3 T4 phage are spread on the surface. Six hours later (a time sufficient for plaque formation, if the phage had been added at the time the bacteria were placed on the agar), no plaques are evident. Why?

11. Phage T4 normally forms small clear plaques on a lawn of *E. coli* strain B. A mutant of *E.coli* called B/4 is unable to adsorb T4 phage particles so that no plaques are formed. T4*h* is a host-range mutant phage capable of adsorbing to *E. coli* B and to B/4 and forms normal-looking plaques. If *E. coli* B and the mutant B/4 are mixed in equal proportions and used to generate a lawn, what will be the appearance of plaques made by T4 and T4*h*?

12. In a broth containing glucose and yeast extract, *E. coli* grows with a doubling time of 30 minutes. Phage T7 has a life cycle of 20 minutes under these conditions and a burst size of 200 phage per infected cell. If a culture of 2×10^7 *E. coli* per milliliter is growing exponentially and 5000 plaque-forming units per milliliter of T7 phage are added, when will the culture lyse (assuming that phage adsorption is rapid and that multiply infected bacteria give the same burst as singly infected bacteria)?

13. One milliliter of a bacterial culture at 5×10^8 cells/ml is infected with 10^9 phage. After sufficient time for greater than 99% adsorption, phage antiserum is added to inactivate all unadsorbed phage. The infected cell is mixed with indicator cells in soft agar, and plaques are allowed to form. If 200 cells are put in each petri dish, how many plaques will be found?

14. P2 and P4 are bacteriophages of *E. coli*. They have the following properties: (1) When one P2 phage infects a bacterium, the bacterium usually bursts, giving about 100 P2 progeny; (2) when a P4 phage infects a bacterium, the bacterium survives because P4 is a defective phage; (3) when P2 phage and P4 phage coinfect the same bacterium, lysis of the bacterium gives 100 P4 progeny and no P2 progeny (because P4 inhibits the growth of P2). If 3×10^8 P2 and 2×10^8 P4 are added to 10^8 bacteria, then:

    a. How many bacteria will not be infected?
    b. How many bacteria will survive?
    c. How many bacteria will produce P2 progeny?
    d. How many bacteria will produce P4 progeny?

15. How does a temperate phage differ from a virulent phage?

REFERENCES

Adams, M. 1959. *Bacteriophages*. Interscience Publishers, New York.

Calendar, R. 1988. *The Bacteriophages*. Plenum Press, New York.

Casjens, S. 1985. *Virus Structure and Assembly*. Jones and Bartlett Publishers, Boston.

Hendrix, R., J. Roberts, F. Stahl, and R. Weisberg. 1983. *Lambda II*. Cold Spring Harbor Laboratory, New York.

Ptashne, M., A.D. Johnson, C.O. Pabo. 1982. A genetic switch in a bacterial virus. *Scientific Am.* November, p. 128.

*Ptashne, M. 1992. *A Genetic Switch*, Second edition. Blackwell Scientific Publications, MA.

Stent, G. S. 1963. *Papers on Bacterial Viruses*. Little, Brown, and Co., Boston.

Wilson, G., and N. Murray. 1991. Restriction and modification systems. *Ann. Rev. Genet.* 25: 585.

Young, R. 1993. Bacteriophage lysis: mechanism and regulation. *Microbiol. Rev.* 56: 412.

*Resources for additional information.

MOLECULAR ASPECTS OF GENE EXPRESSION

Gene Expression

How is the information contained in genes converted to molecules that determine the structure and function of bacteria and phages? Gene expression is accomplished through a sequence of events in which the information contained in the base sequence of DNA is first copied into an RNA molecule, which is used to determine the amino acid sequence of a protein molecule. RNA molecules are synthesized by using the base sequence in a region of one of the DNA strands as a **template** to make the complementary RNA. This reaction is catalyzed by an enzyme called an **RNA polymerase**. The process by which the segment corresponding to a particular gene is selected and an RNA molecule is synthesized is called **transcription**. Protein molecules are then synthesized by using the base sequence of this RNA molecule to direct the sequential joining of amino acids in a particular order, so the amino acid sequence is determined by the DNA base sequence. The production of an amino acid sequence from an RNA base sequence is called **translation**. Some RNA molecules directly serve structural or catalytic roles and hence are not translated. This process is called the central dogma:

$$\text{DNA} \xrightarrow{\textit{transcription}} \text{RNA} \xrightarrow{\textit{translation}} \text{Protein}$$

In this chapter, we describe transcription and two features of translation—the genetic code and protein synthesis.

TRANSCRIPTION

The essential chemical features of the enzymatic synthesis of RNA are as follows:

1. The precursors in the synthesis of RNA are the four 5'-triphosphates of the ribonucleosides adenosine, guanosine, cytosine, and uridine.
2. In forming an RNA molecule, a 3'-OH group of the ribose at the 3' end of the growing RNA molecule reacts with the innermost phosphate of a precursor nucleoside-5'-triphosphate. The two terminal phosphate groups are released as inorganic pyrophosphate (PPi), and a sugar-phosphate bond results, extending the RNA molecule by one nucleotide unit (Figure 6-1a).
3. The sequence of bases in an RNA molecule is determined by the base sequence of the DNA template. Each base added to the growing end of the RNA chain is chosen for its ability to base-pair with the DNA template

strand; thus, the bases C, T, G, and A in a DNA strand cause G, A, C, and U to be added to the growing end of an RNA molecule.

4. Nucleotides are added only to the 3'-OH end of the growing chain. Thus, the 5' end of a growing RNA molecule is a triphosphate. The growing RNA strand and the DNA template strand are antiparallel to one another, similar to the two strands of a DNA molecule.

A common feature of RNA synthesis is that *the DNA molecule being copied is double-stranded, yet in any particular region of the DNA only one strand serves as a template.* The implications of this statement are shown in Figure 6-1b.

The synthesis of RNA consists of five discrete stages: **promoter recognition**, **local unwinding**, **chain initiation**, **chain elongation**, and **chain termination**. These have the following characteristics:

1. RNA polymerase binds to double-stranded DNA within a specific base sequence (typically 20 to 40 bases long), called a **promoter**. RNA polymerase bound to double-stranded DNA is called a **closed promoter complex**.

2. After the initial binding step, local unwinding of the DNA occurs, and the RNA polymerase is said to have formed an **open promoter complex**.

3. The RNA polymerase recognizes a **transcription start site**, which is

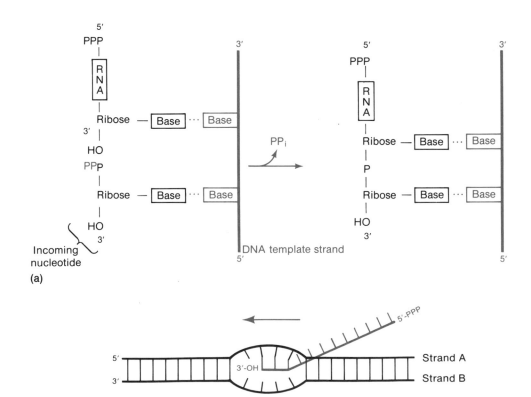

(a)

(b)

Figure 6-1. RNA synthesis. (a) The polymerization step in RNA synthesis. The incoming nucleotide forms hydrogen bonds (three dots) with a DNA base. The OH group in the upper nucleotide (the nucleotide at the 3' end of the RNA molecule) reacts with the black P in the triphosphate group, leading to removal of the red phosphates (PPi). (b) Geometry of RNA synthesis. RNA is copied only from strand A of a segment of a DNA molecule. It is not usually copied from strand B in that region of the DNA. Elsewhere, however,—for example, in a different gene—strand B might be copied; in that case, strand A would not usually be copied in that region of the DNA. The RNA molecule is antiparallel to the DNA strand being copied.

very close to the initial binding site. The first nucleoside triphosphate is added at this site, and synthesis begins.

4. RNA polymerase then moves along the DNA in the 5' → 3' direction, adding nucleotides to the 3'-OH of the growing RNA chain.

5. When RNA polymerase reaches a termination sequence, both the newly synthesized RNA and the RNA polymerase are released.

The existence of promoters was first demonstrated by the isolation of *Escherichia coli* (P⁻) mutations that eliminate activity of the *lac* genes. These mutations did not map within any of the *lac* genes but were located adjacent to them. Furthermore, complementation tests showed that *lac* gene expression was inhibited only if the mutation was located directly adjacent to the genes *in the same DNA molecule*: That is, the mutations were *cis*-acting in a partial diploid cell with two copies of the *lacZ* gene, for example, in a merodiploid with one copy of *lacZ* on the chromosome and one copy on an independent DNA molecule called an F'*lacZ* plasmid. Expression of the *lacZ* gene is required for the cell to synthesize the enzyme β-galactosidase. Table 6-1 shows that a wild-type *lacZ* gene is inactive when a P⁻ mutation is present on the same DNA molecule (either the chromosome or an F' plasmid), but the P⁻ mutation does not affect *lacZ* when it is located on a different DNA molecule. Later molecular studies show that the P⁻ mutation prevents transcription of *lacZ* when it is located in *cis* to the *lacZ* gene. The P⁻ mutations are called **promoter "down" mutations**. It is also possible to isolate **promoter "up" mutations** that overexpress downstream genes because they are better sites for RNA polymerase.

Promoter sequences were first identified in vitro by mixing RNA polymerase with DNA and then treating the DNA with nucleases that digest free DNA but not the region bound to the RNA polymerase. After enzymatic digestion, the RNA polymerase was removed and the base sequence of the DNA determined. Examination of a large number of promoter sequences for different genes and from different bacteria has shown that they have many features in common. Typically *E. coli* and *Salmonella typhimurium* promoters consist of two regions: The sequence TATAAT (or a similar sequence) is called the –10 region because it occurs about 10 base pairs before the transcription start site; the –35 region is commonly located about 35 base pairs before (or "upstream of") the transcription start site. These recognition sequences instruct RNA polymerase where to start transcription. The strength of the binding of RNA polymerase to different promoters varies greatly; this variation is a fundamental mechanism for regulating gene expression. Some promoters have binding sites for other proteins (called activators) that are required for RNA polymerase to bind correctly. For example, many bacterial promoters require binding of the cyclic AMP (cAMP)–cAMP receptor protein (CRP) complex. Activity of these promoters is regulated by the intracellular concentration of cAMP. This regulation is described in more detail in Chapter 7.

Two kinds of termination events are known: those that are dependent on the DNA base sequence only and those that require the presence of termination

Table 6-1 Effect of promoter mutations on transcription of the *lacZ* gene

Genotype	Transcription of *lacZ*⁺ gene	Rationale
1. p⁺lacZ⁺	Yes	– –
2. p⁻lacZ⁺	No	no promoter
3. p⁺lacZ⁺/p⁺lacZ⁻	Yes	– –
4. p⁻lacZ⁺/p⁺lacZ⁻	No	no promoter *cis* to *lacZ*⁺
5. p⁺lacZ⁺/p⁻lacZ⁻	Yes	– –

protein called **Rho**. Both types of events occur at specific but distinct base sequences. Rho-independent termination usually occurs at base sequences in the template DNA strand that consists of a nucleotide palindrome usually interrupted by a few bases followed by a number of adjacent adenines. When this sequence is transcribed into RNA, the RNA can fold back on itself to form a stem-and-loop structure followed by a run of adjacent uracils (Figure 6-2). The stem-loop structure and the run of uracils act in concert to cause termination of RNA synthesis. Rho-dependent terminators have no distinguishing features that have yet been recognized except Rho seems to bind to and act on single-stranded RNA molecules that are not being translated.

Initiation of a second round of transcription need not await completion of the first: Another molecule of RNA polymerase can bind to the promoter once the previous molecule of RNA polymerase has polymerized 50 to 60 nucleotides. For a rapidly transcribed gene, such reinitiation occurs repeatedly, and a gene can be cloaked with numerous RNA molecules in various degrees of completion.

RNA polymerase from *E. coli* and *S. typhimurium* consists of five protein subunits. Thus, it is one of the largest enzymes known and can be easily seen by electron microscopy (Figure 6-3). Four of the subunits constitute the **core**

Figure 6-2. Base sequence of (a) the DNA of the *E. coli trp* operon at which transcription termination occurs and of (b) the 3' terminus of the mRNA molecule. The inverted-repeat sequence is indicated by reversed orange arrows. The mRNA molecule is folded to form a stem-and-loop structure. The relevant regions are labeled in orange; the terminal sequence of U's in the mRNA is shaded in orange.

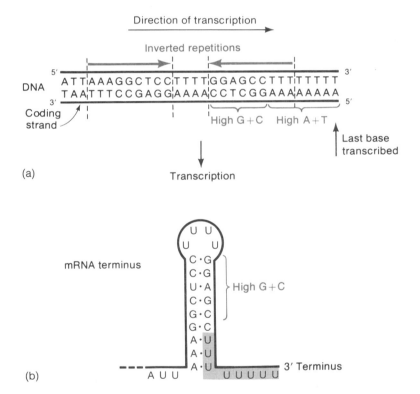

(a)

(b)

Figure 6-3. Electron micrographs of *E. coli* RNA polymerase. (a) Molecules bound to DNA. (b) The holoenzyme viewed by negative-contrast electron microscopy. (× 270,000). (Courtesy of Robley Williams.)

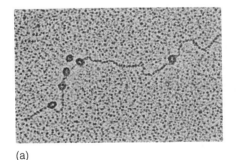

(a)

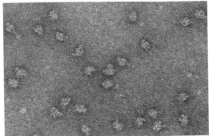

(b)

enzyme, which catalyzes the polymerization of nucleoside triphosphates into RNA. The fifth subunit, called the σ **subunit**, is required for promoter binding. The complex including the fifth subunit is called the **holoenzyme**. Once polymerization begins, the σ subunit dissociates from the core enzyme. When transcription is completed, the core enzyme binds another σ subunit and is then ready to bind to a promoter again.

MESSENGER RNA

Amino acids do not directly recognize DNA. Thus, intermediate steps are needed for arranging the amino acids in a polypeptide chain in the order determined by the DNA base sequence. This process begins with transcription of the base sequence of one of the DNA strands (the **coding** or **sense strand**) into the base sequence of an RNA molecule. This RNA molecule, called **messenger RNA** (mRNA), is used directly in polypeptide synthesis. The mRNA is translated into amino acids by the protein-synthesizing machinery of the cell. In prokaryotes, mRNA molecules often contain information for the amino acid sequences of several different polypeptide chains; in this case, such a molecule is called **polycistronic mRNA**. (Cistron is used synonymously with gene—a base sequence encoding a single polypeptide chain.) The genes contained in a polycistronic mRNA molecule often encode the different proteins of a metabolic pathway. For example, in *S. typhimurium*, the 10 enzymes needed to synthesize histidine are encoded in one mRNA molecule. The use of polycistronic mRNA is an economical way for a cell to regulate synthesis of related proteins in a coordinated way. For example, in prokaryotes, the most common way to regulate synthesis of a particular protein is to control the synthesis of the mRNA molecule that encodes it (see Chapter 7). With a polycistronic mRNA molecule, the synthesis of several related proteins can be regulated by a single signal, so appropriate quantities of each protein are made at the same time; this is termed **coordinate regulation**.

Not all base sequences in an mRNA molecule are translated into the amino acid sequences of polypeptides. For example, translation of an mRNA molecule rarely starts exactly at the 5' end of the mRNA molecule and proceeds to the other end; initiation of polypeptide synthesis may begin hundreds of nucleotides from the 5' end of the mRNA. A section of untranslated RNA before the region encoding the first polypeptide chain is called a **leader**, which in some cases contains regulatory sequences that influence the rate of protein synthesis (an example, the tryptophan biosynthetic genes, is described in Chapter 7). Untranslated sequences are also found at the 3' end. In addition, polycistronic mRNA molecules often contain **spacer sequences** tens of bases long, which separate the **coding sequences**; each coding sequence corresponds to a polypeptide chain.

Generally the coding sequence of each gene is obtained by transcription of only one DNA strand. (It is relatively rare for the two complementary base sequences in a particular gene to be transcribed, although a few exceptions are known in which two genes are transcribed from different strands of the same DNA sequence.) Some genes, however, may be transcribed from one strand of the DNA, and other genes may be transcribed from the opposite strand, so if an extended segment of a DNA molecule is examined, mRNA molecules may be seen growing in either of two directions (Figure 6-4), depending on which DNA strand functions as a template.

In bacteria, most mRNA molecules are degraded within a few minutes after synthesis. Thus continued transcription of a particular mRNA is required to continue to produce its gene product. The short half-life of the mRNA allows cells to rapidly stop making gene products that are no longer needed simply by

Figure 6-4. Schematic drawing showing that complementary DNA strands can be transcribed but not usually from the same region of DNA. Promoters are indicated by black arrowheads and termination sites by black bars. Promoters are present in both strands. Termination sites are usually located such that transcribed regions do not overlap, but this is not always the case.

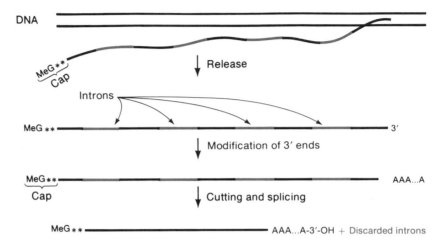

Figure 6-5. Schematic drawing showing production of eukaryotic mRNA. The primary transcript is capped before it is released. Then its 3'-OH end is modified, and finally the introns are excised. MeG = 7-methylguanosine; two asterisks = two nucleotides whose riboses are methylated.

stopping transcription of the mRNA. In contrast, most mRNA molecules in eukaryotes have a much longer half-life.

There are major differences between prokaryotes and eukaryotes in the relation between the transcript and the mRNA used for polypeptide synthesis. In prokaryotes, the immediate product of transcription (called the **primary transcript**) is mRNA; in contrast, in eukaryotes, the primary transcript must be converted to mRNA. This conversion, which is called **RNA processing**, consists of two types of events—modification of the termini and excision of untranslated sequences embedded within coding sequences. These events are illustrated diagrammatically in Figure 6-5. Untranslated sequences within genes are called **intervening sequences** or **introns**. They are common in eukaryotic transcripts but relatively rare in bacteria. (Introns in genes from the *E. coli* phage T4 are well-studied exceptions. Some Archae genes also have introns.) Introns are excised from the primary transcript, and the remaining fragments are rejoined to form the mRNA molecule (see Figure 6-5). Intron excision and the joining of coding sequences (**exons**) to form an mRNA molecule is called **RNA splicing**. Because introns are absent in many bacteria, proper RNA processing is a problem that must be considered when cloning eukaryotic genes in bacteria (see Chapter 20).

TRANSLATION

The synthesis of every protein molecule in a cell is directed by an mRNA intermediate, which is copied from DNA (except in the case of some RNA viruses). Synthesis of proteins from mRNA is called **translation**. The translation system consists of four major components:

1. ***Ribosomes***. These are particles on which the mechanics of protein synthesis is carried out. They contain the enzyme needed to form a peptide bond between amino acids, a site for binding one mRNA molecule, and sites for bringing in and aligning the amino acids in preparation for assembly into the finished polypeptide chain.

2. ***Transfer RNA***. Amino acids do not bind to mRNA, but the order of amino acids in a particular protein is determined by the base sequence in the mRNA molecule. This ordering is accomplished by a set of adaptor molecules, called transfer RNAs (tRNAs). A tRNA molecule "reads" the base sequence of mRNA. Each type of tRNA specifically recognizes three adjacent bases on an mRNA molecule and is charged with a specific amino acid. The amino acids that correspond to each specific three-base sequence in the mRNA define the genetic code.

3. ***Aminoacyl tRNA synthetases***. This set of enzymes catalyzes the attachment of a specific amino acid to its corresponding tRNA.

4. ***Initiation, elongation, and release factors***. These molecules are proteins needed at particular stages of polypeptide synthesis.

In prokaryotes, all of these components are present throughout the cell. In outline, the mechanism of protein synthesis can be depicted as in Figure 6-6. A ribosome binds to a mRNA molecule. Appropriate tRNA–amino acid complexes (formed by the aminoacyl tRNA synthetases) bind sequentially, one by one, to the mRNA molecule that is attached to the ribosome. Peptide bonds are made between successively aligned amino acids, each time joining the amino group of the incoming amino acid to the carboxyl group of the amino acid at the growing end. Finally, the chemical bond between the tRNA and its attached amino acid is broken, and the completed protein is released.

An important feature of the translation is that it proceeds in a particular direction, obeying the following rules (Figure 6-7):

1. RNA is translated from the 5' end of the molecule toward the 3' end but not from the 5' terminus itself nor all the way to the 3' end.

2. Polypeptides are synthesized from the amino terminus toward the carboxyl terminus, by adding amino acids one by one to the carboxyl end. For example, a protein with the sequence NH_2-Met- Pro- . . . -Gly-Ser-COOH, would have started with methionine, and serine would be the last amino acid added to the chain.

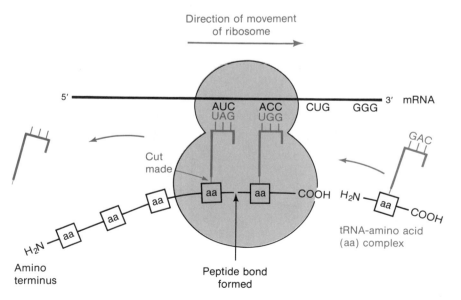

Figure 6-6. A diagram showing how a protein molecule is synthesized.

Figure 6-7. Direction of synthesis of RNA with respect to the coding strand of DNA and synthesis of protein with respect to mRNA.

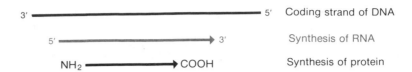

It is conventional to place the 5' terminus at the left when writing nucleotide sequences and to place the NH_2 terminus at the left when writing amino acid sequences. Thus, polynucleotides are generally written to show both synthesis and translation from left to right, and polypeptides are also written to show synthesis from left to right. This convention is used in all of the following sections concerning the genetic code.

GENETIC CODE

There must be distinct combinations of three adjacent bases in the mRNA to specify each of the 20 amino acids in proteins. In fact, more than 20 distinct combinations are needed for polypeptide synthesis because signals are required for starting and stopping the synthesis of the polypeptide chains. An RNA base sequence corresponding to a particular amino acid is called a **codon**, and the sequences that specify start and stop sites are called **start codons** and **stop codons**. Because each of the three bases in a codon can be one of the four nucleotides, there are a total of $4^3 = 64$ possible combinations. All 64 possible codons carry information of some sort. In most cases, several different codons designate the same amino acid. Furthermore, in translating mRNA molecules, the codons do not overlap but are used sequentially (Figure 6-8).

Codons and Features of the Code

The same genetic code is used by almost all biological systems and hence is said to be universal (the exceptions are mitochondria and a few unusual microorganisms). The complete code is shown in Table 6-2. The following features of the code should be noted:

1. Sixty-one codons correspond to amino acids.
2. The start codon is usually AUG. AUG corresponds to the amino acid methionine. In rarer cases, certain other codons (e.g., GUG) initiate translation.
3. There are three stop codons: **UAG**, **UAA**, and **UGA**.
4. In most cases, several different codons direct the insertion of the same amino acid into a protein chain; that is, the code is highly **redundant**. Only tryptophan and methionine are specified by a single codon.
5. The redundancy is not random; except for Ser, Leu, and Arg, all codons corresponding to the same amino acid are in the same box of the codon table (see Table 6-2). That is, *synonymous codons usually differ only in the third base.* For example, GGU, GGC, GGA, and GGG all code for glycine.

In every case in which a mutant protein differs by a single amino acid from the wild-type form, the amino acid substitution can be accounted for by a single

Figure 6-8. Bases in an RNA molecule are read sequentially in the 5' → 3' direction, in groups of three.

Table 6-2   The "universal" genetic code

First position (5' end)	Second position				Third position (3' end)
	U	C	A	G	
U	Phe	Ser	Tyr	Cys	U
	Phe	Ser	Tyr	Cys	C
	Leu	Ser	Stop	Stop	A
	Leu	Ser	Stop	Trp	G
C	Leu	Pro	His	Arg	U
	Leu	Pro	His	Arg	C
	Leu	Pro	Gln	Arg	A
	Leu	Pro	Gln	Arg	G
A	Ile	Thr	Asn	Ser	U
	Ile	Thr	Asn	Ser	C
	Ile	Thr	Lys	Arg	A
	[Met]	Thr	Lys	Arg	G
G	Val	Ala	Asp	Gly	U
	Val	Ala	Asp	Gly	C
	Val	Ala	Glu	Gly	A
	[Val]	Ala	Glu	Gly	G

NOTE: The boxed codons are used for initiation.

base change between the codons corresponding to the two different amino acids. For example, substitution of proline by serine, which is a common mutational change, can be accounted for by the single base changes CCC → UCC, CCU → UCU, CCA → UCA, and CCG → UCG.

Transfer RNA and the Aminoacyl Synthetases

Decoding the base sequence within an mRNA molecule to an amino acid sequence of a protein is accomplished by the tRNA molecules and a set of enzymes, the **aminoacyl tRNA synthetases**. The tRNA molecules are small, single-stranded nucleic acids ranging in size from 73 to 93 nucleotides. Similar to all RNA molecules, they have a 3'-OH terminus, but the opposite end terminates with a 5'-monophosphate rather than a 5'-triphosphate. Internal complementary base sequences form short double-stranded regions, causing the molecule to fold into a structure in which open loops are connected to one another by double-stranded stems (Figure 6-9). In two dimensions, tRNA molecules are often drawn as a planar cloverleaf; however, their three-dimensional structure is more complex.

Three regions of each tRNA molecule are used in the decoding operation. One of these regions is the **anticodon**, a sequence of three bases that can base-pair with a codon sequence in the mRNA. *No normal tRNA molecule has an anticodon complementary to the stop codons UAG, UAA, or UGA; thus, these codons are stop signals.* A second site is the **amino acid attachment site**, the 3' terminus of the tRNA molecule; the amino acid corresponding to the particular mRNA codon that base-pairs with the tRNA anticodon is covalently linked to this terminus. A specific aminoacyl tRNA synthetase matches the amino acid with the anticodon; to do so, the enzyme must be able to distinguish one tRNA molecule from another. The necessary distinction is provided by an ill-defined region encompassing multiple parts of the tRNA molecule and called the **recognition region**. (The bound amino acids are joined together during polypeptide synthesis.)

The different tRNA molecules and synthetases are designated by stating the name of the amino acid that can be linked to a particular tRNA molecule by a specific synthetase; for example, leucyl-tRNA synthetase attaches leucine to tRNA^{Leu}. When an amino acid has become attached to a tRNA molecule, the tRNA is said to be **acylated** or **charged**. An acylated tRNA molecule is designated in several ways. For example, if the amino acid is glycine, the acylated tRNA^{Gly} would be written glycyl-tRNA^{Gly} or Gly-tRNA^{Gly}. The term **uncharged tRNA** refers to a tRNA molecule lacking an amino acid and **mischarged tRNA** to one acylated with an incorrect amino acid (for example, Leu-tRNA^{Gly}).

Accurate protein synthesis, the placement of the "correct" amino acid at the appropriate position in a polypeptide chain, requires (1) attachment of the correct amino acid to a tRNA molecule by the synthetase and (2) fidelity in codon-anticodon binding.

Redundancy and the Wobble Hypothesis

Several features of the genetic code and of the decoding system suggest that something is missing in the explanation of codon-anticodon binding. First, the code is highly redundant. Second, the identity of the third base of many codons appears to be unimportant; that is, XYU, XYA, XYG, and XYC, in which XY denotes any sequence of first and second bases, often correspond to the same amino acid. (Codons, similar to all RNA sequences, are by convention written with the 5' end at the left; thus, the first base in a codon is at the 5' end, and the third base is at the 3' end.) Third, the number of distinct tRNA molecules present in a single organism is less than the number of codons; because all codons are used, the anticodons of some tRNA molecules must be able to pair with more than

Figure 6-9. A tRNA cloverleaf with its bases numbered. A few bases present in almost all tRNA molecules are indicated. The dots between arms represent H-bonds between complementary bases.

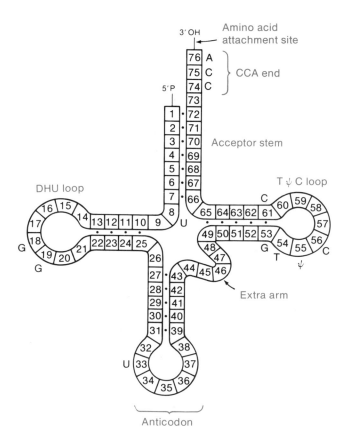

one codon. Experiments with several purified tRNA molecules showed this to be the case.

These observations have been explained by the **wobble hypothesis**, which also provides insight into the pattern of redundancy of the code. Wobble refers to the less stringent requirement for base pairing at the third position of the codon than at the first two positions. That is, the first two bases must form pairs of the usual type (A with U or G with C), but the third base pair can be of a different type (for example, G with U). This observation was derived from the discovery that the anticodon of yeast tRNA^{Ala} contains the nucleoside **inosine (I)**, in the position that pairs with the third base of the codon (the first position of the anticodon). Later analyses of other tRNA molecules showed that inosine was common in this position, although not always present. Inosine can form hydrogen bonds with A, U, and C. In the wobble hypothesis, all pairs of bases that can form hydrogen bonds are considered to be possible in the third position of the codon except purine-purine base pairs, which would cause excessive distortion in the region of the pairing. These possible base pairs are listed in Table 6-3. The wobble hypothesis was later confirmed by direct sequencing of many tRNA molecules. It explains the pattern of redundancy in the code in that certain anticodons (for example, those containing U, I, or G in the first position of the anticodon) can pair with several codons.

Unusual bases, of which inosine is one example, are found in several positions in tRNA. In all cases, they are not incorporated as such but are formed by modification of a standard base that is already in the tRNA. For example, an adenosine in the first position of the anticodon is always enzymatically converted to inosine. Adenosines in other positions are unaffected.

OVERLAPPING GENES

When discussing coding and signal recognition earlier in the chapter, an implicit assumption has been that the mRNA molecule is scanned for start signals to establish the reading frame and that reading then proceeds in a single direction within the reading frame. The idea that several reading frames might exist in a single segment was not considered for many years. The notion of overlapping reading frames was rejected on the grounds that severe constraints would be placed on the amino acid sequences of two proteins translated from the same portion of mRNA.

If multiple reading frames were used, a single DNA segment would be used with maximal efficiency. A disadvantage, however, is that evolution could be slowed because single-base-change mutations would be deleterious more often than if there were a unique reading frame. Nonetheless, overlapping genes do exist in bacteria, and some small viruses and phages have considerable overlapping reading frames.

The *E. coli* phage φX174 contains a single strand of DNA consisting of 5386 nucleotides of known base sequence. If a single reading frame were used, at most

Table 6-3 Allowed pairings
according to the wobble hypothesis

Third position codon base	First position anticodon base
A	U,I
G	C,U
U	G,I
C	G,I

Figure 6-10. The map of *E. coli* phage φX174, showing the start and stop points for mRNA synthesis and the boundaries of the individual protein products. The solid regions are spacers.

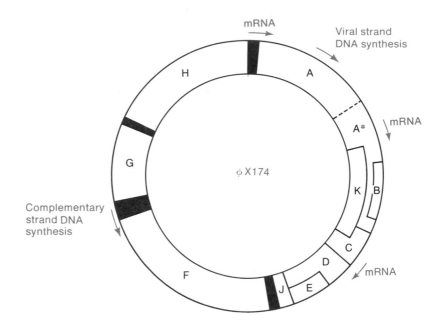

1795 amino acids could be encoded in the sequence, and with an average protein size of about 400 amino acids, only four to five proteins could be made. However, φX174 makes 11 proteins containing a total of more than 2300 amino acids. This paradox was resolved when it was shown that translation occurs in several reading frames from three mRNA molecules (Figure 6-10). For example, the sequence for protein B is contained totally in the sequence for protein A but translated in a different reading frame. Similarly, the protein E sequence is totally within the sequence for protein D. Protein K is initiated near the end of gene *A*, includes the base sequence of B, and terminates in gene *C*. Synthesis is not in phase with either gene *A* or gene *C*. Of note is protein A' (also called A*), which is formed by reinitiation within gene *A* and in the same reading frame, so it terminates at the stop codon of gene *A*. Thus, the amino acid sequence of A' is identical to a segment of protein A. In total, five different proteins obtain some or all of their primary structure from shared base sequences in φX174. The essential features of overlapping genes are the location of each AUG initiation sequence (because these sequences establish the reading frame) and the absence of stop codons in overlapping reading frames.

PROTEIN SYNTHESIS

Protein synthesis can be divided into three stages: **initiation**, **elongation**, and **termination**. The main features of the initiation step are (1) binding of ribosomes to the mRNA, (2) recognition of the initiation codon, and (3) binding of acylated tRNA bearing the first amino acid to the ribosome. The main features of the elongation step are (1) joining together two amino acids by peptide bond formation and (2) moving the mRNA and the ribosome with respect to one another so the codons can be translated successively. In the termination step, the completed protein is dissociated from the complex, and the ribosomes are released to begin another cycle of synthesis.

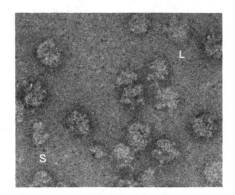

Figure 6-11. Ribosomes. An electron micrograph of 70S ribosomes from *E. coli*. A few ribosomal subunits are identified by the letters S and L, to indicate the small and large subunits.

Ribosomes

Ribosomes are multicomponent ribonucleoprotein particles that bring together a single mRNA molecule and charged tRNA molecules in the proper position and orientation, so the base sequence of the mRNA molecule is translated into an amino acid sequence (see Figure 6-6). All ribosomes contain two subunits (Figure 6-11). For historical reasons, the intact ribosome and the subunits have been given numbers that describe how fast they sediment when centrifuged. In eubacteria (such as *E. coli*), the intact particle is called a **70S ribosome**. The 70S ribosome consists of one 30S subunit and one 50S subunit. Both the 30S and the 50S particles contain RNA (called **rRNA** for ribosomal RNA) and more than 50 protein molecules (Figure 6-12). The 30S subunit contains one 16S rRNA molecule. The 50S subunit contains two RNA molecules: one 5S rRNA molecule and one 23S rRNA molecule.

Stages of Protein Synthesis in Bacteria

An important feature of initiation of protein synthesis in both prokaryotes and eukaryotes is the use of a specific initiating tRNA molecule. In prokaryotes, this tRNA molecule is acylated with the modified amino acid N-formylmethionine (fMet); the tRNA is often designated tRNA^{fMet} (Figure 6-13). Both tRNA^{fMet} and tRNA^{Met} recognize the codon AUG, but only tRNA^{fMet} is used for initiation. The tRNA^{fMet} molecule is first acylated with methionine, then an enzyme (found only in prokaryotes) adds a formyl group to the amino group of the methionine.

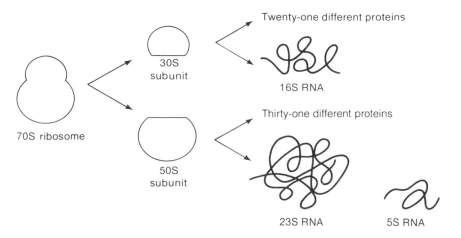

Figure 6-12. Dissociation of a prokaryotic ribosome. The configuration of two overlapping circles is used throughout this chapter, for the sake of simplicity. The correct configuration is shown in Figure 6-11.

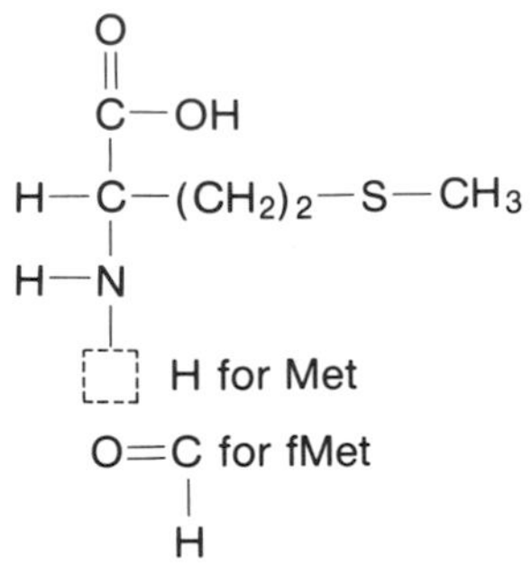

$$
\begin{array}{c}
O \\
\parallel \\
C-OH \\
\mid \\
H-C-(CH_2)_2-S-CH_3 \\
\mid \\
H-N \\
\mid
\end{array}
$$

⌐ ¬ H for Met
└ ┘

O=C for fMet
  |
  H

Figure 6-13. Comparison of the chemical structures of methionine and N-formyl-methionine.

(In eukaryotes, the initiating tRNA molecule is charged with methionine also, but formylation does not occur.) The use of these initiator tRNA molecules means that while being synthesized, prokaryotic proteins have fMet at the amino terminus. This amino acid, however, is frequently deformylated or removed by a specific peptidase later.

Polypeptide synthesis in bacteria begins by the association of one 30S subunit (not the entire 70S ribosome), an mRNA molecule, fMet-tRNA, three proteins known as **initiation factors**, and guanosine 5'-triphosphate (GTP). These molecules constitute the **30S preinitiation complex** (Figure 6-14). Because polypeptide synthesis begins at an AUG start codon and AUG codons are also found within coding sequences (that is, methionine occurs within a polypeptide chain), some signal must be present in the base sequence of the mRNA molecule to identify which particular AUG codon to use as a start signal. The means of selecting the correct AUG sequence differs in prokaryotes and eukaryotes. In prokaryotic mRNA molecules a particular base sequence, called the **ribosome binding site** or a **Shine-Dalgarno sequence**, is usually located about 7 base pairs upstream of the AUG codon used for initiation. The Shine-Dalgarno sequence forms base pairs with a complementary sequence in the 16S rRNA molecule of the ribosome. Although the precise DNA sequence varies considerably for different genes, an example of a Shine-Delgarno sequence is:

$$5'----AGGAGG----3'$$

After formation of the 30S preinitiation complex, a 50S subunit joins with this complex to form a **70S initiation complex** (see Figure 6-14).

The 50S subunit contains two tRNA binding sites. These sites are called the **A (aminoacyl) site** and the **P (peptidyl) site**. When joined with the 30S preinitiation complex, the position of the 50S subunit in the 70S initiation complex is such that the fMet-tRNA^{fMet}, which was previously bound to the 30S preinitiation complex, occupies the P site of the 50S subunit. Placement of fMet-tRNA^{fMet} in the P site fixes the position of the fMet-tRNA anticodon such that it pairs with the AUG initiator codon in the mRNA. Thus, *the reading frame is unambiguously defined on completion of the 70S initiation complex.*

Once the P site is filled, the A site of the 70S initiation complex becomes available to any tRNA molecule whose anticodon can pair with the codon adja-

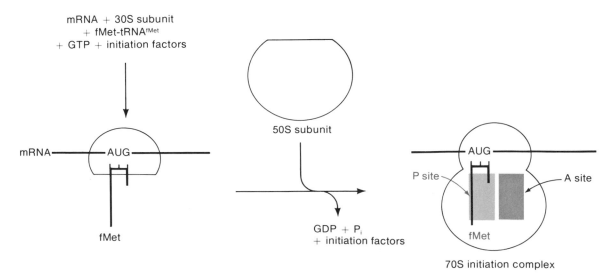

Figure 6-14. Early steps in protein synthesis: in prokaryotes, formation of the 30S preinitiation complex and of the 70S initiation complex.

cent to the initiation codon. After occupation of the A site, a peptide bond between fMet and the adjacent amino acid is formed by a ribosomal enzyme complex called **peptidyl transferase**. As the bond is formed, the fMet is cleaved from the fMet-tRNA in the P site.

After the peptide bond forms, an uncharged tRNA molecule occupies the P site and a dipeptidyl-tRNA occupies the A site. At this point, three movements occur: (1) The tRNA^{fMet} in the P site, now no longer linked to an amino acid, leaves this site; (2) the peptidyl-tRNA moves from the A site to the P site; and (3) the ribosome moves a distance of three bases to position the next codon at the A site (Figure 6-15). This step, called **translocation**, requires the presence of an elongation protein **EF-G** and GTP. After the ribosome has moved, the A site is again available to accept a charged tRNA molecule having a correct anticodon.

When a chain termination codon (UAA, UAG, or UGA) is reached, no tRNA exists that can fill the A site, so chain elongation stops. The polypeptide chain, however, is still attached to the tRNA occupying the P site. Release of the protein is accomplished by proteins called **release factors**, which also cause dissociation of the 70S ribosome into its 30S and 50S subunits, completing the cycle.

If the mRNA molecule is polycistronic and the AUG codon initiating the second polypeptide is not too far from the stop codon of the first, the 70S ribosome will not always dissociate but will re-form an initiation complex with the second AUG codon. The probability of such an event decreases with increasing separation of the stop codon and the next AUG codon. In some genetic systems, the separation is sufficiently great that more protein molecules are always translated from the first gene than from subsequent genes, providing a mechanism for maintaining particular ratios of gene products (see Chapter 7).

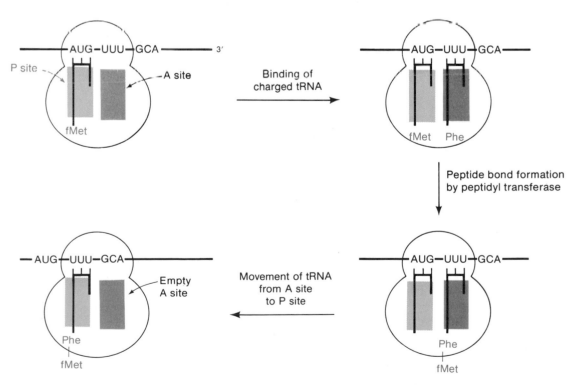

Figure 6-15. Elongation phase of protein synthesis: binding of charged tRNA, peptide bond formation, and movement of mRNA with respect to the ribosome.

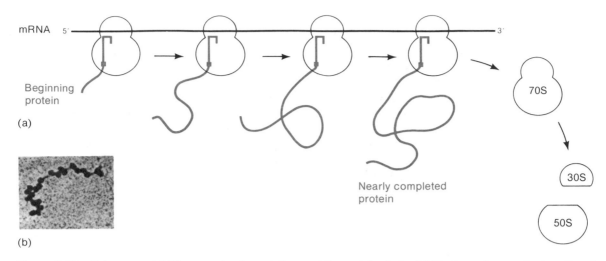

Figure 6-16. Polysomes. (a) Diagram showing relative movement of the 70S ribosome and the mRNA and growth of the protein chain. (b) Electron micrograph of an *E. coli* polysome. (Courtesy of Barbara Hamkalo.)

COMPLEX TRANSLATION UNITS

After about 25 amino acids have been joined together in a polypeptide chain, an AUG initiation codon is completely free of the ribosome, and a second initiation complex can form. This results in two 70S ribosomes moving along the mRNA at the same speed. Likewise, when the second ribosome has moved a suitable distance, a third ribosome can attach to the initiation site. The process of movement and reinitiation continues until the mRNA is covered with ribosomes at a density of about one ribosome per 80 nucleotides. This large translation unit is called a **polyribosome** or a **polysome**. Most mRNA molecules seem to be translated as polysomes. An electron micrograph and a diagram of a polysome are shown in Figure 6-16. Polysomes allow each mRNA to be translated at the maximal rate.

COUPLING OF TRANSCRIPTION AND TRANSLATION

An mRNA molecule being synthesized has a free 5' terminus. Because translation occurs in the 5' → 3' direction, the mRNA is synthesized in a direction appropriate for immediate translation. That is, the ribosome-binding site is transcribed first, followed in order by the initiating AUG codon, the region encoding the amino acid sequence, and finally the stop codon. Thus, the 70S initiation complex can re-form before the mRNA is released from the DNA. This process is called **coupled transcription-translation**. It does not occur in eukaryotes because the mRNA is synthesized and processed in the nucleus and later transported through the nuclear membrane to the cytoplasm, where the ribosomes are located. Transcription and translation, however, are almost always coupled in prokaryotes. Coupling of transcription and translation has many important consequences on the regulation of gene expression, which are discussed in Chapter 7.

ANTIBIOTICS AND ANTIBIOTIC RESISTANCE

Hundreds of antibiotics are known that inhibit gene expression. Some of these that have been valuable tools in microbial genetics are listed in Table 6-4 and described here.

Table 6-4 Mechanisms of sensitivity and resistance to some common antibiotics

Antibiotic	Mechanism
Ampicillin	inhibits crosslinking of peptidoglycan chains in the cell wall of eubacteria. Cells growing in the presence of ampicillin synthesize weak cell walls, causing them to burst due to the high internal osmotic pressure. Amp^r encoded by Mu derivatives and pBR plasmids is due to a periplasmic β-lactamase that breaks the β-lactam ring of ampicillin.
Chloramphenicol	inhibits protein synthesis by binding to the 50s ribosomal subunit and blocking the peptidyl transferase reaction. Cml^r encoded by pBR328 is due to a cytoplasmic chloramphenicol acyltransferase which inactivates chloramphenicol by covalently acetylating it.
Kanamycin	inhibits protein synthesis by binding to the 30s ribosomal subunit and preventing translocation. Kan^r is usually due to a cytoplasmic aminoglycoside phosphotransferase that inactivates kanamycin by covalently phosphorylating it. Kan^r requires phenotypic expression. Neomycin is a structural analog of kanamycin that functions by the same mechanism and is inactivated by the same mechanism.
Tetracycline	inhibits protein synthesis by preventing aminoacyl tRNA from binding to ribosomes. Tet^r encoded by Tn10 and pBR plasmids is due to a membrane protein that actively transports tetracycline out of the cell. When Tn10 is present in multiple copies, cells are less resistant to tetracycline than when only one copy of Tn10 is present.
Streptomycin	inhibits protein synthesis by binding to the S12 protein of the 30s ribosomal subunit and inhibiting translation. A high level of Str^r can result from chromosomal mutations in the gene for the S12 protein (rpsL) which prevent streptomycin from binding to the ribosome. Since only mutant ribosomes are Str^r, resistance to streptomycin is recessive to streptomycin sensitivity. Str^r requires phenotypic expression.

Some antibiotics specifically inhibit transcription in bacteria. Rifampicin binds to one of the subunits of the RNA polymerase core enzyme and prevents initiation of RNA synthesis. Thus, rifampicin has no effect on an RNA polymerase that has already initiated transcription. Rifampicin has been extensively used in experiments in which specific inhibition of mRNA synthesis is desired. Other classes of antibiotics (for example, streptolydigin) bind to RNA polymerase and inhibit the elongation step.

Most of the known antibiotics inhibit protein synthesis. For example, streptomycin (Str) and kanamycin (Kan) bind to the 30S ribosome and inhibit binding of mRNA, tetracycline (Tet) inhibits binding of the aminoacyl tRNA to the 30S ribosome, chloramphenicol (Cam) inhibits peptidyl transferase, and puromycin (Pur) causes premature chain termination. These antibiotics have been used in biochemical studies to inhibit protein synthesis. In microbial genetics, alleles for sensitivity and resistance to these antibiotics are common genetic markers. Other antibiotics, such as penicillin, do not affect transcription or translation but interfere with synthesis of the bacterial cell wall.

Some bacteria are naturally resistant to certain antibiotics; others can acquire resistance by mutation. There are two main ways bacteria develop antibiotic resistance: (1) acquiring an enzymatic activity that directly inactivates the antibiotic or (2) acquiring a mutation that modifies the target site of the antibiotic, which prevents the antibiotic from interfering with the normal function of the target site. For example, penicillin resistance is often due to an enzymatic activity, and streptomycin resistance is often due to modification of the antibiotic's target site. Natural resistance to penicillin is conferred by a gene that synthesizes β-lactamase, an enzyme that cleaves the lactam ring in penicillin, in ampicillin, and in the cephalosporin antibiotics. This gene is carried on many plasmids (see Chapter 11) and transposable elements (see Chapter 12). (We have called this gene amp^r in this book, but it is also often called bla, for β-lactamase.) Streptomycin acts by binding to one of the ribosomal proteins. Streptomycin-resistant mutations can occur that alter the structure of the protein such that it cannot bind streptomycin but without eliminating its ability to function in the ribosome.

SYNTHESIS OF RIBOSOMAL RNA AND TRANSFER RNA

rRNA and tRNA are also transcribed from genes. The production of these molecules is not as direct as synthesis of bacterial mRNA. The main difference is that these RNA molecules are excised from large primary transcripts. Some of these transcripts include both rRNA and tRNA molecules. For example, one of the *E. coli* rRNA transcripts contains one copy each of 5S, 16S, and 23S rRNA as well as four different tRNA molecules (Figure 6-17). Other transcripts include tRNA but not rRNA sequences. Highly specific RNases excise rRNA and tRNA from these large transcripts, and other enzymes produce the modified bases in tRNA. The rRNA and tRNA transcripts do not include regions that are translated.

Figure 6-17. A schematic diagram of one of the *E. coli* rRNA transcripts from which 5S, 16S, and 23S rRNA molecules are excised. The regions containing the 16S rRNA and 23S rRNA molecules are shown in expanded form above the line designating the transcript. The arrows indicate the termini of the 16S rRNA and 23S rRNA molecules.

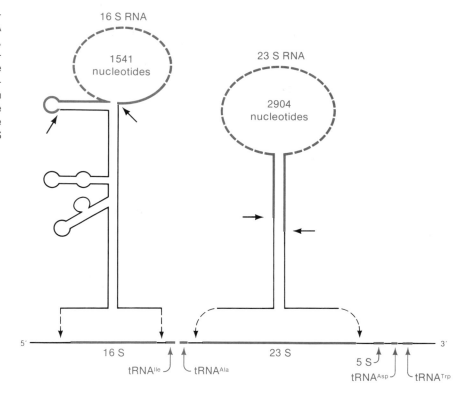

KEY TERMS

aminoacyl tRNA synthetases
coordinate regulation
exon
intron
leader
messenger RNA
polycistronic
polysome
primary transcript
promoter
   closed complex
   open complex
release factors
   rho dependent
   rho independent
ribosome binding site
   (Shine-Delgarno sequence)

ribosomes
   aminoacyl site
   peptidyl site
   peptidyl transferase
RNA polymerase
   holoenzyme
   core enzyme
   σ subunit
RNA processing
RNA splicing
template
transcription
transcription termination
transfer RNA
translation
   start codons
   stop codons
wobble

QUESTIONS AND PROBLEMS

1. What is the direction of synthesis of RNA?

2. What are the three stop codons?

3. What is the principal start codon, and to what amino acid does it correspond?

4. Which chain-termination codon could be formed by a single base change from UCG, UUG, and UAU?

5. Ribonuclease contains 124 amino acids. What is the least number of nucleotides you would expect to find in the gene encoding the protein?

6. The central region of a polypeptide containing a single lysine has the amino acid sequence Phe-Leu-Tyr-Ala-Lys-Gly-Glu . . . . A mutation is found that causes the polypeptide to terminate with Phe-Leu-Tyr-Ala. Which of the lysine codons was used in synthesis of the wild-type protein?

7. Which of the following amino acid changes can result from a single base change: (1) Met → Arg, (2) His → Glu, (3) Gly → Ala, (4) Pro → Ala, (5) Tyr → Val.

8. What is the direction of synthesis of a polypeptide chain?

9. At what stage of polypeptide synthesis do 70S ribosomes form?

10. If the synthetic polynucleotide 5'-A G G U U A U A G G A A A A A-3' is translated in vitro in a system that does not require a start codon, what polypeptide is synthesized? Indicate the NH_2 and COOH termini.

11. Translation has evolved in a particular polarity with respect to the mRNA molecule. What would be the disadvantages of having the reverse polarity?

REFERENCES

Barrell, B. G., G. M. Aire, and C. A. Hutchison III. 1976. Overlapping genes in bacteriophage φX174. *Nature*, 264, 34.

Belfort, M. 1991. Self-splicing introns in bacteria: migrant fossils? *Cell*, 64, 9.

Cold Spring Harbor Laboratory. 1966. *The Genetic Code*. Vol. 31, Symposium on Quantitative Biology, Cold Spring Harbor Laboratory Press.

Crick, F. H. C., et al. 1961. General nature of the genetic code for proteins. *Nature*, 192, 1227.

Ferat, J., and F. Michel. 1993. Group II self-splicing introns in bacteria. *Nature*, 364, 358.

Hatfield, D., and A. Diamond. 1993. UGA: a split personality in the universal genetic code. *Trends Genet.*, 9, 69.

Jukes, T. 1983. Evolution of the amino acid code: inferences from mitochondrial codes. *J. Mol. Evol.*, 19, 219.

Khorana, H. G. 1968. Nucleic acid synthesis in the study of the genetic code. In *Nobel Lectures: Physiology or Medicine*. Vol. 4. American Elsevier.

Losick, R., and M. Chamberlin. 1976. *RNA Polymerase*. Cold Spring Harbor Laboratory Press.

*Nirenberg, M. 1963. The genetic code. *Scientific American*, March, p. 80.

*Noller, H. F. 1984. Structure of ribosomal RNA. *Ann. Rev. Biochem.*, 53, 119.

*Nomura, M. 1973. Assembly of bacterial ribosomes. *Science*, 179, 864.

Nomura, M., A. Tissiageres, and P. Lengyel (eds.). 1974. *Ribosomes*. Cold Spring Harbor.

Pribnow, D. 1975. Nucleotide sequence of an RNA polymerase binding site at an early T7 promoter. *Proc. Natl. Acad. Sci.*, 72, 784.

*Rich, A., and S. Kim. 1978. The three-dimensional structure of transfer RNA. *Scientific American*, January, p. 52.

Roberts, J. 1969. Termination factor for RNA polymerase. *Nature*, 224, 1168.

*Russell, A., and I. Chopra. 1990. *Understanding Antibiotic Action and Resistance*. Ellis Horwood Limited.

Shine, J., and L. Dalgarno. 1974. The 3' terminal sequence of *E. coli* 16S ribosomal RNA: complementarity to nonsense triplets and ribosomal binding sites. *Proc. Natl. Acad. Sci.*, 71, 1342.

Soll, D., J. Abelson, and P. Schimmel (eds.). 1980. *Transfer RNA*. Cold Spring Harbor Laboratory Press.

Steitz, J. A., and K. Jakes. 1975. How ribosomes select initiation regions in mRNA: base pair formation between the 3' terminus of 16S rRNA and the mRNA during initiation of protein synthesis in *E. coli*. *Proc. Natl. Acad. Sci.*, 72, 4734.

Youderian, P., S. Bouvier, and M. Susskind. 1982. Sequence determinants of promoter activity. *Cell*, 30, 843.

Youderian, P., A. Vershon, S. Bouvier, R. T. Saurer, and M. M. Susskind. 1983. Changing the DNA-binding specificity of a repressor. *Cell*, 35, 777.

*Youderian, P. 1988. Promoter strength: more is less. *Trends Genetics*, 4, 327.

---

*Resources for additional information.

7

Regulation of Gene Expression

The number of protein molecules produced per unit time by different genes varies from gene to gene, satisfying the needs of a cell and sometimes also avoiding wasteful synthesis. For some genes, the different rate of gene expression is simply due to the efficiency of transcription and translation. Many gene products, however, are needed only under certain conditions, and regulatory mechanisms that function like an on-off switch allow such products to be made only when demanded by the cell. In addition, more subtle regulatory mechanisms can make minor adjustments in the intracellular concentration of a particular protein in response to needs imposed by the environment.

Most prokaryotes are free-living unicellular organisms that grow and divide indefinitely as long as environmental conditions are suitable and the supply of nutrients is adequate. Thus, their regulatory systems are geared to provide the maximum growth rate in a particular environment except when such growth would be detrimental. This strategy seems to apply also to the free-living unicellular eukaryotes, such as yeast, algae, and protozoa. Phages are less responsive to environmental fluctuations, probably because it is unlikely that a significant change would occur during the short life cycle of a typical phage. Instead their systems are temporally regulated so particular proteins are made only at specific stages in the life cycle.

This chapter discusses the basic mechanisms of genetic regulation and presents several examples of well-understood regulated systems. An enormous variety of regulatory mechanisms are used by bacteria, but most are variations on a few themes. Only a few well-characterized regulatory systems are described in this chapter, each representing a common strategy that serves as a paradigm for many other regulatory systems.

COMMON MODES OF REGULATION

Most metabolic reactions are catalyzed by a sequential series of reactions, each determined by a specific enzyme. In bacteria, regulation of the activity of such a pathway is often accomplished by synthesis (or lack of synthesis) of the entire set of enzymes and accessory proteins; either all of the proteins are synthesized, or none are made. This phenomenon, which is called **coordinate regulation**, often results from control of the synthesis of a single polycistronic mRNA molecule that encodes all of the gene products (control of translation does occur, but it is much less common). Transcription is rarely turned off completely. When

Figure 7-1. The distinction between negative and positive regulation. In negative regulation, an inhibitor, bound to the DNA molecule, must be removed before transcription can occur. In positive regulation, an effector molecule must bind to the DNA. A system may also be regulated both positively and negatively; in such a case, the system is "on" when the positive regulator is bound to the DNA and the negative regulator is not bound to the DNA.

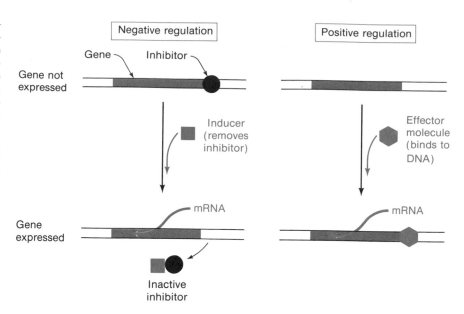

transcription is in the off state, a basal level almost always remains, often consisting of only one or two transcriptional events per cell generation; hence very little synthesis of the gene product occurs. For convenience, the term "off" is used, but it should be kept in mind that usually what is meant is "very low."

Several mechanisms for regulation of transcription are common; the particular one used often depends on whether the enzymes being regulated act in degradative or synthetic metabolic pathways. For example, in a degradative pathway, the availability of the substrate frequently determines whether the enzymes in the pathway will be synthesized. In contrast, in a biosynthetic pathway, the final product is often the regulatory molecule. The molecular mechanisms for each of the regulatory patterns vary quite widely but usually fall in one of two major categories: **negative regulation** and **positive regulation** (Figure 7-1). In a negatively regulated system, a repressor is present in the cell that prevents transcription. In a positively regulated system, an activator is present in the cell that is required for transcription. Negative and positive regulation are not mutually exclusive. Some systems are both positively and negatively regulated, using two regulators to respond to different conditions in the cell, and some proteins can act as both repressors and activators.

A degradative system may be regulated either positively or negatively. In a biosynthetic pathway, the final product often negatively regulates its own synthesis.

LACTOSE OPERON

Metabolic regulation was first studied in detail by Monod, who thoroughly characterized the *Escherichia coli* genes responsible for degradation of the sugar lactose. Most of the terminology used to describe regulation has come from genetic analysis of this system.

In *E. coli*, two proteins are necessary for the metabolism of lactose: the enzyme β-**galactosidase**, which cleaves the disaccharide lactose (a β-galactoside) into the monosaccharides galactose and glucose, and **lactose permease**, a protein required for transport of lactose across the cell membrane into a cell. The requirement for these two proteins for lactose utilization was first shown by a combination of genetic experiments and biochemical analysis.

First, hundreds of Lac⁻ mutants that are unable to use lactose as a carbon source were isolated. Some of the mutations were transferred by genetic recombination from the *E. coli* chromosome to a plasmid called F' *lac*, which carries the genes for lactose utilization. Partial diploids having the genotypes F'*lac*⁻/*lac*⁺ or F'*lac*⁺/*lac*⁻ were constructed by conjugation (see Chapter 14). It was observed that these diploids always had a Lac⁺ phenotype, so none of the mutants produced an inhibitor that inactivated either the enzymatic activity or the function of the wild-type gene (in other words, the *lac*⁻ mutations were recessive to *lac*⁺). Other partial diploids were then constructed in which both the F'*lac* plasmid and the chromosome carried *lac*⁻ genes; these were tested for the Lac⁺ phenotype, with the result that all of the mutants initially isolated could be placed into two complementation groups, *lacZ* and *lacY*. The partial diploids F'*lacY*⁻ *lacZ*⁺/*lacY*⁺ *lacZ*⁻ and F' *lacY*⁺ *lacZ*⁻/*lacY*⁻ *lacZ*⁺ had a Lac⁺ phenotype, but the partial diploids F' *lacY*⁻ *lacZ*⁺/*lacY*⁻ *lacZ*⁺ and F'' *lacY*⁺ *lacZ*⁻/*lacY*⁺ *lacZ*⁻ had a Lac⁻ phenotype. The existence of two complementation groups was good evidence that the *lac* system consisted of at least two genes ("at least" because it was possible that mutations had not yet been obtained in other genes).

Further experimentation was needed to establish the precise function of each gene. Experiments in which cells were placed in a medium containing [¹⁴C]-labeled (radioactive) lactose showed that [¹⁴C]lactose could not enter a *lacY*⁻ mutant, whereas it readily penetrated a *lacZ*⁻ mutant. This result indicated that the *lacY* gene is probably required for transport of lactose across the cell membrane into the cell and is the structural gene for lactose permease. Enzymatic assays showed that β-galactosidase is present in *lac*⁺ but not *lacZ*⁻ cells. These results indicated that the *lacZ* gene is probably the structural gene for β-galactosidase, a conclusion that was confirmed by immunological tests that demonstrated that an altered but inactive protein is present in *lacZ*⁻ cells. In addition, genetic mapping showed that the *lacY* and *lacZ* genes are adjacent.

Regulation of the *lac* System: Repression

Regulation of the *lac* genes is mediated by an "on-off" switch. The on-off nature of the lactose-utilization system is evident in the following observations.

1. If a culture of Lac⁺ *E. coli* is growing in a medium lacking lactose or any other β-galactoside, the intracellular concentrations of β-galactosidase and lactose permease are very low—roughly one or two molecules per bacterium. If lactose is present in the growth medium, however, the number of each of these proteins is about 10^5-fold higher.

2. If lactose is added to a Lac⁺ culture growing in a lactose-free medium (for example, growing on succinate as a carbon source), both β-galactosidase and lactose permease are synthesized nearly simultaneously (Figure 7-2). Analysis of the total mRNA present in the cells before and after addition of lactose shows that no *lac* mRNA (the mRNA that encodes β-galactosidase and lactose permease) is present before lactose is added, but the addition of lactose induces synthesis of *lac* mRNA. This analysis was done by growing cells in a radioactive medium in which newly synthesized mRNA is radioactive, then isolating the mRNA, and finally allowing it to renature with the DNA of a λ phage that carries the *E. coli lac* genes (a λ*lac* transducing phage). Because the only radioactive mRNA that will renature to the *lac*-containing DNA is *lac* mRNA, the amount of radioactive DNA-RNA hybrid molecules is a measure of the amount of *lac* mRNA.

These two observations led to the view that the lactose system is **inducible** and that lactose is an **inducer**. Lactose itself is rarely used in experiments to study

Figure 7-2. The "on-off" nature of the *lac* system. Lac mRNA appears soon after lactose is added; β-galactosidase and permease appear at the same time but are delayed with respect to mRNA synthesis because of the time required for translation. When lactose is removed, no more *lac* mRNA is made, and the amount of *lac* mRNA decreases owing to the usual degradation of mRNA. Both β-galactosidase and lactose permease are stable. Their concentrations remain constant even though no more can be synthesized.

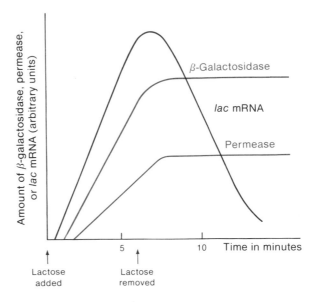

the induction of the *lac* genes for several reasons; one important reason is that the *lacZ* gene product, β-galactosidase, catalyzes the degradation of lactose, resulting in a continual decrease in lactose concentration, which complicates the analysis of many experiments (for example, kinetic experiments). Instead a sulfur-containing analog of lactose is usually used, isopropylthiogalactoside (IPTG); this analog is an inducer but not a substrate of β-galactosidase. Such a substance is said to be a **gratuitous inducer**.

In some mutants, *lac* mRNA is synthesized (hence also β-galactosidase and lactose permease) in both the presence and the absence of an inducer. These **constitutive** mutants provided the key to understanding induction because they eliminated regulation. (The term constitutive is now used to describe the expression genes that are not regulated.) Complementation tests—again with partial diploids carrying two constitutive mutations, one in the chromosome and the other in a plasmid—showed that the mutants fall into two groups, *lacI* and *lacO^c* (Table 7-1). The *lacI^-* mutants are recessive (entries 3, 4): (1) *lacI^+* cells make *lac* mRNA only if the inducer is added; (2) *lacI^-* mutants always make *lac* mRNA whether or not an inducer is added; (3) similar to *lacI^+* cells, *lacI^+/lacI^-* partial diploids make *lac* mRNA only if the inducer is added. These results indicated that the *lacI* gene is apparently a regulatory gene *whose product is an inhibitor that keeps the system turned off.* A *lacI^-* mutant lacks the inhibitor and hence is constitutive. Wild-type copies of the *lacI* gene product made in a *lacI^+/lacI^-* partial diploid can turn off both copies of the *lac* genes. The *lacI* gene product is a protein molecule called the *lac* **repressor**. Genetic mapping experiments place the *lacI* gene adjacent to the *lacZ* gene and establish the gene order *lacI-lacZ-lacY*.

Dominance of *lacO^c* Mutants: The Operator

In contrast to the *lacI* mutants, the *lacO^c* mutants are dominant (entries 1, 2, and 5 in Table 7-1), but the dominance is evident only in certain combinations of *lac* mutations, as can be seen by examining the partial diploids (entries 6 and 7). Both combinations are Lac^+ because a functional *lacZ^+* gene is present. In the combination shown in entry 6, however, synthesis of β-galactosidase is inducible even though a *lacO^c* mutation is present. The difference between the two combinations in entries 6 and 7 is that in entry 6 the *lacO^c* mutation is carried on a DNA molecule that also has a *lacZ^-* mutation, whereas in entry 7, *lacO^c* and *lacZ^+* are car-

Table 7-1 Characteristics of partial diploids having several combinations of *lacI* and *lacO* alleles

Genotype	Constitutive or inducible synthesis of *lac* mRNA
1. F'*lacO^c lacZ^+/lacO^+ lacZ^+*	Constitutive
2. F'*lacO^+ lacZ^+/lacO^c lacZ^+*	Constitutive
3. F'*lacI^- lacZ^+/lacI^+ lacZ^+*	Inducible
4. F'*lacI^+ lacZ^+/lacI^- lacZ^+*	Inducible
5. F'*lacO^c lacZ^+/lacI^- lacZ^+*	Constitutive
6. F'*lacO^c lacZ^-/lacO^+ lacZ^+*	Inducible
7. F'*lacO^c lacZ^+/lacO^+ lacZ^-*	Constitutive

ried on the same DNA molecule. Thus, a *lacO^c* mutation causes constitutive synthesis of β-galactosidase only when the *lacO^c* and *lacZ^+* alleles are located *on the same DNA molecule*; the *lacO^c* mutation is said to be **cis-dominant** because only genes *cis* to the mutation are expressed in dominant fashion. Confirmation of this conclusion comes from an important biochemical observation: The mutant enzyme (encoded in the *lacZ^-* sequence) is synthesized constitutively in a *lacO^c lacZ^-/lacO^+ lacZ^+* partial diploid (entry 6), whereas the wild-type enzyme (encoded in the *lacZ^+* sequence) is synthesized only if an inducer is added. (The mutant enzyme is identified by an immunological test that detects the presence of a protein similar in structure to the active enzyme.) All *lacO^c* mutations are located between the *lacI* and *lacZ* genes; thus, the gene order of the four elements of the *lac* system is *lacI-lacO-lacZ lacY*. An important feature of all *lacO^c* mutations is that they cannot be complemented (a feature of all *cis*-dominant mutations). That is, a *lacO^+* allele cannot alter the constitutive activity of a *lacO^c* mutation. Thus, *lacO* does not encode a diffusible product and must define a site or a noncoding region of the DNA rather than a gene. This site determines whether synthesis of the product of the adjacent *lacZ* gene is inducible or constitutive. The *lacO* region is called the **operator**.

Operon Model

The regulatory mechanism of the *lac* system, which was elucidated by the elegant genetic analysis of Jacob and Monod, has the following features (Figure 7-3):

1. The lactose-utilization system consists of two kinds of components— **structural genes** needed for transport and metabolism of lactose and **regulatory elements** (the *lacI* gene, the *lacO* operator, and the *lac* promoter). Together these components comprise the **lac operon**.

2. The products of the *lacZ* and *lacY* genes are encoded in a single polycistronic mRNA molecule. (This mRNA molecule contains a third gene, denoted *lacA*, which encodes the enzyme transacetylase. Mutants in the *lacA* gene, however, do not affect lactose degradation.)

3. The promoter for the *lacZ lacY lacA* mRNA molecule is immediately adjacent to the *lacO* region. Promoter mutants (*lacP^-*) are unable to make either β-galactosidase or lactose permease because no *lac* mRNA is made (these mutants were discussed in Chapter 6).

4. The *lacI* gene product, the *lac* repressor, binds to the operator site.

5. When the repressor is bound to the operator, initiation of transcription of *lac* mRNA by RNA polymerase is prevented.

6. The inducer stimulates *lac* mRNA synthesis by binding to and inactivating the *lacI* repressor, a process called either **induction** or **derepression**. Thus, in the presence of an inducer, the operator is unoccupied, and the promoter is available for initiation of mRNA synthesis.

Note that regulation of the operon requires that the *lacO* operator is adjacent to the structural genes of the operon (*lacZ*, *lacY*, *lacA*), but the proximity of the *lacI* gene is not necessary because the *lacI* repressor is a soluble protein and is therefore diffusible throughout the cell.

The operon model is supported by a wealth of experimental data and explains many of the features of the *lac* system as well as numerous other negatively regulated genetic systems. The operon model explains the phenotypes of each of the types of *lac* mutants discussed above and predicts the existence of other types of mutants as well. Another predicted type of mutant is noninducible—mutants that fail to increase gene expression in the presence of inducer. Many noninducible mutants are defective for *lacZ* or *lacY* activity due to mutations in these genes, but noninducible mutants contain fully functional *lacZ* and *lacY* genes that are not expressed. One class of noninducible $lacZ^+$ $lacY^+$ mutants maps in the $lacI^-$ gene (called $lacI^s$) and is dominant over wild-type—$lacI^s/lacI^+$ partial diploids make no *lac* mRNA in the presence or absence of inducer. The model predicts that the $lacI^s$ mutant makes a repressor that either fails to bind inducer or that binds inducer and fails to dissociate from the operator DNA, and both

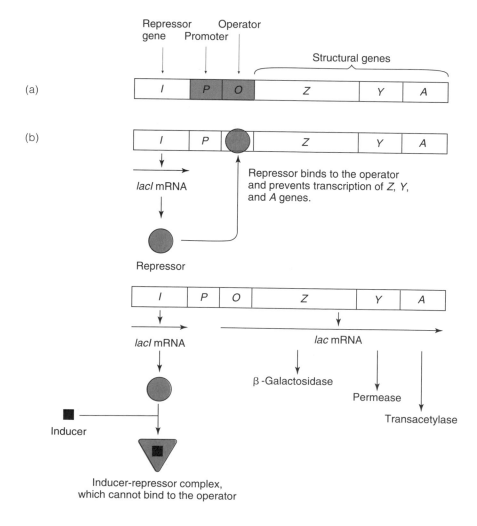

Figure 7-3. (a) Cartoon of the *lac* operon, not drawn to scale: The P and O sites are actually much smaller than the genes. (b) Diagram of the *lac* operon in repressed and induced states. The inducer alters the shape of the repressor, so the repressor can no longer bind to the operator.

types of mutant have been found. The second class of noninducible mutants are *cis*-dominant and affect the promoter for the *lac* operon.

Isolation of Mutations in the *lac* Operon

Many of the details of the organization and regulation of the *lac* operon were learned from the phenotypes of various mutants, but how were the mutants isolated? In initial studies of most operons, mutants are collected merely by looking for a simple defective phenotype (in this case, Lac⁻). The phenotype can be identified by growth on color-indicator plates. For example, on MacConkey lactose plates, a Lac⁺ colony is red and a Lac⁻ colony is white (see Chapter 4). In later stages of analysis of an operon, however, it is invariably necessary to collect large numbers of mutations that affect different aspects of the operon (for example, *lacI* mutations). Once the basic features are understood, it is usually possible (applying a certain amount of cleverness) to design selection procedures for isolating specific types of mutants. A few examples of selections and screens for *lac* mutants are outlined in Table 7-2 and described as follows:

1. *Constitutive mutants*. X-gal is a colorless β-galactoside that produces an intense blue color when cleaved with β-galactosidase (a derivative of the dye indigo is released). X-gal is not an inducer of the *lac* operon, so both Lac⁺ and Lac⁻ cells yield white colonies on plates containing X-gal in the absence of an inducer. Thus, plates containing X-gal can be used to screen for *lac* constitutive mutants. If the cells are constitutive, β-galactosidase is always made, and the colonies are blue. Both *lacI⁻* colonies and *lacOᶜ* colonies are blue, but *lacI⁻* colonies are usually deep blue, and *lacOᶜ* colonies are usually light blue, suggesting that some residual repression remains in most operator mutants.

2. *lacY mutants*. The molecule *o*-nitrophenyl-β-D-thiogalactoside (TONPG) is toxic to cells. It is transported into cells by lactose permease, but it is not an inducer of the *lac* operon. If IPTG (a gratuitous inducer of the *lac* operon described earlier) and TONPG are both included in agar, *lacY⁺* cells will be killed, so only *lacY* mutants will form colonies.

3. *lacZ mutants*. Mutants lacking one of the enzymes in the pathway for utilization of the sugar galactose, *galE* mutants, lyse in the presence of galactose. Recall that the cleavage of lactose by β-galactosidase yields glucose and galactose, so *lac⁺ galE⁻* strains will not grow on a glycerol-lactose plate (glycerol can be used as a carbon source), whereas *lac⁻ galE⁻* strains will. This procedure can yield both *lacZ⁻* and *lacY⁻* mutants because if the lactose never enters the cell, it will not be degraded to galactose. By using phenyl-β-D-galactoside, however, which does not require lactose permease for transport into the cell, only *lacZ* mutants are obtained. (Note that this selection could also yield promoter mutants, but these are quite rare because of the small size of the promoter compared with the *lacZ* gene.)

4. *Noninducible mutants*. To isolate *trans* dominant, noninducible mutants, selection or screening for *lacZ⁻* or *lacY⁻* mutants is done as described above in a partial diploid strain carrying two fully functional copies of the *lac* operon. In the partial diploid strain it takes two mutations to destroy *lacZ⁻* or *lacY⁻* function (because there are two copies of the DNA encoding each protein). However, only one mutation is needed to convert one of the copies of the *lacI* gene to a *lacIˢ* mutant. Since *lacIˢ* is dominant over *lacI⁺*, the second copy of *lacI* has no effect so the strain will have a noninducible phenotype.

Table 7-2 Substrates, indicators, and analogs used to detect expression of the *lac* operon and select for *lac* mutants

Chemical	Action and use
Lactose	Lactose is transported into the cell by the *lacY* gene product, lactose permease, and cleaved by the *lacZ* gene product, β-galactosidase. Growth on minimal medium with lactose as a sole carbon source requires that cells express high levels of both the *lacZ⁺* and *lacY⁺* genes.
2,3,5-Tetrazolium chloride	Tetrazolium is used to make indicator plates that differentiate Lac⁺ from Lac⁻ colonies. When fully reduced, tetrazolium forms an insoluble compound with a deep red color. At low pH, however, the tetrazolium is colorless. Thus, under appropriate conditions, tetrazolium can be used as a redox indicator or as an acid-base indicator. If oxidized tetrazolium is used and the pH of the medium is buffered, Lac⁺ colonies reduce the tetrazolium and turn red, whereas Lac⁻ colonies remain white. If reduced tetrazolium is added and the medium is not buffered, Lac⁺ colonies lower the pH by fermenting lactose so the colonies are white, whereas Lac⁻ colonies turn red.
Phenol red	Phenol red is used as an acid-base indicator in MacConkey lactose plates. At low pH, phenol red turns red colored, so as Lac⁺ colonies lower the pH by fermenting lactose, the colonies turn red, whereas Lac⁻ colonies remain white. The color of the colonies is proportional to the pH, which is in turn proportional to the amount of lactose fermented: Full expression of the *lac* genes results in dark red colonies, and weak expression of the *lac* genes results in pink colonies on MacConkey lactose plates.
5-Bromo-4-chloro-3-indoyl-β, D-galactoside (X-gal)	X-gal is used as an indicator for *lacZ* expression. X-gal is cleaved by β-galactosidase to form galactose and the blue-colored dye, 5-bromo-4-chloro-indigo. Thus, Lac⁺ colonies turn blue on X-gal indicator plates, whereas Lac⁻ colonies remain white. The amount of blue color formed is proportional to the β-galactosidase activity expressed by the cells. X-gal is very sensitive, allowing detection of light blue–colored colonies that express low levels of *lacZ*.
Isopropyl-β, D-galactoside (IPTG)	IPTG is a gratuitous inducer of the *lac* operon: It inactivates the *lac* repressor, but it is not cleaved by β-galactosidase. IPTG can enter cells without lactose permease. Thus, IPTG allows induction of the *lac* operon in the absence of functional *lacZ* or *lacY* gene products.
Phenyl-β,D-galactoside (PG)	PG is cleaved by β-galactosidase into galactose and phenol, but PG cannot induce the *lac* operon. Thus, *gal⁺* cells can use PG only as a sole carbon source if the *lac* operon is induced. Therefore, if no inducer is added, PG can be used to select for *lac* constitutive mutants.
Phenylethyl-β, D-galactoside (TPEG)	TPEG is a competitive inhibitor of β-galactosidase. In the presence of TPEG, cells must make more β-galactosidase to degrade lactose efficiently. Therefore, TPEG can be used to select for increased expression of the *lacZ* gene.
o-Nitrophenyl-β,D-thio-galactoside (TONPG)	TONPG is transported into the cell by lactose permease, but once inside the cell, it poisons cellular metabolism. Thus, TONPG kills cells that are *lacY⁺* but not cells that are *lacY⁻*. Therefore, TONPG can be used to select for *lacY⁻* mutants.
o-Nitrophenyl-β, D-galactoside (ONPG)	ONPG is used as a substrate for colorimetric β-galactosidase assays in vitro. β-galactosidase cleaves the colorless ONPG into galactose and o-nitrophenol, which is yellow colored. The amount of β-galactosidase in a cell suspension can be determined by measuring the rate of o-nitrophenol production using a spectrophotometer.

For a practical description of how to use these compounds, see Silhavy et al. (1984) or Miller (1992).

Differential Translation of the Genes in *lac* mRNA

The ratios of the number of copies of β-galactosidase, lactose permease, and transacetylase are 1.0:0.5:0.2. These differences are due to differential translation of the three genes. There are two main reasons for this:

1. The *lacZ* gene is translated first (Figure 7-4). Frequently ribosomes are released from the *lac* mRNA following chain termination. The frequency with which this occurs is a function of the probability of reinitiation at each subsequent AUG codon. Thus, there is a gradient in the amount of polypeptide synthesis from the 5' end to the 3' end of the mRNA molecule; this effect occurs with most polycistronic mRNA molecules.
2. Most bacterial mRNA is degraded to nucleotides after several rounds of translation. Degradation of *lac* mRNA is initiated more frequently in the *lacA* gene than in the *lacY* gene and more often in the *lacY* gene than in the *lacZ* gene. Hence, at any given instant, there are more complete copies of the *lacZ* gene than of the *lacY* gene and more copies of the *lacY* gene than of the *lacA* gene.

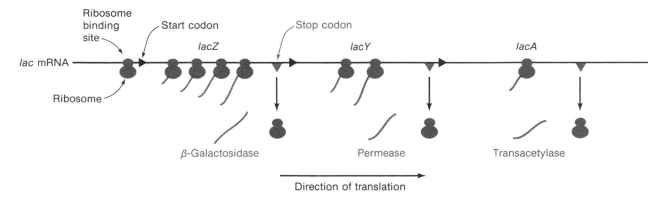

Figure 7-4. Polarity in the *lac* operon. All ribosomes attach to the mRNA molecule at the ribosome binding site. At each stop codon, some ribosomes detach. Thus, the number of ribosomes translating each gene segment decreases for each subsequent gene.

In prokaryotes, the overall expression of an operon is often regulated by controlling transcription of a polycistronic mRNA, and the relative concentrations of the proteins encoded in the mRNA are determined by controlling the frequency of initiation of translation of each gene.

Positive Regulation of the *lac* Operon

The function of β-galactosidase in lactose metabolism is to form glucose by cleaving lactose. (The other cleavage product, galactose, is also ultimately converted to glucose by the enzymes of the galactose operon.) Thus, if both glucose and lactose are present in the growth medium, activity of the *lac* operon is not needed, and, indeed, no β-galactosidase is formed until virtually all of the glucose in the medium is consumed. The lack of synthesis of β-galactosidase is a result of lack of synthesis of *lac* mRNA. No *lac* mRNA is made in the presence of glucose because in addition to an inducer to inactivate the *lacI* repressor, another element is needed for initiating *lac* mRNA synthesis; the activity of this element is regulated by the concentration of glucose. The inhibitory effect of glucose, however, on expression of the *lac* operon is indirect.

The small molecule **cyclic AMP (cAMP)** is universally distributed in animal tissues, and in multicellular eukaryotic organisms, it is important in regulating the action of many hormones (Figure 7-5). It is also present in *E. coli* and many other bacteria. cAMP is synthesized by the enzyme **adenyl cyclase**, and its concentration is regulated indirectly by glucose transport and metabolism. When bacteria are growing in a medium containing glucose, the cAMP concentration in the cells is quite low. In a medium containing glycerol or any carbon

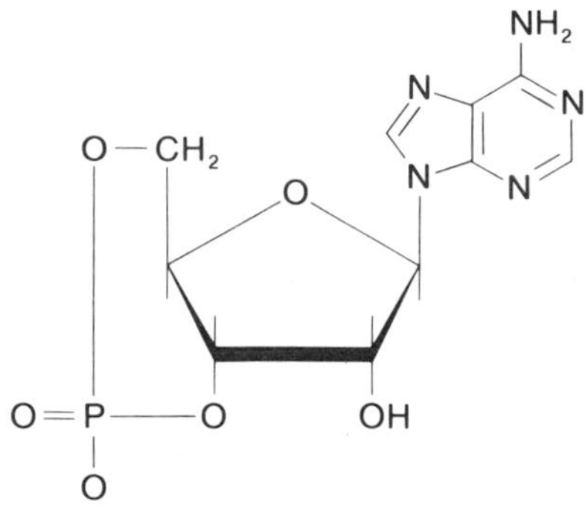

Figure 7-5. Structure of cAMP.

Table 7-3 Concentration of cyclic AMP in cells growing in media having the indicated carbon sources

Carbon source	cAMP concentration
Glucose	Low
Glycerol	High
Lactose	High
Lactose + glucose	Low
Lactose + glycerol	High

source that cannot enter the biochemical pathway used to metabolize glucose (the glycolytic pathway) or when the bacteria are otherwise starved of an energy source, the cAMP concentration is high (Table 7-3). The mechanism by which glucose controls the cAMP concentration is poorly understood; the important point is that *cAMP regulates the expression of the* lac *operon* (and many other operons as well).

E. coli and *Salmonella typhimurium* (and many other bacterial species) contain a protein called the **cAMP receptor protein (CRP)**, which is encoded in a gene called *crp*. A class of Lac⁻ mutations has been isolated that map quite far from the *lac* gene cluster. These mutations affect either the *crp* or the adenyl cyclase (*cya*) gene. Biochemical analysis showed that such mutants are unable to synthesize *lac* mRNA, indicating that both CRP function and cAMP are required for *lac* mRNA synthesis. CRP and cAMP bind to one another, forming a **cAMP-CRP** complex, which has been shown in biochemical experiments with purified components to be required for transcription of the *lacZ-lacA* region. The requirement for cAMP-CRP is independent of the *lacI* repression system because *crp* and *cya* mutants are unable to make *lac* mRNA even if a *lacI⁻* or a *lacOᶜ* mutation is present. The reason is that the cAMP-CRP complex must be bound to a base sequence in the DNA in the promoter region in order for transcription to occur (Figure 7-6). When bound to this DNA base sequence, CRP protein can stabilize binding of RNA polymerase by a protein-protein contact. This contact allows RNA polymerase to efficiently initiate transcription on what would otherwise be an inefficient promoter. Thus, the *cAMP-CRP complex is a **positive regulator** or **activator***, in contrast to the repressor, and the *lac* operon is independently regulated both positively and negatively.

crp⁻ and *cya⁻* mutants are not only Lac⁻, but also defective in their ability to use maltose, galactose, arabinose, and many other carbon sources. The operons responsible for utilization of each of these compounds (called **catabolite-**

Figure 7-6. Three states of the *lac* operon showing that *lac* mRNA is made only if cAMP-CRP is present and repressor is absent.

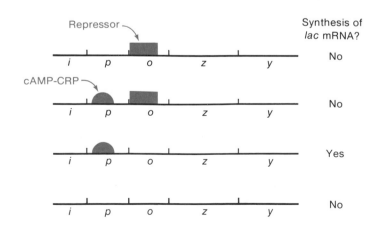

sensitive operons) all require cAMP-CRP complex. In fact, the easiest way to isolate a *crp⁻* or *cya⁻* mutant is to plate the mutants on a color-indicator medium containing both lactose and maltose and screen for a Lac⁻ Mal⁻ double mutant. These double mutants arise at roughly the same frequency as single Lac⁻ or Mal⁻ mutants and hence are likely to be the result of a single mutation. Confirmation of this conclusion comes from the observation that the double mutant is also Gal⁻ and defective in the utilization of the other catabolite-sensitive carbon sources.

REGULATION OF A BIOSYNTHETIC PATHWAY: THE TRYPTOPHAN OPERON

The tryptophan (*trp*) operon of *E. coli* is responsible for the synthesis of the amino acid tryptophan. Regulation of this operon occurs in such a way that when excess tryptophan is present, the *trp* operon is not expressed. That is, excess tryptophan causes transcription of the *trp* operon to be turned off or "repressed"; however, when the concentration of tryptophan is insufficient, transcription of the *trp* operon is turned on. The *trp* operon is quite different from the *lac* operon: Tryptophan acts as a corepressor rather than as an inducer. Furthermore, because the *trp* operon encodes biosynthetic rather than degradative enzymes, cAMP-CRP is not required for its expression.

Tryptophan is synthesized in five steps, each requiring a particular enzyme. In the *E. coli* chromosome, the genes encoding these enzymes are translated from a single polycistronic mRNA molecule. The order of the genes is *trpED-CBA*. Between the promoter and the operator are two regions called the **leader** (*trpL*) and the **attenuator** (Figure 7-7). A repressor gene *trpR* is located quite far from this gene cluster.

Mutations in either the *trpR* gene or in the operator site increase expression of *trp* mRNA, as in the *lac* operon. In contrast to the *lacI* protein, however, the *trpR* protein does not bind to the operator unless tryptophan is present. Hence, it is called an aporepressor. When the *trpR* aporepressor binds a tryptophan molecule, its conformation changes into an active repressor, which binds to the operator DNA:

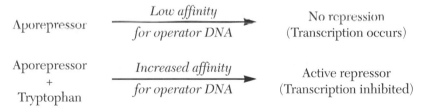

Tryptophan is called the corepressor because *trpR* repressor inhibits transcription only when tryptophan is present. When the concentration of tryptophan decreases, less active repressor is present, the operator is unoccupied, and

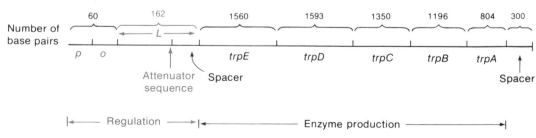

Figure 7-7. The *E. coli trp* operon. For clarity, the regulatory region is enlarged with respect to the coding region. The proper size of each region is indicated by the number of base pairs. L is the leader. The regulatory elements are shown in orange.

transcription begins. Although the details are different, the basic mechanism of repression is similar to the "on-off" switch that regulates the *lac* operon.

A simple on-off system, as in the *lac* operon, is not optimal for an essential biosynthetic pathway; a situation may arise in nature in which some tryptophan is available but not enough to allow normal growth if synthesis of tryptophan were totally shut down. Tryptophan starvation when the supply of the amino acid is inadequate is prevented by a system that subtly modulates the amount of transcription of the biosynthetic genes in proportion to the concentration of tryptophan. This mechanism is found in many operons responsible for amino acid biosynthesis.

When the *trp* operon is not repressed, finer control allows regulation of the *trp* biosynthetic genes in proportion to the amino acid concentration. This control, called attenuation, is due to (1) premature termination of transcription before the first structural gene is reached and (2) regulation of the frequency of this termination by the internal concentration of tryptophan.

A 162-base leader sequence is present at the 5' end of the *trp* operon. In the presence of tryptophan, most of the *trp* mRNA molecules terminate before the *trpE* gene, resulting in a 140-nucleotide transcript that ends within the leader sequence. Furthermore, a deletion mutation that removes bases 123 to 150 of the leader sequence makes the *trp* enzymes at six times the normal rate in both the presence and absence of tryptophan. These results indicate that the *trp* leader is involved in regulation of the *trp* genes. The 28-base region, where regulation of transcription termination occurs, is called the **attenuator**. The base sequence of the region in which termination occurs contains the usual features of a termination site (Figure 7-8): a stem-and-loop structure in the mRNA followed by a sequence of eight AT pairs.

The leader sequence has several important features:

1. It encodes a polypeptide containing 14 amino acids, the **leader polypeptide** (Figure 7-9).
2. Two adjacent tryptophan codons are located in the leader polypeptide at positions 10 and 11.
3. Four segments of the leader RNA—denoted 1, 2, 3, and 4—are capable of base-pairing in two different ways: forming either the base-paired regions 1-2 and 3-4 or just the region 2-3 (Figure 7-10).

This arrangement allows regulation of transcription termination to occur in the *trp* leader region. Regulation of transcription termination is determined by translation of the leader peptide. The two adjacent tryptophan codons in this sequence make translation of the sequence sensitive to the concentration of charged tRNA^Trp. If the concentration of tryptophan is low, much of the tRNA^Trp remains uncharged. Thus, if the intracellular tryptophan concentration is inadequate, translation is slowed at the tryptophan codons. Translation of the leader peptide mediates transcription termination for three reasons: (1) Transcription and trans-

Figure 7-8. The terminal region of the *trp* leader mRNA (right end of L in Fig. 7-7). The base sequence given is extended past the termination site at position 140 to show the long stretch of U's. The orange bases form an inverted repeat sequence that could lead to the stem-and-loop configuration shown (segment 3-4, Fig. 7-10).

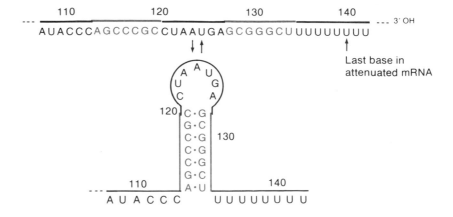

	Leader polypeptide														TrpE protein			
	Met	Lys	Ala	Ile	Phe	Val	Leu	Lys	Gly	Trp	Trp	Arg	Thr	Ser—	Stop	Met	Gln	Thr	Gln
pppAAG...(23)...	AUG	AAA	GCA	AUU	UUC	GUA	CUG	AAA	GGU	UGG	UGG	CGC	ACU	UCC	UGA	...(91)...	AUG	CAA	ACA	CAA

Figure 7-9. The sequence of the *trp* leader mRNA showing the leader polypeptide, the two Trp codons (shaded red), and the beginning of the TrpE protein. The numbers (23 and 91) refer to the number of bases whose sequences are omitted for clarity.

lation are coupled, as is usually the case in bacteria (see Chapter 6); (2) when the sequences 1-2 and 3-4 are paired in the mRNA, the sequence 2-3 cannot be paired; and (3) all base-pairing is eliminated in the segment of the mRNA that is in contact with the ribosome.

The end of the *trp* leader peptide is in segment 1 (see Figure 7-10). A translating ribosome is in contact with about 10 bases in the mRNA past the codons being translated, so when the final codons of the leader are being translated, segments 1 and 2 are not paired. When transcription and translation are coupled, the leading ribosome is not far behind the RNA polymerase. Thus, if the ribosome is in contact with segment 2 when synthesis of segment 4 is being completed, segments 3 and 4 are free to form the 3-4 stem without segment 2 competing for segment 3. The presence of the 3-4 stem-and-loop configuration allows transcription termination to occur after synthesis of the seven uridines in the mRNA. In contrast, if there is limiting tryptophan, the concentration of charged tRNA^{Trp} becomes inadequate, and the translating ribosomes stall at the Trp codons. These codons are located 16 bases before the beginning of segment 2. Thus, segment 2 is free before segment 4 has been synthesized and region 2-3 (the **antiterminator**) can form. In the absence of the 3-4 stem and loop, termination does not occur, and the complete mRNA molecule is made, including the coding sequences for the *trp* genes.

Hence, if tryptophan is present in excess, termination occurs, and little enzyme is synthesized; if tryptophan is absent, termination does not occur, and the enzymes are made. At intermediate concentrations, the fraction of initiation

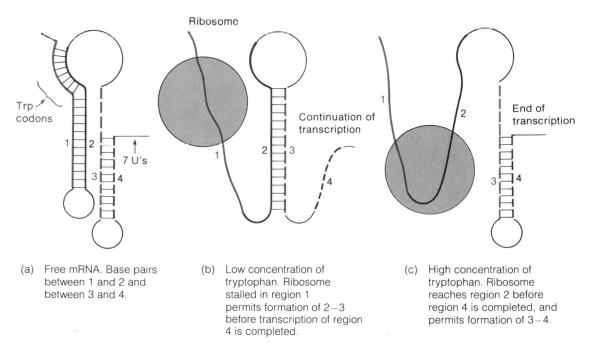

(a) Free mRNA. Base pairs between 1 and 2 and between 3 and 4.

(b) Low concentration of tryptophan. Ribosome stalled in region 1 permits formation of 2−3 before transcription of region 4 is completed.

(c) High concentration of tryptophan. Ribosome reaches region 2 before region 4 is completed, and permits formation of 3−4.

Figure 7-10. The model for the mechanism of attenuation in the *E. coli trp* operon.

events that result in completion of *trp* mRNA depends on how often translation is stalled, which in turn depends on the concentration of tryptophan.

Repression versus Attenuation

The *trp* repressor-operator system does not operate as a simple on-off switch but can yield intermediate levels of operon expression. The concentration of the *trp* enzymes is 10-fold greater in cells with a mutant (inactive) *trp* repressor than in wild-type cells, indicating that the synthesis of *trp* mRNA is partially repressed at all times in a cell growing in the absence of added tryptophan. This observation implies that in wild-type cells, if the internal concentration of tryptophan fluctuates for any reason, the equilibrium between active and inactive repressor shifts to maintain a usable supply of tryptophan.

Repression has been studied independently of attenuation in a cell containing a gene fusion linking the *lacZ* gene to the *trp* promoter-operator region, lacking the attenuator. The activity of β-galactosidase as a function of concentration of tryptophan in the growth medium is a measure of the response of the repressor-operator system. Comparison of the behavior to that of an intact *trp* operon yields the contribution of the attenuation system. It was found that repression is responsible for an 80-fold regulation and attenuation is responsible for sixfold to eightfold regulation of expression of the *trp* operon, for a total variation of 500-fold to 600-fold. Furthermore, repression is the dominant regulatory mechanism at higher concentrations of tryptophan; attenuation is not relaxed until starvation for tryptophan becomes severe (implying that charging of tRNA^{Trp} occurs even when the concentration of tryptophan is quite low).

Many operons responsible for amino acid biosynthesis are regulated by attenuators that are regulated by a mechanism analogous to the *trp* operon. For example, attenuation has been shown to regulate the histidine, threonine, leucine, isoleucine-valine, and phenylalanine operons of a wide variety of gram-negative enteric bacteria. In many cases, the operons lack a repressor-operator system and are regulated solely by attenuation. Because these operons are regulated adequately (although the range of expression is not as great as that of the *trp* operon), one might ask why some amino acid biosynthetic operons have a dual regulatory system. One reason may be that those operons with both regulatory mechanisms also control expression of related biosynthetic operons. For example, the *trp* repressor also regulates expression of the *aroH* gene. A common intermediate in the synthesis of the aromatic amino acids is made by three distinct enzymes, each having the same enzymatic activity. All three enzymes are needed only when all three aromatic amino acids must be synthesized. Thus, each enzyme is independently regulated by an amino acid–specific repressor. It is possible that in the distant past, the *trp* operon was regulated only by attenuation, and *aroH* was regulated as it is now, by a tryptophan-activated *aroH* repressor. The existence of a tryptophan-sensitive repressor allowed the evolution of a sequence in the *trp* promoter into an operator that is regulated by this repressor.

AUTOREGULATION

Many proteins bind to specific sites in the cell. Thus, the amount of such proteins needed depends on the number of available binding sites, not simply the presence or absence of a small molecule. Expression of such proteins is sometimes regulated by **autoregulation**. In the simplest autoregulated systems, the gene product is also a repressor: It binds to an operator site adjacent to the promoter. When the gene product fills all the operator sites, it binds an operator in front of its own gene and represses its own transcription. Generally the affinity for its own operator is less than the affinity for the other operator sites. Thus, as the cell grows and the concentration of repressor decreases, the operator in front of the repressor will be free, so repressor can be made, maintaining the concentration of repressor at a constant steady state. For example, the *trpR* gene is autogenously regulated by the Trp repressor.

GENE AND OPERON FUSIONS

To study regulation of gene expression, it is necessary to have not only a supply of mutations, but also a way to measure expression of the genes. Sometimes this can be done by directly measuring the enzyme activity of the gene products or quantitating the amount of gene-specific mRNA made. In many cases, however, the assays are tedious and time-consuming. One reason the *lac* operon was such a good model for studying gene expression was the ease of measuring β-galactosidase activity and the availability of indicator plates and analogs for isolating mutants. By constructing fusions that join other genes with the *lac* genes, these approaches can be applied to many other genes as well.

Such genetic fusions were first isolated from deletions that joined *lac* genes with neighboring genes on the chromosome (Figure 7-11), but fusions are now usually constructed with special transposon derivatives (see Chapter 12). Two types of fusions can be isolated: operon (or transcriptional) fusions and gene (or translational) fusions (Figure 7-12). Operon fusions join two operons into a single

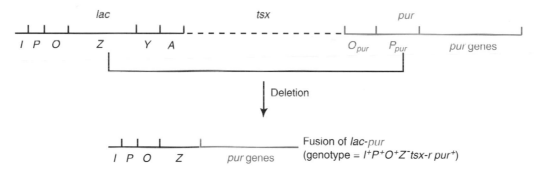

Figure 7-11. Formation of the *lac-pur* fusion.

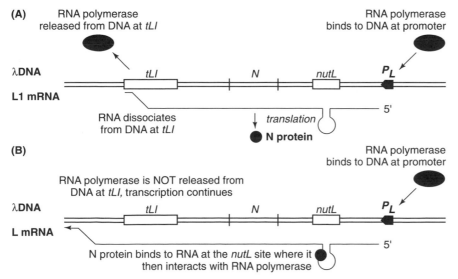

Figure 7-12. (a) Operon and (b) gene fusions. Operon fusions result in two proteins made from the same promoter but different translational start sites. Gene fusions result in a single hybrid protein made from a single promoter and a single translational start site.

transcriptional unit but do not create hybrid proteins. Gene fusions also join two operons into a single transcriptional unit but combine translation of two genes, resulting in a hybrid protein.

Operon fusions retain the translation start sites for *lacZ*, so expression of β-galactosidase is proportional to the rate of transcription of the gene upstream of the *lacZ* insertion. In contrast, gene fusions lack both the *lacZ* translational and the *lacZ* transcriptional start sites, so the expression of β-galactosidase is proportional to both the rate of transcription and the rate of translation of the upstream gene. Thus, operon and gene fusion allow the transcriptional and translational regulation of other genes to be studied by measuring β-galactosidase expression. Furthermore, it is possible to use *lac* indicator plates and analogs to isolate regulatory mutants of such gene and operon fusions. Some sophisticated tools have been developed that make it simple to isolate operon or gene fusions with any gene. These are described in Chapter 19.

ALTERNATIVE TRANSCRIPTION FACTORS

Many developmental processes require expression of a large number of gene products in a carefully timed sequence. For example, many phage require expression of "early genes" required for phage replication and gene expression soon after infection, and "late genes" required to make the structural components of the phage particles (heads and tails). To properly coordinate expression of the right genes at the right time, phages have complex developmental pathways that are regulated in many different ways. Some phage genes are regulated by repressor or activator proteins as described for the *lac* operon earlier in this chapter. Such a repressor-operator mechanism is important for establishing and maintaining the prophage state in temperate phage (discussed in Chapter 17). Many complex pathways that require proper timing of gene expression are regulated by alternative σ factors that change the way that RNA polymerase binds to DNA. Two examples of how alternative promoters regulate gene expression in phages are described here, but alternative σ factors are also important in regulating many complex processes in bacteria as well (for example, sporulation, flagella synthesis, and nitrogen regulation).

E. coli phage T4 has a system for regulating the timing of synthesis of numerous classes of mRNA molecules. Early in the life cycle, the bacterial RNA polymerase initiates transcription at a single class of promoters that are recognized by the host's RNA polymerase. Some of these transcripts encode proteins that modify the host RNA polymerase. The modified polymerase can no longer bind to the original promoter but gains the ability to initiate at other promoters with different base sequences than that of the preceding class. A successive series of modifications occurs by covalent attachment of small molecules and binding of phage proteins to the modified RNA polymerase. Each successive modification causes the polymerase to ignore the earlier promoters and to initiate at new promoters. The net effect is an orderly control of the timing of synthesis of many species of mRNA: Not only are new mRNA species made at the correct time, but also mRNA species made at earlier times that are no longer needed are no longer synthesized.

E. coli phage T7 makes three transcripts, each from the same DNA strand. Transcript I is initiated by *E. coli* RNA polymerase acting on promoter I. The other two promoters (II and III) are not recognized by the *E. coli* enzyme. Transcript I encodes two important proteins. One of these proteins is a new RNA polymerase that (1) does not recognize any promoters in the bacterial DNA, (2) does not recognize promoter I, and (3) initiates transcription at promoters II and III. The second protein inactivates the *E. coli* RNA polymerase. Thus, shortly after infection, the phage has succeeded in preventing all synthesis of bacterial mRNA and has begun to synthesize important phage proteins (e.g., a T7 DNA polymerase) encoded in transcript II. The third transcript encodes structural proteins of the phage particle and a lysis enzyme, both of which, for the sake of efficiency, should be synthesized late in the life cycle. T7 has a unique mode of delaying synthesis of late mRNA: The phage DNA that initiates the infection is injected so slowly that it takes about 12 minutes before promoter III has entered the cell. Thus, the overall sequence of the T7 life cycle is (1) make T7-specific RNA polymerase and inactivate host enzyme, (2) make phage DNA, (3) assemble phage particles, and (4) lyse the cell.

Antitermination

An important step in regulation of the life cycle of *E. coli* phage λ is controlled by modulating transcription termination, but the mechanism is quite different from that used in the *trp* attenuation system. In this case, the λ N protein interacts with sites in the DNA and *inhibits* normal termination. This phenomenon is called **antitermination**. In λ, antitermination occurs at several sites, one of which is described.

Early in both the lytic and the lysogenic pathways of λ, transcription is initiated from the promoter *pL* and yields a short transcript called L1 (Figure 7-13). This transcript includes a gene *N*, whose product is an antitermination protein. Transcription by the unmodified host RNA polymerase stops shortly after the end of the *N* gene because the polymerase encounters the normal transcription-termination sequence *tL1*. Following transcription of the *N* gene, translation of L1 occurs and the *N* gene product is synthesized. The *N* gene product (aided by an *E. coli* protein called Nus A) binds to a short RNA sequence transcribed from the DNA located nearby, called the *N*-utilization site or (*nutL*). As the concentration of the N protein increases, some N protein binds to *nutL*; the polymerase acquires the N protein, and the RNA polymerase–N protein complex is able to transcribe through the termination site. Transcription continues until a second termination site is encountered. The RNA polymerase–N protein complex then stops, forming the longer transcript. Hence, the synthesis of the proteins encoded

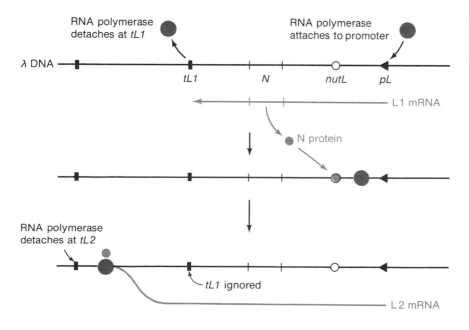

Figure 7-13. Antitermination of λ L1 mRNA induced by the binding of the λ N protein to the *nutL* site in the RNA.

in the longer transcript is delayed by the amount of time required to synthesize a sufficient amount of the *N* protein necessary for antitermination. Transcription of other λ genes is also regulated by antitermination, but the mechanism is different. For example, subsequent to antitermination by N protein, the product of the gene *Q* is made. The Q protein is an antiterminator of the R4 transcript. Downstream from the promoter for R4 is a site called *qut*. Q protein binds to the DNA at the *qut* site. When RNA polymerase encounters Q protein at the *qut* site, it picks up Q and thereby is able to ignore the termination signal for R4. The resulting transcript is a large late mRNA molecule that encodes the head, tail, DNA packaging, and lysis proteins. This series of delays—the time required to make N, make the secondary mRNA, make Q, make the late mRNA, and make the "late" proteins—does not occur until ample DNA has been synthesized to produce a large number of progeny particles. This system is examined further in Chapter 16.

KEY TERMS

adenyl cyclase	inducer
antitermination	noninducible mutant
attenuator	operator
autoregulation	operon
constitutive mutant	operon fusions
corepressor	regulation
cAMP	coordinate
CRP	negative
derepression	positive
gene fusions	repressor
gratuitous inducer	

QUESTIONS AND PROBLEMS

1. What is meant by coordinate regulation?

2. Other than the ability to turn on and off a set of genes in an operon by a single regulatory element, what else is accomplished by having a set of genes contained in one polycistronic mRNA molecule?

3. Which type of regulation, positive or negative, involves removal of an inhibitor?

4. Would synthesis of an enzyme that is needed continually be regulated?

5. What is the biochemical action of an inducer?

6. What is the physical consequence of binding of the *lac* repressor to the *lac* operator?

7. Which enzymes of the *lac* operon are regulated by the repressor?

8. Is the partial diploid F' *lacI*$^+$/*lacI*$^-$ inducible or constitutive?

9. Is the partial diploid F' *lacO*$^+$/*lacO*^{c} inducible or constitutive?

10. Why are all constitutively synthesized proteins not made at the same rate?

11. Is it necessary for a repressor gene to be adjacent to the operator?

12. Is it necessary for the operator to be adjacent to promoter?

13. When glucose is present, is the concentration of cAMP high or low?

14. Can a mutant with either an inactive *cya* gene or an inactive *crp* gene synthesize β-galactosidase?

15. Does the binding of cAMP-CRP to DNA affect the binding of a repressor to the operator?

16. Are all proteins translated from a single polycistronic mRNA necessarily made in the same quantity?

17. Is the attenuator a protein-binding site?

18. Antitermination and attenuation are both concerned with termination of transcription. How do they differ with respect to the role of RNA polymerase?

19. How do lactose molecules first enter an uninduced $lacI^+$ $lacZ^+$ $lacY^+$ cell to induce synthesis of β-galactosidase?

20. For each of the following diploid genotypes, indicate first whether β-galactosidase can be made; second, whether synthesis of β-galactosidase is inducible (I) or constitutive (C); and, finally, whether or not each cell could grow with lactose as sole carbon source. (I, P, O, Z, Y are used for $lacI$, $lacP$, $lacO$, $lacZ$, $lacY$, for simplicity.)
   a. $I^+ Z^- Y^+ / I^- Z^+ Y^+$
   b. $I^+ Z^- Y^+ / O^c I^+ Z^- Y^+$
   c. $I^+ Z^- Y^+ / O^c Z^+ Y^+$
   d. $I^+ Z^+ Y^- / I^- Z^- Y^+$
   e. $I^- Z^+ Y^- / I^- Z^+ Y^+$
   f. $I^- Z^+ Y^+ / I^+ O^c Z^- Y^+$
   g. $I^+ P^- Z^+ / I^- Z^-$
   h. $I^+ O^c Z^- Y^+ / I^+ Z^+ Y^-$
   i. $I^+ P^- O^c Z^- Y^+ / I^+ Z^+ Y^-$
   j. $I^- P^- O^c Z^+ Y^+ / I^- Z^+ Y^-$

21. A cell that is wild-type with respect to the lac operon ($^+$ for all alleles) is grown in a medium without glucose or lactose; that is, it is using another carbon source. What proteins bind the DNA in the lac regulatory region? What if glucose were present?

22. You have isolated a Lac⁻ mutant and found by genetic analysis that its genotype is $lacZ^+$ $lacY^+$. The mutation, which you call $lacI*$, is in the $lacI$ gene. The partial diploid $lacI*$ $lacZ^+ lacY^+ / lacI^+ lacZ^+ lacY^+$ is constructed, and its phenotype is found to be Lac⁻ ($lacI*$ is dominant). The diploid $lacI* lacZ^+ lacY^+ / lacO^c lacI^+ lacZ^+ lacY^+$ is Lac⁺ (β-galactosidase is made). Suggest a property of the mutant repressor that would yield this phenotype. Would $lacI* lacZ^+ lacY^+ / lacO^c lacI^+ lacZ^- lacY^+$ make β-galactosidase?

23. A mutant strain of *E. coli* is found that produces both β-galactosidase and lactose permease whether lactose is present or not.
   a. What are two possible genotypes for this mutant?
   b. Another mutant is isolated that produces no β-galactosidase at any time but produces lactose permease if lactose is present in the medium. If a partial diploid is formed from these two mutants, in the absence of lactose, neither β-galactosidase nor permease is made. When lactose is added, the partial diploid makes both enzymes. What are the genotypes of the two mutants?

24. How many proteins are bound to the trp operon when (a) tryptophan and glucose are present, (b) tryptophan and glucose are absent, or (c) tryptophan is present and glucose is absent?

25. An *E. coli* mutant is isolated that is simultaneously unable to use a large number of sugars as sources of carbon. Genetic analysis, however, shows that none of the operons responsible for metabolism of the sugars is mutated. What are two potential genotypes of this mutant?

26. An operon has the gene sequence A B C D E. Neither the promoter nor the operator has been located. The repressor gene maps far away from the structural genes. Various deletion mutants have been isolated. Some deletions of gene E but none for any of the other genes result in constitutive production of the mRNA of the operon. Where do you think the operator and the promoter are?

27. An operon responsible for using a sugar Q is regulated by a gene called kyu. when Q is added to the growth medium, Qase is made; otherwise, the enzyme is not made. If the gene kyu is deleted (denoted Δkyu), no Qase can be made. The partial diploid $kyu / \Delta kyu$ is inducible. Two types of point mutants of kyu are found: kyu^-1, which never makes Qase, and kyu-2, which makes the enzyme constitutively. The partial diploids $kyu^+/kyu1$ and kyu^+/kyu-2 are inducible and constitutive. What is the likely mode of action of the protein encoded by the kyu gene?

28. The regulation of an operon responsible for synthesis of X depends on a repressor, a promoter, and an operator. In the presence of X, the system is turned off; an interaction between X and the repressor forms a complex that can bind to the operator.

a. What kinds of mutations might occur in the repressor? What is their phenotype (e.g., is the operon on or off)?

b. Describe the phenotype of a partial diploid with one wild-type and one mutant gene for each mutant gene.

REFERENCES

Beatriz, A., P. Olfson, and M. Casadaban. 1984. Plasmid insertion mutagenesis and *lac* gene fusion with mini-Mu bacteriophage transposons. *Proc. Natl. Acad. Sci. USA,* 158, 488.

Bertrand, K., C. Squires, and C. Yanofsky. 1976. Transcription termination *in vitro* in the leader region of the tryptophan operon of *E. coli. J. Mol. Biol.,* 103, 319.

Casadaban, M. J., and S. N. Cohen. 1980. Analysis of gene control signals by DNA fusion and cloning in *E. coli. J. Mol. Biol.,* 138, 179.

Collado-Vides, J., B. Magasanik, and J. Gralla. 1991. Control site location and transcriptional regulation in *Escherichia coli. Microbiol. Rev.,* 55, 371.

Gilbert, W., and A. Maxam. 1973. The nucleotide sequence of the *lac* operator. *Proc. Natl. Acad. Sci.,* 70, 3581.

Gilbert, W., and B. Muller-Hill. 1966. Isolation of the *lac* repressor. *Proc. Natl. Acad. Sci.,* 58, 2415.

Gottesman, S., and M. Maurizi. 1993. Regulation by proteolysis: energy dependent proteases and their targets. *Microbiol. Rev.,* 56, 592.

Greenblatt, J., J. Nodwell, and S. Mason. 1993. Transcriptional antitermination. *Nature,* 364, 401.

Ishihama, A. 1993. Protein-protein communication within the transcription apparatus. *J. Bacteriol.,* 175, 2483.

Jacob, F., and J. Monod. 1961. Genetic regulatory mechanisms in the synthesis of proteins. *J. Mol. Biol.,* 3, 318.

Keller, E. B., and J. M. Calvo. 1979. Alternative secondary structure of leader RNAs and the regulation of the *trp, phe, his, thr,* and *leu* operons. *Proc. Natl. Acad. Sci.,* 76, 6186.

Kutsu, A., K. North, and D. S. Weiss. 1991. Prokaryotic transcriptional enhancers and enhancer-binding proteins. *Trends Biochem. Sci.,* 16, 397.

*Little, J. 1993. LexA cleavage and other self-processing reactions. *J. Bacteriol.,* 175, 4943.

*Magasanik, B., and F. C. Neidhardt. 1987. Regulation of carbon and nitrogen utilization. In F. C. Neidhardt, J. L. Ingraham, K. B. Low, B. Magasanik, M. Schaechter, and H. E. Umbarger (eds.), Escherichia coli *and* Salmonella typhimurium. *Cellular and Molecular Biology.* American Society for Microbiology.

Maloy, S., and V. Stewart. 1993. Autogenous regulation. *J. Bacteriol.,* 175, 307.

Matthews, K. 1992. DNA looping. *Microbiol. Rev.,* 56, 123.

McKnight, S., and K. Yamamoto. 1992. *Transcriptional Regulation.* Cold Spring Harbor Laboratory Press.

Miller, J. H. 1992. *A Short Course in Bacterial Genetics.* Cold Spring Harbor.

*Miller, J. H., and W. S. Reznikoff (eds.). 1978. *The Operon.* Cold Spring Harbor.

Neidhardt, F. 1987. Multigene systems and regulons. In F. C. Neidhardt, J. L. Ingraham, K. B. Low, B. Magasanik, M. Schaechter, and H. E. Umbarger (eds.), Escherichia coli *and* Salmonella typhimurum. *Cellular and Molecular Biology.* American Society for Microbiology.

Oxender, D. L., G. Zurawaski, and C. Yanofsky. 1979. Attenuation in the *E. coli* tryptophan operon: role of RNA secondary structure involving the tryptophan coding region. *Proc. Natl. Acad. Sci.,* 76, 5524.

Pardee, A. B., F. Jacob, and J. Monod. 1959. The genetic control and cytoplasmic expression of inducibility in the synthesis of β-galactosidase by *E. coli. J. Mol. Biol.,* 1, 165.

*Platt, T. 1978. Regulation of gene expression in the tryptophan operon in *E. coli.* In J. H. Miller and W. S. Reznikoff (eds.), *The Operon.* Cold Spring Harbor.

*Resources for additional information.

Postma, P., J. Lengeler, and G. Jacobson. 1993. Phosphoenolpyruvate:carbohydrate phos-photransferase systems of bacteria. *Microbiol. Rev.*, 57, 543.

Reznikoff, W. 1992. The lactose operon-controlling elements: a complex paradigm. *Mol. Microbiol.*, 6, 2419.

*Reznikoff, W. 1992. Catabolite gene activator protein activation of *lac* transcription. *J. Bacteriol.*, 174, 655.

Reznikoff, W., and L. Gold. 1986. *Maximizing Gene Expression*. Butterworths.

*Schleif, R. 1988. DNA-binding by proteins. *Science*, 241, 1182–1187.

Silhavy, T., M. Berman, and L. Enquist. 1984. *Experiments with Gene Fusions*. Cold Spring Harbor Laboratory.

*Yanofsky, C. 1981. Attenuation in the control of expression of bacterial operons. *Nature*, 289, 751.

PART
3

MAINTENANCE OF GENETIC INFORMATION

8

DNA Replication

Genetic information is transferred from parent to progeny organisms by a faithful replication of the parental DNA molecules. Usually the information resides in one or more double-stranded DNA molecules. Some bacteriophage species contain single-stranded DNA instead of double-stranded DNA. In these systems, replication consists of several stages in which single-stranded DNA is first converted to a double-stranded molecule, which then serves as a template for synthesis of complementary single-stranded DNA. Viruses containing single-stranded and double-stranded RNA molecules are also known; these organisms use several different modes of replication, some of which (in eukaryotes) include double-stranded DNA as an intermediate. The modes of replication of each of these types of molecules differ in detail, although certain fundamental features are common to each. This chapter, which is an overview of DNA replication, examines only a few general properties of the replication process.

BASIC RULES FOR REPLICATION OF DNA

The primary role of any mode of replication is to duplicate the base sequence of the parent molecule. The specificity of base pairing—adenine with thymine and guanine with cytosine—provides the mechanism used by all replication systems.

1. Nucleotide monomers are added one by one to the end of a growing strand by an enzyme called a **DNA polymerase**.
2. The sequence of bases in each new or **daughter strand** is complementary to the base sequence in the original template or **parent strand** being copied—that is, if there is an adenine in the parent strand, a thymine nucleotide will be added to the end of the growing daughter strand when the adenine is being copied.

In the following section, we consider how the two strands of a daughter molecule are physically related to the two strands of the parent molecule.

GEOMETRY OF DNA REPLICATION

The production of daughter DNA molecules from a single parental molecule gives rise to several topological problems, which result from the helical structure and enormous size of typical DNA molecules and the circularity of many DNA molecules.

Semiconservative Replication of Double-Stranded DNA

In the semiconservative mode of replication, each parental DNA strand serves as a template for one new or daughter strand, and as each new strand is formed, it is hydrogen-bonded to the parental template (Figure 8-1). Thus, as replication proceeds, the parental double helix unwinds and then rewinds again into two new double helices, each of which contains one of the original parental strands and one newly formed daughter strand.

A classic experiment by Meselson and Stahl provided evidence that DNA replicates semiconservatively. They grew *Escherichia coli* for many generations in a medium containing ^{15}N as the sole source of nitrogen and then transferred the cells to a medium containing the less dense isotope ^{14}N. DNA was isolated before the shift to the low-density medium and sedimented to equilibrium in CsCl (see Chapter 2). The density of the DNA from cells grown in ^{15}N medium was about 1.722 g/cc, compared with 1.708 for cells grown only in ^{14}N medium. After one generation of growth in the ^{14}N medium, all of the DNA isolated from the cells had a density of (1.708 + 1.722)/2 = 1.715, exactly the density expected if one single strand contained ^{15}N and the complementary strand contained ^{14}N. Denaturation of this "hybrid" DNA yielded two components having the density of single-stranded $[^{15}N]$DNA and $[^{14}N]$DNA. In a second round of replication in ^{14}N medium, the $[^{14}N^{15}N]$DNA was converted to equal amounts of two species, $[^{14}N^{15}N]$DNA and $[^{14}N^{14}N]$DNA, as expected for semiconservative replication.

Unwinding a double helix during semiconservative replication presents a mechanical problem. Either the two daughter branches at the Y-fork must revolve around one another, or the unreplicated portion must rotate. If the molecule were fully extended in solution, there would be no problem because rotation

Figure 8-1. The replication of DNA according to the mechanism proposed by Watson and Crick. The two replicas consist of one parental strand (black) plus one daughter strand (orange). Each base in a daughter strand is selected by the requirement that it form a base pair with the parental base.

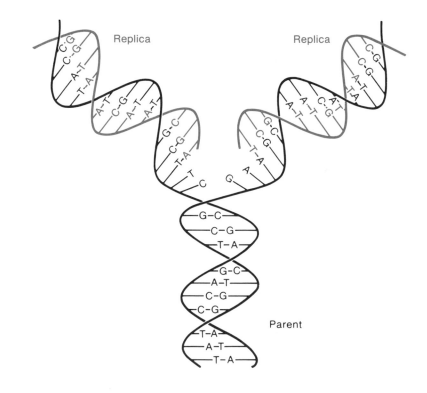

of the unreplicated portion would occur readily. Because the chromosome in *E. coli* is 600 times longer than the cell that contains it (so it must be repeatedly folded as described in Chapter 4), such rotation is unlikely.

In bacteria, most DNA molecules, both bacterial and phage, replicate as circular structures. This introduces a geometric problem that is even more severe than that just described.

Replication of Circular DNA Molecules

The first direct evidence that *E. coli* DNA replicates as a circle came from an autoradiographic experiment by Cairns (earlier genetic-mapping experiments described in Chapter 14 had suggested that the chromosome is circular). Cells were grown in a medium containing [³H]thymine so all DNA synthesized would be radioactive. The DNA was gently isolated to avoid fragmenting it and placed on photographic film. Each ³H-decay exposes one grain in the film. After several months, enough grains were exposed to visualize the DNA by observing the pattern of black grains on the film under a microscope. One of the classic autoradiograms from this experiment is shown in Figure 8-2. Electron micrographs of replicating circular molecules of plasmids, phages, and viruses have also been obtained (Figure 8-3).

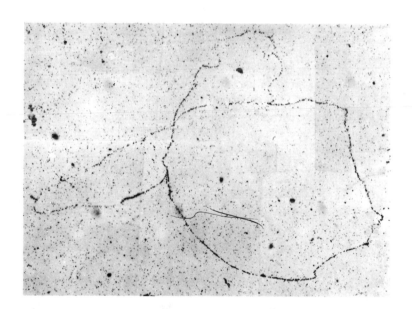

Figure 8-2. Autoradiogram of the intact replicating chromosome of an *E. coli* bacterium that has been allowed to incorporate [³H]thymine into its DNA for slightly less than two generations. The continuous lines of dark grains were produced by electrons emitted during a 2-month storage period by decaying ³H atoms in the DNA molecule. (From J. Cairns. Cold Spring Harbor Symp. Quant. Biol., 28:44, 1963.)

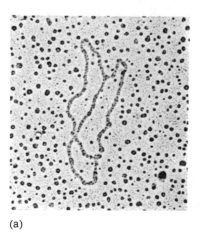

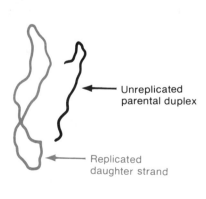

Unreplicated parental duplex

Replicated daughter strand

Figure 8-3. θ-replication. (a) Electron micrographs of a ColE1 DNA molecule (molecular weight = 4.2 × 10⁶) replicating by the θ mode. (b) Interpretive drawing showing parental and daughter segments. (Courtesy of Donald Helinski.)

(a)   (b)

Figure 8-4. Drawing showing that the unwinding motion (curved arrows) of the daughter branches of a replicating circle lacking positions at which free rotation can occur causes overwinding of the unreplicated portion.

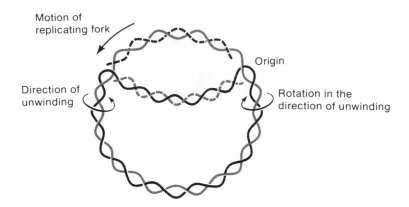

A replicating circle is schematically like the Greek letter θ, so this mode of replication is usually called **θ-replication**.

The unwinding problem in θ-replication is formidable because lack of a free end makes rotation of the unreplicated portion impossible. As replication of the two daughter strands proceeds along the helix, in the absence of some kind of swiveling, the nongrowing ends of the daughter strands would cause the entire unreplicated portion of the molecule to become overwound (Figure 8-4). This in turn would cause positive supercoiling (see Chapter 2) of the unreplicated portion. This supercoiling obviously cannot increase indefinitely because if it were to do so, the unreplicated portion would become coiled so tightly that no further advance of the replication fork would be possible. As discussed in Chapter 2, most naturally occurring circular DNA molecules are negatively supercoiled. Thus, initially the overwinding motion is no problem because it can be taken up by the underwinding already present in the negative supercoil. After about 5% of the circle is replicated, however, the negative superhelicity is used up, and the topological problem arises.

All organisms contain one or more enzymes called **topoisomerases**. These enzymes can produce a variety of topological changes in DNA; the most common are production of negative superhelicity and the removal of superhelicity. In *E. coli*, the enzyme **DNA gyrase**, which is able to produce negative superhelicity, is responsible for removing the positive superhelicity generated during replication. That is, positive superhelicity is removed by gyrase introducing negative twists by binding ahead of the advancing replication fork.

Termination also poses a topological problem. When double-stranded circular DNA replicates semiconservatively, the result is a pair of circles that are linked as in a chain. Such a structure is called a **catenane**. Catenated molecules have been observed in numerous systems, and evidence is accumulating to indicate that they result from replication. Apparently they are a precursor to the separated circles that ultimately result. Figure 8-5 shows that DNA gyrase is capable of decatenating two circles, and it is believed that a topoisomerase is the enzyme

Figure 8-5. The processes of catenation and decatenation, catalyzed by DNA gyrase.

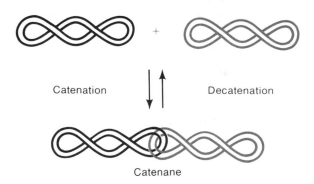

responsible for separating the two daughter molecules. This topoisomerase is probably not DNA gyrase, but a related protein called topoisomerase IV. Support for this hypothesis comes from a study of replication of the *E. coli* nucleoid (see Chapter 4) in a bacterial mutant that makes a temperature-sensitive topoisomerase IV. Nucleoids isolated from these cells grown at temperatures at which topoisomerase IV was active (permissive temperature) appeared in a microscope as a spherical object. When the cells were instead grown at a temperature at which topoisomerase IV was nearly inactive, paired spheres accumulated, whose size and appearance was consistent with the presence of two completely synthesized *E. coli* chromosomes.

ENZYMOLOGY OF DNA REPLICATION

The enzymatic synthesis of DNA is a complex process, primarily because of the need for high fidelity in copying the base sequence and for physical separation of the parental strands. About 20 proteins are required for DNA replication. The enzymes that form the sugar-phosphate bond (the phosphodiester bond) between adjacent nucleotides in a nucleic acid chain are called **DNA polymerases**.

Polymerization

Three principal requirements must be fulfilled before DNA polymerases can catalyze synthesis of DNA.

1. The 5' triphosphates of the four deoxyribonucleosides, deoxyadenosine, deoxyguanosine, deoxycytidine, and thymidine, are required. Synthesis does not occur with the 3'-triphosphates or 5'-diphosphates or if one of the four 5'-triphosphates is lacking.
2. Single-stranded template DNA is required.
3. A short nucleic acid primer, hydrogen-bonded to a template DNA strand, is required. The primer may be very short and either DNA or RNA (Figure 8-6). *None of the known DNA polymerases are able to initiate a DNA synthesis without a primer.* Thus, an oligonucleotide primer with a free 3'-OH group is absolutely necessary for initiation of replication. (In this way, DNA polymerases differ from RNA polymerases, which can initiate RNA synthesis without a primer).

The reaction catalyzed by a DNA polymerase is the formation of a phosphodiester bond between the free 3'-OH group of the primer and the innermost phosphorus atom of the nucleoside triphosphate being incorporated at the new primer terminus (see Figure 8-6). Thus, DNA synthesis always occurs by the elongation of primer chains, *in the 5' to 3' direction*. Recognition of the appropriate incoming nucleoside triphosphate during growth of the primer chain depends on base-pairing with the opposite nucleotide in the template chain. A DNA polymerase usually catalyzes the polymerization reaction, incorporating the new nucleotide at the primer terminus only when the correct base pair is present within the active site; in this reaction, the two terminal phosphate groups of the nucleoside triphosphate are released as a pyrophosphate (PP_i) unit.

Three DNA polymerases have been purified from *E. coli*. DNA **polymerase III (Pol III)** is the major replication enzyme. DNA **polymerase I (Pol I)** plays a secondary role: it is responsible for removing RNA primers and replacing them with DNA. It is also required for repairing certain types of DNA damage (see Chapter 9). DNA **polymerase II** is a minor enzyme that plays a role in DNA repair.

Error Correction

Pol I and Pol III both have the job of selecting a deoxynucleoside 5'-triphosphate that can hydrogen-bond to the template strand and of carrying out the polymerization reaction. Because of the need for faithful replication of a DNA base sequence, selection of the correct base must be extremely accurate. Errors do occur on occasion, however, and systems have evolved for correcting these errors. A major error-correcting process is carried out by the polymerases themselves. Pol I and Pol III of *E. coli* both have an exonuclease activity that acts from the 3' terminus (a 3' to 5' exonuclease activity). This **proofreading** or **editing function** excises a nucleotide from the 3'-OH end of the growing chain if it is not correctly base-paired to the corresponding nucleotide in the template chain.

The editing function of the polymerases is exceptionally efficient, but the integrity of the base sequence of DNA is so important that a second system exists for correcting the occasional error missed by the editing function. This correction system is called **mismatch repair**. In mismatch repair, a pair of non–hydrogen-bonded bases that is not at the 3' end of a growing strand is recognized as incorrect, and a polynucleotide segment is excised from one strand by an endonuclease, thereby removing one member of the unmatched pair. The resulting gap is filled in by Pol I.

To correct replication errors, the mismatch repair system must be able to distinguish the correct base in the parental strand from the incorrect base in the daughter strand. If it were unable to do this, the correct base might sometimes be replaced by the complement of the incorrect base, thereby producing a mutation. The key to understanding the correction process came from the discovery of rare methylated adenines in DNA and from studies with *dam⁻* (methylation-defective) mutants of *E. coli*. The *dam* gene product methylates adenines located in the sequence GATC. For any genetic locus, the mutation frequency in a *dam⁻* mutant is much higher than in a *dam⁺* bacterium. This indicates that incorrectly incorporated bases are less frequently corrected in a *dam⁻*

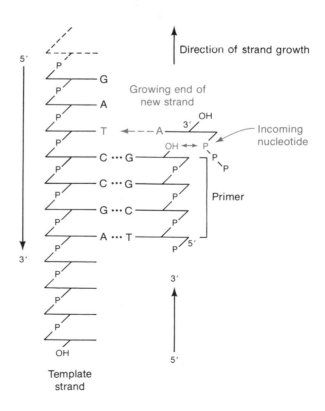

Figure 8-6. Addition of nucleotides to the 3'-OH terminus of a primer. The recognition step is shown as the formation of hydrogen bonds between the orange A and the orange T. The chemical reaction is between the orange 3'-OH group and the orange phosphate of the triphosphate.

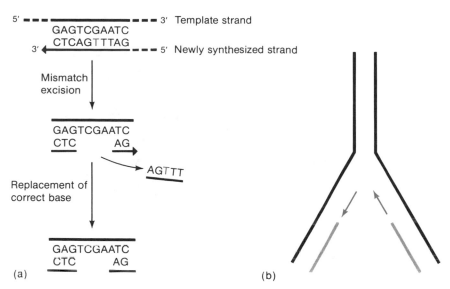

5' ▬ ▬ ▬ ▬ ▬ ▬ 3' Template strand
    GAGTCGAATC
3' ◄——— CTCAGTTTAG ▬ ▬ ▬ 5' Newly synthesized strand

Mismatch
excision

GAGTCGAATC
CTC   AG

——► AGTTT

Replacement of
correct base

GAGTCGAATC
CTC   AG

(a)   (b)

Figure 8-7. Mismatch repair. (a) Excision of a short segment of a newly synthesized strand and repair synthesis. (b) Methylated bases in the template strand direct the excision mechanism to the newly synthesized strand containing the incorrect nucleotide. The regions in which methylation is complete are light orange; the regions in which methylation may not be complete are orange.

mutant than in the wild-type. The reason for this is that the mismatch repair system recognizes parental (fully methylated) and daughter strand (undermethylated) and *preferentially excises nucleotides from the daughter strand* (Figure 8-7) The daughter strand is always the undermethylated strand because methylation lags somewhat behind the moving replication fork; the parental strand is fully methylated at rare GATC sites, having been methylated in the previous round of replication.

DISCONTINUOUS REPLICATION

During bidirectional DNA replication, *one of the daughter strands is made in short fragments, which are then joined together*. All known DNA polymerases can add nucleotides only to a 3'-OH group. If both daughter strands grew in the same direction (for example, both clockwise), only one of these strands would have a free 3'-OH group; the other strand would have a free 5' end because the two strands of DNA are antiparallel. The solution to this geometric problem is that both strands grow in the 5' to 3' direction at the growing fork. Thus, one strand of the DNA is made as short fragments (called Okazaki fragments; Figure 8-8). This results in a single-stranded region of the parental strand on one side of the replication fork because the discontinuous strand is initiated only periodically. In fact, the 3'-OH terminus of the continuously replicating strand is always ahead of the discontinuous strand. This had led to the use of the convenient terms **leading**

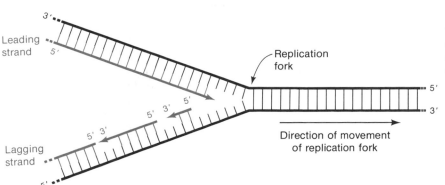

Leading
strand

Lagging
strand

Replication
fork

Direction of movement
of replication fork

Figure 8-8. Short fragments in the replication fork. For each tract of base pairs, the lagging strand is synthesized later than the leading strand.

Figure 8-9. (a) θ-replication of phage λ DNA. The arrows show the two replicating forks. The segment between each pair of thick lines at the arrows is single stranded; note that it appears thinner and lighter. (b) An interpretive drawing. (Courtesy of Manuel Valenzuela.)

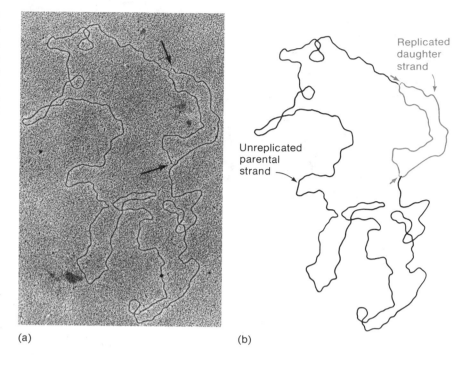

Replicated daughter strand

Unreplicated parental strand

(a)   (b)

strand and **lagging strand** for the continuously and discontinuously replicating strands (see Figure 8-8). Such regions have been seen in high-resolution electron micrographs of replicating DNA molecules (Figure 8-9).

How is the lagging strand made? To answer this, we must first consider (1) initiation and priming of replication, (2) how fragments are attached to one another, and (3) the role of the DNA Pol I in replication of *E. coli* DNA.

Initiation of DNA Replication

Recall that DNA Pol III cannot provide the first nucleotide to initiate chain growth but requires a primer. Thus, another enzyme must synthesize an oligonucleotide primer, which can be extended by Pol III.

In *E. coli*, initiation of synthesis of the leading strand and of the precursor fragments of the lagging strand occurs by somewhat different mechanisms, possibly because initiation of leading-strand synthesis begins with a double-stranded DNA template, whereas in initiation of the lagging strand, fragments begin with a single-stranded DNA template (that is, the strand to be copied is already unwound). In both cases, the primer is a short RNA oligonucleotide. The size of the RNA primer varies considerably, depending on whether the lagging or leading strand is being primed and on the particular organism. This RNA primer is synthesized by copying a complementary base sequence from one DNA strand. It differs from a typical RNA molecule in that after its synthesis, *the primer remains hydrogen-bonded to the DNA template*. In bacteria, two different enzymes synthesize primer RNA molecules. RNA polymerase, which is the same enzyme used for synthesis of most RNA molecules, primes the leading strand in some phage systems, once for each round of replication. **Primase**, the product of the *dnaG* gene, primes the precursor fragments of the lagging strand and may also prime leading-strand synthesis. In all cases, the growing end of the RNA primer is a 3'-OH group to which Pol III can easily add the first deoxynucleotide; the 5' end of the RNA chain, which remains free and has a 5'triphosphate group. Thus, a precursor fragment has the following structure while it is being synthesized:

Ligation of Okazaki Fragments

The fragments in the lagging strand (called Okazaki fragments) are ultimately joined to yield a continuous strand. This strand contains no ribonucleotides, so assembly of the lagging strand requires removal of the primer ribonucleotides, replacement with deoxynucleotides, and then joining of the DNA fragments. In *E. coli*, the first two processes are accomplished by DNA Pol I, and joining is catalyzed by the enzyme DNA ligase, which can link adjacent 3'-OH and 5'-P groups at a nick (Figure 8-10). Pol III extends the growing strand until the RNA nucleotide of the primer of the previously synthesized precursor fragment is reached. It then dissociates from the 3'-OH terminus, leaving a nick between the 3'-OH of the DNA and 5'-triphosphate of the RNA primer. *E. coli* DNA ligase cannot seal the nick because a triphosphate is present (it can only link a 3'-OH and a 5'-monophosphate and it cannot join RNA to DNA). Pol I has an exonuclease activity, however, that can remove a nucleotide from the 5' end of a base-paired fragment. It is effective with either DNA or RNA fragments. This activity is called its **5' to 3' exonuclease** activity. Thus, Pol I acts at the 3'-OH terminus left by Pol III and moves in the 5' to 3' direction, removing ribonucleotides and adding deoxynucleotides to the 3' end. When the RNA primer has been completely removed (probably with some DNA as well), DNA ligase joins the 3'-OH group to the terminal 5'-phosphate of the precursor fragment. Each Okazaki fragment is assimilated into the lagging strand by this sequence of events.

Advance of the Replication Fork and the Unwinding of the Helix

DNA replication also requires a means of unwinding the parental double helix. Pol III is unable to unwind a helix. (Pol I can, but it is the only known polymerase that can do so.) Helix-unwinding is accomplished by enzymes called **helicases**. The helicase active in *E. coli* DNA replication is the DnaB protein.

In *E. coli*, the Pol III enzyme synthesizing the leading strand is not immediately behind the advancing DnaB protein (Figure 8-11). Thus, behind the DnaB

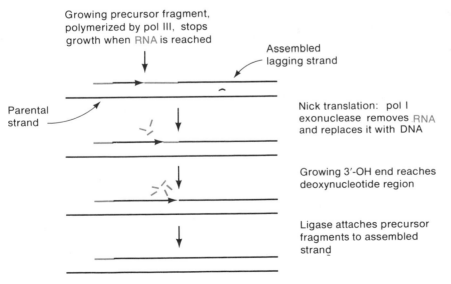

Figure 8-10. Sequence of events in assembly of precursor fragments. RNA is indicated in orange. The replication fork (not shown) is at the left.

Figure 8-11. The unwinding events in a replication fork.

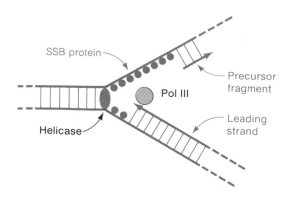

protein are two single-stranded regions: a large, single-stranded region on the lagging strand and a smaller, single-stranded region just ahead of the leading strand. To prevent the single-stranded regions from reannealing or from forming intrastrand hydrogen bonds, the single-stranded DNA is coated with a single-stranded DNA binding protein (**SSB protein**). As Pol III advances, it must displace the SSB protein in order that base-pairing of the nucleotide being added can occur.

BIDIRECTIONAL REPLICATION

Somewhat after initiation of synthesis of the leading strand at the replication origin, the first precursor fragment is synthesized. This is shown in part I of Figure 8-12, in which the overall direction in which the replicating fork moves is counterclockwise. In the discussion of lagging-strand replication just presented, it was noted that synthesis of each precursor fragment is terminated when the growing end reaches the primer of the previously synthesized fragment. In the case of the first precursor fragment, however, there is no earlier-made fragment. Thus, the precursor fragment becomes a leading strand for a second replication fork, moving clockwise, as shown in the figure. Clockwise replication requires the synthesis of Okazaki fragments in the second replication fork, but this can be achieved by the standard mechanism. The result of these events is that the DNA molecule will have two replication forks moving in opposite directions around the circle. This is called **bidirectional replication**, and it is an almost universal phenomenon. In a few systems, replication is unidirectional, but bidirectional replication is advantageous in that, com-

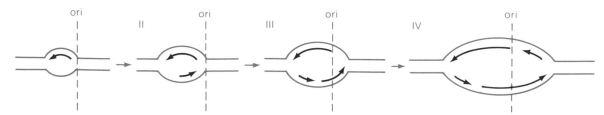

Figure 8-12. The formation of a bidirectionally enlarging replication bubble. (I) The leftward-leading strand starts at ori. (II) The leading strand has progressed far enough that the first rightward precursor fragment begins. (III) The leftward-leading strand has progressed far enough that the second rightward precursor fragment has begun. The first rightward precursor fragment has passed ori and has become the rightward-leading strand. (IV) The rightward-leading strand has moved far enough that the first leftward precursor fragment has begun. There are now two complete replication forks.

pared with unidirectional replication, it halves the time required to replicate a circle.

In either mode of replication the replication fork or forks must start in a specific segment of the chromosome called the **origin** of DNA replication. If DNA replication began in random regions of the chromosome daughter cells with only partial chromosomes would result. The origin of chromosomal replication for the *E. coli* chromosome is called **oriC**. *oriC* is a small (260 bp) segment of DNA found at 84 min on the *E. coli* genetic map that lacks genes which encode proteins. Chromosome initiation at *oriC* requires several adjacent DNA sequences that specifically bind the DnaA protein. The bound DnaA protein opens up the DNA and allows the DnaB helicase to begin the two replication forks. The two forks then proceed around the circular chromosome until they encounter two termination (**ter**) sites located in the DNA halfway around the chromosome from *oriC*. A protein (called Tus) binds to the *ter* sites and stops the DnaB helicase, resulting in termination of DNA replication. The completed chromosomes are then partitioned into two daughter cells during cell division.

REGULATION OF BACTERIAL CHROMOSOME REPLICATION

In contrast to eukaryotic cells, which have a cell cycle that involves a temporal series of coordinated events before cell division, in *E. coli* the time required for a single cell to double in size and divide depends on the rate of production of useful energy and of precursor molecules. If glucose is provided as the sole carbon source and all other nutrients are simple inorganic compounds (that is, if the cells are grown in a glucose-minimal medium), it takes about 45 minutes at 37°C for a cell to double. If succinate is the sole carbon source, ATP is synthesized more slowly, and the doubling time is about 70 minutes. With even poorer carbon sources, the doubling time may be increased to 10 hours. In a glucose medium, the time required to replicate the bacterial DNA is 40 minutes; that is, initiation is delayed by a few minutes after completion of a round of replication. In succinate medium, the replication time is still 40 minutes, so the time between successive rounds of replication is 30 minutes. In a medium in which the doubling time is 10 hours, the replication time is increased by only a few minutes. These observations indicate that the rate of DNA synthesis and termination is constant, so regulation of the rate of DNA replication must be controlled by the rate of initiation.

When *E. coli* is grown in a nutrient broth, however, the doubling time may be as short as 20 minutes, yet even under these conditions, replication still takes 40 minutes (Figure 8-13). This apparent paradox is explained by the phenomenon of **premature initiation** (also called **dichotomous replication**): In rich media, initiation of a second round of DNA replication begins before replication is complete. Figure 8-13c shows how a second initiation event at the time replication is half complete allows segregation of two daughter molecules to occur at twice the normal rate. Thus, the rate of initiation is carefully controlled relative to the growth rate. A major factor in control of this process seems to be the level of DnaA protein.

A consequence of the ordered and dichotomous replication of the chromosome is that there are more DNA copies per cell of those genes close to *oriC* than of genes far away from *oriC* (especially those close to *ter*). Since more DNA copies should give rise to more gene product, we might expect that genes whose products are needed in large quantities would be located close to *oriC*. Indeed,

Figure 8-13. Stages of replicating DNA at various times in the *E. coli* life cycle when the doubling time is 22, 40, and 60 minutes. Colors alternate with round of replication.

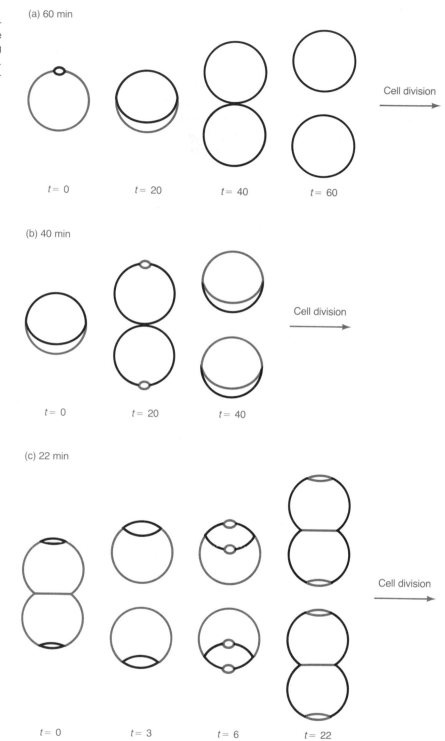

(a) 60 min

$t= 0$ $t= 20$ $t= 40$ $t= 60$

Cell division

(b) 40 min

$t= 0$ $t= 20$ $t= 40$

Cell division

(c) 22 min

$t= 0$ $t= 3$ $t= 6$ $t= 22$

Cell division

the seven genes that encode the ribosomal RNAs (cell components needed in very large quantities) are all located close to *oriC*.

ROLLING CIRCLE REPLICATION

There are numerous instances in which, in the course of replication, a circular DNA molecule gives rise to linear daughter molecules in which the base sequence

XYZABC	XYZABC	XYZABC	XYZABC

Figure 8-14. A concatemer consisting of the repeating unit ABC . . . XYZ. Note that the definition of concatemer does not make any requirements about the terminal sequences.

of the circular DNA is repeated many times, forming a **concatemer** (Figure 8-14). These concatemers are often an essential intermediate in phage production. Likewise, in bacterial mating, a linear DNA molecule is transferred by a replicative process from a donor cell to a recipient cell, as described in Chapter 14. Both phenomena are consequences of initiation of a replication mode known as **rolling circle replication**.

Consider a duplex circle in which a nick is made having 3'-OH and 5'-P termini (Figure 8-15). Under the influence of a helicase and SSB protein, a replication fork can be generated. Synthesis of a primer is unnecessary because of the 3'-OH group, so leading-strand synthesis can proceed by elongation from this terminus. At the same time, the parental template for lagging-strand synthesis is displaced. The polymerase used for this synthesis is usually Pol III (although some phages use other enzymes). The displaced parental strand is replicated in the usual way by lagging-strand synthesis. The result of this mode of replication is a circle with a linear branch. There are four significant features of rolling circle replication:

1. The leading strand is covalently linked to the parental template for the lagging strand.
2. Before precursor fragment synthesis begins, the linear branch has a free 5'-P terminus.
3. Rolling circle replication continues unabated, generating a long, linear concatemer.
4. The circular template for leading-strand synthesis never leaves the circular part of the molecule.

A variant of the rolling circle mode, called **looped rolling circle replication**, generates a single-stranded circle from a double-stranded circular template. For *E. coli* phage φX174, this occurs in the following way (Figure 8-16). A phage protein (the A protein) nicks the viral-strand replication origin and becomes covalently linked to the newly formed 5'-P terminus. Using the Rep and SSB proteins and Pol III, chain growth occurs from the 3'-OH group, displacing the broken parental strand, called the (+) strand. This strand becomes coated with SSB protein and does not serve as a template for synthesis of precursor fragments. Synthesis continues until the origin is reached. At this point, the A protein binds to the 3'-OH group of the (+) strand and joins the 3'-OH and 5'-P groups of the (+) strand, dissociates, and attaches to the newly synthesized (+) strand. This process can continue indefinitely, generating numerous circular (+) strands. Note that in looped rolling circle replication, the displaced strand never exceeds the length of the circle, in contrast with ordinary rolling circle replication. This mode of DNA replication is common during certain stages in the life cycles of phages and plasmids (see Chapter 11).

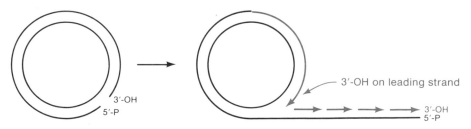

Figure 8-15. Rolling circle or σ-replication. Newly synthesized DNA is shown in orange.

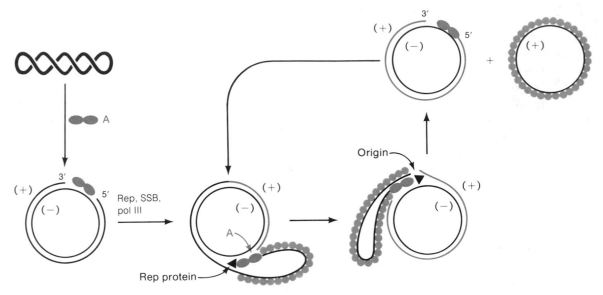

Figure 8-16. A diagram of looped rolling-circle replication of phage φX174. The gene A protein nicks a supercoil and binds to the 5' terminus of a strand, known as the (+) strand, whose base sequence is the same as that of the DNA in the phage particle. Rolling circle replication ensues to generate a daughter strand (orange) and a displaced (+) single strand that is coated with SSB protein and still covalently linked to the A protein. When the entire (+) strand is displaced, it is cleaved from the daughter (+) strand and circularized by the joining activity of the A protein. The cycle is ready to begin anew. Note that the (–) strand is never cleaved.

KEY TERMS

3' → 5' exonuclease	methylation
5' → 3' exonuclease	mismatch repair
catenane	Okazaki fragments
concatemer	premature initiation
discontinuous replication	primase
leading strand	proofreading
lagging strand	θ-replication
DNA gyrase	rolling circle replication
DNA polymerase	semiconservative replication
helicase	topoisomerase

QUESTIONS AND PROBLEMS

1. In semiconservative replication, what fraction of the DNA consists of one of the original parental strands and one daughter strand after one, two, and three rounds of replication?

2. Will a ^{15}N-labeled circular DNA replicating in ^{14}N medium using the rolling circle mode ever achieve the density of $^{14}N^{14}N$ DNA?

3. What are three enzymatic activities of DNA Pol I?

4. In what direction does a DNA polymerase move along a template strand?

5. What are the precursors for DNA synthesis?

6. DNA polymerization occurs by addition of a nucleotide to what chemical group?

7. What are the roles of the 5'→3' and 3'→5' exonuclease activities of DNA Pol I?

8. What are the roles of DNA Pol I and III in DNA replication?

9. What is the chemical difference between the groups joined by a DNA polymerase and by DNA ligase?

10. What is the fundamental difference between the initiation of θ-replication and of rolling circle replication?

11. Does the chemistry of polymerization by RNA polymerases differ from that by DNA polymerases?

REFERENCES

Baker, T., and S. Wickner. 1992. Genetics and enzymology of DNA replication in *Escherichia coli*. *Ann. Rev. Genet.*, 26, 447.

*Cairns, J. 1966. The bacterial chromosome. *Scientific American*, January, p. 36.

DeLucia, P., and J. Cairns. 1969. Isolation of an *E. coli* strain with a mutation affecting DNA polymerase. *Nature*, 224, 1164.

Konrad, E. B., and I. R. Lehman. 1974. A conditional lethal mutant of *E. coli* defective in the 5'→3' exonuclease associated with DNA polymerase I. *Proc. Natl. Acad. Sci.*, 71, 2048.

Kornberg, A. 1960. Biological synthesis of deoxyribonucleic acid. *Science*, 131, 1503.

*Kornberg, A., and T. Baker. 1992. *DNA Replication, Second Edition*. W. H. Freeman.

Lehman, I. R., and D. Uyemura. 1976. DNA polymerase I: essential replication enzyme. *Science*, 193, 963.

Meselson, M., and F. W. Stahl. 1957. The replication of DNA in *E. coli*. *Proc. Natl. Acad. Sci.*, 44, 671.

Okazaki, R. T., et al. 1968. Mechanism of DNA chain growth. I. Possible discontinuity and unusual secondary structure of newly synthesized chains. *Proc. Natl. Acad. Sci.*, 59, 598.

Prescott, D. M., and P. L. Keumpel. 1972. Bidirectional replication of the chromosome in *E. coli*. *Proc. Natl. Acad. Sci.*, 69, 2842.

Scheuermann, R. H., and H. Echols. 1984. A separate editing exonuclease for DNA replication: the ε subunit of *E. coli* DNA polymerase III holoenzyme. *Proc. Natl. Acad. Sci.*, 81, 7747.

Schmid, M., and J. Sawitzke. 1993. Multiple bacterial topoisomerases: specialization or redundancy? *Bioessays*, 15, 445.

*Schmid, M. 1988. Structure and function of the bacterial chromosome. *Trends Biochem. Sci.*, 18, 131.

*Smith-Keary, P. 1991. *Molecular Genetics: A Workbook*. Guilford.

Valenzuela, M., et al. 1976. Lack of a unique termination site in lambda DNA replication. *J. Mol. Biol.*, 102, 569.

Watson, J. D., and F. H. C. Crick. 1953. Genetic implications of the structure of desoxyribonucleic acid. *Nature*, 171, 964.

Yoshikawa, H., and N. Ogasawara. 1991. Structure and function of DnaA and the DnaA-box in eubacteria: evolutionary relationships of bacterial replication origins. *Mol. Microbiol.*, 5, 2589.

Zyskind, J., and D. Smith. 1986. The bacterial origin of replication, *oriC*. *Cell*, 46, 489.

*Resources for additional information.

DNA Damage and Repair

Maintenance of the base sequence of DNA from one generation to the next is one of the primary goals of all biological systems. Nevertheless, sequence alterations can arise in a variety of ways. For example, in Chapter 8 we saw that incorrect nucleotides are occasionally added during DNA replication. Two error correcting systems, proofreading and mismatch repair, serve to eliminate most of the misincorporated nucleotides. DNA, however, is also subject to environmental damage from chemicals and radiation. In this chapter, we describe mechanisms that repair this type of damage.

BIOLOGICAL INDICATIONS OF DAMAGE TO DNA

When bacteria are exposed to radiation or various chemicals, they lose the ability to form colonies. Similarly, phage lose plaque-forming ability. This loss of viability can be expressed graphically by plotting the fraction of the initial population that survives various exposures to the radiation or the chemicals versus some measure of exposure (Figure 9-1). The most detailed studies have been with radiation, in which the exposure is simply the total amount of radiation, or the radiation dose. Such a dose-response graph is called a **survival curve**. For bacteria, such curves are obtained in the following way. Samples are removed at intervals from a population of bacteria that is being irradiated, for example, with ultraviolet light or x-rays. The samples are plated, normally on a nutrient agar, and the colonies that form are counted. The fraction of the initial number of cells that remain able to produce colonies is plotted as a function of the dose. For phages, the irradiated phage are plated on a lawn of bacteria so plaques can form.

Analysis of survival curves has provided considerable insight into the various mechanisms of radiation damage and also provided the first suggestion that environmental damage to DNA is often repaired. A simple mathematical theory, called **target theory**, has been useful in analyzing survival curves.

Target Theory

Survival curves for various types of populations of N identical organisms exposed to a dose D of radiation (or some other external agent) that causes damage of some kind can be represented by mathematical equations. The simplest case assumes that each organism possesses only one sensitive site (a target) that, if

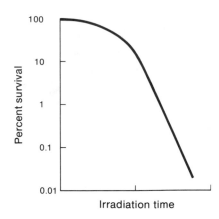

Figure 9-1. A typical ultraviolet-light survival curve for a bacterium. Initially the curve is fairly flat because initial damage does not cause killing. Note that the y-axis is logarithmic.

damaged or "hit" by a photon (a "particle" of light), inactivates the organism. The number dN damaged by a dose dD is proportional to the number N that existed before receiving that dose; that is, $-dN/dD = kN$, in which the constant k is a measure of the effectiveness of the dose and is proportional to the fraction of incident photons that causes an inactivating hit—in other words, the probability that a single photon can cause such a hit. Integrating this equation from $N = N_0$ at $D = 0$ yields:

$$(1)\ N = N_0\, e^{-kD}$$

The surviving fraction $S = N/N_0$ is:

$$(2)\ S = N/N_0 = (e^{-kD})^n$$

so that a plot of $\ln S$ versus D gives a straight line with a slope of $-k$. Curves of this type are called **exponential** or **single-hit curves**. An example of this type of curve is typically observed when phages are irradiated with ionizing radiation such as x-rays (Figure 9-2).

Now consider a population of different organisms in which each organism contains n sites, each of which must be hit (damaged) if the organism is to be inactivated. In this case, inactivation requires at least n hits ("at least" because statistically some sites will be hit twice, and we assume that two hits in one site are not more effective than one hit on that site). The probability of one unit being hit by a dose D is $1 - e^{-kD}$, so the probability P_n that all n units become inactivated is:

$$(1)\ P_n = (1 - e^{-kD})^n$$

Figure 9-2. Survival curves for two X-irradiated populations, one of the bacterium *E. coli* and the other of the phage T4.

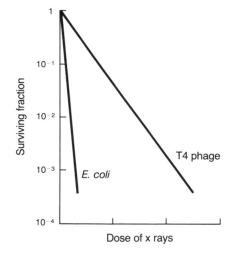

The surviving fraction S of the population is $1 - P_n$ or:

$$(2)\ S = 1 - (1 - e^{-kD})^n$$

Expansion of this equation yields:

$$(3)\ S = 1 - (1 - ne^{-kD} + \ldots + e^{-nkD})$$

At large values of D, the higher order terms become negligible compared with ne^{-kD}, so that at high dose, $S = ne^{-kD}$, or:

$$(4)\ \ln S = \ln n - kD$$

A plot of Equation 3 for $k = 1$ and various values of n shows that for small values of D, $\ln S$ changes slowly (Figure 9-3). At large D, Equation 4 predominates, and the curve becomes linear. Extrapolation of the linear part (high-dose region) of a curve yields $S = n$ at the y intercept. Thus, if experimental data for sufficiently large values of D can be obtained, the number of targets n can be estimated.

As might be expected, straight lines are observed for x-ray inactivation of phages. The results for bacteria are not always straightforward for two reasons. First, populations of bacteria include cells at different stages of DNA replication—that is, cells that have just divided have only one chromosome, but cells that are almost ready to divide have nearly two complete copies of the chromosome. Second, many types of radiation damage are readily repaired, so small numbers of potentially lethal hits may not be detected. Ionizing radiation causes three types of damage to DNA: single-strand breaks, double-strand breaks, and alterations of bases. (1) Single-strand breaks are for the most part resealed by DNA ligase and do not contribute to lethality. (2) Double-strand breaks are often lethal because the free ends initiate degradation of the DNA by nucleases. (3) Damage to bases, which is an oxidative process requiring molecular O_2, is often lethal, probably because the damaged bases constitute a replication block.

It was pointed out in the derivation of Equation 1 that the constant k is in some way a measure of the probability of a hit. The relation between k and this probability can be seen by looking at the effect of radiation in a slightly different way. Radiation such as x-rays produces ionizations in matter. The number of ionizations per unit volume is proportional to dose, and a fixed fraction of the ionizations produce a lethal hit. Thus, if V is the volume of the sensitive target molecule (V is called the **target volume**), the average number of hits within the

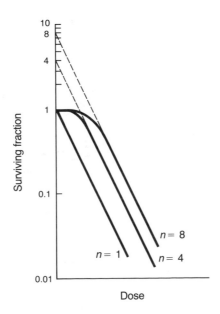

Figure 9-3. Survival curves for various values of n, showing that at high doses each curve becomes linear and that extrapolation to the y-axis yields n as the intercept.

target volume is cVD, in which c is a proportionality constant relating the number of ionizations and the number of hits. If the hits are random and independent, the probability $P(n)$ that n hits occur within the volume is given by the Poisson distribution, or:

$$P(n) = e^{-cVD}(cVD)^n n!$$

For a single-hit mechanism (that is, survivors must have zero hits), the surviving fraction $S = 1 - P(n)$ is:

$$S = e^{-cVD}$$

Comparison with Equation 2 shows that $k = cV$, or k is proportional to the volume of the target. This is not hard to understand: If one DNA molecule A has twice the number of nucleotides as a second DNA molecule B, the dose required to damage a nucleotide pair (for example, either a double-strand break or base damage) in A is half that required to damage one in B. That is, A is twice as sensitive as B. This phenomenon can easily be seen by examining curves for x-ray inactivation of phages having DNA molecules of different sizes, as shown in Figure 9-4. Phage T5, which has the largest DNA molecule of the three, is most sensitive, and T7, which is the smallest, is the least sensitive. Furthermore, the ratio of the k values is simply the ratio of the DNA molecular weights.

Ultraviolet Radiation

Ultraviolet radiation (UV) also causes inactivation (killing) of bacteria and phages. Nucleic acids and proteins absorb light in nearly the same range of wavelengths: 260 nm is the absorption maximum for nucleic acids, and 280 nm is the maximum for proteins. Analysis of UV survival curves for a variety of bacteria and phages, however, makes it clear that the target molecule is DNA. The experiment consists of irradiating several identical phage samples with different wavelengths of UV. The survival curves all show similar kinetics, but the slopes depend on the wavelength used. Such experiments show that the most effective wavelength is 260 nm, and that 280 nm radiation is quite ineffective. In fact, *the most effective wavelength for killing and mutagenesis by UV irradiation matches the absorption spectrum of DNA*, which suggests that DNA, not protein, is the target molecule. (The absorption spectrum of RNA is quite similar to that of DNA, but because of the large number of RNA molecules in cells and because of the similarity in the action spectra of bacteria and phages that have no RNA, the possibility of an RNA target was never seriously considered.)

Chemical analyses of UV-irradiated bacteria and phages as well as of irradiated purified DNA have shown that the major photoproduct is an intrastrand dimer formed by two adjacent pyrimidines as a result of UV. The most important

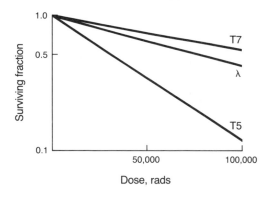

Figure 9-4. Loss of plaque-forming ability of the three phages irradiated with x-rays. The molecular weights of the phage DNA molecules are: T7 = 25 × 10⁶; λ = 31 × 10⁶; T5 = 76 × 10⁶.

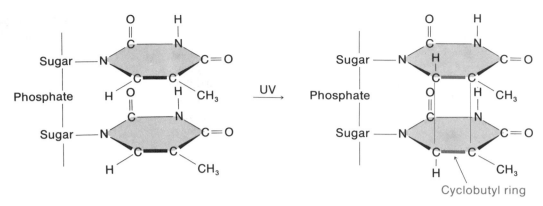

Figure 9-5.   Structure of a cyclobutylthymine dimer. Following UV irradiation, adjacent thymine residues in a DNA strand are joined by formation of the bond, shown in red. Although not drawn to scale, these bonds are considerably shorter than the spacing between the planes of adjacent thymines, so the double stranded structure becomes distorted. The shape of the thymine ring also changes as the C=UC double bond (heavy horizontal line in left panel) of each thymine is converted to a C=TC single bond (horizontal orange lines in right panel) in each cyclobutyl ring.

Figure 9-6.   Distortion of the DNA helix caused by two thymines moving closer together when joined in a dimer. The dimer is shown as two joined lines.

dimer is apparently the thymine dimer, shown in Figure 9-5. The significant effects of the presence of thymine dimers are the following: (1) The DNA helix becomes distorted as the thymines, which are in the same strand, are pulled toward one another (Figure 9-6), and (2) as a result of the distortion, hydrogen-bonding to adenines in the opposing strand, although possible (because the hydrogen-bonding groups are still present), is significantly weakened; this structural distortion blocks the growing replication fork.

Why do thymine dimers block DNA replication? When DNA polymerase III (Pol III) reaches a thymine dimer, the replication fork is temporarily stalled. A thymine dimer is still capable of forming hydrogen bonds with two adenines because the chemical change in dimerization does not alter the groups that engage in hydrogen bonding. The dimer introduces a distortion into the helix, however, and when an adenine is added to the growing chain, Pol III reacts to the distorted region as if a mispaired base had been added; the editing function (see Chapter 8) then removes the adenine. The cycle begins again—an adenine is added and then it is removed; the net result is that the polymerase is stalled at the site of the dimer. A cell in which DNA synthesis is permanently stalled cannot complete a round of replication, so a colony cannot form. Stalling is only temporary, however, for there are several different ways by which DNA synthesis can restart: (1) The dimer can be directly repaired by photoreactivation; (2) the dimer can be excised and the correct bases replaced by DNA polymerase I; (3) DNA synthesis can reinitiate on the other side of the dimer, and then the dimer can be repaired by recombination repair; and (4) induction of SOS repair can allow trans-dimer synthesis (error prone repair).

EVIDENCE FOR REPAIR SYSTEMS

From the beginning, analysis of UV survival curves was paradoxical. First, it was found that bacteria almost never yielded a single-hit curve. One might have concluded that damage was required in both DNA strands, so two hits are required. The survival curves, however, were not two-hit either. Furthermore, when survival curves were extended to very high doses, a truly linear portion was never observed: The curve continued to bend downward. The significance of this phenomenon was not understood at first, although we now know that it is a result of repair of UV damage. That is, if fairly efficient repair occurs, small doses are not effective in killing bacteria or phages. As the dose increases, however, the putative repair system may become saturated, or some types of damage may be nonreparable, or the repair system itself may become damaged. Any one of these effects would result in a survival curve that has an initial plateau region followed by continued downward curvature.

The most striking observations that led to the concept of DNA repair were two discoveries: (1) an increase in the surviving fraction of bacteria resulting from certain postirradiation treatments and (2) mutant strains that were more sensitive to radiation than wild-type strains. Furthermore, the survival of plaque-forming ability of irradiated phage was also affected by use of these bacterial mutants as a lawn.

Photoreactivation

Repair of UV damage was first recognized by a chance observation that the survival level of irradiated bacteria was increased when the cells were left in a window and exposed to sunlight before being allowed to form colonies. A quantitative analysis of this phenomenon, which is called **photoreactivation**, is shown in Figure 9-7. Panel (a) shows that exposure to visible light increases the survival to a level that is determined by the UV dose. In panel (b), the data of panel (a) are combined into two survival curves, with and without exposure to visible light. These curves clearly indicate that visible light eliminates some of the damage introduced by the UV light. Biochemical analysis has indicated that photoreactivation is an enzymatic reaction in which an enzyme, called **photolyase**, cleaves T-T dimers, restoring them to the monomeric state. The enzyme is inactive unless exposed to visible light (300–600 nm). A folic acid cofactor associated with the enzyme absorbs the light, then the enzyme uses the energy of the absorbed light to perform the cleavage. Mutants (Phr⁻) that cannot photoreactivate and lack photolyase have been isolated. They show no other phenotypic properties, suggesting that photoreactivation is the sole function of photolyase.

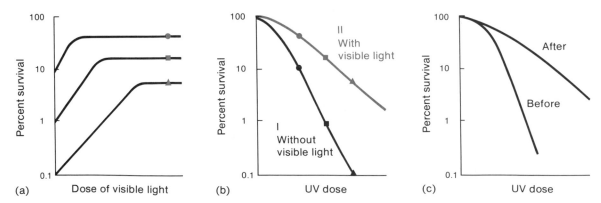

Figure 9-7. Types of repair. (a) Increase in survival of three different samples of ultraviolet light–irradiated bacteria as a function of the dose of visible light. This is called photoreactivation. (b) A pair of survival curves showing the effect of postirradiation with visible light. Curve I consists of the points taken from the y-axis of part (a), and curve II is a plot of the orange points taken from the plateaus in part (a). (c) Survival curves before and after incubation in buffer following ultraviolet-light irradiation. This is called liquid-holding recovery.

UV-irradiated phages are not photoreactivated by exposure to light because they do not possess photolyase. If examined in the correct way, however, irradiated phage can be photoreactivated. Phage are normally plated by mixing them with an excess number of bacteria and adding the mixture to soft agar. The same number of plaques result if the phage are preadsorbed to bacteria in a buffer that does not allow cell growth (to inhibit phage development temporarily), and then these phage-bacterium complexes (infective centers) are placed in soft agar. If irradiated phage are preadsorbed to bacteria, however, and a portion of the infective centers is exposed to visible light before plating, the infective centers that have been irradiated with visible light produce more plaques. This is a result of light activation of photolyase in the cells and subsequent photoreactivation of the UV-irradiated phage.

Dark Repair

Another response to damage that pointed to the existence of a second repair system was liquid-holding recovery. If UV-irradiated cells are held in a nonnutrient buffer for several hours before plating, the surviving fraction for a particular dose is increased (see Figure 9-8c). When first observed, it was suggested that delaying cell growth or perhaps merely DNA replication allowed additional time for some repair process to occur. The Phr⁻ mutant can undergo liquid-holding recovery. Furthermore, because liquid-holding recovery takes place without light from any source, it was hypothesized that *Escherichia coli* possesses two distinct repair processes: a light-dependent one (photoreactivation) and a light-independent one. Proof of the existence of two repair systems came from the isolation of an extraordinarily UV-sensitive *E. coli* mutant called Bs (Figure 9-8). This mutant does not show higher viability after UV irradiation when held in a buffer in the dark (that is, it does not exhibit dark repair), but it does exhibit normal photoreactivation. Liquid-holding recovery is now known to be a manifestation of a general phenomenon called **dark repair**, which is accomplished in several ways.

E. coli Bs exhibits loss of another repair phenomenon, called **host-cell reactivation** of phage. That is, the survival curve for UV-irradiated phage of certain types (for example, *E. coli* phage T1) is steeper if the phage are plated on *E. coli* Bs than on the wild-type *E. coli* strain B, indicating that strain B can repair some of the UV damage in the phage but Bs cannot.

Mutants of the Dark-Repair Systems

Insight into the mechanism of dark repair came from a study of UV-sensitive mutants. Although the UV-sensitive mutant Bs had been isolated, at the time this

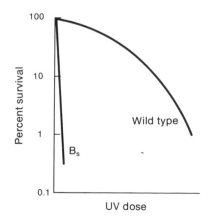

Figure 9-8. Survival curves showing the great sensitivity to ultraviolet light of the mutant *E. coli* Bs.

mutant was isolated, techniques had not yet been developed for carrying out genetic analysis with *E. coli* strain B. Thus, similar mutants in the more genetically accessible *E. coli* K12 strain were sought. The search for mutants made use of the phenomenon of host-cell reactivation described earlier. Several hundred thousand cells of a mutagenized sample of *E. coli* were spread on agar along with about 10^7 UV-killed T1 phage. The concentration of phage on the plate was such that before colonies became visible, each microcolony had been infected with several UV-killed phage. If the microcolony consisted of wild-type cells, a fraction of the UV-killed phage was reactivated, and these went on to infect and lyse the microcolony. Colonies of mutants unable to engage in host-cell reactivation were also infected, but they did not release progeny phage and hence survived to produce visible colonies. These colonies were streaked on agar to isolate the mutant cell from free phage and wild-type cells (see Chapter 4), bacterial cultures were prepared, and the radiosensitivity of the cultures was tested. Many of these colonies were exceedingly sensitive to UV.

Complementation tests showed that the mutations fell into three classes, which defined the genes *uvrA, uvrB,* and *uvrC.* Biochemical analysis showed that the *uvrA, uvrB,* and *uvrC* mutants are defective in an endonuclease required for excision of thymine dimers (as well as in repair of many types of chemical damage). Survival curves for some of these mutants are shown in Figure 9-9. Several other classes of mutants were also found to be UV sensitive.

In studies of genetic recombination in *E. coli* (described in Chapter 14), which were totally unrelated to the repair phenomenon, recombination-deficient mutants were isolated. These mutations mapped in three genes, designated *recA, recB,* and *recC.* When the phenotypes of the *rec* mutants were examined, it was discovered that they are sensitive to UV radiation (see Figure 9-10). Biochemical analysis, however, showed that these mutants excised thymine dimers normally, indicating that a system that uses recombination (or at least requires the *rec* genes) is responsible for another type of repair.

In the course of studying DNA replication in *E. coli,* a mutation was isolated in the gene *polA,* which encodes DNA Pol I. The mutant was viable, which suggested that Pol I is not the major replication enzyme. This finding provided the impetus for seeking other DNA polymerases in *E. coli,* and in fact Pol III was isolated from the *polA* mutant. Detailed examination of the *polA* mutant ultimately showed that it retained normal 5' → 3' exonuclease activity and had a residual polymerizing activity of about 2% of the wild-type, which was sufficient for it to play its essential role in the removal of RNA primers and the joining of precursor fragments (see Chapter 8). An important observation about the *polA* phenotype was that the mutant had somewhat increased sensitivity to UV, suggesting that DNA Pol I might be involved in DNA repair.

Figure 9-9. Survival curves of *E. coli* showing the sensitization to ultraviolet light resulting from the *uvrA⁻, recA⁻,* and *recA⁻ uvrA⁻* mutations.

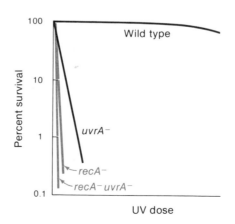

REPAIR OF THYMINE DIMERS

The four major pathways for dealing with thymine dimers in DNA can be subdivided into two classes: photoreactivation and light-independent pathways. Dark repair can be accomplished by three distinct mechanisms: (1) excision of the damaged bases (**excision repair**), (2) reconstruction of a functional DNA molecule from undamaged fragments (**recombinational repair**), and (3) tolerance of the damage (**SOS repair**).

Excision Repair

A major mechanism for several types of dark repair is the elimination of thymine dimers from DNA. Dimers are not cleaved and converted to monomers, however, as in photoreactivation, but instead *the dimer is completely excised from the DNA*. Evidence comes from the following experimental result. A population of bacteria is UV-irradiated and then incubated for various periods of time in a non-nutrient buffer (to allow liquid-holding recovery). During this period, the number of thymine dimers present in the DNA, determined by direct biochemical analysis, continually decreases, and at the same time, thymine dimers appear both in the intracellular fluid and in the buffer.

Excision repair is a multistep enzymatic process (Figure 9-10). Two distinct mechanisms have been observed for the first step, an **incision** step. In *E. coli*, a repair endonuclease recognizes the distortion produced by a thymine dimer and

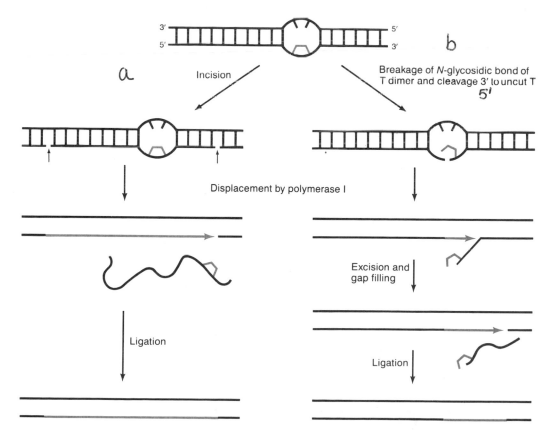

Figure 9-10. Two modes of excision repair. (a) The *E. coli* mechanism. Two incision steps are followed by gap-filling and displacement by polymerase I. (b) The *M. luteus* mechanism. A pyrimidine dimer glycosylase breaks an N-glycosidic bond and makes a single incision. DNA polymerase I displaces the strand, which is removed by an exonucleolytic event. In both mechanisms, the final step is ligation.

makes two cuts in the sugar-phosphate backbone: one is eight nucleotides to the 5' side of the dimer, and another is four to five nucleotides to the 3' side (see Figure 9-11a). At the 5' incision site, a 3'-OH group is produced, which DNA Pol I uses as a primer and synthesizes a new strand while displacing the DNA segment that carries the thymine dimer. The final step of the repair process is joining of the newly synthesized segment to the original strand by DNA ligase. The excised fragment is ultimately degraded to single nucleotides plus a thymine dimer dinucleotide by the combined activity of numerous scavenging exonucleases and endonucleases. Note that the role of DNA polymerase was anticipated by the observation described in the previous section that *polA* mutants are UV-sensitive.

The incision activity of *E. coli* is determined by a complex of the products of the three genes *uvrA*, *uvrB*, and *uvrC*, which are the three subunits of the excision endonuclease. The UvrA protein binds the helix distortion, the UvrB protein then binds to the UvrA-DNA complex, followed by binding of the UvrC protein that cuts the DNA.

The Uvr system is able to repair lesions other than thymine dimers. These lesions have in common either the displacement of bases, as in thymine dimer formation, or the addition of bulky substituents on the bases. It is thought that the incision enzyme recognizes a helix distortion.

In several other systems (for example, the bacterium *Micrococcus luteus* and *E. coli* phage T4), the incision step occurs in two distinct stages (see Figure 9-10b). The first step is an enzymatic cleavage of the *N*-glycosidic bond in the 5' thymine nucleotide of the dimer. Incision of the strand is completed by an endonuclease activity that recognizes a deoxyribose lacking a base; the enzyme makes a single cut at the 5' side of the remaining thymine in the dimer site. Then the deoxyribose is removed, and Pol I acts at the new 3'-OH group, displacing the strand and filling the gap. The displaced strand is excised by one of several different enzymes.

Recombinational Repair

If excision repair accounted for all dark repair, one might expect that a UV dose yielding one or a small number of thymine dimers per cell would be a lethal event for a *uvrA* mutant. A large number of thymine dimers, however, are required to kill a *uvrA* mutant (about 300 unexcised dimers), suggesting that the cells possess another repair system. Evidence for such a system came from the observation described earlier that *recA* mutants, a gene that is essential for genetic recombination in *E. coli*, are very UV-sensitive (see Figure 9-9). Excision of thymine dimers occurs in a *recA* mutant, so RecA-mediated repair clearly differs from excision repair. The existence of two repair systems is confirmed by a quantitative analysis of the survival curve of a *uvrA recA* double mutant, which is more UV-sensitive than either of the single mutants (see Figure 9-9).

Recall that the thymine dimers induced by UV block DNA replication. One way to deal with a thymine-dimer block is to bypass it and initiate chain growth beyond the block (Figure 9-11). This process, called *postdimer initiation*, appears to involve restarting DNA synthesis by an unknown mechanism, perhaps similar to that used in lagging strand synthesis. The result of postdimer initiation

Figure 9-11. Blockage of replication by thymine dimers (represented by joined lines) followed by re-starts several bases beyond the dimer. The black region is a segment of ultraviolet light–irradiated parental DNA. The orange region represents synthesis of a daughter molecule from right to left. The daughter strand contains gaps.

is that the daughter strands have large gaps, one for each unexcised thymine dimer. There is no way to produce viable daughter cells by continued replication alone because the strands having the thymine dimer will continue to turn out gapped daughter strands, and the first set of gapped daughter strands will be fragmented when the growing fork enters a gap. By a recombination mechanism involving **sister-strand exchange**, however, an intact double-stranded molecule can be made.

The essential idea in sister-strand exchange is that a single-stranded segment free of any defects is excised from the homologous DNA segment and inserted into the gap created by excision of a thymine dimer (Figure 9-12). This genetic recombination event requires the RecA protein. DNA Pol I then synthesizes the complementary strand, and DNA ligase joins this inserted piece to adjacent DNA, thus filling in the gap. The gap formed by excision of one strand from the donor molecule is also repaired by DNA Pol I and DNA ligase. If each thymine dimer is repaired this way, two complete daughter single strands can be formed, and each can serve in the next round of replication as a template for synthesis of normal DNA molecules. Note that the system fails if two dimers in opposite strands are near one another because then no undamaged sister-strand segments are available to recombine with the other strand. Many molecular details of recombinational repair are still not known, so the model shown in Figure 9-12 is simply a model.

Recombinational repair is an important mechanism because it eliminates the necessity for delaying replication for the many hours that would be needed for excision repair to remove all thymine dimers. Furthermore, recombination repair may correct some kinds of damage that cannot be corrected by excision repair—for example, alterations that do not cause helix distortion but do stop DNA synthesis.

Recombinational repair can also occur with UV-irradiated phages. If a population of a phage that fails to engage in excision repair is UV-irradiated, fewer phage plate on a *recA⁻* host than on a *recA⁺* host.

In contrast to excision repair, recombinational repair occurs after DNA replication; hence, recombination repair has been called **postreplicational repair**. Recombination repair is also called **daughter-strand gap repair** because only the gaps formed by opposite dimers, rather than the dimers themselves, are repaired.

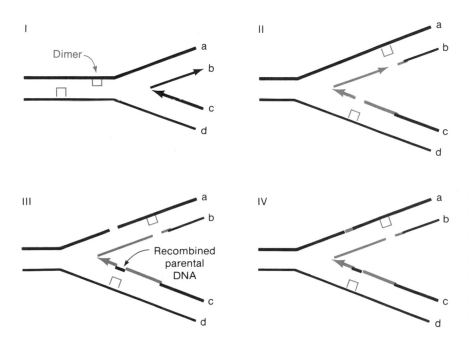

Figure 9-12. Recombination repair. (I) A molecule containing two thymine dimers (orange boxes) in strands a and d is being replicated. (II) By postdimer initiation, a molecule is formed whose daughter strands b and c have gaps. If repair does not occur, in the next round of replication, strands a and d would yield gapped daughter strands, and strands b and c would again be fragmented. (III) A segment of parental strand is excised and inserted into strand c. (IV) The gap in strand b is similarly filled in by repair synthesis. Such a DNA molecule would probably engage in a second exchange in which a segment of c would fill the gap in b. DNA synthesized after irradiation is shown in orange. Heavy and thin lines are used for purposes of identification only.

SOS Repair

Recall that UV light is a powerful mutagen. The repair processes we have discussed thus far, however, are not mutagenic; photoreactivation, excision repair, and recombinational repairs all result in faithful repair of the damage. A clue to UV mutagenesis is that an amount of mutagenesis is not a linear function of UV dose: Mutagenesis requires high doses of UV. This finding suggests that when the amount of UV damage (chiefly thymine dimers) exceeds the capacity of the faithful repair systems to correct the DNA damage, another process can allow cell survival but at the cost of mutagenesis. This process is called SOS repair (derived from the international distress signal) because it is a "last ditch" attempt to allow DNA replication—and hence cell survival—to proceed. It therefore seems that unrepaired DNA damage somehow induces the SOS response.

Thus, SOS repair is a **bypass system** that allows DNA chain growth across damaged segments at the cost of fidelity of replication. It is an error-prone process; that is, even though intact DNA strands are formed, the strands often contain incorrect bases. SOS repair is not yet thoroughly understood, but one of the results seems to be a relaxation of the editing system to allow polymerization to proceed across a dimer (transdimer synthesis) despite the distortion of the helix. SOS repair is the major cause of mutagenesis by UV and many chemical mutagens.

Regulation of the SOS Response

The SOS response involves the coordinate turn-on and turn-off of a large number of genes (about 20) following extensive DNA damage. The genetics and physiology of the SOS response were a confusing puzzle until a connection was made between a function of the RecA protein and the induction of phage.

Recall that phage λ has two potential lifestyles: either lytic growth producing infective virus or lysogenic growth integrated into the chromosome of *E. coli*. The integrated phage DNA is called a prophage, and the host cell carrying a prophage is called a lysogen. The prophage produces a protein called the *cI* repressor, which prevents the phage from expressing lytic functions (this is discussed in detail in Chapter 17). UV damage induces a λ prophage to go from its integrated state in the bacterial chromosome to a lytic state, where the phage replicates in the host cell kills and lyses the host cell, releasing progeny phage (**lysogenic induction**). This response is analogous to "rats deserting a sinking ship": The prophage senses that the host cell is severely damaged and is likely to die. If the host cell dies and the phage DNA remains in the host DNA, the phage

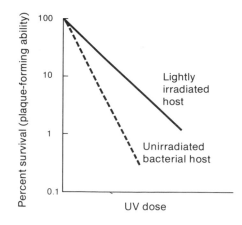

Figure 9-13. W-reactivation of ultraviolet light–irradiated phage λ. The dashed line shows the survival curve (for plaque-forming ability) obtained when λ phage irradiated with various doses of ultraviolet light are plated on unirradiated bacteria. The solid line represents survival of plaque-forming ability, when ultraviolet light–irradiated λ are plated on lightly irradiated bacteria.

also dies. If the phage goes into lytic growth, however, it can replicate and escape the damaged cell and, it is hoped, find an undamaged cell to infect. Lysogenic induction occurs only at UV doses sufficient to trigger the SOS response. Thus, it seemed that induction of the λ lysogen might be part of the SOS response.

A second clue came from the behavior of *recA* mutants. Recall that *recA* strains are extremely sensitive to UV. Part of this sensitivity is due to the requirement of RecA protein for recombinational repair. *recBC* mutants, however, which are also defective in recombination, are not nearly as sensitive to UV irradiation as are *recA* strains. This result suggested that RecA protein may play another role in UV repair in addition to its role in recombinational repair. Further experiments showed that the UV was not mutagenic in *recA* mutants: More *recA* cells died after UV irradiation, but the survivors did not have an increased number of mutations. Hence, the RecA protein is required for the SOS response. The SOS response cannot be turned on in a *recA* mutant, resulting in the extreme sensitivity to UV and the absence of UV mutagenesis. Moreover, in a *recA* strain, λ prophages are not induced by UV light. This could not be attributed to effects of UV on phage production because λ prophages could be induced in *recA* host strains by other means (see Chapter 17). Because induction of the prophage requires inactivation of the *cI* repressor protein, encoded by λ, it seemed likely that the combination of UV damage and functional RecA protein somehow affected the λ *cI* protein.

These clues were tied together by Roberts and Roberts, who examined the fate of λ repressor following UV irradiation. They showed that the native *cI* repressor is a 38 kDa protein, but following UV treatment it is cleaved in half. When the prophage was in a *recA* mutant, no cleavage of the *cI* repressor protein was seen. This result suggested that RecA protein might be a protease as well as a recombination protein. The in vivo experiments were confirmed in vitro using purified sources of *cI* repressor and RecA protein. The in vitro system required two additional components, a source of single-stranded DNA fragments and ATP. These two components had to be bound to RecA protein before cleavage of the repressor protein occurred. Thus, it seemed that binding of single-stranded DNA and ATP to RecA protein converted the protein into a protease form (RecA*) capable of cleaving *cI* repressor. Subsequent work showed that RecA* is not a protease. The protease activity is actually within the *cI* repressor itself. RecA* facilitates the self-cleavage of *cI* repressor by interacting with the repressor protein and causing a conformational change in the repressor protein. This brings the protease active site close to the region of the repressor protein which is cleaved, resulting in autoproteolysis. RecA* also facilitates the self-cleavage of other proteins of the SOS repair pathway.

The RecA protein has several functions in SOS repair. It directly inhibits the editing function of DNA Pol III by binding tightly in the region of the distortion resulting from a pyrimidine dimer. When Pol III encounters a dimer site to which RecA is bound, RecA interacts with the Pol III subunit responsible for the $3' \rightarrow 5'$ proofreading and inhibits the editing function, allowing the replication fork to advance. Because most UV damage is due to thymine dimers, most of the time Pol III randomly places two adenines in the daughter strand. Mispairing is enhanced by the distortion, however, which normally would activate the editing response. The presence of RecA at the dimer site inhibits editing and thus allows the mispaired base to remain in the daughter strand as a mutation.

The *umuDC* gene products are required for error-prone repair. Activated RecA protein facilitates proteolytic cleavage of the UmuD protein, producing UmuD', the active C-terminal fragment. The biochemical mechanism of the UmuD' and UmuC proteins is not known, but it seems likely that the UmuD'-UmuC complex directly interacts with RecA protein and the stalled DNA polymerase to promote error-prone replication. Not all bacteria seem to have functional *umuDC* genes, but functional homologs of the *umuDC* genes are often present on plasmids. Bacteria that lack functional *umuDC* genes are poorly mutated by UV irradiation. For example, *Salmonella typhimurium* is poorly mutated by UV unless a *umuD* homolog is provided on a plasmid (for example, the *mucAB* genes derived from pKM101).

Another important feature of SOS repair is that the system is induced as a result of damage to the DNA. The best evidence for this point comes from an analysis of the survival of UV-inactivated phage on UV-irradiated bacterial host cells. When λ is heavily UV irradiated, the phage titer decreases owing to DNA damage that cannot be repaired. If the UV-irradiated λ particles infect UV-irradiated *E. coli* cells, however, the survival of the phage is much greater than when they infect cells that have not been UV irradiated. That is, UV-irradiated λ produce more infective centers with irradiated *E. coli* than with an unirradiated host (Figure 9-13). This phenomenon, which is called **UV-reactivation** or **W-reactivation** (for Weigle, who discovered it), seemed to involve induction of a repair process in the cells by the UV damage. UV reactivation occurs only in RecA⁺ cells and does not require the *uvr* genes. Furthermore, although more phage survive, the surviving phage contains a higher proportion of phage mutants when the irradiated host is used. The SOS system has been turned on by the UV irradiation of the host.

Because SOS repair allows the frequency of replication errors to increase when necessary, it must be regulated in an on-off fashion to keep the normal error frequency low. Repair is needed only following certain types of DNA damage; thus, it seems reasonable that some feature of the damage would be the inducer. The same is true of other types of UV repair, such as recombinational repair and excision repair, and indeed these systems are inducible and are controlled by the same elements that regulate SOS repair. These repair systems plus several other operons compose the SOS regulon (the term **regulon** refers to a set of operons that are coordinately regulated).

The SOS regulatory system has two components, the *lexA* and *recA* gene products (Figure 9-14). These gene products have three essential features:

1. The *lexA* gene encodes a repressor of all SOS operons. The LexA repressor binds to a common operator sequence adjacent to each gene or operon. The LexA gene is itself regulated by binding of *lexA* to an operator site adjacent to the *lexA* gene, so LexA is autoregulated.
2. The RecA protein turns on the SOS response by facilitating proteolysis of the LexA repressor, as described for the λ *cI* repressor earlier.
3. DNA damage causes a conformational change in the RecA protein (converting it to RecA*) that promotes proteolysis of LexA and λ *cI* repressors.

In the absence of DNA damage, the SOS repair proteins are not required. Following expression of DNA damage, the proteins must be expressed, but once the repair is complete, the proteins should optimally be rapidly turned off. The three features of the SOS regulon that allow these regulatory events to occur are (1) the damage-induced activation of the RecA protease facilitator activity (recA*), (2) the sensitivity of LexA protein to proteolysis, and (3) the autogenous regulation of the *lexA* gene.

In an undamaged cell, the RecA protein lacks the ability to facilitate self-cleavage of the LexA repressor. In a UV-irradiated cell, RecA binds to single-stranded DNA (possibly to the short segment of single-stranded DNA generated by the T-T distortion). Binding of RecA protein to DNA causes a conformational change in the protein, which stimulates its protease facilitator activity. The activated RecA protein facilitates cleavage of the LexA repressor, which allows transcription of all operons of the SOS regulon to increase about 50-fold. This results

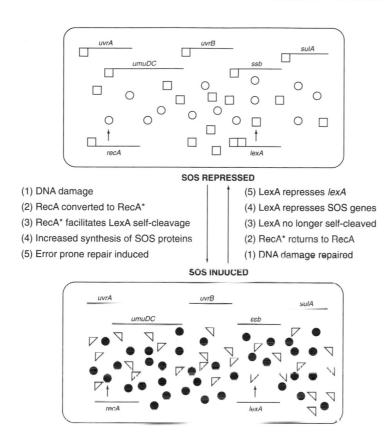

Figure 9-14. The SOS response. In uninduced cells the RecA protein [○] is not activated and thus does not facilitate self-proteolysis of the LexA protein [□]. The LexA protein functions as a repressor, turning off transcription of many different genes, including the *recA* gene and the *lexA* gene itself (i.e., it is autoregulatory). DNA-damaging agents induce the SOS response by activating RecA protein to RecA* [●], which facilitates the self-cleavage of LexA protein [◹] and several other proteins including the λ repressor. After LexA protein is cleaved, it cannot function as a repressor, resulting in transcription of all of the genes regulated by the LexA protein. As the SOS genes repair the DNA damage, RecA* returns to RecA, LexA is no longer cleaved, and accumulation of LexA represses the SOS genes.

SOS REPRESSED

(1) DNA damage

(2) RecA converted to RecA*

(3) RecA* facilitates LexA self-cleavage

(4) Increased synthesis of SOS proteins

(5) Error prone repair induced

(5) LexA represses *lexA*

(4) LexA represses SOS genes

(3) LexA no longer self-cleaved

(2) RecA* returns to RecA

(1) DNA damage repaired

SOS INDUCED

in high levels of each of the SOS proteins required to repair the DNA damage. Because *lexA* is autoregulated, the LexA protein is also made in high levels. The large amount of activated RecA* protein, however, continues to facilitate cleavage of LexA, so all proteins of the SOS regulon continue to be made as long as the DNA damage persists. Once the DNA repair is completed, RecA loses its proteolytic activity, and LexA is no longer cleaved. Once the RecA* protease activity decreases, LexA rapidly accumulates, binds to the SOS operators, and turns off the SOS operons, and the state of the cell existing before DNA damage occurred is reestablished.

Overview of Repair Systems

E. coli and *S. typhimurium* have four distinct systems to repair DNA damaged by UV light. At first thought it may seem strange that organisms that live in a dark environment, the intestine, need to defend their DNA against UV light. However, these organisms are also found in bodies of water where light can be intense. Moreover, the dark repair systems also function to repair damage to DNA caused by other agents, such as chemicals. The only repair system specific to UV damage, photoreactivation, uses visible light to repair the damage—thus broad spectrum light such as that from the sun contains both the wavelengths causing DNA damage (UV) and the wavelengths needed to repair the damage. The SOS repair system should be thought of as a "last ditch" attempt to repair the genomic DNA. The heavily irradiated cell seems willing to tolerate mutagenesis to replicate its chromosome. Are these all of the repair systems *E. coli* and *S. typhimurium* have to repair UV damage to DNA? It seems likely that this is the case since a *phr⁻ recA⁻ uvrA⁻* mutant of *E. coli* is killed by a UV dose that gives *one* thymine dimer per cell. Therefore, a strain unable to repair its DNA due to defects in photoreactivation (*phr⁻*), excision repair (*uvrA⁻*), recombinational repair (*recA⁻*), and SOS repair (*recA⁻*) gives a survival curve (see Figure 9-1) that is a straight line.

WHY DO DAMAGED CELLS DIE?

We have discussed a variety of repair systems that seem to be able to fix many types of DNA damage, yet we have not explained why cells are still killed by high doses of radiation and toxic chemicals. There are many possible explanations. One trivial reason is that there are probably some alterations that are not reparable—for instance, damaged regions may be opposite one another or tightly clustered so an undamaged template is not available for repair. A second possibility is that the repair systems themselves can be damaged. A third possibility is that a lethal event may often take place before repair occurs.

KEY TERMS

dark repair	RecA* protein
excision repair	SOS repair
LexA protein	survival curve
photolyase	thymine dimers
photoreactivation	UV-reactivation

QUESTIONS AND PROBLEMS

1. Are thymine dimers formed between adjacent thymines in the same strand or between opposite thymines in complementary strands?

2. If UV-irradiated bacteria are incubated in a buffer before plating, will the number of colonies formed be increased or decreased by the treatment?

3. If UV-irradiated phage are incubated in a buffer or exposed to light before plating, how will the efficiency of plaque formation change?

4. Which repair system cleaves thymine dimers?

5. Which enzymes are required for excision repair in *E. coli*?

6. What is the difference between incision and excision?

7. What two features of SOS repair distinguish it from all other repair systems?

8. Would you expect photoreactivation to occur with equal efficiency over a wide range of wavelengths, or will it exhibit a spectrum with maximum effectiveness in a small range of wavelengths?

9. A cell possesses a repair system capable of removing half of the damage produced by some agent. Which of the following statements would be true of a survival curve: (1) A dose yielding x percent survival when repair does not occur would yield $2x$ percent survival if repair occurs; (2) the dose required for a particular percentage of survival in the absence of repair is doubled when repair occurs.

10. The ability of UV-irradiated T4 phage to form plaques is almost the same on both Uvr⁺ and Uvr⁻ bacteria, with the survival curve on the Uvr⁻ bacteria being only slightly steeper than the one obtained with Uvr⁺ bacteria.
    a. How might you explain this observation?
    b. A T4 mutant is UV-irradiated and plated on Uvr⁺ and Uvr⁻ bacteria. The survival curves are much steeper than with wild-type phage and considerably steeper with Uvr⁻ bacteria than with Uvr⁺ bacteria. Suggest a defect for the mutant.

11. A bacterial repair system called X removes thymine dimers. You have in your bacterial collection the wild-type (X^+) and an X^- mutant. Phage 1, when UV-irradiated and then plated, gives a larger number of plaques on X^+ than on X^- bacteria. It has been proposed based on survival curve analysis that the X enzyme is inducible. To test this proposal, UV-irradiated 1 phage are adsorbed to both X^+ and X^- bacteria in the presence of the antibiotic chloramphenicol (which inhibits protein synthesis). No thymine dimers are removed in the X^- cell, and 50% are removed in the X^+ cell. In the absence of chloramphenicol, the same results were obtained.

   a. Is X an inducible system?
   b. Suppose 5% of the thymine dimers are removed in the presence of chloramphenicol and 50% in its absence; how would your conclusion be changed?

12. A Uvr$^-$ bacterial culture that has been grown for many generations in medium containing [^{3}H]thymidine is transferred to nonradioactive medium. The culture is UV-irradiated; because the cells are Uvr$^-$, no radioactive material appears in the medium. A phage sample is used to infect the irradiated culture, and several minutes later radioactive thymine dimers are found in the medium. The phage does not cause degradation of bacterial DNA from unirradiated bacteria. Explain the appearance of the dimers.

13. It is generally observed that if the UV survival curves of two different phages are compared, the phage with the larger DNA molecule has the steeper curve.

   a. Explain this phenomenon.
   b. You have isolated a new phage whose DNA is very large but which has the radiosensitivity of a phage with a small DNA. Suggest a possible explanation.

REFERENCES

Cox, M. M. 1991. The RecA protein as a recombinational repair system. *Mol. Microbiol.*, 5, 1295.

*Echols, H., and M. Goodman. 1991. Fidelity mechanisms in DNA replication. *Ann. Rev. Biochem.*, 60, 477.

Eisenstadt, E. 1987. Analysis of mutagenesis. In F. Neidhardt, J. Ingraham, K. B. Low, B. Magasanik, M. Schaechter, and H. E. Umbarger (eds.), Escherichia coli *and* Salmonella typhimurium: *Cellular and Molecular Biology.* American Society for Microbiology, Washington, D.C.

Kunz, B., and S. Kohalmi. 1991. Modulation of mutagenesis by deoxyribonucleotide levels. *Ann. Rev. Genet.*, 25, 339.

*Howard-Flanders, P. 1981. Inducible repair of DNA. *Scientific Am.* March, p. 72.

Lin, J., and A. Sancar. 1992. (A)BC excinuclease: the *Escherichia coli* nucleotide excision repair enzyme. *Mol. Microbiol.*, 6, 2219.

Little, J. W., et al. 1980. Cleavage of the *E. coli lexA* protein by the *recA* protease. *Proc. Natl. Acad. Sci. USA*, 77, 3225.

Livneh, Z., and I. R. Lehman. 1982. Recombinational bypass of pyrimidine dimers by the RecA protein of *E. coli. Proc. Natl. Acad. Sci. USA*, 79, 3171.

Modrich, P. 1991. Mechanisms and biological effects of mismatch repair. *Ann. Rev. Genet.*, 25, 229.

*Peterson, K., N. Ossanna, A. Thliveris, D. Ennis, and D. Mount. 1988. Derepression of specific genes promotes DNA repair and mutagenesis in *Escherichia coli. J. Bacteriol.*, 170, 1.

Roberts, J. J. 1978. The repair of DNA modified by cytotoxic, mutagenic, and carcinogenic chemicals. *Adv. Radiat. Biology*, 6, 212.

Setlow, R. B. 1966. Cyclobutane-type pyrimidine dimers in polynucleotides. *Science*, 153, 379.

Tessman, M., S. Liu, and M. Kennedy. 1992. Mechanism of SOS mutagenesis of UV-irradiated DNA: Mostly error-free processing of deaminated cytosine. *Proc. Natl. Acad. Sci. USA*, 89, 1159.

*Resources for additional information.

*Walker, G. C. 1984. Mutagenesis and inducible responses to deoxyribonucleic acid damage in *Escherichia coli*. *Microbiol. Rev.*, 4, 60.

Weigle, J. 1953. Induction of mutations in a bacterial virus. *Proc. Natl. Acad. Sci. USA*, 39, 628.

Willets, N., and A. J. Clark. 1969. Characteristics of some multiply recombinant-deficient strains. *J. Bacteriol.*, 10, 231.

Winston, F., D. Botstein, and J. Miller. 1979. Characterization of amber and ochre suppressors in *Salmonella typhimurium*. *J. Bacteriol.*, 137, 433.

Woodgate, R., and S. G. Sedgwich. 1992. Mutagenesis induced by bacterial UmuDC proteins and their plasmid homologues. *Mol. Microbiol.*, 6, 2213.

Mutagenesis, Mutations, and Mutants

Many uses of mutants have been discussed in previous chapters. But what is a mutant, how are mutations formed, and how are mutants detected?

BIOCHEMICAL BASIS OF MUTATIONS

The term **mutant** refers to an organism in which the base sequence of DNA has been changed. The mutation may or may not cause a phenotype. The chemical and physical properties of proteins are determined by their amino acid sequence. An amino acid substitution can change the structure and hence the biological activity of a protein. Even a single amino acid change is capable of altering the activity of, or even completely inactivating, a protein. For instance, consider a hypothetical protein whose three-dimensional structure is determined entirely by an interaction between one positively charged amino acid (for example, lysine) and one negatively charged amino acid (aspartic acid). A substitution of methionine, which is uncharged, for the lysine would clearly destroy the three-dimensional structure. Similarly, a protein might be stabilized by a hydrophobic cluster, in which case substitution of a polar amino acid for a nonpolar one in the cluster would also be disruptive.

An amino acid substitution does not always lead to a mutant phenotype. For instance, a hydrophobic cluster might be virtually unaffected by a replacement of a leucine by another nonpolar amino acid such as isoleucine. When an amino acid substitution has no detectable effect on phenotype, it is called a **silent mutation**. A base change without an amino acid alteration (for example, in the third position of a condon) is also a silent mutation.

The shapes of proteins are determined by such a variety of interactions that sometimes an amino acid substitution is only partially disruptive. This could cause a reduction, rather than a complete loss, of activity of an enzyme. For example, a bacterium carrying such a mutation in the enzyme that synthesizes an essential substance might grow very slowly (but it would grow), unless the substance is provided in the growth medium. Such a mutation is called a **leaky mutation**.

Several common alterations other than amino acid substitutions may also eliminate activity of a protein:

1. Deletion of 3n bases (where *n* is a positive integer), which causes one or more amino acids to be absent in the completed protein.
2. A deletion or insertion that causes shift in the reading frame such that all the codons after the mutation are changed.

179

3. A chain termination mutation, in which a base change generates a stop codon. Such mutations result in a shorter polypeptide lacking the carboxyl terminus of the protein.

SPONTANEOUS MUTATIONS

Mutations are random events, and there is no way of knowing when or in which cell a mutation will occur. Every gene, however, mutates spontaneously at a characteristic rate, making it possible to assign probabilities to particular mutational events. Thus, there is a definite probability that a given gene will mutate in a particular cell and likewise a definite probability that a mutant allele of the gene will occur in a population of a particular size. Most mutations are considered random in the sense that their occurrence is not related to any adaptive advantage they may confer on the organism in its environment.

Random and Nonadaptive Nature of Mutation

The idea that mutations are spontaneous random events unrelated to adaptation was not accepted by many microbiologists until the late 1940s. Before that time, it was believed that mutations occur in bacterial populations in response to particular selective conditions. When antibiotic-sensitive bacteria are spread on medium containing an antibiotic, some colonies can be isolated that consist of cells having an inherited resistance to the drug. The initial interpretation of such observations was that the mutations were adaptive variations induced by the selective agent itself. In 1943, however, Luria and Delbruck designed an experiment that demonstrated the spontaneous, nonadaptive nature of mutation, an experiment that marked the birth of microbial genetics. In this experiment, the origin of mutations in *Escherichia coli* that confer resistance to phage T1 (T1^r mutations) was investigated. The approach was to compare the number of T1^r mutant cells arising in different cultures of T1^s (T1 sensitive) cells with the number found in repeated samples of the same size taken from a single culture. A statistical test called a fluctuation test was used to analyze the results of the experiment. The data shown in Table 10-1 were obtained from one experiment in which twenty 0.2-ml cultures and one 10-ml culture, each containing a T1^s bacterial strain at an initial concentration of 10^3 cells/ml, were grown to a concentration of 2.8×10^9 cells/ml (21 generations). Each of the small cultures and ten 0.2-ml samples from the large culture were plated on individual plates covered uniformly with about 10^{10} T1 phage (enough to kill all T1^s cells), and the number of colonies that grew (due to T1^r mutants) were counted. Each plate received the same number of bacterial cells (5.6×10^8), but the number of T1^r colonies formed depended on whether the cells had been grown in the small individual cultures or in the large bulk culture. No T1^r cells were detected in 11 of the 20 small cultures, and the numbers in the other 9 of these cultures ranged from 1 to 10^7; in contrast, the 10 samples from the large culture each had about the same number. The alternatives expected in such an experiment were the following: (1) If T1^r bacteria arise in response to the phage, there should be about equal numbers in all populations of the same size; (2) if T1^r cells arise by spontaneous mutation at different times in the growth of the cultures in the absence of the phage, the numbers in different cultures may vary greatly. The results of the experiment shown in Table 10-1 and others like it were consistent with spontaneous mutation, as shown in Figure 10-1. Panel (a) shows four separate bacterial cultures and the pedigrees by which the cells could be derived from a single ancestor. A mutation to phage resistance is shown to have occurred during a dif-

Table 10-1   The number of T1 phage-resistant *E. coli* mutants in small individual cultures and in samples from a large bulk culture.

Small individual cultures		Samples from large culture	
Culture	T1^r colonies per plate	Sample	T1^r colonies per plate
1	1	1	14
2	0	2	15
3	3	3	13
4	0	4	21
5	0	5	15
6	5	6	14
7	0	7	26
8	5	8	16
9	0	9	20
10	6	10	13
11	107		
12	0	Mean	16.7
13	0		
14	0	Variance	15
15	1		
16	0	Variance/Mean	0.9
17	0		
18	64		
19	0		
20	35		
Mean	11.4		
Variance	694		
Variance/Mean	60.8		

Source: Data from S. F. Luria and M. Delbrück. *Genetics* 28 (1943): 491.

ferent generation in each lineage (line of descent). The two extreme cases are cultures 1 and 2. Culture 1 depicts a mutation occurring just before sampling, resulting in only a single mutant colony. Culture 2 depicts a mutation occurring soon after addition of bacteria to the medium, yielding an entirely mutant population. Cultures 3 and 4 are intermediate cases. Thus, separate cultures contain significantly different numbers of mutant cells. Panel (b) shows the result of growing the four lineages in panel (a) in a single large culture. With this arrangement, the mutants from each lineage become uniformly dispersed in the medium, so sample-to-sample variation is greatly reduced. This experiment strongly suggested that resistance to T1 was due to spontaneous mutations. Although the experiment was elegant and convincing, however, the results were indirect.

Direct evidence that T1^r mutations occurred spontaneously in *E. coli* cells before exposure to the phage was obtained in 1952 by a procedure developed by Lederberg and Lederberg called replica plating (Figure 10-2). In this procedure, bacteria are plated, and after colonies have formed, a piece of sterile velvet mounted on a solid support is pressed onto the surface of the plate (the master plate). Some bacteria from each colony stick to the fibers, as shown in panel (a). Then the velvet is pressed onto a fresh plate (the replica plate), transferring the cells and giving rise to new colonies in the same position as those on the first plate. Panel (b) shows how the method was used to demonstrate the spontaneous origin of T1^r mutants. Master plates containing 10^7 colonies growing on nonselective medium (with no T1) were replica-plated onto a series of plates that had been spread with T1 phage. After incubation, a few colonies of T1^r bacteria appeared in the same positions on each of the replica plates. This meant that the T1^r cells that formed the colonies must have been transferred from corresponding colonies

(a) Individual cultures

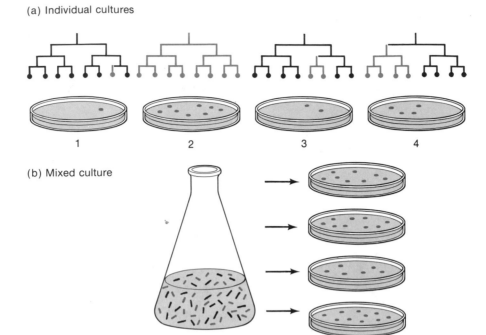

(b) Mixed culture

Figure 10-1. The fluctuation test. (a) Pedigrees showing the source of the mutants (orange colonies) found in each of four different samples. In each pedigree, the occurrence of a mutation is indicated by a shift of the path to red. (b) Typical results of sampling from a single mixed culture. Black and orange rods indicate wild-type and mutant bacteria; they are not drawn to scale, being roughly 10^8 times too large compared with the colonies.

on the master plate. Because the colonies on the master plate had never been exposed to the phage, the mutations to resistance must have occurred by chance in cells not induced by exposure to the phage.

Mutation Rates

The mutation rate is the probability that a gene will be mutated in a single generation. Measurement of mutation rates is important in population genetics, in studies of evolution, and in analyzing the effect of environmental mutagens.

Estimation of mutation rates in bacteria is complicated by the fact that a mutation can occur at any time during the growth of a culture, and division of the mutant bacteria usually results in an increase in their number at a rate the same as that of the increase of the population as a whole. Thus, as shown in the fluctuation test, the occurrence of phage-resistant mutants in different cultures may vary dramatically. With a mutation rate per generation of μ, the probability P_0 of obtaining no mutants in a culture of N cells can be estimated by the Poisson distribution as $e^{-\mu N}$. (The total number of divisions of individual cells required to produce N cells from 1 cell is $N - 1$, but because N is a large number, $N - 1 = N$.) Thus, $-\mu N = \ln P_0$, or $\mu = -(1/N) \ln P_0$. In the fluctuation test data in Table 10-1, 11 of the 20 small cultures contained no phage-resistant mutants, and the average number of cells per culture (N) was 5.6×10^8. Thus, the mutation rate could be estimated as $\mu = -(1/5.6 \times 10^8) \ln(11/20) = 1.1 \times 10^{-9}$ per cell per generation. The fluctuation test is one important method for estimating mutation rates in bacteria. Another method is the following.

Consider a culture of wild-type *E. coli* obtained by inoculating a single colony into liquid medium. At various times, samples are removed and tested for a par-

(a) The transfer process

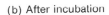

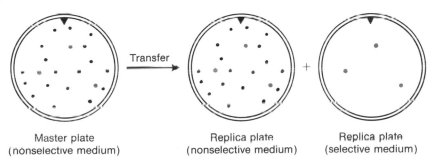

Velvet is pressed on the master plate and then onto fresh medium

Master plate

Replica plate

(b) After incubation

Master plate
(nonselective medium)

Transfer

Replica plate
(nonselective medium)

+

Replica plate
(selective medium)

Figure 10-2. Replica plating. (a) The transfer process. A velvet-covered disc is pressed onto the surface of a master plate to transfer cells from colonies on that plate to a second medium. (b) Detection of mutants. Cells are transferred onto two plates containing either a nonselective medium (on which all form colonies) or a selective medium (for example, one spread with T1 phage). Colonies form on the nonselective plate in the same pattern as on the master plate. Only mutant cells (for example, T1^r) can grow on the selective plate; the colonies that form correspond to certain positions on the master plate. Colonies containing mutant cells are shown in orange.

ticular phenotype. In the experiment to be described, the wild-type cells are sensitive to phage T5, and the mutants are T5^r. Data for this experiment are shown in Figure 10-3. Note that the culture was grown in the absence of any phage, so neither T5^r nor T5^s bacteria were selected, and changes in allele frequency were determined entirely by mutation. The figure shows that the frequency of T5^r cells increased linearly over the course of the experiment. The reason for the increase was that almost all of the cells in the initial population were T5^s; in each generation, some of these T5^s cells underwent spontaneous mutation to become T5^r, and thus T5^r cells accumulated in the population as time progressed. The T5^r cells are rare, so the reverse mutation from T5^r to T5^s may be ignored.

The linearity of the curve can be deduced from a simple equation. The frequencies of T5^s and T5^r cells in the population can be represented by p and q, with subscripts indicating the generation (for example, P_n = proportion of T5^s cells at generation n). Because reverse mutation can be neglected, the frequency of T5^s cells in the nth generation equals the proportion of T5^s bacterial cell lineages that escaped mutation for n consecutive generations. All other lineages will have mutated to T5^r at some prior time. If mutation from T5^s to T5^r occurs at the rate μ per generation, the probability that a particular T5^s lineage will escape mutation in each of n generations must equal $(1-\mu)^n$. Consequently:

$$P_n = (1-\mu)nP_0$$

Making the approximation $(1-\mu)n = 1 - n\mu$, which is valid when μ is very small, and substituting $1 - q = p$ yields $1 - q_n = (1 - n\mu)(1 - q_0) = 1 - n\mu - q_0 + n\mu q_0$. The term $n\mu q_0$, which is the product of two small numbers, can be ignored, so rearrangement of the equation yields:

$$q_n = q_0 + n\mu$$

This is the equation of a straight line, and it explains why the curve in Figure 10-3 is linear. Furthermore, the slope of the line is μ, the mutation rate. Thus, Figure 10-3 shows that the mutation rate of $T5^s$ to $T5^r$ is 7.8×10^{-8} mutations per generation. It should be noted that the mutation rates for $T5^s$ to $T5^r$ and $T1^s$ to $T1^r$ are quite different (by a factor of 70). This should not be surprising because the mutation rates depend on the size and nucleotide sequence of the gene and on the amino acid sequence and three-dimensional structure of the gene product.

Origin of Spontaneous Mutations

Mutations are due to modification of DNA. Several mechanisms for such modification are known. The two most common causes of spontaneous mutagenesis are (1) errors occurring during replication and (2) spontaneous alteration of a nucleotide.

Errors in nucleotide incorporation during DNA replication occur with sufficiently high frequency that the information content of a daughter DNA molecule would differ significantly from that of the parent were it not for two mechanisms for correcting such errors: proofreading by DNA polymerase and mismatch repair (described in Chapter 8).

One reason for such incorporation errors is tautomerism of the nucleotide bases. Some of the bases exist in alternative forms with different base-pairing properties. Examples are shown in Figure 10-4. For example, the rare form of adenine can pair with cytosine, and the enol form of thymine can pair with guanine. Thus, if such tautomers occur during DNA replication, an incorrect base will be correctly hydrogen-bonded to the template strand, so it is not recognized by the proofreading function as incorrect. If the base in the template strand later assumes its normal structure, a mismatched base pair will be present, which may be corrected by the mismatch repair system. If the region of the daughter strand containing the incorrect base becomes methylated before it is repaired, however, the mismatch repair system will be unable to distinguish parental and daughter strands, and the mutation is retained. Both proofreading and mismatch repair are efficient; however, they are not perfect, and mutations do occur.

Another source of spontaneous mutation is an alteration of 5-methylcytosine (MeC), a methylated form of cytosine that pairs with guanine. About 5% of

Figure 10-3. Accumulation of cells resistant to phage T5 resulting from recurrent mutation in a population of *E. coli* growing in conditions that allow continuous growth at constant cell density. (After Kubitschek, H. E. 1970. Introduction to Research with Continuous Cultures, p. 33. Reprinted by permission of Prentice-Hall, Inc., Englewood Cliffs, NJ.)

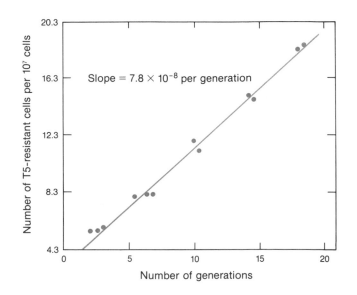

the cytosines in the DNA of many bacteria and viruses are MeC. Both cytosine and 5-methylcytosine can occasionally lose the amino group. For cytosine, this loss yields uracil (Figure 10-5). Because uracil pairs with adenine instead of guanine, replication of a molecule containing a GU base pair ultimately leads to substitution of an AT pair for the original GC pair (by the process GU → AU → AT in successive rounds of replication). Cells, however, possess an enzyme (uracil glycosylase) that specifically removes uracil from DNA, so the C-to-U conversion rarely leads to mutation. Loss of the amino group of 5-methylcytosine yields 5-methyluracil, or thymine (see Figure 10-5). Because thymine is a normal DNA base, no thymine glycosylase exists, so the GMeC pair becomes a GT pair. A GT pair is subject to correction by mismatch repair. Because MeC is a methylated base, however, and therefore is present in a methylated strand, the mismatch repair system does not recognize the thymine as incorrect. The direction of correction is random, sometimes yielding the correct GC pair and sometimes an

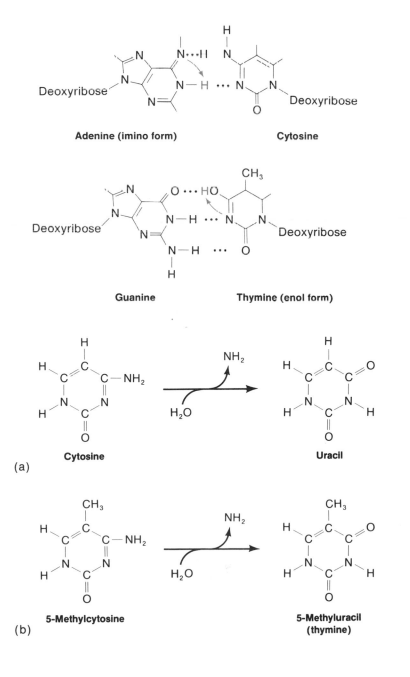

Figure 10-4. Base-pairing between the rare imino form of adenine and cytosine and the enol form of thymine and guanine. The orange H is the one that has moved from the more common position. Compare with Figure 2-5, which shows the standard base pairs.

Figure 10-5. Spontaneous loss of the amino group of (a) cytosine to yield uracil and (b) 5-methylcytosine to yield thymine.

Figure 10-6. A portion of the T4 *rII* gene showing the number of mutations isolated at each site. (From Benzer, S. 1961. Proc. Nat. Acad. Sci., 47, 410.)

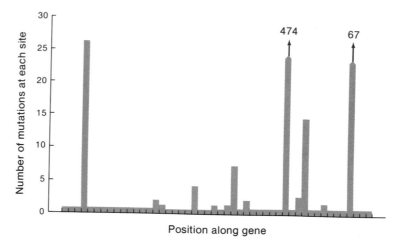

incorrect AT pair. Thus, MeC sites, which exist at only a few locations in a gene, constitute highly mutable sites, and the mutations are always GMeC → AT changes.

Sites within a gene at which mutations occur with much higher frequency than at other sites were first observed in the fine structure mapping experiments of the T4 *rII* locus (described in Chapter 15), in which several thousand independently isolated *rII* mutations were mapped (Figure 10-6). These sites were called hot spots. In later years, determination of the base sequence of several bacterial genes and of mutant alleles has shown that many hot spots are GMeC base pairs.

Some hot spots yield large deletions rather than point mutations. The base sequences of these regions indicate that the deletion is often bounded by a repeated sequence. Two possible mechanisms for the production of a deletion in molecules having repeated nucleotide sequences are illustrated in Figure 10-7. These mechanisms are recombinational excision and a particular type of replication error.

ISOLATION OF MUTANTS

Because mutants are so rare in the natural population, most mutant "hunts" begin with mutagenesis of the bacterium or phage. Usually the incidence of a mutation in a given gene is 10^6 or less. That is, at most only one in a million organisms is the mutant being sought. A few of the selections discussed later are sufficiently powerful to isolate such rare mutants, but selections lack this discrimination. Screening methods cannot cope with mutation frequencies that are this low. For these reasons, the bacteria or phage are treated with a mutagen. This can be a chemical mutagen (for example, nitrous acid), radiation (for example, ultraviolet), or a mutator gene. Bacteria are generally mutagenized by treatment of cell suspensions with the mutagen but can also be mutagenized by treatment of isolated DNA (or transducing phages) followed by introduction into the bacteria. Phages can be mutagenized by treatment of suspensions of phage particles or by infection of a bacterial host that is treated with a mutagen (or contains a mutator gene).

After mutagenesis, the bacteria or phage are grown for several generations. This period of growth, called "**phenotypic expression**," allows the products of mutagenesis to segregate. For example, usually only one of the two DNA strands of a given gene is altered by the mutagenic treatment; the other strand has the normal sequence. If the normal strand is the one used for mRNA transcription, the mutation will be silent. Moreover, the mutated cell still contains a supply of the gene product. Thus, even if the mutagenized strand is transcribed, the mutated bacterial cell remains physiologically normal until the gene product is diluted by cell division. The period of growth thus allows DNA replication and cell division such that the mutated strand will be copied and segregated into a daughter cell (where both strands of the gene will be mutant). Cell division also results in progressive dilution of the supply of normal gene product present in the original mutagenized cell. Phenotypic expression is done under conditions where the mutant sought will grow.

Given a culture of mutated bacteria that have undergone phenotypic expression, a method is needed to isolate the mutants of interest. Many ingenious methods have been developed to isolate mutants, but most of the strategies used can be lumped under a few headings: screening, enrichment, and selection.

(a)

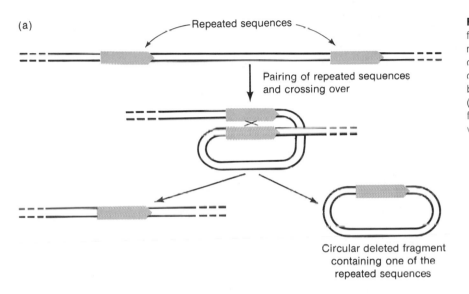

(b)

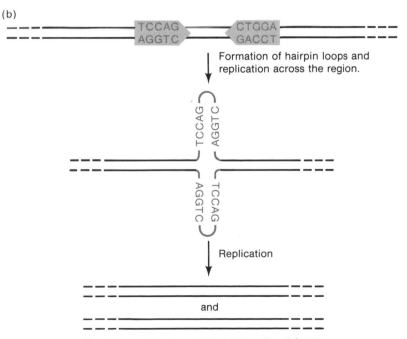

Daughter molecules with the red region deleted.

Figure 10-7. Two mechanisms for the production of deletions in molecules with repeated nucleotide sequences. (a) Crossing over (homologous recombination) between paired direct repeats. (b) Replication across an aberrant form of a molecule with an inverted repeat.

1. *Screening.* Screening methods are the most obvious means of isolating mutants but can be labor-intensive (leading to the term "brute force"). In the simplest screens, mutants can be recognized by colony appearance. For example, mutants defective in sugar fermentation can be detected on agar plates of various indicator media where colonies of cells that ferment the sugar have a different color (for example, red versus white colonies on MacConkey plates) than colonies of nonfermenting cells. In some cases, 10^4 to 10^5 cells can be screened on a single plate, and thus given enough plates, many mutants can be isolated. This is the exception, however—an indicator medium is not available for many mutants, and some types of indicator media work only with few colonies (<100) per plate because diffusion of products between closely spaced colonies affects the results.

   Another approach is to look for colonies that fail to grow on a given medium. For example, if one wishes to isolate *E. coli* mutants unable to synthesize leucine, phenotypic expression of the mutagenized *leu*$^+$ strain is done in the presence of leucine followed by plating for single colonies on medium containing leucine. Both *leu*$^+$ and *leu*$^-$ cells form colonies on these plates, but if a few cells from each colony are transferred to a second plate lacking leucine, only *leu*$^+$ cells form a colony. This is usually done by replica plating (see Figure 10-2): (1) the original plate (called the "master plate") is gently pressed onto a sterile velvet pad that is supported by a round "block" the size of a petrie dish—cells from each colony become attached to the velvet; (2) the velvet is then gently pressed onto a sterile plate to transfer the cells—the pattern of the cells is identical to the pattern on the original plate; (3) this can be repeated for many plates (up to about 15 plates); (4) after the plates are incubated, the resulting colonies at each specific position on each of the plates will have arisen from the same original colony on the master plate.

2. *Enrichment.* Replica plating is a powerful approach for finding mutants, but even given the advantages of replica plating, screening can be a tremendous amount of work. Many plates must be replicated and compared for each mutant isolated. For this reason, various procedures to enrich for mutant organisms in a population have been developed. Such enrichment procedures must generally be followed by a screening procedure, but now many fewer colonies must be screened.

   An example of an enrichment procedure is the "penicillin selection" technique. Penicillin kills bacteria by interfering with the synthesis of the bacterial cell wall. When penicillin is added to a culture of growing bacteria, cell wall synthesis stops, but growth (enlargement) of the cell continues until the cell explodes and dies. Thus, penicillin does not kill nongrowing cells. For example, a Leu$^-$ cell in a growth medium lacking leucine cannot synthesize protein so it cannot grow; hence, it is not killed by penicillin. Consider a culture containing one Leu$^-$ cell per 10^6 Leu$^+$ cells growing in medium containing leucine. The culture is filtered, and the cells collected on the filter are resuspended, in a growth medium containing penicillin but lacking leucine. The Leu$^+$ cells continue to grow, and about 99% of them are killed by the penicillin; in contrast, the Leu$^-$ cells, which cannot grow, survive. After about 1 to 2 hours, incubation of the culture is stopped because the killed Leu$^+$ cells release their contents to the medium, and the pool of leucine released from these cells would allow the Leu$^-$ cells to grow and therefore also to be killed. The treated culture now consists of one Leu$^-$ cell per 10^4 cells. The penicillin enrichment is then repeated, yielding about 1 Leu$^-$ cell per 10^2 cells, allowing the mutants to be easily detected by replica-plating.

For some organisms, penicillin selection cannot be used. For example, some bacteria synthesize β-lactamase, an enzyme that inactivates penicillin, so an unrelated chemical that also inhibits peptidoglycan synthesis must be used (for example, D-cycloserine). Another example involves the *Archaea* (previously called *Archaeobacteria*), which lack peptidoglycan. Auxotrophs of various *Archaea* species have been selected by using ogaguanine, a toxic nucleic acid base analog.

Another enrichment method is incorporation of a radioactive metabolic precursor. The presence of high levels of radioisotopically labeled cellular molecules results in cell death if the cells are stored under nongrowing conditions. During storage, the β particles released by disintegration of the incorporated isotope cause cell death, probably by triggering a chain of free-radical reactions. The amounts of radioactivity needed (about 1 disintegration/minute/cell) are large, and thus the approach is limited to isotopes with a high specific activity or a short half-life (for example, ^{3}H, ^{35}S, or ^{32}P), which are readily incorporated into cellular components which comprise a major fraction of cell mass (protein, nucleic acid, or lipid). This approach, often termed "radioactive-suicide" because the cell must actively incorporate the radioactive precursor, can be used as a general enrichment for nongrowing cells (such as penicillin enrichment), but it has the advantage that it can be targeted to a particular metabolic step. For example, if a radioactive lipid precursor is used, among the surviving cells, mutants defective in lipid synthesis will dominate. These selections are often done for temperature-sensitive mutants (see later), and thus the radioactive labeling is done at the nonpermissive temperature, and plating for surviving cells is done at a low (permissive) temperature. Mutants defective in transport of the radioactive precursor would also be enriched among the survivors.

3. *Selection.* A selection method is defined as an experimental condition in which only mutant cells grow; wild-type cells either die or survive but fail to grow. The most straightforward example is selection for resistance to a metabolic inhibitor (such as an antibiotic). Wild-type cells are killed by the inhibitor, and only mutant cells survive. Such mutants can be due to a large variety of metabolic alterations. A few examples include:

   a. The mutant cell may produce an enzyme that inactivates the antibiotic by hydrolysis or modification of the inhibitor molecule.

   b. The intracellular molecule to which the inhibitor binds may be altered such that the inhibitor no longer binds or binds but is unable to inhibit. Examples include mutations in ribosomal protein S12, which prevent binding of streptomycin, and mutations in DNA gyrase, which prevent binding of naladixic acid or coumermycin.

   c. Accumulation of the inhibitor in the mutant may be prevented because the inhibitor cannot enter the cell or enters but is actively pumped out of the cell. Selection for resistance to amino acid analogs often gives mutants defective in amino acid transport. Tetracycline resistance can be due to either preventing antibiotic entry or active efflux of the antibiotic.

   d. The mutant may dilute out the inhibitor. For example, the amino acid analog 5-methyltryptophan inhibits the growth of *E. coli* by competing with tryptophan for incorporation into proteins (5-methyltryptophan-containing proteins function poorly). Mutants resistant to 5-methyltryptophan are generally altered in the regulation of the tryptophan biosynthetic operon, resulting in high levels of intracellular tryptophan and dilution of the inhibitor such that less 5-methyltryptophan is incorporated into cell protein. Cells can also become resistant by overproducing the target protein. Suppose that a target protein is an enzyme present in 1000 copies per cell and that enough inhibitor enters the cell to bind and inactivate 950 enzyme molecules. The remaining 50 enzyme molecules cannot support growth. If, however, a mutation arises that doubles the amount of enzyme expressed to 2000 molecules, the mutant will be resistant to the inhibitor. For these reasons, selection for resistance to various metabolic inhibitors often results in mutants altered in the regulation of a pathway.

Another type of selection is for the use of molecules that cannot be used by wild-type organisms. For example, *E. coli* cannot grow on short-chain fatty acids (with fewer than 12 carbon atoms) as the sole carbon source, although it readily grows on longer chain fatty acids. Plating cells on media containing a 10 carbon fatty acid (decanoic acid) as the sole carbon source results in growth of a few (10^{-7}) mutant colonies. The mutant colonies are *fadR* mutants in which the normally inducible β-oxidation pathway is now expressed constitutively. Analogous constitutive mutants can be isolated in other pathways. In other cases, selection gives rise to mutants that are not expressed under normal physiological conditions but become activated by mutation. For example, *E. coli* K-12 has two acetohydroxyacid synthases, the enzyme responsible for the first step of valine synthesis and the second step of isoleucine biosynthesis, but both are feedback inhibited by valine. Thus, *E. coli* cannot grow on a minimal medium containing valine because the cells are unable to synthesize isoleucine (addition of isoleucine to the medium restores growth). Selection for mutants that grow on valine-containing medium results in selection for mutants of the *ilvG* gene. The *ilvG* gene encodes acetohydroxyacid-synthase II, an enzyme that is not inhibited by valine. The *ilvG* gene is expressed in *E. coli*, but the translated protein lacks enzyme activity owing to a naturally occurring frameshift mutation within the structural gene. The valine-resistant mutants therefore are mutants of *ilvG* having a 1 bp deletion or 2 bp insertion that restores the reading frame encoding the active acetohydroxyacid-synthase II. (Note that the genetic nomenclature of such strains becomes confusing. By definition, the wild-type strain is *ilvG*⁺, so the mutation that results in a functional protein is *ilvG*.)

Isolation of Conditionally Lethal Mutations

In the examples given, the mutant strains isolated grow normally if the proper medium is provided (i.e., a required nutrient or a usable carbon source is supplied). These approaches suffice for mutants defective in many steps of small molecule metabolism but are useless in the study of the metabolism of most macromolecules. For example, a mutant lacking a key enzyme in DNA synthesis would be unable to grow because no DNA could be produced for the daughter cells. This is a lethal mutation. Most mutants in DNA, RNA, protein, and membrane synthesis would also be lethal. These are essential molecules that are too large and complex to be provided in the growth medium. So how can you isolate mutants in macromolecular metabolism? The trick is to use conditional mutations.

Temperature-sensitive mutations are one common type of conditional mutations. The idea is that the folding of proteins (hence enzymes) is sensitive to temperature and that alteration of the amino acid sequence of a protein can make the mutant protein unfold (denature) under conditions in which the wild-type protein remains folded. Specifically a mutant protein may be able to fold more or less normally and be active in cells grown at 30°C but will be unfolded and inactive in cells grown at 42°C. The wild-type protein is folded and active at both temperatures. Note that the gene is not altered at 42°C, rather the *product* of the gene is inactive at 42°C. This trick allows study of lethal mutations; the mutant strain can be isolated and grown at 30°C, but the mutation is not expressed until the cells are shifted to 42°C.

In practice, temperature-sensitive lethal mutations can be isolated in the same manner as conventional mutants except that the permissive conditions during phenotypic expression are 30°C and the selections or screens are done at 42°C. For example, to isolate mutants in DNA synthesis, one could mutagenize a culture of cells, grow the surviving cells at 30°C for phenotypic expression, then shift the culture to 42°C, and enrich for mutants defective in DNA synthesis by a radiation-suicide selection using ^{3}H-thymidine incorporation. The survivors of the suicide selection would then be plated at 30°C to allow growth of the mutants and the individual colonies screened for defective DNA synthesis at 42°C. Indeed, many mutants in DNA synthesis were isolated by similar experimental protocols. It should be noted that most temperature-sensitive mutants are missense mutants in which a single amino acid residue of the protein has been changed. This amino acid change alters folding at 42°C, but the behavior of the various mutant proteins differs in that some spontaneously renature on shift back to 30°C and others remain denatured. In the latter case, growth cannot resume until new protein is synthesized (and hence *properly* folded) into an active conformation at 30°C.

Another class of conditional mutations are cold-sensitive mutations. Mutants of *E. coli* can be isolated that grow normally at 37°C but cease growth at 20°C (wild-type strains grow well at both temperatures). The chemical basis for the loss of function at high temperature seen in temperature sensitive mutants is straightforward. The chemical bonds that maintain the proper secondary and tertiary structures of proteins are weakened at high temperature, and thus the mutant proteins unfold and become nonfunctional. Loss of function at low temperature is more subtle because most chemical bonds (the exception being hydrophobic interactions) are strengthened at low temperature. Probably for these reasons cold-sensitive mutations are restricted to a smaller number of genes than temperature-sensitive mutants. Cold-sensitive mutants are often defective in molecular interactions required for function rather than function per se. The change in primary structure and the weakened hydrophobic interaction at low temperatures often alter the associations of small molecules (such as allosteric effectors) with proteins and association of proteins with one another. For example, a well-understood example of a cold-sensitive mutation that affects the interaction of a protein with an allosteric effector is a mutant in histidine biosynthesis—the enzyme functions properly at 42°C, but at 20°C, the mutant enzyme is strongly feedback inhibited by its allosteric effector, histidine, resulting in insufficient histidine synthesis for growth. Examples of cold-sensitive mutants that affect protein-protein interactions include a number of mutants defective in the assembly of ribosomes at low temperatures—many mutants of ribosomal proteins alter the conformation of the protein at 20°C such that it cannot properly associate with other ribosomal proteins.

Isolation of Bacteriophage Mutants

Some classes of bacteriophage mutants are conceptually similar to various bacterial mutants. For example, mutants in plaque morphology or size can be readily isolated by simply looking at a lot of plaques. The clear (*c*) mutants of phage λ and the *rII* mutants of phage T4 were isolated in this manner. Mutants in most phage genes, however, prevent viral growth and hence production of a plaque. Thus more general approaches must be used. One of these is the isolation of temperature-sensitive phage mutants; that is, phages that form plaques on a host grown at 30°C but not on the same host grown at 42°C. The chemical rationale of such mutants is the same as that for bacterial temperature-sensitive mutants: The mutant phage protein denatures at 42°C where the protein encoded by a wild-type strain is stable. Cold-sensitive phage mutants have also been isolated, and as expected from the bacterial examples, these are usually defective in assembly of multiprotein complexes.

MUTAGENESIS

The production of a mutation requires a change in the nucleotide sequence. This change can be stimulated to occur in five main ways in vivo: (1) failure to remove an incorrectly inserted base; (2) tautomerization of a base, which allows a substitution to occur in subsequent replication; (3) a previously inserted base chemically altered to a base having different base-pairing specificity; (4) one or more bases skipped during replication; or (5) one or more extra bases inserted during replication. Mutations may be due to spontaneous, uncorrected errors or may be induced by specific mutagens (Table 10-2). Site-specific mutagenesis is described in Chapter 20.

Base-Analog Mutagens

A base analog is a compound sufficiently similar to one of the four DNA bases that it can be incorporated into a DNA molecule during normal replication. Such a compound must be able to hydrogen-bond with a base in the template strand,

Table 10-2   Types of mutagens

Mutagen	Mode of action	Example	Consequence
Base analog	Substitutes for a standard base during replication and causes a new base pair to appear in daughter cells in a later generation	5-Bromouracil	$A \cdot T \rightarrow G \cdot C$, and $G \cdot C \rightarrow A \cdot T$
		2-Aminopurine	$A \cdot T \rightarrow G \cdot C$
Chemical mutagen	Chemically alters a base so that a new base pair appears in daughter cells in a later generation	Nitrous acid	$G \cdot C \rightarrow A \cdot T$, and $A \cdot T \rightarrow G \cdot C$
		Hydroxylamine	$G \cdot C \rightarrow A \cdot T$
		Ethyl methane sulfonate (EMS)	$G \cdot C \rightarrow A \cdot T$, $G \cdot C \rightarrow C \cdot G$, and $G \cdot C \rightarrow T \cdot A$
		Ultraviolet light	All single base-pair changes are possible.
Intercalating agents	Addition or deletion of one or more base pairs	Acridines	Frameshifts
Mutator genes	Excessive insertion of incorrect bases or lack of repair of incorrectly inserted bases	_____	All single base-pair changes are possible.
None	Spontaneous deamination of 5-methylcytosine (MeC)	_____	$G \cdot MeC \rightarrow A \cdot T$

Note: Italicized changes in base pairs are transversions; those that are not italicized are transitions.

or the editing function removes it. If the base analog has two modes of hydrogen-bonding, however, it will be mutagenic.

The substituted base 5-bromouracil (BU) is a close analog of thymine because the bromine is about the same size as the methyl group of thymine (Figure 10-8). In subsequent rounds of replication, BU functions like thymine and primarily pairs with adenine. In discussing tautomerism earlier in this chapter, it was pointed out that thymine occasionally assumes a form that can pair with guanine. The mutagenic activity of 5-bromouracil stems in part from a shift in the keto-enol equilibrium caused by the bromine atom; that is, the enol form exists for a greater fraction of time for BU than for thymine. Thus, if BU replaces a thymine, in subsequent rounds of replication, it may pair with a guanine, which in turn specifies cytosine, resulting in formation of a GC pair (Figure 10-9). BU can also induce a change from GC to AT. The enol form is actually sufficiently prevalent that BU is sometimes (but infrequently) incorporated into DNA in that form. When that occurs, BU acts as an analog of cytosine rather than thymine. Even though it may become part of the DNA, while temporarily having the base-pairing properties of cytosine, the keto form is the predominant form; hence in subsequent rounds of replication, BU usually pairs like thymine. Thus, a GC pair, which as a result of an incorporation error, is converted to a GBU pair, ultimately becomes an AT pair, as shown in Figure 10-9b.

Experiments suggest that BU is also mutagenic in another way. Because of complex regulatory pathways for nucleotide synthesis, the BU nucleoside triphosphate inhibits production of dCTP. The ratio of TTP to dCTP then becomes quite high, and the frequency of misincorporation of T opposite G increases. The rate of misincorporation then exceeds the capacity of the editing and mismatch repair systems, and a persistent, incorrectly incorporated thymine will pair with adenine in the next round of DNA replication, yielding a GC → AT change in one of the daughter molecules.

Both base-pair changes induced by BU maintain the original purine (Pu)–pyrimidine (Py) orientation. That is, the original and the altered base pairs

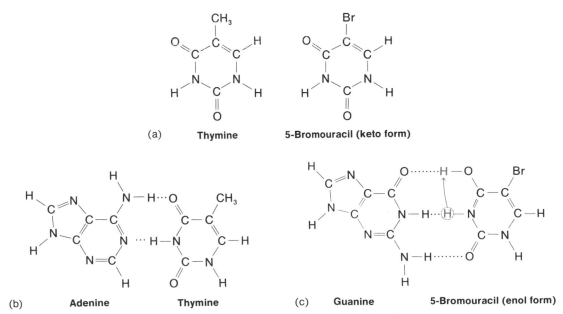

Figure 10-8.   Mutagenesis by 5-bromouracil. (a) Structural formulas of thymine and 5-bromouracil. (b) A standard adenine-thymine base pair. (c) A base pair between guanine and the enol form of 5-bromouracil. The orange H in the dashed circle shows the position of the H in the keto form. When tautomerization occurs, the orange double bond forms.

both have the orientation PuPy—for example, AT and GC. If the original pair was TA, the altered pair would be CG—that is, PyPu for both the original and the altered pairs. A base change that does not change the PyPu orientation is called a **transition**. Base-analog mutations are always transitions. Later we see changes from PuPy to PyPu and from PyPu to PuPy; when such a change of orientation occurs, the mutation is called a **transversion**. Note that BU induces transitions in both directions: AT → GC by the tautomerization route and GC → AT by the misincorporation route.

Chemical Mutagens

A chemical mutagen is a substance that can alter a base that is already incorporated in DNA and thereby change its hydrogen-bonding specificity. Three commonly used chemical mutagens are nitrous acid (HNO_2), hydroxylamine (HA), and ethylmethane sulfonate (EMS). The chemical structures of these mutagens are shown in Figure 10-10.

Nitrous acid primarily converts amino groups to keto groups by oxidative deamination. For example, cytosine and adenine are converted to uracil (U) and hypoxanthine (H), which form the base pairs UA and HC. Therefore the changes are GC → AT and AT → GC as cytosine and adenine are deaminated.

Hydroxylamine (NH_2OH) is often used to mutagenize DNA in vitro. When used in vitro, hydroxylamine reacts specifically with cytosine, converting it to a modified base (N_4-hydroxycytosine) that pairs with adenine instead of guanine. This has two consequences: (1) hydroxylamine produces only GC → AT transitions, and (2) mutations induced by hydroxylamine cannot be reverted by

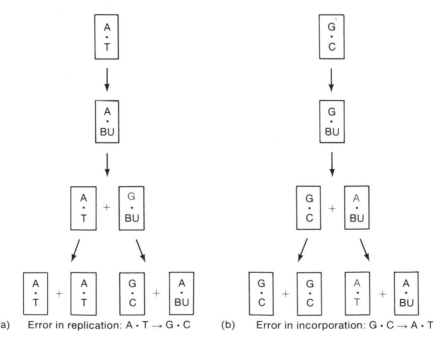

(a)   Error in replication: A · T → G · C

(b)   Error in incorporation: G · C → A · T

Figure 10-9. Two mechanisms of 5-bromouracil (BU)–induced mutagenesis. (a) During replication, BU, in its usual keto form, substitutes for T, and the replica of an initial AT pair becomes an ABU pair. In the first mutagenic round of replication, the BU, in its rare enol form, pairs with G. In the next round of replication, the G pairs with a C, completing the transition from an AT pair to a GC pair. (b) During replication of a GC pair, a BU, in its rare enol form, pairs with a G. In the next round of replication, the BU is again in the common keto form, and it pairs with A, so the initial GC pair becomes an AT pair. The replica of the ABU pair produced in the next round of replication is another AT pair.

hydroxylamine. (When hydroxylamine is used in vivo, it produces free radicals that damage the DNA, and this DNA damage induces the SOS system, resulting in a wide variety of types of mutations.)

EMS is an alkylating agent. Many sites in DNA are alkylated by these agents; the major effect of EMS is the addition of an alkyl group to the hydrogen-bonding oxygen of guanine and thymine. These alkylations impair the normal hydrogen-bonding of the bases and cause mispairing of G with T, leading to the transitions AT → GC and GC → AT (the latter markedly predominates). EMS also reacts with adenine and cytosine.

Another phenomenon resulting from alkylation of guanine is depurination, or loss of the alkylated base from the DNA molecule by breakage of the bond joining the purine nitrogen and deoxyribose. Depurination is not always mutagenic because the gap left by loss of the purine can be efficiently repaired. Sometimes, however, the replication fork may reach the apurinic site before repair has occurred. When this happens, replication stops just before the apurinic site, the SOS system is activated, and replication proceeds, almost always putting an adenine nucleotide in the daughter strand opposite the apurinic site. Because the original parental base (which was removed) was a purine, the base pair at that site will be a mismatch (PuA), and after replication, the base pair at that site will change orientation from PuPy to PyPu, the first example we have seen of a transversion. Treatment of phages with buffers at pH 4 also produces depurinations. On replication of phages treated in this way, numerous transversions occur, in agreement with the adenine-insertion mechanism just suggested.

Mutagenesis by Intercalating Agents

Acridine orange, proflavine, and acriflavine (Figure 10-11) are planar, heterocyclic molecules whose dimensions are roughly the same as those of a purine-pyrimidine pair. In aqueous solution, these substances can insert into DNA between the adjacent base pairs, a process called **intercalation**. When DNA containing intercalated acridines is replicated, additional bases appear in the sequence (Figure 10-12). The usual addition is a single base, although occasionally

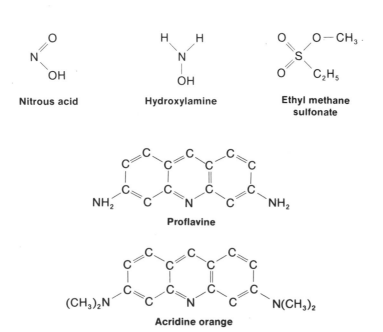

Figure 10-10. Structures of three chemical mutagens.

Figure 10-11. Structures of two mutagenic acridine derivatives.

two bases are added. Deletion of a single base also occurs, but this is far less common than base addition. Mutations of this sort are called **frameshift mutations**. Although the DNA sequence is correct beyond the point of the insertion or deletion, because the base sequence is read in groups of three bases during translation into an amino acid sequence, the addition or deletion of one or two bases changes the reading frame (Figure 10-13). Thus, downstream of the frameshift mutation, out-of-frame codons will be read until translation is terminated by a nonsense codon in the improper reading frame.

Streisinger and coworkers proposed the "strand-slippage" model to explain how frameshift mutations are formed. Frameshift mutations usually occur in sequences with monotonous repeats of one or a few base pairs. The strand-slippage model proposes that during DNA replication, the strands may separate, slip, then re-pair in such a way that one or two bases are looped out (see Figure 10-12). As the DNA is elongated, extra bases would be added to one strand of the DNA. Intercalating agents may induce frameshift mutations by stabilizing the looped out structure.

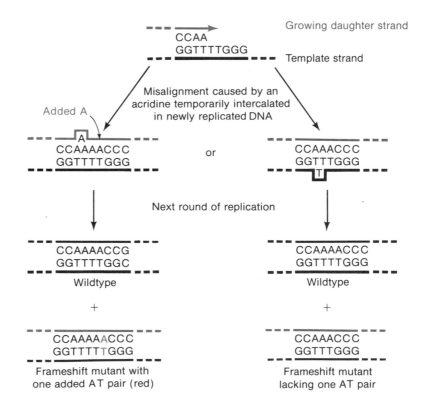

Figure 10-12. Proposed mechanism for misalignment mutagenesis generated by intercalation of an acridine molecule in the replication fork. Acridine is present only in the first round of replication; the second round serves to produce a true-breeding mutant. The left and right paths generate a base addition and deletion. The growing daughter strand is shown in orange.

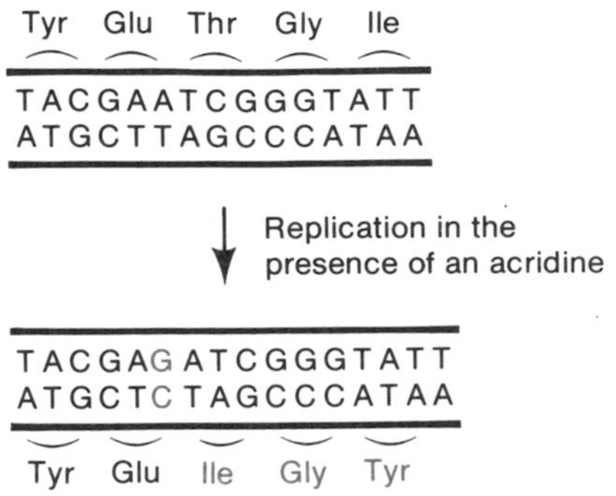

Figure 10-13. A base addition (red) resulting from replication in the presence of an acridine. The change in amino acid sequence read from the upper strand in groups of three bases is also shown in orange.

Mutator Genes

Specific mutations can also cause other mutations to appear frequently in other genes throughout the chromosome. This is called a mutator phenotype, and the corresponding genes are called *mut*. The normal function of *mut* genes is to keep the mutation frequency low; that is, it is only when the product of a mutator gene is itself defective that the frequency of spontaneous mutations increases. For example, *mutD* is a mutation in the proofreading function of DNA polymerase III (*dnaQ*). Another is a mutant *dam* gene, the gene responsible for the methylation of DNA and directing the mismatch repair system to the correct template strand (see Chapter 8, Figure 8-8). In a *dam*⁻ mutant, there is little or no methylation of the parent (or any other) strand, so about half the time, the mismatch repair system excises the parental (correct) base and inserts a base that can pair with the daughter (incorrect) base. The products of the mutator genes *mutH*, *mutL*, and *mutS* also participate in mismatch repair. However, the function of some mutator genes is still poorly understood.

REVERSION

So far, we have discussed changes from the wild-type to the mutant state. The reverse process, in which the wild-type phenotype is regained, also occurs. This process is called back mutation, or **reversion**, and the resulting mutant is called a revertant. Reversion may occur in two ways: (1) a "**true reversion**" is due to a back mutation that exactly restores the original DNA sequence; (2) a "**pseudo-reversion**" or "**suppressor mutation**" is due to an additional mutation at a second site that restores the original phenotype. Reversion can occur spontaneously, or it can be induced by mutagens.

Reversion Frequency

The probability of obtaining a revertant depends on the nature of the original mutation. Thus, the reversion frequency is sometimes used as a criterion for identifying the type of mutation present in a mutant. The **reversion frequency** is the fraction of cells in a population of mutants that regain the original phenotype per generation. Point mutations revert at the highest frequency because a single base change in the DNA is sufficient to restore the original sequence, and spontaneous mutations that yield true revertants occur at some measurable frequency (typically about 10^{-8}). In contrast, deletion mutations cannot revert by repairing the original DNA sequence because the probability that the missing DNA will be replaced with material having an equivalent base sequence, thereby restoring a functional gene, is virtually zero.

With bacteria, many reversion events can be selected by measuring the ability of a population of bacteria to form colonies on solid growth medium. For example, of 10^9 cells of a *pro*⁻ bacterial strain are placed on a solid medium lacking proline, about 10 colonies arise; these colonies are formed by spontaneous Pro⁺ revertants (cells able to grow without an external supply of proline). The reversion frequency in this case is $10/10^9 = 10^{-8}$. The production of a spontaneous revertant is a random process, so the reversion of a double mutant would require two independent events. If each were to occur at a frequency of 10^{-8}, the frequency of reversion of a double mutant would be $(10^{-8})^2 = 10^{-16}$—so rare that it would probably never be detected.

Second-Site Intragenic Revertants

True revertants regain the wild-type phenotype by restoring the wild-type genotype (that is, the wild-type DNA sequence of the gene). It is also possible, however, to regain the wild-type phenotype by acquiring an additional mutation in the gene.

The frequency of spontaneous reversion of a typical point mutation suggests that spontaneous reversion rarely results in a restoration of the wild-type base sequence. The typical frequency of finding a mutation in a specific gene is about 10^{-7}. For example, if 10^9 cells of a Leu$^-$ mutant are plated on medium lacking leucine, about 100 colonies will grow. If there are about 500 sites in the gene that can give the Leu$^-$ phenotype, however, the probability of a mutation in a specific site is $^1/_{500} \times 10^7$ or $^1/_5 \times 10^9$ cells.

Furthermore, a precise reversion requires a specific mutation in a specific site. Because any base can be changed to three other bases, the original base would be restored at a frequency of one-fourth of the frequency of a mutation at that one specific position. It is usually found, however, that in a population derived from a single Leu$^-$ mutant, about 1 in 10^8 cells is Leu$^+$. For any particular mutation, the reversion frequency can vary widely, but it is frequently observed that the fraction of the mutant population with the revertant phenotype is much too high to be explained by a precise return to the wild-type base sequence. The explanation is simply that reversion events occurring at many different sites in a gene can produce the same phenotype. Reversion due to an additional mutation in a gene that does not change the mutation at the original mutant site is called a **second site intragenic mutation** or an **intragenic suppressor mutation**. This has been confirmed by biochemical data in which it has been shown that the amino acid sequence in a revertant is rarely the wild-type sequence: The original mutant amino acid substitution is usually still present. Some suppressor mutations are due to a base change at the original site, but with insertion of a non–wild-type amino acid at that site.

Consider a hypothetical protein containing 97 amino acids whose structure is determined by an ionic interaction between a positively charged (+) amino acid at position 18 and a negative one (–) at position 64 (Figure 10-14). If the (+) amino acid were replaced by a (–) amino acid, the protein would clearly be inactive. Three kinds of reversion events could restore activity (Figure 10-14a): (1) The original (+) amino acid could be replaced. (2) A different (+) amino acid could be put at position 18. (3) The (–) amino acid at position 64 could be replaced by a (+) amino acid; this second site mutation would restore the interaction between (+) and (–) charges. In addition, sometimes insertion of a (+) amino acid at position 17 or 19 would work.

Figure 10-14b shows another more complicated example of **intragenic reversion**. In this case, the structure of a protein is maintained by a hydrophobic interaction. The replacement of an amino acid with a small side chain by a bulky phenylalanine changes the shape of that region of the protein. A second amino acid substitution providing space for the phenylalanine could restore the protein structure.

Second-Site Intragenic Revertants of Frameshift Mutations

Reversion of frameshift mutations is often due to another mutation at a second site. It is of course possible that a particular added base could be removed or a particular deleted base could be replaced, but such true revertants are usually quite rare. Instead reversion of frameshift mutations is usually due to a compensating frameshift mutation in the gene: For example, reversion of a +1 frameshift mutation can occur by a −1 frameshift mutation. Second site reversion of a frameshift mutation has two requirements, illustrated in Figure 10-15: (1) The reversion event must be near the original site of mutation, so few amino acids are altered between the two sites; and (2) the segment of the polypeptide chain in which both changes occur must be able to withstand substantial alterations.

Reversion as a Test of Cause and Effect

Mutations are usually introduced to enable some biological system to be understood. As pointed out in Chapter 1, if the properties of a mutant are to be

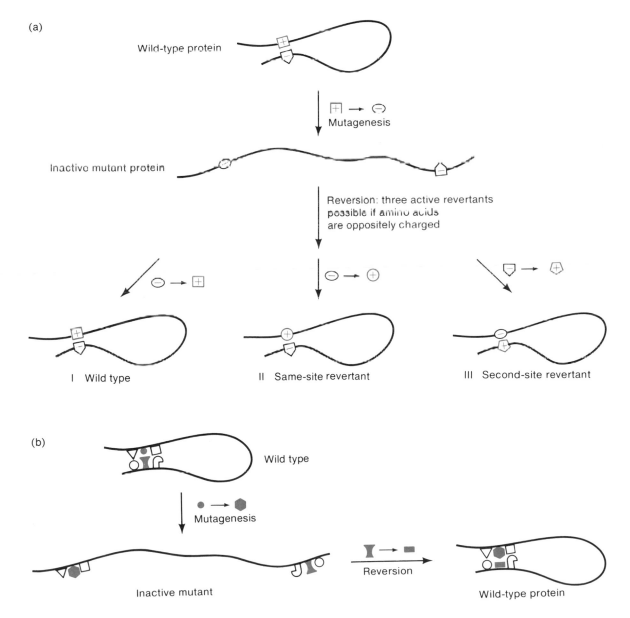

Figure 10-14. Several mechanisms of reversion. (a) The charge of one amino acid is changed, and the protein loses activity. The activity is returned by (I) restoring the original amino acid, or (II) by replacing the (−) amino acid by another (+) amino acid, or (III) by reversing the charge of the original (−) amino acid. In each case, the attraction of opposite charges is restored. (b) The structure of the protein is determined by interactions between six hydrophobic amino acids. Activity is lost when the small circular amino acid is replaced by the bulky hexagonal one and is restored when space is made by replacing the concave amino acid by the small rectangle.

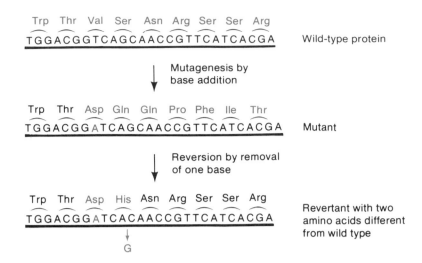

Figure 10-15. Reversion by base deletion from an acridine-induced, base-addition mutant.

compared with those of a wild-type bacterium or phage, it is essential that only a single mutation be introduced. That is, the mutant and the wild-type organisms must be isogenic except for a single mutation because otherwise it is not possible to know what really caused a change in properties. This problem is important in mutagenesis because some mutagens tend to produce multiple mutations in clusters. Reversion is a useful test to determine whether the phenotype is due to a specific mutation because if an observed mutant phenotype were the result of two mutations, reversion of the phenotype would occur at an exceedingly low frequency.

Consider a proline auxotroph, isolated from a mutagenized population, that grows slowly even when proline was added to the medium. One might hypothesize that the mutation affected a protein that was common to the proline biosynthetic pathway and to some other important pathway. Alternatively, the mutant might carry mutations in two distinct genes. The hypothesis could be distinguished by plating a large number of mutants on medium lacking proline and selecting a Pro⁺ revertant. If the revertant grew at a normal rate, one would conclude that both the proline requirement and the slow growth rate resulted from the same mutation. If the Pro⁺ revertant grew very slowly, however, two mutations in distinct genes were probably present in the original mutant.

Reversion as a Means of Detecting Mutagens and Carcinogens

With the increased number of chemicals used and accumulated as environmental contaminants, it is important to have quick, simple tests to determine whether a chemical is a carcinogen. Most carcinogens are also mutagens, so assaying mutagenicity can be used for the initial screening for these hazardous agents. One simple method for screening large numbers of substances for mutagenicity is a reversion test using auxotrophic mutants of bacteria. In the simplest type of reversion test, a compound that is a potential mutagen is added to solid growth media, known numbers of a mutant bacterium are plated, and the number of revertant colonies that arise is counted. A significant increase in the reversion frequency above that obtained in the absence of the compound tested would identify the substance as a mutagen. Simple tests of this type, however, fail to demonstrate the mutagenicity of a number of potent carcinogens. The explanation for this failure is that some substances are not directly mutagenic (or carcinogenic) but are converted to active compounds by enzymatic reactions that occur in the liver of animals. The normal function of these enzymes is to protect the organism from various noxious substances that occur naturally by chemically converting them to nontoxic substances. When the enzymes encounter certain compounds, however, they convert these substances, which may not be themselves directly harmful, into mutagens or carcinogens. The enzymes are contained in a component of liver cells called the microsomal fraction. Addition of the microsomal fraction of the rat liver to the growth medium as an activation system has been used to extend the sensitivity and usefulness of the reversion test system. Ames pioneered the use of the microsomal fraction in the **Ames Test** for carcinogens.

In the Ames test, histidine requiring (His$^-$) mutants of *Salmonella typhimurium*, containing either a base substitution or a frameshift mutation, is used to test for reversion to His$^+$. The strain also contains a plasmid that enhances SOS repair (see Chapter 9). In addition, the bacterial strains have been made more sensitive to mutagenesis by the incorporation of mutations that inactivate the excision repair system (see Chapter 9) and make the cells more permeable to foreign molecules. Because some mutagens act only on replicating DNA, the solid medium used contains enough histidine to support a few rounds of replication but not enough to permit formation of a colony. The procedure is the following. Rat-liver microsomal fraction is spread on the agar surface, and bacteria are plated. Then a paper disc saturated either with distilled water (as a control) or a solution of the compound being tested is placed in the center of the plate. The test compound diffuses outward from the disc, forming a concentration gradient. If the substance is a mutagen or is converted to a mutagen, colonies form. With a highly effective mutagen, colonies will be present all over the surface of the medium, including far from the disc where the concentration is low; in contrast, with a weak mutagen, colonies will form only very near the disc, where the concentration is high. The procedure is highly sensitive and permits the detection of weak mutagens. A quantitative analysis of reversion frequency can also be carried out by incorporating known amounts of the potential mutagen in the medium. The reversion frequency depends on the concentration of the substance being tested and, for a known carcinogen or mutagen, correlates roughly with its known effectiveness.

The Ames test has now been used with thousands of substances and mixtures (such as industrial chemicals, food additives, pesticides, hair dyes, and cosmetics), and numerous unsuspected substances have been found to stimulate reversion in this test. A high frequency of reversion does not mean that the substance is definitely a carcinogen but only that it has a high probability of being so. As a result of these tests, many industries have reformulated their products: for example, the cosmetic industry has changed the formulation of many hair dyes and cosmetics to render them nonmutagenic. Ultimate proof of carcinogenicity is determined from testing for tumor formation in laboratory animals. The Ames test and several other microbiological tests (e.g., the Devoret test; see Chapter 17) are used to reduce the number of substances that have to be tested in animals because to date only a few percent of more than 300 substances known from animal experiments to be carcinogens failed to increase the reversion frequency in the Ames test. Thus, these microbial tests greatly reduce the number of animals used for this type of testing.

SUPPRESSION

Intergenic reversion may also be due to a mutational change in a second gene that eliminates or "suppresses" a mutant phenotype. Intergenic suppression can occur in several ways.

1. *Informational suppressors.* It is possible to isolate suppressors that change the cell's translational machinery in such a way that the mutation is misread, making a functional protein from a mutant gene. Such suppressor mutations are called **informational suppressors** because they change the way the cell reads the "information" in the mRNA. Most informational suppressors are due to mutations in tRNA genes, as described in detail later. Some informational suppressors, however, are due to mutations in ribosomal genes that cause the ribosomes to translate the mRNA incorrectly at a high frequency. One of these mutations is called *ram* for ribosomal ambiguity. Because these ribosomal mutations cause errors in all proteins, such mutants are usually very unhealthy.

2. *Interaction suppressors.* When two proteins interact, a mutation in one of the genes may have a mutant phenotype simply because it disrupts proper protein-protein interactions. In such cases, it is possible to isolate **interaction suppressors** in the second gene that restore the interaction. Interaction suppressors are usually "**allele specific**"—that is, a specific interaction suppressor suppresses only a small subset of mutations in a gene. Often such suppressor mutations define sites on the second protein that directly interact with the mutant site on the mutant protein. Interaction suppressor mutants are sometimes used to identify other proteins that interact with a mutant protein or as a genetic approach for studying protein-protein interaction. This approach is especially useful for studying complex interactions that are difficult to define biochemically. For example, DNA synthesis requires many different gene products that must interact with each other in precise but poorly understood ways. Maurer and Botstein selected for interaction suppressors that define specific protein-protein interactions between several of these gene products.

3. *Overproduction suppressors.* Because many mutants produce a protein product with lower activity than is required for growth, it is sometimes possible to overcome the mutant phenotype by simply producing more of the mutant protein. Suppressors that increase the amount of the mutant protein are often due to mutations in regulatory genes or regulatory sites that increase the synthesis of the mutant protein. For example, promoter-up mutations can be selected as suppressors of a *lacZ* missense mutant. Another way of increasing the amount of a mutant protein is to decrease its degradation. Mutant proteins are often more sensitive to cellular proteases than the wild-type protein, so mutations that decrease the amount of a protease may also increase the intracellular concentration of the mutant protein and thereby suppress the mutant phenotype.

4. *Bypass suppressors.* In the aforementioned classes of suppressors, the suppressor mutation restores only the original phenotype of certain mutations in a gene—the suppressor mutation will not restore the function of every mutation in a gene. For example, a tRNA suppressor that suppresses a nonsense mutation in the *proA* gene would not suppress a deletion mutation in the *proA* gene. In contrast, **bypass suppressors** turn on a new pathway that eliminates the need for the mutant gene. For example, it is possible to isolate *argD* mutations that suppress *proA* or *proB* mutations: Mutations in the *proA* or *proB* gene are unable to synthesize glutamate semialdehyde (an intermediate in proline biosynthesis), but *argD* mutations allow the synthesis of glutamate semialdehyde from N-acetylglutamate semialdehyde (an intermediate in arginine biosynthesis). Thus, the *argD* suppressor mutation completely bypasses the first two enzymatic steps of proline biosynthesis. Alternatively, some bypass suppressors turn on an inactive ("cryptic") gene that has a similar function.

5. *Physiological suppressors*. Some mutations may be suppressed by general changes in cell physiology. For example, many missense mutations produce proteins that are still functional but are unstable. These mutants may be suppressed by mutations that increase the intracellular concentration of molecules that stabilize the proteins or other proteins that help the mutant protein fold properly ("chaparonins").

tRNA Suppressors

Because they directly affect the translation of specific codons, tRNA suppressors not only suppress the phenotype of the original mutation, but also can suppress certain mutations in many other genes as well. Such "**suppressor-sensitive mutations**" are a type of conditional mutation: The mutation has the wild-type phenotype in cells that produce the tRNA suppressor but has a mutant phenotype in cells that do not produce the tRNA suppressor. (Recall that temperature-sensitive mutations are another type of conditional mutation.) For example, a phage with a suppressor sensitive mutation in an essential gene will grow in a strain of bacteria that produces a tRNA suppressor (denoted *sup*) but will not grow in other strains that do not produce the tRNA suppressor (*sup*°). This type of suppression is due to changes in the translation system.

Suppressor-sensitive mutations are of two main types: nonsense or chain termination mutations and missense or amino acid substitution mutations. Nonsense suppressors are considered first.

Genetic Detection of a Nonsense Suppressor

Nonsense mutations are base substitutions that introduce a premature translational stop codon—UAG ("amber"), UAA ("ochre"), or UGA ("opal")—in a gene. Such chain termination mutations are common. For example, a single base substitution in any of the codons AAG, CAG, GAG, UCG, UUG, UGG, UAC, and UAU can give rise to the chain termination codon UAG. If such a mutation occurs within a gene, a mutant protein results because there is no tRNA molecule whose anticodon is complementary to UAG (see Chapter 6). Thus, only a fragment of the wild-type protein is produced, and this truncated protein usually has little or no biological function (unless the mutation is near the carboxyl terminus of the wild-type protein).

In the appropriate suppressor mutants, a nonsense mutation does not cause chain termination. For example, if a specific Tyr residue in the *lacZ* gene is mutated to a nonsense codon, the protein will terminate at that site, making the bacterium genotypically and phenotypically Lac⁻ (Figure 10-16). Lac⁺ revertants can be isolated by mutagenizing a population of the Lac⁻ cells, allowing several generations of growth to express the reversion, and plating the culture on lactose-minimal agar, to select for Lac⁺ colonies. Three classes of revertants can be found. In the first class of revertants, the chain termination mutation is reversed by a base substitution mutation that converts UAG back to UAC. In this class of revertant, the complete protein chain is present, the tyrosine is restored at the correct position, and the protein has the wild-type amino acid sequence. Such "true revertants" are the rarest class of revertants. In the second class of revertants, a new amino acid is substituted in the complete protein, and the base sequence is likewise altered from the original wild-type sequence. In this case, the chain termination mutation has been changed to a mutation that has a wild-type or near wild-type phenotype. For example, in the figure this has happened as a result of a base substitution mutation converting UAG to UCG, the serine codon. The substitution of serine

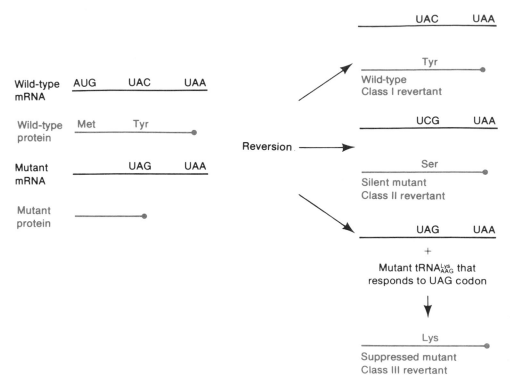

Figure 10-16.   Three types of revertants of a chain-termination mutation.

for tyrosine does not markedly alter function of the protein; that is, serine is an acceptable amino acid at this position in this protein. Both of these two classes of revertants are due to intragenic mutations. The third class of revertants is due to intergenic mutations. These revertants produce a complete polypeptide chain, usually by supplying some other amino acid at the mutant site without in any way altering the original mutant base sequence—that is, the codon remains UAG. Such intergenic revertants often suppress nonsense mutations in other genes as well. The revertant cell has gained the ability to ignore, or suppress, UAG-type chain termination mutations by translating the mutant UAG codon into an amino acid.

The molecular explanation for this type of suppression is that the class III revertant, which is called a suppressor mutant, contains an altered tRNA molecule—one that has the anticodon CUA, which can pair with UAG. This mutant tRNA is called a **suppressor tRNA**.

How can such a mutant tRNA molecule arise? Because a tRNA molecule is the product of a tRNA gene, these mutants occur in exactly the same way as any other mutant—by a replication error. Clearly it must alter a tRNA gene. Thus, in the example given previously, a $tRNA^{Lys}$ molecule whose anticodon is CUU has been altered to have the anticodon CUA, which can hydrogen-bond to the codon UAG.

Because a single base change is sufficient to alter the complementarity of an anticodon and a codon, there are (at most) eight tRNA molecules with a complementary anticodon that, with a single base change, can recognize a UAG codon. Thus, suppressor tRNAs can substitute the following amino acids at the site of a chain termination codon: Lys (AAG), Gln (CAG), Glu (GAG), Ser (UCG), Trp (UGG), Leu (UUG), and Tyr (UAC and UAU). Note that these are the same amino acid codons that can be altered by mutation to form a UAG site.

Suppressors also exist for chain termination mutations of the UAA and UGA type. These too are mutant tRNA molecules whose anticodons are altered by a single base change.

In conventional notation, suppressors are given the genetic symbol *sup* followed by a letter or number that distinguishes one suppressor from another. The genotype of cells lacking a suppressor is designated sup° or sup^+.

Several features of nonsense suppression are important:

1. A particular UAG suppressor will not suppress all UAG chain termination mutations because the amino acid it inserts will not always produce an active protein.
2. Often the suppression is incomplete: The new proteins may have sufficient activity for colony formation, but the specific activity of the protein may be less than the wild-type protein.
3. A cell can survive the presence of a suppressor only if the cell also contains two or more copies of that tRNA gene. If there were only one copy of the $tRNA^{ser}$ gene and it was mutated so $tRNA^{ser}$ recognized only the UAG codon, all normal UCG codons would be read as stop codons. There are, however, multiple copies of most tRNA genes, so in any living cell containing a suppressor tRNA, there will always be an additional copy of a wild-type tRNA that can function in normal translation.

Normal Termination in the Presence of a Suppressor tRNA

You would predict that if a cell contains a UAG suppressor, genes that end with a single UAG codon would not terminate, so the existence of a suppressor tRNA would be lethal. Two features of the normal chain termination process allow the survival of a suppressor-containing cell:

1. Protein factors active in translation termination respond to chain termination codons even when a tRNA molecule that recognizes the codon is present; thus suppression is weak. For example, if termination were suppressed only 10% of the time, the amount of the mutant protein made would be 10% of the wild-type, but 90% of other proteins would properly terminate.
2. Normal chain termination often uses pairs of distinct termination codons such as the sequence UAG-UAA. Thus, the existence of a UAG suppressor would not prevent termination of a double terminated protein.

It is likely that both of these mechanisms are responsible for the existence of viable suppressor-containing organisms. Suppression of UAG and UGA mutations can be efficient; some nonsense suppressors prevent termination more than 50% of the time. (Note that the activity of the suppressed mutant may not always be as high as 50% because the particular amino acid insertion may affect the structure and function of the protein rather than because of inefficient suppression of chain termination.) A partial explanation for the viability of organisms containing UAG and UGA suppressors is that when either UAG or UGA is used as a natural stop signal, it usually occurs adjacent to or either 3 or 6 base pairs away from a second, different stop codon in the same reading frame.

Natural chain termination is accomplished most commonly by a single UAA codon. This is consistent with the observation that UAA suppressors are typically inefficient, preventing termination at UAA codons only between 1% and 5% of the time. Furthermore, cells containing UAA suppressors are generally unhealthy and grow more slowly than cells lacking any suppressor or containing either a UAG or UGA suppressor. Presumably some critical proteins are damaged by not being terminated properly in UAA suppressor mutants.

Missense Suppressors

Suppression can also occur for missense mutations. For example, a protein in which valine (nonpolar) has been mutated to aspartic acid (polar), resulting in loss of activity, can be restored to the wild-type phenotype by a missense suppressor that substitutes alanine (nonpolar) for aspartic acid. Such a substitution can occur in three ways: (1) a mutant tRNA molecule may recognize two codons, (2) a mutant tRNA molecule may be incorrectly recognized by an aminoacyl synthetase so it carries the wrong amino acid, and (3) a mutant aminoacyl synthetase may charge an incorrect tRNA molecule. Suppression of missense mutations is necessarily inefficient. If a suppressor that substitutes alanine for aspartic acid worked with 20% efficiency, in virtually every protein molecule synthesized by the cell, at least one aspartic acid would be replaced, which is a situation that a cell could not possibly survive. The usual frequency of missense suppression is about 1%. In this way, a small amount of a functional, essential protein is made, and thereby a mutant cell is able to survive. Missense suppression, however, still introduces a significant number of defective proteins of all other types, and as a result, a cell carrying a missense suppressor usually grows slowly and is generally unhealthy.

Frameshift Suppressors

It is also possible to isolate tRNA suppressors that suppress certain frameshift mutations. Frameshift suppressors often change the tRNA so the anticodon recognizes four bases instead of three bases as normal. These tRNA suppressors shift the translational reading frame at a low frequency, thus restoring translation of the correct amino acid sequence downstream of the frameshift mutation.

KEY TERMS

base-analogs	spontaneous mutations
cold-sensitive mutations	suppressor mutations
conditional lethal mutations	intragenic suppressor
enrichment	intergenic suppressor
frameshift mutations	informational suppressors
intercalating agents	interaction suppressors
leaky mutations	overproduction suppressors
mutagenesis	bypass suppressor
mutator genes	physiological suppressors
phenotypic expression	nonsense suppressors
pseudorevertant	missense suppressors
reversion	frameshift suppressors
screening	suppressor tRNA
selection	temperature-sensitive
silent mutations	mutations

QUESTIONS AND PROBLEMS

1. Is a base-pair change always a mutation?

2. Does a base-pair change necessarily change the phenotype?

3. Which of the following base-pair changes are transitions, and which are transversions? (a) AT → TA; (b) AT → GC; (c) GC → TA.

4. Do base analogs produce transitions or transversions?

5. Assuming that all possible base changes can occur with equal frequency, what would be the ratio of transversions to transitions in a large collection of mutants?

6. Distinguish a missense and a nonsense mutation.

7. What makes a particular mutation temperature sensitive?

8. How can tautomerization cause mutation?

9. How does 5-bromouracil induce mutations?

10. Does a frameshift always cause a phenotypic change?

11. A deletion occurs that eliminates a single amino acid in a protein. How many base pairs were deleted?

12. A mutant is isolated that cannot be reverted. What biochemical type(s) of mutation might it carry?

13. One class of point mutation cannot be reverted by intragenic reversion at a second site. What is this class?

14. Why is a liver microsomal fraction included in the Ames test for mutagens?

15. One type of conditional mutation is the cold-sensitive (Cs) mutation, which has a mutant phenotype below a particular temperature. Bacteria containing the mutation *ess-2*(Ts) can form colonies at 32°C but not at 37°C and 42°C, whereas those containing the mutation *ess-5*(Cs) form colonies at 42°C but not at 32°C or 37°C. What would be the phenotype of an *ess-2*(Ts) *ess-5*(Cs) double mutant?

	32°C	37°C	42°C
ess-2(Ts)	+	–	–
ess-5(Ts)	–	–	+

16. *E. coli* DNA Pol III possesses several enzymatic activities. Two important activities are the polymerizing function and the 3′–5′ exonuclease. Mutant polymerases have been found that either increase or decrease mutation rates in an organism containing the mutant enzyme. A mutant that increases the mutation rate is called a mutator; a mutant that decreases the mutation rate is called an antimutator. The mutator and antimutator activities are usually a result of changes in the ratio of the two enzymatic activities described earlier. How do you think the ratios change in a mutator and in an antimutator?

17. An enzyme has the property that if amino acid 28, which is glutamic acid, is replaced by asparagine in a mutant, all activity is lost. If in this mutant protein, amino acid 76, which is asparagine, is then replaced by glutamic acid, full activity of the enzyme is restored. What can you say about amino acids 28 and 76 in the normal protein?

18. If a cell contains 2000 genes and if the average mutation rate per gene is 1×10^{-5} per generation, what is the average number of new mutations per cell per generation?

19. A fluctuation test was carried out to estimate the mutation rate of an *E. coli* locus conferring resistance to phage T1. If 5 of the 12 small independent cultures contained no phage-resistant mutants after growth of the cultures was completed and the average number of bacterial cells per culture was 5×10^8, what is the estimated rate of mutations to T1 resistance?

20. A fluctuation test is carried out for two different genes *A* and *B*. The following data are obtained. For gene *A*, 22 of 40 cultures had no mutants, with N = 5.6×10^8. For gene *B*, 15 of 37 cultures had no mutants, with N = 5×10^8. What are the mutation frequencies for the two genes?

21. In microorganisms, which mutation rates are more easily measured: ability to inability to synthesize proline (*pro*⁺ to *pro*⁻) or the reverse (*pro*⁻ to *pro*⁺)?

22. Which of the following amino acid substitutions would be likely to yield a mutant phenotype if the change occurred in a fairly critical part of a protein: (a) Pro → His; (b) Arg → Lys; (c) Thr → Ile; (d) Val → Ile; (e) Gly → Ala; (f) His → Tyr; (g) Gly → Phe?

23. Several hundred independent missense mutants have been isolated in the *trpA* gene which encodes *E. coli* tryptophan synthetase, a protein having 268 amino acids. Fewer than 30 of the positions were represented with one or more mutant. Why do you think that the number of different positions represented by amino acid changes is so limited?

24. The molecule 2-aminopurine is an analog of adenine, pairing with thymine. It also pairs on occasion with cytosine. What types of mutations will be induced by 2-aminopurine?

25. Nitrogen mustard reacts efficiently with guanine, causing ring cleavage and subsequent hydrolysis of the N-glycosidic bond. What base-pair change does this cause?

26. Can a mutation induced by nitrous acid be induced to revert at the same site by treatment with nitrous acid?

27. Two hundred Leu$^-$ mutants of a bacterial strain are examined separately to determine reversion frequencies. Of these, 90 revert at a frequency of 10^{-5}, 98 at 3×10^{-6}, 6 at 3×10^{-11}, and 6 at 10^{-10}.

  a. What type of mutant is probably contained in the class whose reversion frequency is 10^{-11}: single-point mutations, double-point mutations, or deletions?

  b. Can you say anything from these frequencies about whether any of the classes of mutations are chain termination mutations?

28. Revertants of temperature-insensitive mutations often prove to be temperature sensitive. That is, they exhibit a wild-type phenotype at low temperature and mutant at a higher temperature. Explain this phenomenon.

29. A mutation of a bacterial Lac$^+$ strain yielding a Lac$^-$ colony was isolated. Several lines of experiments indicate that the mutation resulted from production of a UGA codon. Spontaneous revertants were found at a frequency of 10^{-8} per cell per generation, and 9 of 10 of them were caused by suppressor tRNA molecules. What do you think is the rate of production of suppressor mutations in the original Lac$^+$ culture?

REFERENCES

*Ames, B. W. 1979. Identifying environmental chemicals causing mutations and cancer. *Science*, 204, 587.

Ames, B. W., et al. 1973. Carcinogens are mutagens: a simple test system combining liver homogenates for activation and bacteria for detection. *Proc. Natl. Acad. Sci.*, 70, 2381.

*Bossi, L. 1985. Informational suppression. In *Genetics of Bacteria*. Academic Press.

*Botstein, D., and R. Maurer. 1982. Genetic approaches to the analysis of microbial development. *Ann. Rev. Genet.*, 16, 61.

Coulondre, R., et al. 1978. Molecular basis of base-substitution hotspots in *E. coli*. *Nature*, 274, 775.

*Cox, E. C. 1976. Bacterial mutator genes and the control of spontaneous mutation. *Ann. Rev. Genet.*, 10, 135.

Davis, B. D. 1948. Isolation of biochemically deficient mutants of bacteria by penicillin. *J. Amer. Chem. Soc.*, 70, 4267.

Drake, J. W. 1970. *The Molecular Basis of Mutation*. Holden-Day.

Drake, J. W. 1991. Spontaneous mutation. *Ann. Rev. Genet.*, 25, 125.

Freese, E. 1959. The specific mutagenic effect of base analogues on phage T4. *J. Mol. Biol.*, 1, 87.

Garen, A., and S. Garen. 1963. Complementation in vivo between structural mutants of alkaline phosphatase from *E. coli*. *J. Mol. Biol.*, 7, 13.

Gorini, L., and H. Kaufman. 1960. Selecting bacterial mutants by the penicillin method. *Science*, 131, 604.

Grossman, A., R. Burgess, W. Walter, and C. Gross. 1983. Mutations in the *lon* gene of *E. coli* phenotypically suppress a mutation in the sigma subunit of RNA polymerase. *Cell*, 32, 151.

*Hartman, P., and J. Roth. 1973. Mechanisms of suppression. *Advances in Genetics*, 17, 1.

Hill, C. 1975. Informational suppression of missense mutations. *Cell*, 6, 419.

Hong, J. S., and B. N. Ames. 1971. Localized mutagenesis of any small region of the bacterial chromosome. *Proc. Natl. Acad. Sci.*, 68, 3158.

*Resources for additional information.

Kohno, T., M. Schmid, and J. Roth. 1980. Effect of electrolytes on growth of mutant bacteria. In D. Rains, R. Valentine, and A. Hollaender (eds.), *Genetic Engineering of Osmoregulation*. Plenum.

Kurland, C. 1992. Translational accuracy and the fitness of bacteria. *Ann. Rev. Genet.*, 26, 29.

Lederberg, J., and E. Lederberg. 1952. Replica plating and indirect selection of bacterial mutants. *J. Bacteriol.*, 63, 399.

Luria, S. E., and M. Delbruck. 1943. Mutations of bacteria from virus sensitivity to virus resistance. *Genetics*, 28, 491.

Modrich, P. 1991. Mechanisms and biological effects of mismatch repair. *Ann. Rev. Genet.*, 25, 229.

Novick, A., and L. Szilard. 1951. Experiments with the chemostat on spontaneous mutations in bacteria. *Proc. Natl. Acad. Sci.*, 36, 708.

*Roth, J. R. 1974. Frameshift mutations. *Ann. Rev. Genet.*, 8, 319.

*Roth, J. R. 1981. Frameshift suppression. *Cell*, 24, 601.

Schlesinger, M., and C. Levinthal. 1963. Hybrid protein formation of *E. coli* alkaline phosphatase leading to in vitro complementation. *J. Mol. Biol.*, 7, 1.

*Smith-Keary, P. 1991. *Molecular Genetics: A Workbook*. Guilford.

Streisinger, G., et al. 1966. Frameshift mutations and the genetic code. *Cold Spring Harb. Symp. Quant. Biol.*, 31, 77.

GENETICS OF BACTERIA
AND PHAGES

11

Plasmids

Plasmids are circular, supercoiled DNA molecules (Figure 11-1) present in most species, but not all strains, of bacteria. Most plasmids are small, from about 0.2 to 4% the size of the bacterial chromosome (Table 11-1). Under most conditions of growth, plasmids are dispensable to their host cells. Many plasmids, however, contain genes that have value in particular environments. Often these genes are the main indication that a plasmid is present. For example, R plasmids render their host cells resistant to certain antibiotics, so in nature a cell containing such a plasmid can survive better in environments in which the antibiotic is present.

In many bacterial species, plasmids are responsible for a particular type of gene transfer between cells, a property that accounted for the initial interest in plasmids in the 1950s. Similar to phages, plasmids heavily depend on the metabolic functions of the host cell for their reproduction. They normally use most of the replication machinery of the host and hence have been useful models for understanding certain features of bacterial DNA replication. In addition, they have been valuable to the microbial geneticist in constructing partial diploids (see Chapter 7) and as gene-cloning vehicles in genetic engineering (see Chapter 20).

TYPES OF PLASMIDS

This chapter is concerned primarily with plasmids of *Escherichia coli*. Many types of plasmids have been detected in various *E. coli* strains, but the greatest amount of information has been obtained about three main types—the F, R, and Col plasmids—which share some properties but which are, for the most part, quite different. The presence of an F, R, or Col plasmid in a cell is indicated mainly by the following traits:

1. *F, the fertility or sex plasmids*. These plasmids mediate the ability to transfer chromosomal genes (that is, genes not carried on the plasmid) from a cell containing an F plasmid to one that does not. The F plasmid itself can also be transferred to a cell lacking the plasmid.
2. *R, the drug-resistance plasmids*. These plasmids make the host cell resistant to one or more antibiotics, and many R plasmids can transfer the resistance to cells lacking R.
3. *Col, the colicinogenic plasmids*. Col plasmids synthesize proteins, collectively called **colicins**, that can kill closely related bacterial strains that lack a Col plasmid of the same type. The mechanism of killing is different for different types of Col plasmids.

Figure 11-1. Two supercoiled plasmid DNA molecules.

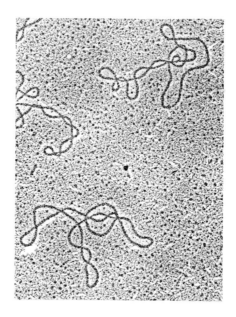

DETECTION OF PLASMIDS

Plasmids can be detected by both genetic and physical experiments. The first plasmid that was discovered was F. An *E. coli* strain (A) with phenotype $Met^- Bio^- Thr^+ Leu^+$ was mixed with a second strain (B) with phenotype $Met^+ Bio^+ Thr^- Leu^-$, and the mixture was plated on minimal agar. At a frequency of about 10^{-7}, colonies formed on the minimal agar; these had the phenotype $Met^+ Bio^+ Thr^+ Leu^+$ and hence were recombinants. If strain A was treated with streptomycin (and then washed free of the antibiotic) before mixing the cells, recombinant colonies still formed. If strain B was first treated with streptomycin, however, no recombinants were found. This experiment indicated that the recombinants were derived from strain B and that mating somehow involved a one-way transfer of genetic information. Another experiment involved a third strain C, which could not transfer any genetic information to B. If A and C bacteria, however, were mixed and allowed to grow together for a long time and then

Table 11-1 Examples of some plasmids and their properties

Plasmid	Size (Kb)	Number of copies per chromosome	Self-transmissible	Phenotypic features
Col plasmids				
ColE1	6.4	10–15	No	Colicin E1 disrupts energy gradient, host immunity to Colicin E1
ColE2	7.6	10–15	No	Colicin E2 is a DNase, host immunity to Colicin E2
ColE3	7.6	10–15	No	Colicin E3 is a ribosomal RNase, host immunity to Colicin E3
F plasmid	94.5	1–2	Yes	F-pilus, conjugation
R plasmids				
R100	106.7	1–2	Yes	$Cam^r Str^r Sul^r Tet^r$
RK2	56.0	5–8	Yes	Broad host range
pSC101	9.0	<5	No	Low copy number, compatible with ColE1-type plasmids, Tet^r
Phage plasmid				
λdv	6.4	50	No	λ genes *cro, cl, O, P*
Recombinant plasmids				
pBR322	4.4	20	No	Medium copy number, ColE1-type replication, Amp^r
pUC18	2.7	200–500	No	High copy number, ColE1-type replication with a mutation that increases the copy number, Amp^r
pACYC184	4.0	10–12	No	$Cam^r Tet^r$

colonies of C were isolated, these colonies (C') could transfer genes to strain B. The interpretation of these results was that strain A contained a fertility element, called F, which mediated transfer of chromosomal genes from bacterium A to bacterium B. Strain C also lacked F. When strains A and C were grown together, however, F was transferred to C, generating strain C', which could then (because it contained F) transfer genetic markers to B cells. Thus, the initial cross was:

$$F^+met^-bio^-thr^+leu^+ \times F^-met^+bio^+thr^-leu^-$$

Because of the one-way nature of the transfer, the cell containing F is said to be a **donor** (or "**male**"), and the cell without F is the **recipient** (or "**female**").

Variants of F, called F', are known that carry chromosomal genes in addition to plasmid genes. One that was studied in great detail carried the *lac* operon and was used in constructing the partial diploids described in Chapter 7. It is designated F' *lac*. Transfer of an F' is recognized easily with minimal lactose plates containing an antibiotic (for example, streptomycin) to which the donor cells are sensitive and the recipients are resistant. For example, in the cross:

$$F'lac^+/lac^- Str^s \times lac^- Str^r$$

in which the donor cell carries the *lac*$^+$ marker on the plasmid and a *lac*$^-$ marker in the chromosome (note the use of / to distinguish plasmid and chromosomal markers), plating the mixture on minimal-lactose agar containing streptomycin yields recombinant colonies (F' *lac*$^+$/*lac*$^-$ Str^r *recombinants*). A control with the donor alone would yield no colonies (because all cells are Str^s), and a control with the recipient alone would not yield colonies (all would be Lac$^-$). Note that the function of the streptomycin marker is to prevent growth of donor cells; such a marker is termed the **counterselection**, or **counterselective marker**. Antibiotic resistance is often used for counterselective purposes, but other phenotypes can be used as well. For example, F' transfer can be detected in the cross F' *lac*$^+$/*met*$^-$ × *lac*$^-$ *met*$^+$ by plating on minimal lactose plates. Donors will not form colonies because methionine is lacking in the agar (absence of methionine is the counterselection), and recipients lacking a transferred F' *lac*$^+$ will not grow because they are Lac$^-$ and thus cannot use lactose as a carbon source. In this case, the **selected marker** is Lac$^+$.

The experiments with the F-containing strain (that is, the cross between strain A and strain B) and the F' strains show significant quantitative differences. If the cells are mixed for about 30 minutes before plating, recombinants arise at a frequency of about 10^{-7} per donor cell in the A × B experiment and about 50% in the F' experiment. The difference is that in the F' *lac* experiment, the plasmid carries the genetic marker, whereas the genetic markers in the other experiment are chromosomal. Thus, in the A × B experiment, even though about half of the recipients receive a copy of F, only a tiny fraction receive chromosomal markers. This process of rare **chromosomal mobilization** is discussed in Chapter 14.

How is the plasmid transferred? The mating

$$F' lac^+ / lac^- Str^s tsx^r \times F' lac^- Str^r tsx^s$$

was carried out, in which *tsx*^{r} indicates inability to absorb phage T6. After a period of mating, a portion of the bacterial mixture was plated on lactose color-indicator plates containing streptomycin, and more than 90% of the colonies were Lac$^+$. Thus, most donors had transferred F' *lac*$^+$. To determine whether donors that had transferred F' *lac*$^+$ still had a copy of the plasmid, a portion of the mating mixture was treated with an excess of T6 phage; a high multiplicity of infection was used so the recipients, which were all Tsx^s, were lysed within a few minutes by the simple action of thousands of phage particles punching holes in the cell wall. The surviving cells, which were the Tsx^r donors, were then plated on lactose-indicator plates, and Lac$^+$ colonies formed. From several other experiments, it was known that most of the original donor cells contained only one copy of F' *lac*⁺. Thus, cells that had transferred the plasmid still retained a copy of the plasmid, which means that transfer is accompanied by DNA replication. Indeed, further tests showed that cultures derived from the colonies remained able to transfer the *lac*⁺ marker.

How can you detect a plasmid if it does not have simple genetic phenotypes? A single colony is taken from a plate, gently lysing the cells as described in Chapter 2, and the DNA analyzed by agarose gel electrophoresis (Figure 11-2). The bacterial chromosome is large and cannot penetrate the gel, but the plasmid DNA is small enough to migrate into the gel. Plasmid DNA, if present, forms a narrow band at a position in the gel characteristic of its molecular weight. The band can be visualized by staining the gel with ethidium bromide (see Figure 2-21), which binds tightly to the DNA and fluoresces on irradiation with ultraviolet light. From the distance moved in a particular time interval relative to that for plasmids of known molecular weight, the molecular weight of the plasmid DNA can be calculated, as shown in Figure 11-3.

PURIFICATION OF PLASMID DNA

Plasmid DNA can be purified from bacteria in a similar way. Plasmid-containing bacteria are gently opened by the detergent treatment (see Chapter 2). The resulting translucent solution (called a **cell lysate**) is centrifuged. The bacterial chromosome complex, which contains protein and RNA, is large and compact and sediments to the bottom of the centrifuge tube; the smaller plasmid DNA remains in the clear supernatant, which is called a **cleared lysate**. Some chromosomal DNA is usually present in the cleared lysate, but because most of the plasmid DNA is covalently circular, the contaminating chromosomal DNA and any sheared plasmid DNA can be removed by several procedures (see Chapter 2).

TRANSFER OF PLASMID DNA

F and F' plasmids were first detected by virtue of their ability to be transferred from a donor cell to a recipient. How is the DNA transferred?

Stages in Transfer Process

One of the earliest discoveries about recombination between donor and recipient cells is that it requires cell-to-cell contact. This was demonstrated by experiments in which F⁺ and F⁻ cultures were separated by a porous filter. Recombinants were not produced when cell-to-cell contact was prevented in this way, so recombination could not simply result from movement of genetic material through the growth medium. This led to the use of the term **bacterial conjugation** or **mating**. Some years later, striking electron micrographs were obtained of conjugating bacteria (Figure 11-4). A variety of experiments have shown that plasmid transfer can be divided into four stages:

1. Formation of specific donor-recipient pairs (**effective contact**).
2. Preparation for DNA transfer (**mobilization**).
3. DNA transfer.
4. Formation of a replicative functional plasmid in the recipient.

When cells containing F or an F' come in contact with a recipient, all four steps occur. Many types of plasmids, however, are genetically unable to carry out all of these processes. Plasmids fall into four groups based on these properties: A **nontransmissible plasmid** lacks genes necessary for effective contact and DNA transfer. A **conjugative plasmid** is a plasmid carrying genes that determine the effective contact function. A **mobilizable plasmid** can prepare its DNA for transfer. A **self-transmissible plasmid**, such as F, is both conjugative and mobilizable.

The conjugative functions are, in many cases, not plasmid specific, and hence one plasmid can assist transfer of a second. For example, a single cell may contain both F and ColE1. F is both conjugative and mobilizable (it is self-transmissible). ColE1 is mobilizable but nonconjugative—hence, a cell containing only ColE1 cannot transfer the plasmid. In a cell containing both plasmids, F can provide the missing conjugative function to ColE1, and ColE1 can thereby be transferred to a recipient that lacks both plasmids. This process, in which a nonconjugative plasmid is transferred via the effective contact provided by a conjugative plasmid, is called **donation**; its hallmark is efficient transfer of the nonconjugative plasmid (10 to 100%). In contrast, the mobilization functions are usually plasmid specific, so a self-transmissible plasmid cannot enable a nonmobilizable plasmid to be transferred by simple complementation. Transfer can still occur, however, in some cases, although at low frequency. The transfer requires recombination between the two plasmids to form a single transferable DNA molecule. That is, the self-transmissible plasmid just carries along the nonmobilizable plasmid. This process is called **conduction** (in contrast to donation). The frequency of transfer of a non self-transmissible plasmid is a useful criterion for determining whether a plasmid is mobilizable by donation or conduction.

In the F-mediated transfer of the chromosome, the chromosome was mobilized by conduction. That is, genetic recombination occurred between sequences in F and in the chromosome, and F carried the chromosome along into the recipient. Transfer by conduction (which requires a low-frequency recombination event between F and the chromosome) accounts for the 10^7-fold lower frequency of transfer of chromosomal genes than of the *lac*+ marker from the F' *lac*+–containing cell.

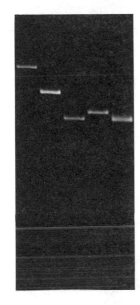

Figure 11-2. An agarose gel showing the migration of five different plasmids. Movement is from top to bottom. Each vertical column ("lane") represents a single plasmid. After electrophoresis, the gel was soaked in a solution of ethidium bromide, washed, and then illuminated with near-ultraviolet light. The ethidium bromide, which is bound to the DNA, fluoresces. The single intense band in each lane contains supercoiled DNA molecules. (Courtesy of Elaine Cocuzzo and Pieter Wensink.)

Effective Contact and Pili

The first step in effective contact is pair formation between a donor and a recipient cell. This requires a hairlike protein appendage, called a **sex pilus**, on the

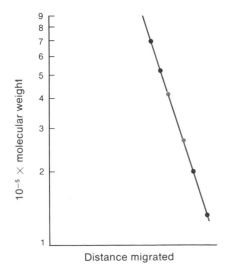

Figure 11-3. Determination of the molecular weights of DNA molecules (orange points) by gel electrophoresis with DNA molecules whose molecular weights are known (black points). The line is a plot of the black points; the molecular weights of the molecules being studied are then determined from the positions of the orange points.

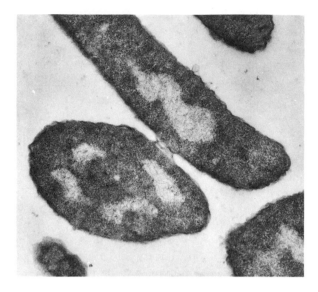

Figure 11-4. Electron micrograph of two *E. coli* cells during conjugation. The small cell is an F⁻ cell; the larger cell contains F'*lac*. (With permission, from Caro, L. 1966. J. Mol. Biol., 16, 269. Copyright: Academic Press Inc.)

donor cell (Figure 11-5). The pili on F-containing and R-containing cells are called **F pili** and **R pili**. On the average, there are 1.4 to 2.7 F pili per cell, depending on the growth conditions.

The F pilus has been purified. It consists of a single, hydrophobic protein, called **pilin**, which forms a hollow tubular structure. Four experimental results indicate that the F pilus is necessary for conjugation. (1) Plasmid mutants that cannot synthesize wild-type pilin cannot transfer the plasmid. (2) Pili can be stripped from cells by violent agitation of a culture. Pili re-form over a period of about 0.5 hour. F is unable to be transferred from such stripped cells until the pili have regrown. (3) Purified pili can bind to recipient cells. (4) Several phages are known that bind to F pili (these are sometimes called "**male-specific phages**"). These phages are of two types, binding either to the tip of the pilus or to the shaft. Addition of those that bind to the pilus tip inhibit pair formation and transfer, whereas those that bind to the shaft do not interfere in a major way with F transfer.

The fact that pili are hollow tubes suggested that plasmid DNA is transferred through the core of the pilus. No experiment, however, has ever given evidence for DNA within a pilus. Other experiments indicate that pili retract into

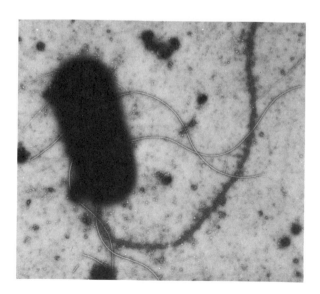

Figure 11-5. An *E. coli* cell showing a single F pilus, which is coated with the F-specific phage R17 to make the thin pilus visible as a rough dark appendage. The five heavy bright fibers are flagella. The very faint thin hairs are called fimbrae. [Courtesy of Barry Eisenstein.]

the donor cell after pairing. Thus, it seems most likely that the pilus serves first to bring the pair into initial contact and then to draw the cells together into close contact. The nature and identity of the actual conjugation tube that occurs at the cell-cell junction are not clear.

Not all mating systems depend on pili. For example, some strains of the bacterium *Streptococcus faecalis* carry a self-transmissible plasmid. Plasmid-free recipients produce a mating protein (analogous to the pheromones of female insects) that is not made in plasmid-containing (donor) cells. The pheromone causes the donor to synthesize a protein called adhesin that coats the donor cells and causes donor-recipient pairs to form. Once the plasmid is transferred, synthesis of the pheromone is inhibited.

Mobilization and Transfer

Mobilization begins when a plasmid-encoded protein, which is probably the nicking protein of the relaxation complex, makes a single-strand break in a unique base sequence called the **transfer origin** or *oriT*. This nick initiates rolling circle replication (see Figure 8-16), and the linear branch of the rolling circle is transferred. It is thought that the nicking protein remains bound to the 5' terminus and that the replication mode is similar to the rolling-circle mechanism used by phage φX174 (see Figure 8-17). The sequence of events during transfer is shown schematically in Figure 11-6.

Note that DNA synthesis occurs both in donor and in recipient cells. The synthesis in the donor, called **donor conjugal DNA synthesis**, serves to replace the single strand that is transferred. Synthesis in the recipient cell (**recipient conjugal DNA synthesis**) converts the transferred single strand to double-stranded DNA.

Usually the transferred strand is simultaneously replaced by donor conjugal synthesis and converted to double-stranded DNA in the recipient. This would indicate that DNA synthesis and transfer are coupled were it not for the following observations. (1) Transfer occurs even if DNA synthesis in the donor is inhibited by appropriate host mutations, such as a temperature-sensitive mutation in the DNA polymerase III gene. (2) Donor synthesis occurs even if transfer is prevented by appropriate plasmid mutations. (3) Transfer occurs even though recipient conjugal synthesis is inhibited by mutations in the recipient cell. These observations raise the question of the identity of the motive force for transfer because clearly it is not DNA replication.

Fertility Inhibition

If a single *E. coli* cell containing the F plasmid is added to a culture of growing F⁻ cells, after 15 to 20 generations of cell growth, a large fraction of the cells contain F. This is a result of transfer of a replica of F from an F⁺ cell to an F⁻ cell without loss of F by the F⁺ cell. After transfer, the F replica remaining in the original cell can replicate again, and another replica can be transferred to a second F⁻ cell. Every recipient acquires F and therefore becomes a donor from which F can be transferred to other F⁻ cells. Transfer can occur once or twice per cell generation, so F quickly spreads throughout a bacterial population.

Such rapid spread is not the case for most transmissible plasmids. For example, only about 0.02% of a population of cells containing most R plasmids are competent donors. That is, if an R^+/lac^- culture is mixed with an $R^- lac^+$ culture and the culture is diluted 1000-fold shortly after mixing so no further pair formation can occur, the number of R^+/lac^+ recombinants is only 0.02%

the number of initial R^+ cells. If after transfer has occurred, however, an excess of R^- recipients is added at a concentration that allows rapid pair formation, the R plasmid quickly spreads through the recipient culture. The kinetics of transfer show that an R^- cell that has just received an R plasmid is able to transfer the R plasmid almost immediately. This phenomenon, which is called **fertility inhibition**, is a result of the activity of a multisubunit repressor (encoded by two *fin* genes), which acts on an operator and prevents transcription of the genes required for transfer. Most R plasmids have an active *fin* repressor, which accounts for the low transfer frequency. Due to an IS3 insertion in *finO*, F lacks one of the *fin* genes, and hence the transfer system is always transcribed constitutively. With most R plasmids, once pair formation occurs, derepression occurs because the repressor is absent in the recipient, so recent recipients can transfer again almost immediately (in time, repressor is made, and fertility decreases). The operator in F apparently can bind the R repres-

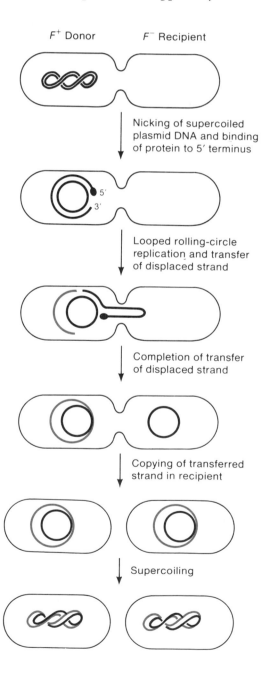

Figure 11-6. A model for transfer of F plasmid DNA from an F+ cell by a looped rolling-circle mechanism. The displaced single strand is transferred to the F− recipient cell, where it is converted to double-stranded DNA. Chromosomal DNA and proteins of the relaxation complex in completed DNA molecules are omitted for clarity.

F⁺ Donor   *F⁻* Recipient

Nicking of supercoiled plasmid DNA and binding of protein to 5′ terminus

5′
3′

Looped rolling-circle replication and transfer of displaced strand

Completion of transfer of displaced strand

Copying of transferred strand in recipient

Supercoiling

sors because transfer of F is markedly reduced by the presence of an R plasmid; as might be expected, if the R plasmid contains a defective *fin* gene, transfer of F is not inhibited.

It should be noted that although fertility inhibition is a common feature in R plasmids, it is not a property of all R plasmids.

tra Genes of F

In the preceding section, we saw that transfer is regulated by the *fin* genes. What are the essential transfer functions that *fin* regulates?

The essential transfer (*tra*) functions are encoded in an operon. Most of the *tra* genes are involved in pili synthesis. For example, the *traA* gene encodes pilin, and the *traB, C, E, F, G, H, K, L, Q, U, V, W* genes are needed for assembly of a functional pilus. Other genes are required for pairing, triggering transfer, actual transfer, nicking at the transfer origin *oriT*, DNA replication from the normal (nontransfer) replication origin *oriV*, surface exclusion (the inability of an F⁺ cell to pair with another F⁺ cell), and plasmid incompatibility (see later section). These and several genes are listed in Table 11-2 and on the map of F shown in Figure 11-7.

Table 11-2   Some F plasmid genes and sites and their functions

Gene	Function
traA	Pilin, major subunit of the pilus
traB, C, E, F, G, H, K, L, Q, U, V, W	Biosynthesis and assembly of the pilus
traG, N	Stabilization of mating pairs
traH	DNA helicase
traJ	Regulation of the *tra* operon
traM	Initiation of plasmid transfer
traS, T	Surface exclusion, inhibition of mating between F-containing cells
traY	Nicking at *oriT*
oriT	Site nicked to initiate rolling circle replication and DNA transfer
oriV	Origin of circular DNA replication
finO, P	Fertility inhibition, *finO*, is mutant in F but the *finO⁺* gene product can be supplied in trans by certain other plasmids
incB, C, E	Plasmid incompatibility

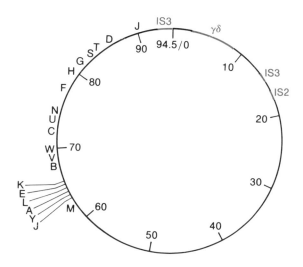

Figure 11-7. A map of the F plasmid. Map positions in kilobases are shown. The single capital letters refer to the midpoints of the locations of the corresponding *tra* genes. The sequences γδ, IS2, and IS3 are shown in orange.

Host Restriction in Transfer

In Chapter 5, host restriction and modification of phages was described. This same phenomenon occurs in plasmid transfer because a plasmid carries the modification pattern of the host cell in which it resides. Thus, F' lac^+ does not appear to be transferred efficiently from *E. coli* strain K12 to strain B and vice versa. In fact, transfer does occur, but the DNA transferred from strain K12 is destroyed by B-specific restriction nucleases when it enters strain B. Just as with phages, the rare B recipient to which successful transfer of F' lac^+ from strain K12 has occurred can then transfer F'lac^+ to a strain-B recipient, but not back to K. Restriction in bacterial conjugation is exercised to a lesser extent than that shown (see Table 5-2) for phages. This may have to do with the fact that a single strand is transferred, and the single strand is not attacked as efficiently by the restriction nuclease. Also, when the completed strand is synthesized, it rapidly becomes methylated. Once the DNA is methylated in one strand, it is not restricted.

IN VITRO PLASMID TRANSFER

Often it is desirable to transfer a nontransmissible plasmid to a specific host cell. It is possible to transfer the purified DNA as long as a genetic selection is available for recipients that possess the plasmid. Uptake of purified DNA is called **transformation**. Some species of bacteria are naturally transformable (see Chapter 13). Many bacteria, however, that are most useful for genetic engineering are not naturally transformable. These bacteria require chemical transformation or electrotransformation. Although the mechanisms are poorly understood, these techniques are the basis for many genetic engineering experiments that require manipulation of DNA in vitro then returning it to a host cell to replicate.

Chemical Transformation

When some bacteria are treated with appropriate chemicals under the proper conditions, they can be transformed. The most common method of preparing competent cells is by hypotonic Ca^{++} shock. To prepare competent cells by this method, an early log phase culture is centrifuged and resuspended in a cold hypotonic $CaCl_2$ solution. When DNA is added to these cells, it forms a calcium-DNA complex that adsorbs to the cell surface. The cells are then briefly warmed (heat shocked), which allows transport of the DNA into the cell. An alternative method for preparing competent cells involves treating the cells with polyethylene glycol (PEG) and dimethylsulfoxide (DMSO). In this technique, a PEG-DNA complex adsorbs to the cell surface. DMSO apparently makes the membrane permeable to the PEG-DNA complex. Both techniques yield transformation efficiencies of about 2×10^8 transformants/µg DNA.

If the DNA can replicate (for example, DNA from a plasmid or phage), it can become permanently established in the recipient cell. (If a linear DNA fragment is used, most of the DNA is digested by exonuclease V in *E. coli*.) A typical experiment would be to transform tetracycline-sensitive (Tet^s) cells with an R plasmid that is Tet^r and to plate the cells on nutrient agar containing tetracycline. All Tet^r colonies will contain the plasmid. At the same time, controls are plated with the cells only or the DNA only—no Tet^r colonies should arise on either control plate. (Spontaneous Tet^r mutants are not found because they are normally present in a Tet^s culture at a frequency of less than 10^{-8}). If an amount of plasmid DNA is added such that all cells take up the DNA, a few percent of the treated cells will become Tet^r.

Electrotransformation

Cells can also take up exogenous DNA by **electroporation**. When cells are exposed to an electric field, the membrane becomes polarized, and a voltage potential develops across the membrane. If the voltage potential exceeds a threshold level, small "pores" transiently form in the membrane, which make the cells permeable to exogenous macromolecules, including DNA. Uptake of exogenous DNA by electroporation is called electrotransformation.

Although electroporation requires a special apparatus, it has several advantages over "chemically induced" transformation. (1) Preparation of the cells is simple and quick. A culture of cells is thoroughly washed (to remove salts from the medium) and resuspended in a low ionic strength solution. (2) The efficiency of electrotransformation is typically several orders of magnitude greater than other methods of transformation. Electroporation of $E. coli$ or $Salmonella \ typhimurium$ typically yields 10^9 to 10^{10} electrotransformants/μg of plasmid DNA. (3) The efficiency of electrotransformation of $E. coli$ and $S. typhimurium$ is so high that plasmids can be directly transferred between cells without purifying the DNA first. (4) Electrotransformation is much less exacting about the size and purity of donor DNA than other transformation methods. Electrotransformation of supercoiled and relaxed plasmids from 2 to 44 kb occurs at high efficiency. (5) A wide variety of bacteria can be transformed by electroporation, including many bacteria that are transformed poorly or not at all using other methods.

The electroporation apparatus generates a rapid, high-voltage electrical pulse across a cuvette that holds the cells and DNA. The voltage is stored in a capacitor and released as an electrical pulse, which decays exponentially. The initial voltage (V_0) decays over time according to the equation:

$$V_t = V_0 \ e^{-t/\tau} \text{ where } \tau = R \times C$$

R is the resistance in ohms (Ω), C is the capacitance in farads, and τ is the time constant in seconds. τ is the time required for the voltage to decay to $1/e$ (about 37%) of the initial voltage. Thus, τ is proportional to the length of the pulse. For the conditions used for electroporation of enteric bacteria, τ is optimally 4 to 5 msec.

The voltage potential across the cell membrane is proportional to the voltage gradient between the electrodes, also called the electric field (E):

$$E = V/d$$

where d is the distance between the electrodes (i.e., the path length of the cuvette). E and τ are the most important electrical variables affecting electroporation. If E is too low, the voltage potential across the membrane is too small to disrupt the membrane, so electroporation does not occur. If E is too high, the cell membrane is irreversibly disrupted. E can be adjusted by varying the voltage or changing the path length of the cuvette. For electroporation of $E. coli$ and $S. typhimurium$, the optimal E is between 12.5 to 20 kV/cm, but the optimal conditions for different bacteria are different.

PLASMID REPLICATION

A plasmid can replicate only within a host cell and each type of plasmid must contain DNA sequences (origins of replication, see Chapter 8) that allow replication of the plasmid in an ordered manner. As with phages, however, there is enormous variation in both the enzymology and mechanics of plasmid DNA replication, as is seen in the following sections.

Use of Host Proteins in Plasmid Replication

Plasmids rely heavily on the host replication proteins for their replication. DNA Pol III is the major replication protein for the *E. coli* chromosome, but some plasmids use DNA Pol I instead. For example, in *polA* mutants, which have very low levels of Pol I, the F plasmid replicates normally, but ColE1 fails to replicate. This is because DNA replication of F uses Pol III, but ColE1 uses Pol I. Some plasmids use host gene products exclusively. For example, ColE1 can be replicated in vitro by adding purified ColE1 DNA to a cell extract prepared from cells that do not contain ColE1 or any other plasmid. Other plasmids require some plasmid-encoded gene products. For instance, F' *lac* plasmids carrying a temperature-sensitive mutation in F have been isolated that fail to replicate at 42°C; at this temperature, the F'(Ts) *lac*+ plasmid fails to replicate and thus is rapidly segregated from the population as the cells divide. All plasmids examined to date replicate semiconservatively and remain circular throughout the replication cycle. There are significant differences, however, in the replication pattern from one plasmid to the next. For example, some plasmids replicate unidirectionally and others bidirectionally. The plasmid RK6 replicates first in one direction and then later in the opposite direction from the same origin. The bidirectionally replicating plasmids terminate replication in two ways. In one type, termination occurs when the growing forks collide. Others have a fixed termination site that is sometimes reached by one growing fork before the other fork reaches it. In the most carefully studied plasmids, replication occurs by the so-called butterfly or rabbit's ears mode (Figure 11-8), in which a partially replicated molecule contains untwisted replicated portions, as is usually the case in θ replication, and a supercoiled unreplicated portion. When the replication cycle is completed, one of the circles is cleaved to separate the daughters. The result after one round of replication is one nicked molecule and one supercoiled molecule. The nicked molecule is then sealed and, somewhat later, supercoiled.

Control of Copy Number

We have just seen that plasmids normally use most of the replication machinery of the host. Each type of plasmid, however, possesses its own genes for regulating

Figure 11-8. Replication of ColE1 DNA. (a) Diagram of a butterfly molecule. The newly synthesized strands are orange. (b) Electron micrographs of butterfly molecules. (c) Nicked butterfly molecules, showing that a nick converts a "butterfly" molecule to a "u" molecule. [Courtesy of Donald Helinski.]

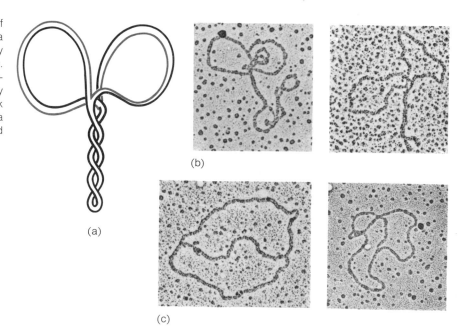

(a)

(b)

(c)

the rate of initiation of replication and hence the number of plasmid copies per cell. Based on their copy number, plasmids may be classified as **stringent** or **low-copy-number** plasmids (with 1 to 2 copies per cell) or **relaxed** or **high-copy-number** plasmids, (with 10 to 100 copies per cell). Transformation of both types of plasmids has been used to show that the copy number is established and regulated by controlling the rate of initiation of DNA synthesis. If a plasmid-free culture is transformed with low-copy-number plasmid DNA at a DNA concentration such that no cell receives more than a single copy of the plasmid, the plasmid DNA replicates only once or twice before cell division. If a cell, however, is transformed with a single DNA molecule of a high-copy-number plasmid, the plasmid DNA replicates repeatedly until the proper copy number is reached.

One mechanism for the regulation of copy number is that the plasmid encodes a repressor that negatively regulates the initiation of replication. How can this idea explain maintenance of copy number from generation to generation? Assume that the activity of the repressor depends on its concentration. Thus, as a cell grows, the volume increases, so the repressor concentration drops, replication is not inhibited, and the number of plasmid DNA molecules doubles. At this point, there are twice the initial number of repressor genes; therefore as a result of protein synthesis, the repressor concentration also doubles, causing replication to stop. A similar sequence of events would occur if there were initially only one copy of a high-copy-number plasmid—that is, replication would continue until there is sufficient repressor to turn off synthesis. How can this explanation account for differences in copy number? A simple model is that for the high-copy-number plasmids, complete repression requires a higher concentration of repressor than is required for low-copy-number plasmids. Thus, only when the number of plasmids per cell is high is the "gene dosage" high enough to cause inhibition.

The following experiment indicates that each plasmid controls its own copy number. A hybrid plasmid, pSC134, was constructed (by recombinant DNA techniques; see Chapter 20) that consists of a complete copy of each of two plasmids, ColE1 and pSC101 (derived from an R plasmid). Under the conditions used for this experiment the copy numbers for these plasmids are roughly 18 and 6 (Figure 11-9).

1. In wild-type cells, plasmid pSC134 replicates from the ColE1 replication origin and has a copy number of 16—roughly equal to that for ColE1. Thus, the higher-copy-number origin is dominant.
2. If pSC134 is moved into a *polA* mutant by transformation (recall that ColE1 cannot replicate in a *polA* mutant), the pSC101 origin is used, and the copy number becomes 6, the value for pSC101. These two results show that the copy number correlates with the replication origin that is used.
3. If pSC101 DNA is moved into a bacterium containing 16 copies of pSC134 by $CaCl_2$ transformation, the pSC101 cannot replicate. This lack of replication shows that the pSC101 inhibitor is being made by pSC134.

The interpretation of these three results is the following. If the copy number were less than 6, both pSC101 and ColE1 origins would be active, and replication from both origins would increase the copy number. If the number were greater than 6, the pSC101 origin would be inactive because the pSC101 inhibitor concentration would be above its inhibitory level. Replication of pSC101 would continue, but if it were self-regulated, the concentration would not exceed that produced by a copy number of 6. The ColE1 origin would remain active until the copy number reached 18. In a *polA* mutant, ColE1 cannot replicate; consequently, the copy number is totally controlled by the concentration of pSC101 inhibitor, and thus it will not exceed its normal value. This interpretation is consistent with several other types of experiments that also support the repressor model of copy-number regulation.

Replication of normal plasmid in *polA*+ or *polA*− cell:

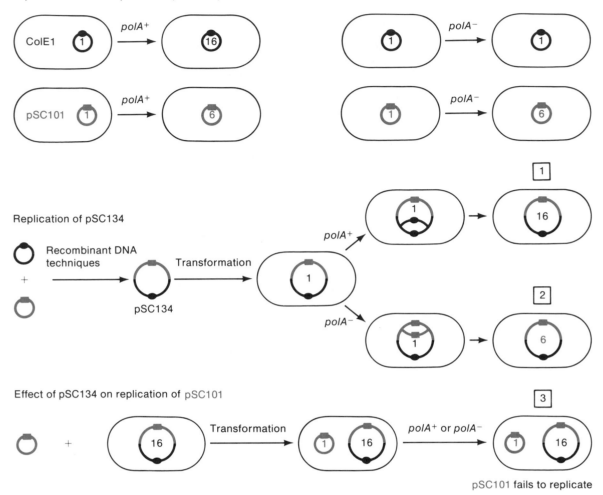

Figure 11-9. Diagram depicting the replication of ColE1, pSC101, and the hybrid plasmid pSC134, starting with one copy per cell. The black solid circle and the orange solid square designate the replication origins of ColE1 and pSC101.

The numbers (1, 6, and 16) in the DNA molecules indicate the number of copies in the cell. The boxed numbers (1, 2, and 3) refer to similarly numbered items in the text.

The following important point must be understood about the repressor model. In a strain containing a single plasmid type, all of the plasmids in a particular cell are identical, so a repressor molecule cannot (for most plasmids) distinguish one DNA molecule from another. Thus, when the concentration of repressor is low enough that all DNA molecules are not repressed, the few that are free are drawn randomly from the population. This means that if one plasmid replicates to form two daughter plasmids, an individual daughter plasmid has the same probability of replicating a second time as any other molecule has of replicating (a first time). That is, at any instant, a DNA molecule is chosen for replication by random selection from the entire population of plasmids.

Plasmid Amplification

Some high-copy-number plasmids exhibit a phenomenon called **amplification**. If an inhibitor of protein synthesis, such as chloramphenicol, is added to a culture of plasmid-containing bacteria, initiation of replication of chromosomal DNA is inhibited, but the plasmid continues to replicate until the number of plasmids per cell increases to 1000 or more. Bacterial DNA synthesis is inhibited because each time chromosomal DNA replication is initiated, new protein synthesis is required. If the plasmid, however, uses only stable bacterial replication proteins or stable plasmid-encoded proteins, plasmid DNA synthesis can continue. In addition, because the plasmid copy-number repressor is a concentration-dependent inhibitor and the repressor cannot accumulate in the absence of protein synthesis, plasmid replication will continue unregulated. Increased availability of bacterial replication proteins, owing to the lack of bacterial DNA synthesis, plus metabolic instability of RNA molecules contribute to the excessive synthesis. (The argument given here assumes a protein repressor, but some plasmids have an RNA inhibitor.)

Amplification is a convenient way to increase the amount of plasmid DNA that can be isolated from a culture and was used extensively in genetic engineering before very high copy number plasmids and improved plasmid purification methods were available.

Incompatibility

Pairs of closely related plasmids usually cannot be stably maintained in a single cell; such plasmids are said to be **incompatible**. The repressor model for initiation of plasmid replication also explains this phenomenon.

Consider a cell that contains two plasmids, say, F and ColE1, that have different repressors. Replication of each type of plasmid proceeds independently of one another because the repressor of one type (e.g., F) does not regulate the replication of the other type (e.g., ColE1). Thus, F and ColE1 are compatible, or another way of saying this is that they belong to different **incompatibility groups**.

The situation is quite different with two plasmids A and B, whose repressors are either identical or are similar enough that the repressor of A can regulate replication of B and vice versa. Consider a cell having one copy of A and one of B, which has enlarged sufficiently that initiation occurs (Figure 11-10). Because the two plasmid copies are selected at random for replication, the result of the first replication event is a cell having either one copy of A and two of B (1A,2B) or one of B and two of A (2A,1B). When the second replication event occurs, each cell will have four plasmids, but depending on the plasmid that is replicated, the plasmid composition may be (1A,3B), (2A,2B), or (3A,1B), as shown in Figure 11-11. At this point, the cell, which has twice the initial number of plasmids, can divide. The plasmid composition of the two daughter cells will be one of the following:

(1A,3B) becomes (1A,1B) + (1B,1B)
(2A,2B) becomes (1A,1A) + (1B,1B) or (1A,1B) + (1A,1B)
(3A,1B) becomes (1A,1A) + (1A,1B)

Note that two possible types of cells, (1A,1A) and (1B,1B), contain only one of the two kinds of plasmids; daughter cells obtained from these cells will, of course, continue to have only one kind of plasmid. In subsequent cell divisions, each cell still containing one A and one B will, with 50% probability, produce daughter cells lacking one of the plasmids. As a single cell initially containing two incompatible plasmids divides, the percentage of the progeny population containing only one plasmid type will increase with each generation. Thus, incompatibility is a result of (1) two plasmids having a similar repressor and (2) the random selection of plasmids for DNA replication.

Hundreds of plasmids have been sorted into incompatibility (Inc) groups. This classification is of some value because members of a single group are evolutionarily related with respect to replication functions and often also with respect to features of the pili. Plasmids can be classified into Inc groups by the following test:

1. Transfer a second plasmid B into cells of a culture already containing a resident plasmid A, and select for the presence of plasmid B.
2. Pick at least 10 colonies, and test these for the traits that identify the two plasmids.
3. If plasmid A is absent from all colonies, carry out a mating in which plasmid A is in the donor and B is in the recipient. If introduction of either plasmid always eliminates the other, the two plasmids are incompatible.
4. If the resident plasmid is still present in the progeny colonies, test the colonies carrying both plasmids for stability and independent replication of the two plasmids. Stability is shown by the continual presence of markers from both plasmids after many generations of growth in a nonselective medium. Separate replication is implied if the two plasmids are separately transferred in a mating.

If two plasmids coexist stably and are separably transferable, they are compatible and hence belong to different incompatibility groups. As many as seven different compatible plasmids can be maintained in a single cell. The limitation may be the amount of space available for excess DNA. For example, the cells with seven different plasmid types had a 25% excess of DNA over that of the chromosome.

Many plasmids can be transferred between various enterobacteria (for example, *E. coli*, *S. typhimurium*, and *Shigella dysenteriae*), and these plasmids

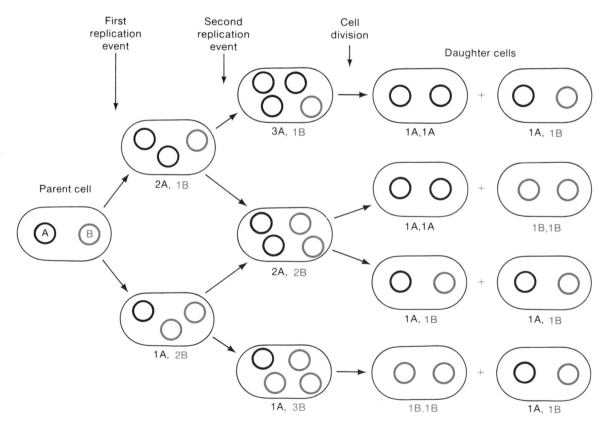

Figure 11-10. Four possible pairs of daughter cells that   the incompatible plasmids A and B. See text for details.
can arise from division of a cell containing one copy each of

Table 11-3 Some incompatibility groups of plasmids that are transmissible to *E. coli**

Group	Sex pili	Group	Sex pili
FI	F	I1	I
FII	F-like	I2	I-like
FIII	F-like	N	N
FIV	F-like	P	RP4

*The F plasmids are all in group F1; R1 and R100, which are mentioned in the text, are in group FII; Col plasmids are mainly in FII, FIII, and I1.

have been classified into about 25 incompatibility groups. Plasmids can also be sorted according to the pili they produce (usually immunological relatedness of pili is detected). The different pili classes are: F, F-like, I, I-like, N, and so forth. A typical Inc group consists of plasmids that not only have a common replication system, but also a common type of pili, and hence the pili class is used, to some extent, in naming the Inc groups. This can be seen in Table 11-3, which describes eight enterobacterial Inc groups.

Replication Inhibition by Acridines

Various agents that can intercalate between the bases of DNA, particularly acridines (e.g., proflavin and acridine orange), inhibit replication of many plasmids without inhibiting chromosomal DNA replication. Such inhibition can lead to loss of the plasmid (**acridine curing**). The phenomenon, which is detected most easily when the plasmid contains host genes, is illustrated in Figure 11-11. If a culture of a bacterium whose genotype is F' *lac+/lac−* is grown in medium

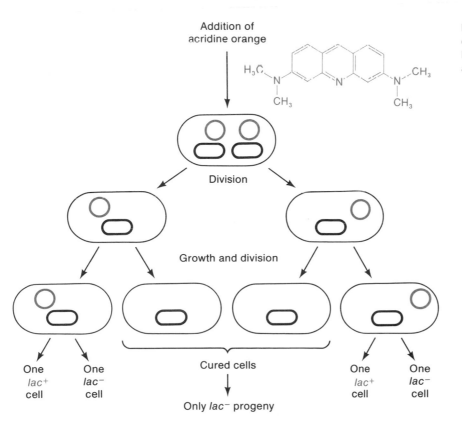

Figure 11-11. Curing of a cell containing F' *lac+* (orange circle) by growth in a medium containing acridine orange.

containing acridine orange, cells continue to grow and divide. The number of Lac⁺ cells in the population, however, remains constant, and the number of Lac⁻ cells increases. Why? At the time of adding acridine orange, most cells contain two copies of F' *lac*. After one generation, each cell contains only one F' *lac* because plasmid replication is inhibited. Thus, in the next cell division, only one plasmid is available for the two daughter cells; as a consequence, half the daughter cells will not contain the plasmid. Acridine curing can be useful for isolating plasmid-free segregrants.

The mechanism of acridine curing has been elusive. It works only in a fairly narrow range of acridine concentrations—too much inhibits chromosome replication, and too little fails to inhibit plasmid replication. Why plasmid DNA replication is more sensitive to acridines than replication of the chromosome is unclear. Furthermore, curing has an extraordinary dependence on the pH of the growth medium, being optimal at pH 7.6 and quite inefficient at pH 7 and pH 8. In addition, it is not known why some plasmids are not inhibited by acridines.

Some plasmid-containing strains (for example, many F⁺ strains) are cured by growth at 45°C, exposure to low concentrations of detergents, or certain antibiotics at concentrations too low to inhibit cell growth.

PARTITIONING OF PLASMIDS AT CELL DIVISION

A variety of experiments indicate that for low-copy-number plasmids partitioning of plasmids between daughter bacterial cells during cell division is carefully regulated. That is, if a cell that is ready to divide contains two copies of plasmid DNA molecules (as is the case with F plasmids), each daughter cell receives one copy. If partitioning were a random process, the fraction of cells lacking a plasmid after division would be determined by the Poisson distribution, and hence roughly 37% of the cells would be plasmid-free. For an F' *lac⁺/lac⁻* cell, however, segregation of Lac⁻ cells occurs at a frequency of only about 1 per 10^4 cells per generation, a fairly typical frequency of plasmid loss for the low-copy-number plasmids. Evidence that a genetically defined system regulates partition of F comes from the isolation of mutations in F that increase the segregation of plasmid-free cells to that expected for random partition. Two distinct systems seem to prevent segregation of plasmid-free cells. One is a system called *par*. The other is a system called *ccd* (control of cell division) that inhibits cell division in cells carrying only one copy of F.

It seems clear that the low-copy-number plasmids would have a partition function to avoid segregation of large numbers of plasmid-free cells. Such a system is not needed for high-copy-number plasmids: Partitioning may simply be random for some high-copy-number plasmids. For example, segregation of ColE1 plasmid occurs at about 10^{-6} to 10^{-5} per generation, essentially equal to the expected value for random segregation from a cell with 20 to 30 plasmids; however, the frequency of segregation of low-copy-number plasmids is about the same, so no conclusions can be drawn from these data alone. If the copy number of the plasmids, however, is reduced to two or three per cell (for example, using a ColE1 temperature-sensitive replication mutant), segregation of plasmid-free cells occurs at a rate of about 0.2 per generation, indicating that these plasmids do not have a Par function.

PROPERTIES OF PARTICULAR BACTERIAL PLASMIDS

Particular plasmids have unique properties that have important uses in bacterial genetics. A few of these are described next.

F Plasmids

An important property of F is its ability to integrate into the bacterial chromosome to generate an Hfr cell. The properties of these cells are discussed in detail in Chapter 14. Integration is a reciprocal exchange similar to that occurring when phage λ lysogenizes a bacterium. Integration of F into the *E. coli* chromosome, however, differs from λ integration because there are several possible exchange sites in F and many sites in the chromosome where integration can occur. More than 20 major sites and nearly 100 minor sites in the chromosome are known. The affinity of F for each site is not the same.

Excision of F also occurs, although this is quite rare (excision occurs, however, at a higher frequency at some locations than at most other locations). Often excision is imperfect, and one cut is made at one of the two termini within the integrated F, and a second cut is in the adjacent chromosomal DNA. In such a case, the new F plasmids contain chromosomal genes. This aberrant excision is the origin of the F' plasmids. In Chapter 14, the genetic procedures used to isolate a strain containing an F' plasmid are described.

Integration of F into certain bacterial mutants that have defects in DNA replication gives rise to a phenomenon called **integrative suppression**. F uses many *E. coli* replication proteins but does not require the bacterial *dnaA* gene product, which is necessary to initiate chromosomal DNA replication. Thus, if F exists as a free plasmid in an *E. coli dnaA*(Ts) mutant and the temperature is raised to 42°C (which inactivates the mutant DnaA protein), initiation of chromosomal replication is no longer possible, but F can still replicate. In contrast, in a *dnaA*(Ts) mutant strain in which F is integrated into the chromosome (an Hfr strain), chromosomal replication can occur indefinitely at the high temperature. Replication, however, is not initiated at the *E. coli* replication origin but instead is initiated at the *oriV* site for F replication. Thus, integration of F suppresses the *dnaA*(Ts) phenotype by providing a DnaA-independent replication origin.

Screening for integrative suppression is one way to isolate a strain in which F is integrated. For example, an F' *lac*+/Str^s strain can be mated with a *lac*⁻ *dnaA*(Ts) Str^r strain, and Lac⁺ Str^r cells can be selected by plating at 42°C on a minimal lactose plate containing streptomycin. All surviving colonies should contain an integrated F' *lac*. Each of these colonies should be an Hfr.

Drug-Resistance Plasmids

The drug-resistance (R) plasmids were originally isolated from the bacterium *S. dysenteriae* during an outbreak of dysentery in Japan and have since been found in *E. coli* and many other bacteria. R plasmids confer resistance on their host cell to a variety of fungal antibiotics and are usually self-transmissible. Most R plasmids consist of two contiguous segments of DNA (Figure 11-12). One of these segments is called **resistance transfer factor (RTF)**; it carries genes regulating DNA replication and copy number, the transfer genes, and sometimes the gene for tetracycline resistance (Tet). The other segment, sometimes called the **R determinant**, is variable in size and carries other genes for antibiotic resistance. R plasmids commonly carry resistance to the drugs ampicillin (Amp), chloramphenicol (Cam), streptomycin (Str), kanamycin (Kan), and sulfonamide (Sul), in a variety of combinations. Small drug-resistance plasmids, lacking the ability to transfer but still containing the Tet gene, are also known; one of these, pSC101, is commonly used in genetic engineering (see Chapter 20). The two-component R plasmids are reminiscent of F' plasmids, but the drug-resistance genes are not acquired by integration of RTF followed by aberrant excision; instead they result from acquisition of transposons that carry antibiotic-resistance genes, as described in Chapter 12.

Evidence for an RTF and an R component of the larger R plasmids is provided by both physical and genetic experiments. Examination of the DNA isolated from cultures of certain R plasmid–containing strains showed the presence of three sizes of circular DNA molecules. The length of the largest was the sum of the lengths of the other two molecules. Genetic studies of R plasmids carrying markers for resistance to chloramphenicol (Cam), streptomycin (Str), and tetracycline (Tet) indicated that strains often arose that were $Tet^r Cam^s Str^s$, $Tet^s Cam^r Str^r$, or sensitive to all three antibiotics. The Cam^r and Str^r markers were never dissociated—either both or none were present. Furthermore, Tet^r-$Cam^s Str^s$ cultures could transfer the Tet^r marker, but $Tet^s Cam^r Str^r$ strains could not transfer any of the markers. Finally, $Tet^r Cam^r Str^r$ strains could transfer either all three markers or just Tet^r. Physical studies showed that a mid-sized circle was present in $Tet^r Cam^s Str^s$ cells and a small circle in $Tet^s Cam^r Str^r$ cells. Thus, it was clear that the Tet^r marker resided on a self-transmissible plasmid (RTF), and another plasmid that was not self-transmissible (an R unit) carried the Cam^r and Str^r markers.

A critical observation was that in some strains and with some plasmids an RTF-R composite plasmid could not be stably maintained: RTF and R segregants continually segregate. Furthermore, strains that initially contain RTF and R units usually produce some cells that contain a composite plasmid. Thus, these plasmids must be able to associate and dissociate. This equilibrium gives rise to a gene amplification when a cell carrying the $Tet^r Cam^r Str^r$ plasmid is challenged with such large amounts of either Cam or Str that even these resistant cells are partially inhibited. With continued growth, the culture develops increased resistance to these antibiotics as a result of the acquisition of multiple copies of the R determinant with a corresponding increase in the size of the plasmid. The amplification does not occur if large amounts of Tet are added. It is unlikely that the antibiotic challenge directly induces the gene amplification. More likely, within the culture, there are a few rare cells in which the number of R determinants on the plasmid have duplicated, and these mutant plasmids are simply selected for by the antibiotic challenge because of the increased growth rate of cells containing these plasmids. These gene amplifications are unstable because if the antibiotic is removed from the growth medium and the cells are grown without antibiotic selection for many generations, the plasmid returns to the original size. Association and dissociation and the gene amplification presumably result from recombination between common sequences present in both RTF and R, such as IS elements (see Chapter 12). Association would result from recombination between IS elements in different circles, and dissociation would result from recombination between two elements in a single circle.

An interesting group of R plasmids are the F-like R plasmids, including R1, R6, and R100. In cells containing both F and one of these plasmids, the R plasmids caused fertility inhibition of F. Complementation analysis indicated that the F-like plasmids contained most of the F *tra* genes. Electron microscopic examination of heteroduplexes formed between F and an F-like R plasmid showed extensive homology in the transfer regions. Detailed analysis of these hetero-

Figure 11-12. The components of an infectious R plasmid.

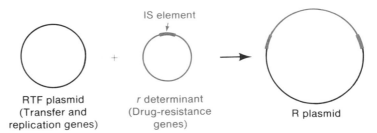

RTF plasmid
(Transfer and
replication genes)

IS element

r determinant
(Drug-resistance
genes)

R plasmid

duplexes shows that the F-like R plasmids consist of a large segment of F linked to a typical R determinant. Numerous inversions, deletions, and substitutions are also present in the F region, which suggests that the plasmids may have arisen a long time ago from a common ancestor.

R plasmids are of considerable medical interest because they can be transferred between bacteria that cause major epidemics, such as *S. typhimurium* and *S. dysenteriae*, and to strains that cause infections in hospitals (various enterobacteria, *Pseudomonas aeruginosa*, and *Staphylococcus aureus*). In fact, it has become clear that since the beginning of the "antibiotic era," R plasmids have increased dramatically in nature. For example, penicillin was introduced to general use in the early 1940s. By 1946, 14% of *S. aureus* strains isolated in hospitals were Pen^r. The fraction was 38% in 1947, 59% in 1969, and nearly 100% by the 1970s. The majority of these resistant strains either carry an R plasmid or a Pen^r gene of the type found in R plasmids. Transfer of bacteria that carry R plasmids also commonly occurs from farm animals to humans. R plasmids are rampant in farm animals owing to the extensive use of penicillin and tetracycline in animal feed. (Use of low levels of antibiotics in animal feed leads to more rapid growth of the animals and hence is economically valuable.) For example, poultry is frequently contaminated with *E. coli* and *S. typhimurium*, which can colonize the human intestine. Handling of raw meat is a common route for transmission to humans; bacteria from the meat gets on utensils and kitchen surfaces, eventually transferring to humans. Numerous epidemics of drug-resistant salmonellosis have occurred since 1960, which were caused by transmission from farm animals to humans. Because the bacteria were resistant to multiple antibiotics, the antibiotics commonly used for *S. typhimurium* infections were not effective. To make matters worse, treatment with broad-spectrum antibiotics often kills off some of the normal bacteria in the host, allowing the pathogenic bacteria to proliferate even more, thus exacerbating the infection.

Colicinogenic Plasmids

Col plasmids are *E. coli* plasmids able to produce colicins, proteins that prevent growth of susceptible bacterial strains that do not contain a Col plasmid. They are one class of a general type of plasmid called bacteriocinogenic plasmids, which produce bacteriocins in many bacterial species. Bacteriocins, of which colicins are one example, are proteins that can interact with sensitive bacteria and inhibit one or more essential processes, such as DNA replication, transcription, translation, or energy metabolism. There are many types of colicins, each designated by a letter (for example, colicin B) and each having a particular mode of inhibition of sensitive cells (Table 11-4).

Colicin production is detected by an assay similar to that for detecting phage. A colicin-producing cell is placed on a lawn of sensitive cells; the colicin inhibits growth of nearby bacteria, producing a clear area, known as a **lacuna**, in the turbid layer of bacteria. The presence of the Col plasmid may have survival value for the host because it may be able to compete more effectively in the wild with colicin-sensitive cells.

Table 11-4 Properties of several colicins

Colicin	Action
Colicin Ib, Colicin B	Damages cytoplasmic membrane
Colicin E1, Colicin K	Uncouples electrochemical potential across membrane
Colicin E2	Degrades DNA
Colicin E3	Cleaves 16S ribosomal RNA

Studies on many purified colicins suggest that colicins may be of two types—true colicins and defective phage particles. Some colicins are simple proteins. Others look like phage tails when examined by electron microscopy. Such colicins may be gene products from remnants of ancient prophages: a defective prophage that has lost the genes for replication and head production, but retained genes encoding a repressor system, the lysis enzyme, and the tail proteins have survived intact. Similar to phage, colicins also bind to specific receptor sites on the cell wall. In addition, expression of some colicins is normally repressed but can be induced by DNA damaging agents, such as UV light.

In most cases, colicins are inactive against a cell that contains a related Col plasmid. This is called immunity. In many cases, immunity is conferred by an excreted small protein that binds to the larger colicin protein. The immunity protein not only confers immunity on Col^+ cells, but also is necessary for killing of Col^- cells. For example, the colicin cloacin DF13 protein consists of three regions (Figure 11-13): a receptor-binding region, an RNase, and a segment that binds the immunity protein. The receptor-binding terminus is hydrophobic and probably also interacts with the cell membrane. The immunity-binding segment has a strong negative charge that is neutralized by the positively charged immunity protein. After binding to the receptor, the colicin is cleaved. The N-terminal region remains outside the cell, and the RNase segment enters the cell, leaving the immunity protein on the cell surface. The RNase acts on the RNA in ribosomes and thereby kills the sensitive cells.

The Col plasmids vary greatly in size: Those that are not self-transmissible tend to be small, and those that are self-transmissible are quite large. The best-studied Col plasmid is ColE1. Its complete sequence of 6646 base pairs is known. It is used extensively in recombinant DNA research (see Chapter 20). Many of the large self-transmissible Col plasmids are hybrids between a small Col plasmid and F or F' plasmids.

ColE1 is a mobilizable but nonconjugative plasmid and has provided the most definitive evidence that transfer requires a plasmid-encoded nuclease (for ColE1, the nuclease is encoded by the *mob* gene), and a specific base sequence called *bom* (*basis of mobility*), which contains the cutting site. The critical experiment is shown in Figure 11-14, in which the compatible plasmids F and ColE1 are present in the same cell, and F donates the conjugal functions that ColE1 lacks, thereby enabling ColE1 to be transferred. Panel (a) shows the state of ColE1 in an F^- cell. The *mob* gene is transcribed, the *mob* product nicks in the *bom* site (the actual cutting site is called *nic*), and the ColE1 supercoiled DNA is converted to a nicked circle. Transfer cannot occur because ColE1 lacks the ability to form pili and hence to form conjugate pairs. In panel (b), the cell contains both F and ColE1. F causes synthesis of the pilus and the transfer apparatus, and ColE1 is transferred. Mutants (*mob^-*) of ColE1 that cannot be transferred when F donates the transfer apparatus have been isolated. Genetic analysis shows that the *mob* mutation is recessive, which indicates that the *mob* gene encodes a diffusible protein. Furthermore, when the mutant plasmid is isolated, it is not in the form of a relaxation complex. The failure of F to help the *mob* mutant to transfer is shown in panel (c), in which the F pilus and transfer apparatus are formed, but nicking of the ColE1 DNA does not occur. Another ColE1 mutant, with the Mob^- phenotype, but which has a *cis*-

Figure 11-13. Regions of the polypeptide chain of cloacin DF13.

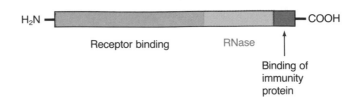

inant mutation, is a deletion that removes the *bom* site. The Mob protein is made, but nicking fails to occur, and there is no transfer, as shown in panel (d).

Agrobacterium Plasmid Ti

A crown gall tumor found in many dictyledonous plants is caused by the bacterium *Agrobacterium tumefaciens*. The tumor-causing ability resides in a plasmid called Ti. When a plant is infected, some of the bacteria enter and grow within the plant cells and lyse there, releasing their DNA in the cell. From this point on, the bacteria are no longer necessary for tumor formation. A small fragment of the Ti plasmid (the T-DNA), containing the genes for replication, becomes integrated into the plant cell chromosomes. The integrated fragment modulates the hormonally regulated system that controls cell division, causing the cell to be converted into a tumor cell. This plasmid has become important in plant breeding because specific genes can be inserted into the Ti plasmid by recombinant DNA techniques, and sometimes these genes can become integrated into the plant chromosome, thereby permanently changing the genotype and phenotype of the plant. New plant varieties having desirable and economically valuable characteristics derived from unrelated species can be developed in this way. This is discussed further in Chapter 21.

Broad Host Range Plasmids

Most plasmids can exist in only a limited number of closely related bacteria; these are called narrow host range plasmids. The self transmissible R plasmids of incompatibility group IncP of *E. coli* and of group IncP1 of *Pseudomonas*

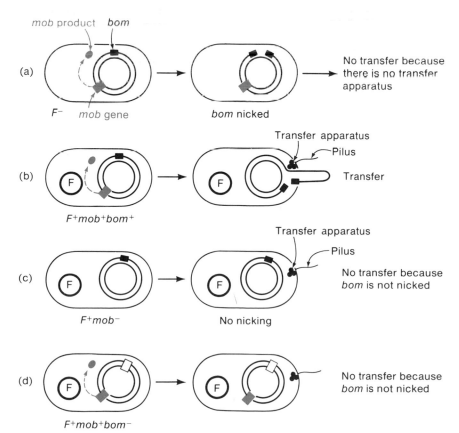

Figure 11-14. Conditions in which F can mobilize a ColE1 plasmid. The *mob* gene and the *mob* product are drawn in orange. The *bom* site in the DNA is drawn as a solid box, when functional, and as an open box, when mutant or deleted. Transfer occurs only when ColE1 makes an active *mob* product that acts on a functional *bom* site and F provides the transfer apparatus.

aeruginosa, however, are notable in that they can be transferred to and maintained in bacteria of a large number of species. These are called **broad host range plasmids**. Why some plasmids have a narrow host range and others a broad host range is unknown. The broad host range plasmids can be used for genetic studies in a wide variety of bacteria. Because they integrate (albeit at low frequency) into the host chromosome of numerous species, they provide a useful tool for genetic mapping in species in which mapping had not previously been possible. In this way, genetic maps have been obtained for several economically important bacteria; such mapping facilitates the construction of strains with desirable characteristics.

Most of these plasmids are able to mobilize the chromosome only at very low frequency (about 10^{-8} per cell). By various genetic and recombinant DNA techniques, however, variants of these plasmids have been constructed in which chromosome mobilization is enhanced by a factor of 10^3 to 10^5.

Other Plasmids

Several plasmids render fairly innocuous plasmid-free bacteria pathogenic. The Ent plasmids of *E. coli* synthesize enterotoxins that are responsible for travelers' diarrhea. A plasmid called Hly (for hemolysis) has been found in *E. coli* strains isolated from pigs. Although the hemolysin destroys red blood cells in blood samples, the Hly plasmid does not seem to cause any pathogenicity. Certain plasmids residing in the human pathogen *S. aureus*, enhance pathogenicity. A penicillinase (penicillin-destroying) *S. aureus* plasmid has been carefully studied.

Many species of *Pseudomonas* can use several hundred organic compounds as carbon sources—in particular, toxic organic compounds such as camphor, toluene, and octane. This metabolic ability resides in a set of plasmids collectively known as **degradation plasmids**. Each plasmid provides one or more metabolic pathways to degrade these compounds. Because many enzymes are needed, the plasmids are fairly large. These plasmids enable bacteria to degrade many *synthetic* compounds and hence are making an important contribution in the removal of man-made environmental pollutants. For example, some strains can degrade persistent chlorinated hydrocarbons, herbicides, and various detergents. Many laboratories are attempting to use genetic engineering to construct "super plasmids," which can be used for pollution control and chemical syntheses.

Plasmids have also been isolated that confer resistance to toxic metal ions. Such plasmids are found in environments containing these ions, such as the sludge produced by industrial reprocessing of photographic film (resistance to the Ag^+ ion). The biochemistry of resistance to Hg^{++} ions has been well studied: It results from a plasmid-encoded reductase that converts Hg^{++} to metalic mercury, which is sufficiently volatile that it evaporates away.

KEY TERMS

amplification	Hfr
broad host range plasmids	incompatibility
chromosomal mobilization	pili
conjugation	plasmid
copy number	recipient
counterselection	relaxed plasmids
curing	self-transmissible
donor	stringent plasmids
electroporation	transfer origin
F'	transformation
fertility inhibition	

QUESTIONS AND PROBLEMS

1. Is there any relation between plasmid size and copy number?

2. What mode of replication is used by plasmids for their reproduction?

3. Are there any circumstances in which plasmids undergo rolling circle replication?

4. During transfer, does replication occur in the donor, the recipient, or both?

5. What early enzymatic step is required in transfer replication but is not required for typical θ replication?

6. Are sex pili present on the donor or recipient cell?

7. What elements encode genes for pili synthesis?

8. What two roles are played by pili in bacterial conjugation?

9. What are the *tra* genes of F?

10. Define the terms conjugative, mobilizable, and self-transmissible.

11. State which of the terms in question 12 describe F and ColE1.

12. If a plasmid is mobilizable but nonconjugative, what function does it lack?

13. In a hybrid plasmid carrying two replication origins, one from a high-copy-number plasmid and the other from a low-copy-number plasmid, which origin will be used and what will the copy number be?

14. If two plasmids cannot be maintained in a single cell, what property is common to the plasmids?

15. What feature of F, not present in ColE1, enables F, but not ColE1, to integrate into the host chromosome?

16. If ColE1 could be altered to contain insertion sequences homologous to sequences in the chromosome, such that it could integrate in the chromosome, would Hfr-like cells arise?

17. What are the two components of R factors, and which carries the genes for replication and transfer?

18. Could you have a plasmid with no genes whatsoever?

19. Lac$^+$ and Lac$^-$ cells form purple and white colonies, respectively, on EMB-lactose agar. A culture consisting of cells whose genotype is F' *lac$^+$/lac$^-$* is mutagenized and plated on EMB-lactose agar at 30°C. Several purple colonies whose color is a little less intense than the others are studied further. These give white colonies at 42°C and light purple colonies at 30°C. Bacteria obtained from white (42°C) colonies remain white when replated at 30°C. What type of mutation has been acquired?

20. If, in a particular cell type, rifampicin were to inhibit DNA transfer, what would you conclude about the transfer mechanism?

21. What are the roles of DNA synthesis in the donor and the recipient?

22. A Lac$^-$ bacterial strain has a *dnaA*(Ts) mutation, which prevents colony formation at 42°C. An F'*lac$^+$* plasmid is introduced into the strain by conjugation. The culture is grown for many generations at 30°C, and then 10^6 cells are plated at 42°C. A few colonies arise, and these are capable of growth at both 30°C and 42°C.
    a. Are these colonies Lac$^+$ or Lac$^-$?
    b. Do the cells still carry the *dnaA*(Ts) mutation?
    c. What feature of the cells has changed that enables them to grow at 42°C?
    d. Can the cells grow at 42°C in the presence of acridine orange?

23. An F' (Ts)*lac$^+$* plasmid has a temperature-sensitive mutation in its replication system.
    a. What is the phenotype of an F'(Ts) *lac$^+$/lac$^-$* cell at 42°C?
    b. An F'(Ts) *lac$^+$/lac$^-$ gal$^+$* strain is grown for many generations and then plated at 42°C. Some Lac$^+$ colonies form at 42°C. How have these formed?
    c. Some of the Lac$^+$ colonies in (b) are Gal$^-$. How have these formed?

24. Plasmids contained in Rec$^+$ cells frequently dimerize. A dimer can then be isolated and transformed into a Rec$^-$ cell, in which it will be stable and not revert to the

monomer. If a monomer plasmid has a copy number of 10, what will the copy number of the dimer be?

25. On EMB-lactose agar Lac$^+$ and Lac$^-$ colonies are purple and white, respectively. If several thousand F' *lac$^+$/lac$^-$* cells are plated, a few sectored colonies appear. This may have a purple half and a white half or a wedge of white in a predominately purple colony. Sectored colonies that are predominately white are not found. Explain the cause of sectoring.

26. An F' *lac$^+$*/Str^s is mated with a *lac$^-$*Str^r recipient that also carries a *dnaG*(Ts) mutation. Mating is at a nonpermissive temperature. After 30 minutes of mating, streptomycin and an inducer of the *lac* operon are added. Will any β-galactosidase be made in the culture?

REFERENCES

Broda, P. 1979. *Plasmids*. W. H. Freeman.

Bukhari, A. I., J. A. Shapiro, and S. L. Adhya. 1978. *DNA Insertion Elements and Plasmids*. Cold Spring Harbor.

Cesareni, F., M. Helmer-Citterich, and L. Castagnoli. 1991. Control of ColE1 plasmid replication by antisense RNA. Trends Genet., 7, 230.

Clewell, D.B. 1993. Bacterial sex pheromone-induced plasmid transfer. Cell., 73:9.

*Clowes, R. C. 1973. The molecule of infectious drug resistance. *Scientific American*, April, p. 18.

Hooykaas, P. J. J., and R. A. Schilperoort. 1985. The Ti plasmid of *Agrobacterium tumefaciens*: a natural genetic engineer. *Trends Biochem. Sci.*, 10, 307.

Kornberg, A., and T. Baker. 1992. DNA Replication, Second Edition. New York: W.H. Freeman.

*Levy, S. 1993. *The Antibiotic Paradox: How Miracle Drugs are Destroying the Miracle*. Plenum Press.

Low, K. B. 1972. *E. coli* K-12 F' factors, old and new. *Bacteriol. Rev.*, 36, 587.

Manen, D., and L. Caro. 1991. The replication of plasmid pSC101. *Mol. Microbiol.*, 5, 233.

*Novick, R. P. 1980. Plasmids. *Scientific American*, December, p. 102.

Rownd, R. H., et al. 1978. Dissociation, amplification, and reassociation of composite R-plasmid DNA. In *Microbiology 1978* (D. Schlessinger, ed.). p. 33. American Society for Microbiology.

Sambrook, J., E. F. Fritsch, and T. Maniatis. 1989. *Molecular Cloning: A Laboratory Manual*. Cold Spring Harbor Laboratory.

Willetts. N., and R. Skurray. 1987. Structure and function of the F factor and mechanism of conjugation. In F. Neidhardt, J. Ingraham, K. B. Low, B. Magasanik, M. Schaechter, and H. E. Umbarger (eds.), Escherichia coli *and* Salmonella typhimurium: *Cellular and Molecular Biology*. American Society for Microbiology, Washington, D.C.

*Resources for additional information.

Transposable Elements

A long-standing assumption about gene organization and chromosome structure was that each gene has a definite and unvarying location in a particular chromosome. This assumption was strengthened over the years by the generation of genetic maps by both genetic and physical techniques. Rearrangements (for example, inversion and duplication of genes), however, occur at low frequency in both bacteria and eukaryotes. In the 1940s, in a genetic analysis of the mottling of the kernels of maize, McClintock discovered regulatory elements that moved from one site to another and thereby affected gene expression. Her work was followed 30 years later by the recognition that bacteria contain mobile segments of DNA that relocate at low frequency. The frequency with which the elements move in both bacteria and eukaryotes is fairly low and depends on the particular element (10^{-7} to 10^{-2} per generation), but nonetheless the concept of a static genome has gradually been replaced by the current view of the "genome in flux."

Movement of mobile elements requires DNA exchange, analogous to recombination. It differs, however, from the more commonly observed recombination systems in which exchange occurs between homologous DNA sequences because homology is not required for the process. In bacteria, homologous recombination always depends on the *recA* gene product or its equivalent (see Chapters 9 and 14). In contrast, movement of these elements occurs at the same frequency in *recA*⁻ and *recA*⁺ cells. The movement discussed in this chapter is called **transposition**, and the mobile segments are called transposable elements or **transposons**.

TRANSPOSON TERMINOLOGY

Many bacterial transposons contain easily recognizable genes, which may or may not exist elsewhere in the genome. For example, antibiotic-resistance genes are commonly carried on transposons. Transposons carrying antibiotic-resistance genes are the ones most frequently studied because they can be identified by simple plating tests. Most antibiotic-resistance transposons were originally designated by the abbreviation Tn followed by a number (e.g., Tn5), which distinguished the different elements. By agreement, all newly discovered bacterial transposons are to be designated in this way even if no recognizable gene is present. When it is necessary to refer to the genes carried on a transposon, these are indicated by standard genotypic designations, for example, Tn*1*(*amp*ʳ), in which *amp*ʳ indicates that the transposon carries the genetic locus for resistance to

239

ampicillin. The transposons first discovered did not contain any known host genes, and for historical reasons, they were called insertion sequences, or IS elements, designated IS*1*, IS*2*, and so forth. Transposons have occasionally been designated in nonstandard ways—for example, the transposon Tn*1000* (an element contained in the F plasmid) is often called γδ.

A transposon is frequently located within a particular gene; this creates a mutation in that gene, which is assigned an allele number. The name of the gene and the allele number are followed by two colons then the name of the transposon. For example, mutation 87 in the *lac*Z gene produced by the transposon Tn3 is designated *lac*Z87::Tn3. The antibiotic-resistance genes present in transposons are usually quite different from antibiotic-resistance genes that arise by simple mutation of bacteria lacking transposons. In most cases, the origin of the transposon genes is unknown. The antibiotic-resistance genes of most R plasmids are carried by transposons present in the plasmid DNA.

INSERTION SEQUENCES

The first transposable elements in bacteria were discovered in 1967. A collection of *gal⁻* and *lac⁻* mutations with unusual properties were found in *Escherichia coli*. The mutations had the following features:

1. They were highly polar mutations. Each mapped in the first gene of an operon, but proteins of the downstream genes were not synthesized. For example, the mutation might be in the *lac*Z gene, yet complementation does not occur with an F' *lac*Z⁺ *lac*Y⁻ plasmid because of lack of *lac*Y expression both from the plasmid and from the chromosome. In each case, the polarity resulted from the presence of a transcription-termination sequence.

2. These mutations could not be reverted by base-analog or frameshift mutagens, so the mutations could not be base substitutions or single-base additions or deletions. (Other polar mutations had been known since 1960, but these were chain-termination mutations that caused decreased transcription of downstream genes; mutagens increased the frequency of reversion of these mutants.)

3. If a plasmid was transferred to a cell containing one of these polar mutations, similar polar mutations occasionally appeared in genes carried on the plasmid. The locations of the chromosomal and plasmid polar mutations could be quite different, for example, in the *gal* operon on the chromosome and the *lac* operon on the plasmid. An example of such an event was the following: F' *lac⁺* was transferred by mating to a *gal⁻* (polar) *lac*Z⁻ cell to form a *gal⁻* (polar) *lac*Z⁻/F' *lac⁺* cell. *Lac⁻* mutants arose, which exhibited the same polar phenotype. The mutation was located on the F', as demonstrated by the observation that when the F' was transferred to an F⁻ *lac*Y⁻ recipient, it could not confer the Lac⁺ phenotype.

4. Physical studies of several plasmids containing such polar mutations in different genes showed that in each case the plasmid was larger than the original plasmid because a segment of DNA had been inserted in the operon. It was hypothesized that these inserted segments were mobile elements that moved from one region of a DNA molecule chromosome to another site.

When various phages infected cells carrying these polar mutations, rare phage mutants could be isolated that contained polar mutations. The DNA of phages carrying polar mutations was examined by electron microscopy. Hybridization of the DNA from a phage bearing the mutation with the DNA of a similar phage lacking a polar mutation showed a loop (Figure 12-1), which indicated that the

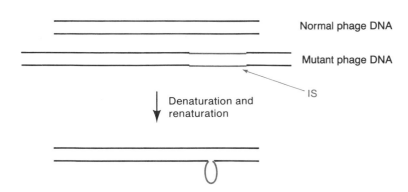

Figure 12-1. A hybridization test showing that an IS element is a segment of inserted DNA. After renaturation, half of the molecules have re-formed the normal and mutant phage DNA molecules (only the heteroduplex has been drawn).

mutation is an insertion. The length of the insertion was the length of the loop. In other experiments, wild-type phage DNA was hybridized with DNA obtained from various mutant phage, each carrying a polar insertion (derived from a single bacterial strain) in different phage genes. In these experiments, the insertion had a constant size and a position that corresponded to the position of the mutant site on the genetic map. Furthermore, if different phage species infected the same bacterial strain, the loop of the DNA from one phage mutant carrying a polar insertion could hybridize with the insertion present in another phage species. These observations together indicated that the mutant bacterium donated the same sequence to all of the phage that acquired a polar mutation.

A large number of phage mutants containing genes from numerous regions of the *E. coli* chromosome into which an **IS element** had been inserted have been compared to plasmids containing **insertion sequences**. This analysis demonstrated the existence of several different IS elements. Many of these elements have been isolated from mutant phage DNA, and their base sequences have been determined. A number of significant features of these elements have become apparent from the sequence analyses:

1. Those elements that produce polar mutations contain transcription-stop signals (except for IS*1*) and also chain-termination mutations in all possible reading frames. These properties are sufficient to account for the polar effects.

2. The termini of each IS element have inverted-repeat sequences consisting of 16 to 41 base pairs, the number depending on the element (Figure 12-2).

3. At many sites of insertion (for example, in the *E. coli gal* operon), the sequences are inserted in either the left-to-right or right-to-left orientation. This is not surprising because of the symmetry of the inverted-repeat configuration of the termini.

4. Different IS elements contain different numbers of bases. Radioactive copies of the elements have been prepared and hybridized with denatured *E. coli* DNA and with denatured F plasmid DNA to determine the number of copies of the elements in the chromosome and in F. The results are summarized in Table 12-1.

Figure 12-2. An example of a terminal inverted repeat in a DNA molecule. The arrows indicate the inverted base sequences. Note that the sequences AGTC and CTGA are not in the same strand. In a direct repeat, the sequence in the upper strand would be AGTC . . . AGTC.

Table 12-1   Properties of some *E. coli* insertion elements

Element	Number of copies and location*	Number of base pairs
IS*1*	5–8 in chromosome	768
IS*2*	5 in chromosome; 1 in F	1327
IS*3*	5 in chromosome, 2 in F	1258
IS*4*	1 or 2 in chromosome	1426
IS*5*	Unknown	1195
Tn 1000 (γδ)	1 or more in chromosome; 1 in F	5980

5. The elements contain at least two apparent coding sequences initiated by an AUG and terminated with an in-phase stop codon. By using in vitro protein-synthesizing systems with isolated IS DNA as template, proteins have been made. Some of these proteins have also been isolated from cells containing the element. Study of a large number of bacterial transposons indicates that each encodes at least two proteins.

DETECTION OF TRANSPOSITION IN BACTERIA

The IS elements just described are a special class of bacterial transposon. In this section, we examine the more general features of transposons and how these elements are observed.

In contrast with IS elements, which are normally detected by polarity effects, the more common class of bacterial transposon contains recognizable genes—for instance, genes for antibiotic resistance—that are often not present elsewhere in the genome. These genes provide the means for detecting transposition. An example is shown in Figure 12-3.

In this experiment, an R plasmid having an Amp^r (ampicillin-resistance) transposon is transferred by conjugation to a *recA*$^-$ recipient bacterium that carries a transposon having the Kan^r (kanamycin-resistance) gene. On continued growth in the presence of both antibiotics, transposition occasionally occurs and yields an R plasmid carrying both drug-resistance markers and hence both transposons. The presence of this two-transposon plasmid is detected by transferring the plasmid by conjugation to a Kan^s Amp^s bacterium and plating on agar containing both antibiotics. As a further test, the plasmid containing the two markers could be hybridized to the original plasmid molecule and examined by electron microscopy; loops such as those in Figure 12-1 would be seen, indicating the presence of an inserted DNA sequence.

When transposition occurs, most transposons can be inserted at a large number of positions. The most direct evidence comes from several types of experiments in which mutants formed by insertion of a particular transposon are isolated. For example, consider a Tet^s *E. coli* culture that is infected with a λ *O*$^-$ *P*$^-$ *int*$^-$ phage carrying the transposon Tn*10* (Tet^r): The phage is unable to replicate or integrate into the chromosome, so Tet^r cells can arise only by transposition of the Tn*10* into the chromosome. Plating the cells on nutrient agar containing tetracycline selects for colonies derived from cells in which transposition occurred. If a sufficiently large number of Tet^r bacterial colonies are examined and tested (by replica plating) for both nutritional requirements and ability to use particular sugars as a carbon source, with most transposons it is observed that a colony can be found bearing a mutation in almost any gene that is tested. This indicates that insertion sites for transposition are scattered throughout the *E. coli* chromosome.

Although there are certainly many sites of insertion in *E. coli*, and even in several phages, the locations clearly are not randomly distributed; certain positions seem to be excluded ("cold spots"), and others ("hot spots") are used repeatedly. This can be seen by examining the insertion sites in small regions whose base sequences are known. Such studies show that there is a rather broad range of insertion specificities from one transposon to the next. At one extreme is IS4; 20 independent *galT* insertion mutants were isolated, and all insertions were at exactly the same position in the gene. At the other extreme is Mu, which can insert almost anywhere. Most transposons exhibit insertion specificities between these extremes. For example, the insertion sites of transposon Tn9 in the final 160 base pairs of the *lacZ* gene have been determined by base-sequencing: Of 28 independent insertions, 16 different positions have been noted, but 5 of these are represented by multiple occurrences of insertion. The transposon Tn10 is even more specific; for example, 22 different sites have been identified in the *his* operon of *Salmonella typhimurium*, but 40% of all the transposition events occurred at a single nucleotide position.

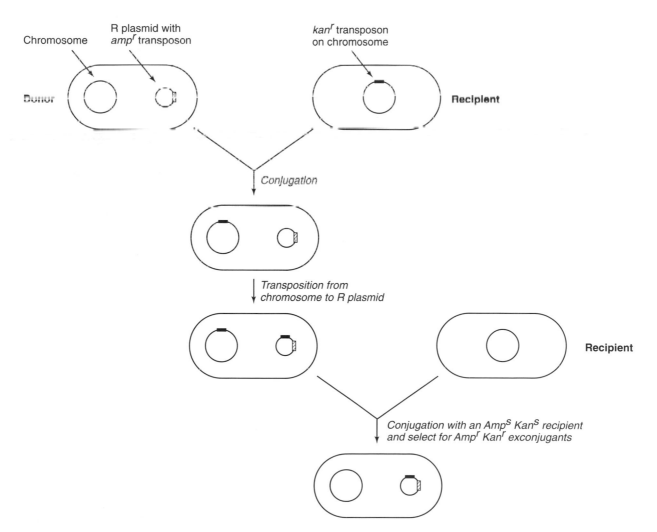

Figure 12-3. An experiment to demonstrate transposition. A donor bacterium (left) containing an Amp^r R plasmid transfers R to a recipient cell (center) containing a Kan^r transposon. In subsequent generations, the Kan^r transposon is transposed to the R plasmid. The presence of the second transposon on R is demonstrated by simultaneous transfer of the Amp^r and Kan^r genes. The Amp^r gene is shown as a transposon, as are most drug-resistant determinants on plasmids, but this fact is not essential to the experimental demonstration of transposition.

Figure 12-4 shows another way that transposition can be detected. In this case, a temperate λ phage acquires an Amp^r gene from a *recA*^− cell containing an Amp^r R plasmid, in which the Amp^r gene is carried on a transposon and transfers the transposon to another cell. A *recA*^− cell (chosen for the experiment to eliminate homologous recombination) is infected with the phage, and in a small number of progeny phage (perhaps 1 in 10^7), the transposon carrying the Amp^r gene has been picked up. The resulting phage population is used to infect an Amp^s culture of bacteria at a ratio of about 1 phage per 10 bacteria in a growth medium containing ampicillin. Uninfected bacteria are killed. More important, cells infected with a phage lacking the Amp^r gene do not yield progeny phage because of premature lysis induced by the drug; however, if the phage carries the Amp^r gene, the bacteria survive, and progeny Amp^r phage are produced. Thus in the presence of ampicillin, only the λ carrying the Amp^r transposon can grow; the lysate therefore consists of a homogeneous population of λ Amp^r phage.

To demonstrate that the Amp^r gene is part of a transposon, the λ Amp^r phage are used to infect a *recA*^− Amp^s (λ) lysogen. Because the lysogen is immune to infection by λ, the phage are unable to develop, and the bacteria survive. The infected cells are then plated on agar containing ampicillin. The Amp^r phage enables the cells to grow, but because the phage cannot replicate (because of the presence of the λ repressor in the lysogen), in each division cycle, one daughter cell does not inherit a λ Amp^r DNA molecule, and a bacterial colony cannot form. In a few bacteria, however, transposition does occur, and the bacterial chromosome acquires the Amp^r transposon, and all daughter cells inherit an Amp^r gene.

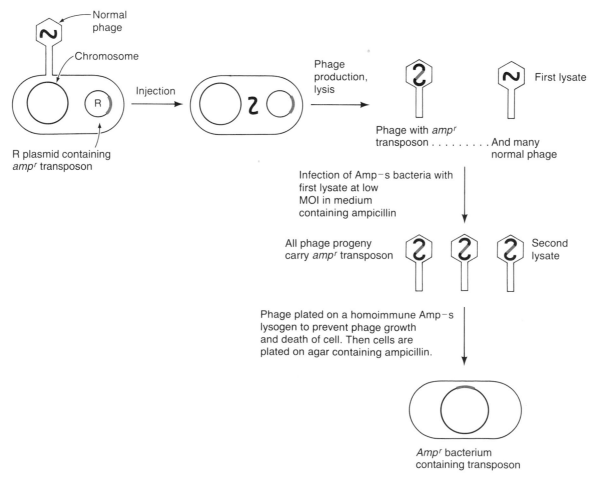

Figure 12-4.   An experiment demonstrating transposition. The transposon is shown in orange. See text for details.

Thus, formation of a colony indicates that transposition has occurred. Note that the resulting Amp^r bacteria could be reinfected again by a normal phage to yield a second population of phage, a few of which would contain the transposon.

TYPES OF BACTERIAL TRANSPOSONS

Because of particular related features, bacterial transposons have been divided into several distinct classes. One class includes the IS elements, which we have already discussed. Three more classes are described here:

1. *Composite transposons.* A variety of transposons carrying antibiotic-resistance genes are flanked by two identical or nearly identical copies of an IS element. These transposons are called composite type-I transposons. The IS elements in these composite units can be in an inverted or direct repeat configuration (Figure 12-5). Because the two ends of an IS element are themselves inverted repeats, the relative orientation (inverted or direct) of the flanking IS elements of a composite transposon does not alter its terminal sequences—that is, they remain the same in the inverted-repeat array. The ability to transpose is determined by the terminal IS elements. Three frequently studied composite transposons are Tn5, Tn9, and Tn10. Some properties of these and several other transposons are shown in Table 12-2.

   Most of the composite type-I transposons were first detected as genetic elements that could be transposed from one plasmid to another or to a phage. When the DNA of some of these plasmids was denatured

Inverted repeat

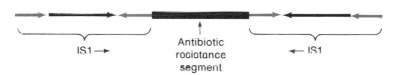

IS1 →   Antibiotic resistance segment   ← IS1

Direct repeat

IS1 →   IS1 →

Figure 12-5. Two composite type-I transposons flanked by IS1 in either inverted or direct repeat. The central black region contains the antibiotic-resistance genes and the genes necessary for transposition.

Table 12-2   Properties of selected composite type I transposons of *E. coli*

Element	Genes carried*	Size in base pairs	Terminal IS element and size in base pairs	Relative directions of terminal IS Elements
Tn5	kan	5818	IS50 (1533)	Inverted
Tn9	cam	2638	IS1 (768)	Direct
Tn10	tet	9300	IS10 (1329)	Inverted
Tn204	cam, fus	2457	IS1 (768)	Direct
Tn903	kan	3094	IS903 (1057)	Inverted
Tn1681	ent	2088	IS1 (768)	Inverted

*cam, chloramphenicol; ent, enterotoxin; fus, fusidic acid; kan, kanamycin; tet, tetracycline.

Figure 12-6. Formation of a stem-and-loop structure by denaturation and intramolecular annealing of a plasmid carrying an inverted repeat. The configurations labeled EM indicate the appearance of such a molecule in an electron microscope; light and heavy lines represent single-stranded and double-stranded regions.

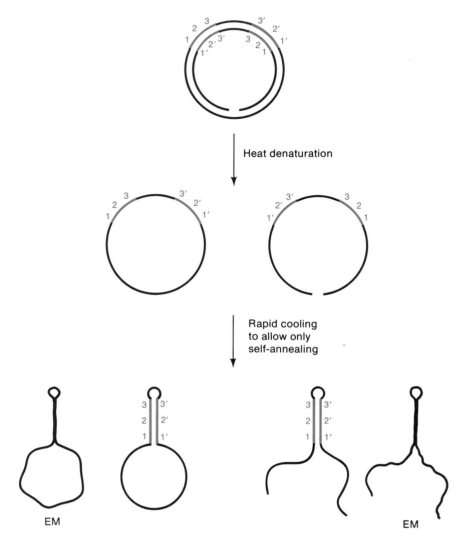

and renatured, single-stranded circles containing a stem-and-loop structure were observed by electron microscopy. Such a structure is indicative of an inverted repeat, as shown in Figure 12-6.

The composite transposons have two interesting features: (1) Sometimes the terminal IS elements are able to transpose by themselves. (2) For many composite transposons, if the transposon resides in a small plasmid, it behaves like two different transposons. This is because, within the frame of reference of the terminal IS elements, the DNA they flank and the DNA that surrounds them are interchangeable. For example, consider a composite transposon having the structure IS-A-IS, in which A is the central sequence. If the transposon inserts in a plasmid whose sequence we call B, the structure of the transposed plasmid will be

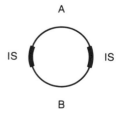

Because the plasmid is circular, both A and B are flanked by the IS elements, and transposition of either IS-A-IS or IS-B-IS can occur.

2. *The Tn3 transposon family*. The Tn3 family of transposons are about 5000 bp. The prototype, Tn3, was first detected by the appearance of an inverted-repeat element in plasmid R1. Each transposon carries three genes, one encoding β-lactamase (which confers resistance to ampicillin) and two others needed for transposition (discussed later). All Tn3-like transposons contain short (38-bp) inverted repeats, and none are flanked by IS-like elements.

3. *The transposable phages*. Two related phages, Mu and D108, include transposition as an essential part of their life cycles: Replication of the phage DNA requires transposition. Mu is described in greater detail later in the chapter.

TRANSPOSITION

The end result of the transposition process is the insertion of a transposon between two nucleotide pairs in a recipient DNA molecule. Base sequence analysis of many transposons and their insertion sites reveals that there is no sequence homology between the transposon and the insertion site, as expected from the lack of a requirement for the *E. coli recA* system. There are, however, several noteworthy features of the DNA sequences of the regions in which the transposon is joined to the recipient DNA.

Duplication of a Target Sequence at an Insertion Site

Insertion of a transposon always is accompanied by the duplication of a short nucleotide sequence (3 to 12 bp) in the recipient DNA molecule, called the target sequence: The inserted transposon is sandwiched between the repeated nucleotides. This arrangement is shown in Figure 12-7. Note that only one copy of the target sequence is present in the recipient DNA before insertion of the transposon, and it is not present in the transposon itself.

The length of the target sequence varies from one transposon to the next but is the same for all insertions of a particular transposon. For example, an inserted IS*1* is always flanked by a 9-bp sequence, whereas another transposon is

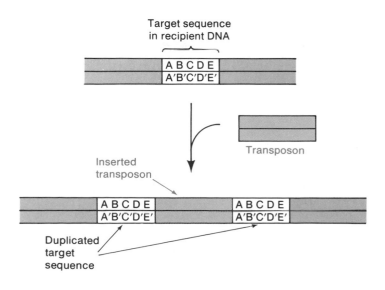

Figure 12-7. Insertion of a transposon generates a duplication of the target sequence. The transposon is never present as a free DNA segment but comes from another region of the genome, as shown in Figure 12-10.

Figure 12-8. A schematic diagram indicating how target sequences might be duplicated. The transposon is never present as a free DNA segment but comes from another region of the genome, as shown in Figure 12-10.

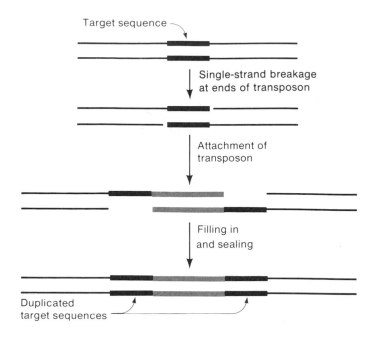

always flanked by a duplication of 5 bp. Note that for a particular transposon, the target sequence is different for each insertion site; only the length of the duplicated sequence is constant.

The basic mechanism shown in Figure 12-8 could account for the duplication of the target sequence. An enzyme, called **transposase** (encoded in the transposon), acts on a target sequence in the recipient DNA by making two single-strand breaks, one in each strand, at the ends of the target sequence. The transposon DNA is then attached to the free ends generated by the nicks, such that one strand of the transposon is joined to one strand of the target sequence, and the other strand of the transposon is joined to the other strand at the opposite end of the target sequence. This joining leaves two gaps, one across from each strand of the target sequence. Each gap is then filled in, and the nicks are sealed; thus, the two strands of the recipient DNA are again continuous, and two copies of the target sequence flank the transposon. Note that no homology is used in joining the terminal sequence of the transposon to the recipient DNA.

Because every time the transposon is integrated into a new site it is flanked by a duplicated target sequence, the question arises: Is the duplication also necessary for transposition from a donor site? This is definitely not the case because DNA sequences have been constructed (using recombinant DNA techniques) in which a transposon is not flanked by a directly repeating sequence, and these transposons are still able to transpose.

Structure of Transposons and Transposon-Target-Sequence Joint

DNA sequence analysis of many transposons has been done. Two important facts have emerged from these analyses:

1. *Transposons have well-defined ends*. The transposon sequences on both sides of the target sequences are identical for every insertion of a particular transposon (Figure 12-9). Each kind of transposon, however, has a specific sequence, so the transposition process requires joining the ends of the transposon to the ends produced by nicking at opposite sides of the target sequence.

2. *The two termini of most transposons consist of a single sequence in an inverted repeat array* (some exceptions are known). This is shown schematically in Figure 12-9. Note that for transposon I, the sequence at the left end of the upper strand and the right end of the lower strand is 1234, reading each time in the $5' \rightarrow 3'$ direction. Depending on the transposon, the number of bases in the inverted repeat is 8 to 38.

A great deal of genetic evidence indicates that the termini of a transposon are essential for transposition. For example, mutations in the transposons that produce single base-pair changes, small deletions, and small insertions in the inverted repeats invariably prevent transposition.

Replicative Transposition

Consider a cell with two plasmids, A and B, one of which contains a transposon. At a frequency of about 10^{-7} events per generation, the phenomenon shown in Figure 12-10 occurs. That is, after many generations of growth, cells are produced in which plasmid B (now called B') also possesses the transposon originally in A. With some transposons (for example, Tn3 and Mu), the transposon is now contained in both plasmids. Thus for these transposons, the original transposon sequence is apparently duplicated in the transposition process—transposition of these elements is not simply due to excision of the element from one site and insertion of the same DNA fragment at a second site. Thus, transposition of these transposons is a replicative process. For other transposons (for example, Tn10 and Tn5), it is clear that replication is not required for transposition.

Transposons that use **replicative transposition** have a unique intermediate. At a frequency also of about 10^{-7} events per cell generation, the two plasmids fuse to form a single plasmid called a **cointegrate**. (This process is sometimes also called replicon fusion.) This hybrid plasmid is not simply a fusion

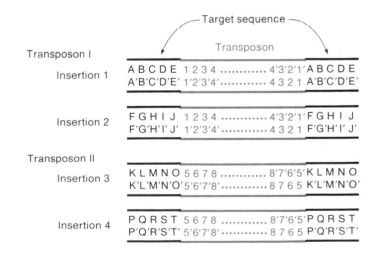

Figure 12-9 Diagram showing that two insertions by the same transposon have the same sequences at the junction of the transposon and the target sequence. That is, 1234 is at the left terminus of all insertions by transposon I. For transposon II, the sequence is always 5678. Each number and each letter represent a base; the prime denotes a complementary base. The identifications are merely symbolic; 5678 designates a sequence different from 1234 and not one that follows 1234 in numerical order.

Figure 12-10. Transposition of a transposon from plasmid A to a plasmid B.

of the two plasmids because it contains two (not just one) copies of the transposon. Both copies are in the same orientation (that is, in direct repeat) and are located precisely at the junctions between the donor and recipient plasmid sequences (Figure 12-11).

The fact that the cointegrate has two copies of the transposon, whereas only one copy was present before cointegration occurred, indicates that the transposon has replicated. Those transposons that transpose via a nonreplicative process do not form cointegrates.

Cointegrate as an Intermediate in Transposition of Tn3

Tn3 and closely related transposons use replicative transposition. Certain experimental results with transposons in the Tn3 family suggest that formation of a cointegrate is an intermediate step in the transposition of these elements.

The structure of Tn3 is shown in Figure 12-12. Tn3 has a pair of terminal inverted repeats that flank three genes. The three gene products have been isolated. From their size and from the base sequence of Tn3, we know that the genes include all of the bases between the inverted repeats. The leftmost gene encodes a large protein (1015 amino acids), TnpA, called transposase; it is responsible for formation of cointegrates. The rightmost gene encodes β-lactamase, an enzyme that inactivates ampicillin. The central gene encodes a small protein (185 amino acids), TnpR, called **resolvase**; it has two functions—repression of the synthesis of TnpA and catalysis of a site-specific exchange that resolves cointegrates in a second step of the transposition process. Near the boundary of the genes encoding TnpA and TnpR is a sequence of DNA called the internal resolution site or **res**; it includes the base sequence where TnpR acts.

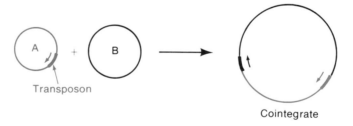

Figure 12-11. Formation of a cointegrate by transposon-mediated fusion of two plasmids A and B. Two copies of the transposon are generated in the process. The arrows denoted an arbitrary direction to show that both copies of the transposon in the cointegrate are in direct repeat.

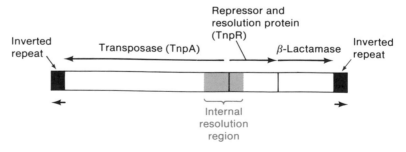

Figure 12-12. The physical map of transposon Tn3. The transposon is 4957 bp long. Each inverted repeat contains 38 bp. The three genes are indicated above the arrows, which show the direction of transcription. The internal resolution region res is discussed in the text.

Mutations in each of these genes indicated the role of the proteins. The *tnpA*⁻ mutants are unable either to transpose or to form cointegrates. In contrast, neither *tnpR*⁻ mutations nor deletions of the internal resolution site interfere with the ability to form cointegrates, but they completely prevent transposition. That is, in a cell containing a plasmid with a *tnpR*⁻ copy of Tn3 and a second plasmid lacking Tn3, the process shown in Figure 12-11 can occur, but the process shown in Figure 12-10 does not occur. These observations suggest that Tn3-mediated transposition proceeds by a two-step mechanism: First, TnpA induces formation of a cointegrate and then TnpR catalyzes a site-specific exchange at the internal resolution site. A schematic diagram of this sequence of events is shown in Figure 12-13.

Some transposons outside of the Tn3 family do not carry a TnpR-like, site-specific recombination system but are still able to transpose. These systems may use a homology-dependent exchange system because once a cointegrate has formed, there are two identical sequences—duplicate copies of the transposon—so homologous recombination process could carry out the exchange shown in Figure 12-13. Because transposition of these elements, however, occurs in *recA*⁻ cells, the homology-dependent recombinase would have to be encoded in the transposon. Evidence that a homology-dependent process could resolve the cointegrate is supported by the observation that a mutant of Tn3 lacking TnpR or the internal resolution site can carry out the transposition process of Figure 12-10 (that is, reduction of a cointegrate to individual plasmids) in a *recA*⁺ host cell.

Mechanism of Replicative Transposition

The mechanism of replicative transposition has been extensively studied in Mu. Four hallmarks of replicative transposition are as follows: (1) One copy of the transposon is retained in the donor at the initial site of the transposon, (2) the intermediate is a cointegrate, (3) a short repeated sequence of target DNA is

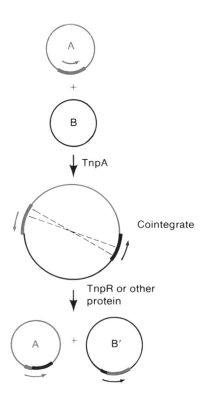

Figure 12-13. A model for transposition using a cointegrate intermediate.

generated on each side of the newly integrated element, and (4) some transposons require an internal resolution site or a homology-dependent exchange.

Two critical steps are required to initiate replicative transposition:

1. Two single strands are joined together (ligated) without base pairing to determine the position where the two DNA molecules are joined (the "join-point").
2. A replication fork is created by the ligation event.

These results can be explained by the model shown in Figure 12-14a. Consider a donor plasmid containing a transposon and a recipient plasmid. In the recipient, there is a short base sequence, the target sequence, at which integration of the transposon is to occur. The following steps, numbered as in Figure 12-14a, have been proposed:

I. A pair of single-strand breaks is made in the target sequence by the transposase. These breaks are staggered; thus, cleavage of the target sequence yields two complementary single strands.

Figure 12-14a. A model for replicative transposition. The model is drawn for transposition between two circular DNA molecules. (I) Nicks are made at sites in the DNA molecule indicated by arrows. (II) Strand separation between nicks and joining of nonhomologous strands. Two replication forks result. (III) DNA synthesis (hollow lines) forms this structure. A site-specific strand exchange occurs between the internal resolution sites (arrows), or possibly a homology-dependent strand exchange occurs anywhere in the transposon. (IV) The DNA molecules separate.

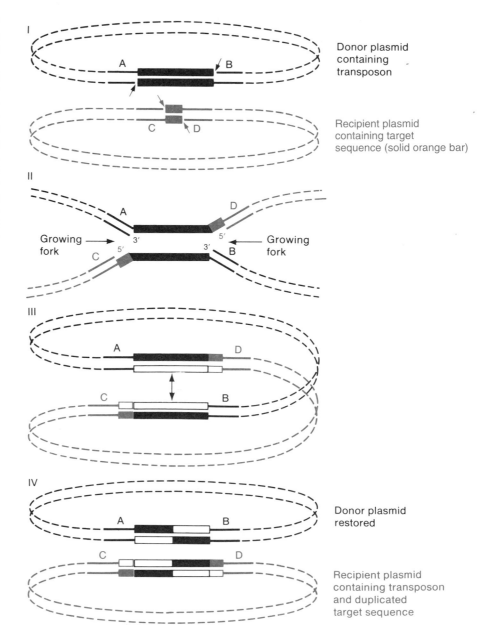

II. A pair of single-strand breaks is made in complementary strands at opposite ends of the transposon; this generates two free ends. Each 3' free end is attached by a single strand to a protruding strand (which carries a 5' group) of the staggered cut in the target site; this generates two replication forks.

III. Replication begins by synthesis of a strand complementary to the protruding end of the target sequence. When replication is completed, a cointegrate is formed containing two copies of the transposon.

IV. Finally, double-strand exchange occurs between the two copies of the transposon. In Tn3, this exchange usually takes place in the internal resolution site. The result of the exchange is to separate the cointegrate into the donor and recipient units, each of which now has a copy of the transposon. In the recipient, the transposon is flanked by a direct repeat of the target site; the length of this repeat is the number of base pairs between the staggered cuts.

Nonreplicative Transposition

Other transposons do not use replicative transposition: Transposition occurs by a conservative "cut and paste" mechanism, which requires DNA synthesis only to fill in the gaps at the ends between transposon and the target sequence (Figure 12-14b).

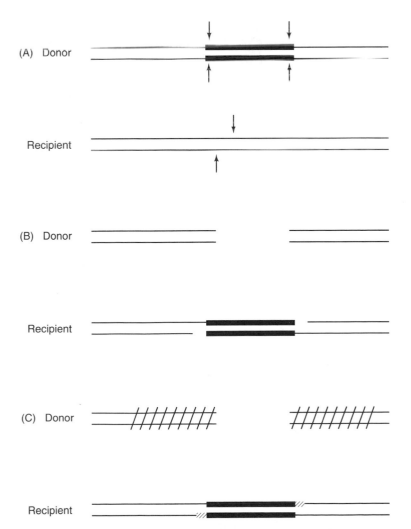

Figure 12-14b. A model for nonreplicative transposition (A) The transposase makes double-stranded cuts at each end of the transposon in the donor DNA molecule and staggered single-stranded cuts in the recipient DNA molecule. (B) The transposon is joined to the 5' ends of the recipient DNA at the target site. (C) The small, single-stranded gaps at each end are filled in with DNA polymerases from the host, resulting in a small duplication at each end of the transposon insertion. The free double-stranded ends of the donor DNA are degraded by host exonucleases.

The clearest evidence for this mechanism came from an experiment by Bender and Kleckner in 1986. They constructed a derivative of Tn10 with an insertion of the *lacZ* gene. They then made a heteroduplex between a copy of the Tn10 carrying the *lacZ*⁺ gene and a copy of Tn10 with a mutation in *lacZ*. They then examined the *lac* phenotype following transposition of the heteroduplex Tn10. If Tn10 transposition occurred by semiconservative replication, on transposition, the cell would get a copy of either *lacZ*⁺ or *lacZ*⁻, *not both*. In contrast, if transposition occurred by a cut and paste mechanism, the cell would get the heteroduplex with *both* copies of *lacZ*: On cell divisions, one of the daughter cells would inherit the *lacZ*⁺ copy of Tn10, and the other daughter cell would inherit the *lacZ*⁻ copy of Tn10. When they examined the resulting colonies on color indicator plates, they found that many of the colonies had the phenotype predicted for nonreplicative transposition.

EXCISION OF TRANSPOSONS

Transposons are sometimes excised from their insertion sites. Excision can be recognized in a variety of ways. For example, how could you detect excision of Tn5 from *lacZ*::Tn5 (which carries a Kan^r insertion in the *lacZ* gene)? In one study, 19 insertions were isolated. These insertion mutants were stable, and mapping experiments showed that the *lac*⁻ and Kan^r mutations were tightly linked. Cultures derived from each of the 19 mutants were prepared and tested for Lac⁺ reversion. Lac⁺ revertants arose at a frequency of 10^{-6} per generation, and 99% of the revertants were Kan^s. Thus, associated with recovery of the Lac⁺ phenotype was complete loss of Tn5 from the cell—a process called **precise excision**. In the 1% of the Lac⁺ cells that were still Kan^r, the Kan^r transposon gene was excised from the *lac* gene and relocated at a new site. A small percentage of these Lac⁺ Kan^r revertants were auxotrophic for various nutrients, as would be expected if reinsertion occurred at random sites.

Excision can occur in *recA*⁻ cells, indicating that excision is not simply the result of homologous exchange between flanking target sequences.

GENETIC PHENOMENA MEDIATED BY TRANSPOSONS IN BACTERIA

Transposons mediate a variety of genetic phenomena, such as gene rearrangements and integration of plasmids into the chromosome.

When a transposon is present in a chromosome or plasmid, the frequency of nearby deletions is increased about 100-fold to 1000-fold. When such a deletion occurs, it is often found that (1) the deletion originates at the end of the element, and (2) the transposon is bounded by only one copy of the target sequence (that is, the deletion extends from one side of the transposon). These features are shown in Figure 12-15.

Figure 12-15. Characteristics of a deletion mediated by a transposon. Segments 2, 3, 4, and 5 are deleted and replaced by a transposon.

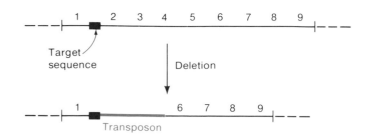

The frequency of inversions is also enhanced in the vicinity of transposons. In this process, beginning with a sequence . . . transposon-*a b c d e f* . . . in which *a–f* are genes adjacent to the transposon, an event occurs that leads to a sequence such as . . . transposon-[5]-*b a*-transposon-[5]-*c d e f* . . . in which [5] denotes an arbitrary 5-bp target sequence. Note that *a b* is inverted and that a new target sequence and the transposon are duplicated and surround the inverted sequence. The molecular mechanisms for both deletion and inversion generation are not known but can be explained by an abortive transposition event, as shown in Figure 12-16.

In this figure, the array shown in Figure 12-14a is repeated—that is, segments A and B are separated by a transposon, and segments C and D surround a target sequence, as shown in panels I and 1 of the two figures. Panel 3 shows

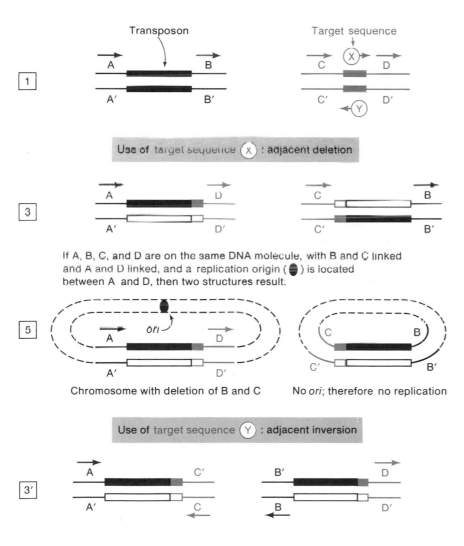

Figure 12-16. Schematic diagram showing how genetic deletions and inversions can form in accord with the model shown in Figure 12-14a. Panel 1 has the same array as I in Figure 12-14; 3 and 3' correspond to III in Figure 12-14 except that two different orientations of a target sequence are used. Solid arrows indicate the direction of transcription. Other symbols are those used in Figure 12-14.

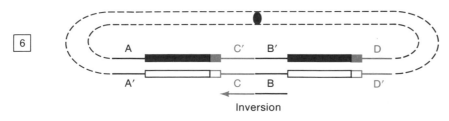

the array surrounding the two copies of the transposon after steps I through III of Figure 12-14 are completed. If each of the segments in panel 1 were circles, the two segments in panel 3 would be fused into one circle, as explained in the preceding section. If both segments are part of a single DNA molecule (for example, a chromosome), however, the transposition process would yield the two molecules shown in panel 5. Note that only one molecule in this panel contains the replication origin and that this molecule has also lost the segment containing B and C; that is, this molecule is deleted for B and C. The smaller circle containing B and C does not have a copy of the replication origin and therefore cannot replicate. Thus, if this entire process occurred in a bacterium, the smaller circle could not replicate, and progeny bacteria would contain only the deleted chromosome. Note that the segment deleted is immediately adjacent to one side of the transposon and includes all material between the transposon and the target sequence.

The lower part of Figure 12-16 shows how genetic inversions could be produced. In this case, the initial gene array is exactly that which yielded the deletion, but the target sequence has an orientation opposite to that of the transposon (that is, the strands ligated in panel I of Figure 12-14 are reversed); this is equivalent to insertion of the transposon in the opposite orientation. The transposition process then gives rise to the arrays shown in panel 3'. Note that the orientations of A and C are inverted with respect to the starting array, as are the sequences C and D. Thus, if the initial segments come from a single DNA molecule, the array shown in panel 6 results. In this case, no DNA is lost (that is, there is no deletion), but the segment between one side of the transposon and the target sequence is inverted with respect to the other segment—this is a genetic inversion.

It is also possible that inversions may arise by homologous recombination between the inverted repeats, although such events are rare. A simple mechanism by which this might occur is shown in Figure 12-17. Here transposition occurs at a target sequence (indicated by a square in the left circle of the figure) in the molecule containing the transposon. Insertion can occur in either a direct-repeat orientation (I in the figure) or an inverted repeat (II in the figure). With

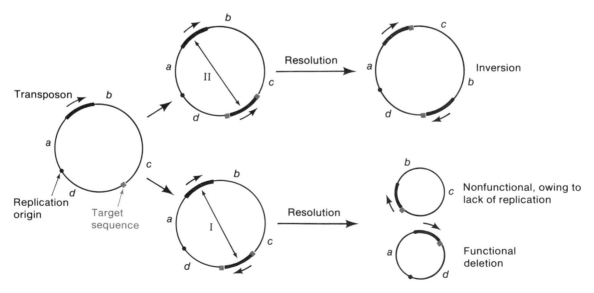

Figure 12-17. Model for production of genetic deletions and inversions. The transposon DNA is inserted into the target sequence in orientation I or II. The circle could be a plasmid or the chromosome. Resolution by a site-specific strand exchange at sites indicated by double-headed arrows or by exchange in homologous sequences yields a deletion from I and an inversion from II.

a transposon capable of resolution, a site-specific exchange could occur within the two transposons; alternatively, a homology-dependent exchange mediated by the RecA protein may occur. If the transposons are in a direct-repeat array, the exchange would generate two circles. As in the model shown in Figure 12-16, because the chromosome or a plasmid generally contains a single replication origin, only the element possessing the origin can persist; this element has lost a DNA segment and thus is a deletion mutant. If the transposons are in the inverted-repeat array, the result is instead an inverted sequence, as shown in the right portion of the figure. Note that the end results of the schemes shown in Figures 12-16 and 12-17 are the same. Other models also explain deletions and inversions.

PHAGE MU

E. coli phage Mu is a temperate phage capable of both lytic and lysogenic growth. Its name, Mu, is short for "mutator" because about 3% of all Mu lysogens are mutant for some easily recognizable gene. The mutations arise because Mu inserts its DNA at randomly distributed sites in the *E. coli* chromosome. Insertion into the host chromosome is an obligatory stage in the lytic life cycle of Mu. Furthermore, insertion always results in a duplication of a target sequence. Thus, Mu is a giant transposon that has acquired phage functions that enable it to be packaged in a phage coat, lyse its host, and thereby leave the cell.

Mu DNA

Mu contains about 38 kb of linear DNA. In a population of Mu phage particles, however, the ends of each DNA molecule are different. When a sample of Mu DNA is denatured and then renatured, electron microscopy of the renatured double helices (which invariably contain two single strands derived from different double-stranded molecules) yields the two structures shown in Figure 12-18. All single-stranded regions consist of sequences that are noncomplementary. Further experiments have indicated the following:

1. The noncomplementary sequences at the termini are bacterial DNA segments, which differ from one DNA molecule to the next.
2. If a segment of Mu DNA in the α region is increased in length (by an insertion of foreign DNA), the nonhomologous region at the left (c) end retains the wild-type length, whereas the segment at the right (S) end becomes shorter.

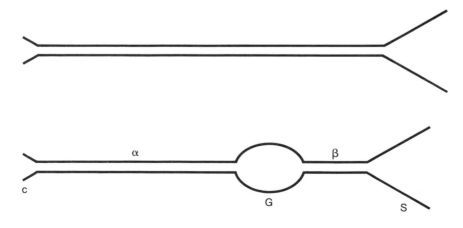

Figure 12-18. The major structures of Mu DNA seen after denaturation and renaturation. The symbols c, α, G, β, and S refer to distinguishable regions of the heteroduplexes.

3. Renatured molecules may have an unpaired region called the G loop. The two strands of the G loop, which are phage DNA, are identical, but one is reversed in direction (inverted) with respect to the other. That is, a G loop has the structure shown in Figure 12-19. Occasionally in the phage life cycle the G segment becomes inverted by genetic recombination. The normal and inverted configurations are called G (+) and G (-): Annealing of a strand containing G (-) with a complementary strand containing G (+) yields a G loop. The G loop affects the host range of Mu, but it is not relevant to transposition.

Replication and Maturation of Mu DNA

Points 1 and 2 in the preceding section are explained by the fact that transposition is obligatory in Mu DNA replication. Shortly after infection, the incoming Mu DNA becomes inserted at a random position in the E. coli chromosome. The terminal bacterial sequences are not inserted: Only the Mu segment is inserted, as is usually the case in transposition. Replication of Mu DNA proceeds by repeated transposition to various sites in the chromosome, which explains the association of Mu DNA with various pieces of bacterial DNA described in point 1.

Numerous physical experiments show that insertion of progeny Mu DNA at various sites occur throughout the lytic cycle. This was first indicated, however, by genetic experiments. F' lac–containing cells were infected with a Mu mutant that was unable to kill the cells. Shortly after infection, the cells were mated with appropriate females, and F' lac transfer was scored. The transferred F' lac often contained integrated Mu DNA, and the insertion site was variable. Use of various F' plasmids indicated that Mu could insert in various parts of the DNA.

Mu DNA is never found free in the cell. The DNA in the phage particles always contains terminal bacterial DNA. These sequences arise from the packaging process. Packaging of the DNA occurs as follows (Figure 12-20): The phage-DNA maturation system recognizes the c end of the integrated Mu DNA and makes a cut in the host DNA about 100 bases to the left of the phage DNA. Pack-

Figure 12-19. The structure of a G loop (see Figure 12-18) of Mu DNA showing inversion of the G sequence denoted by ABCD.

Figure 12-20. Diagram of inserted Mu DNA (orange) showing how the cutting process generates host sequences at the termini.

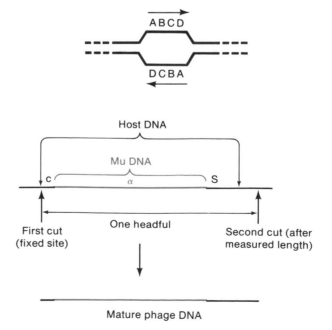

aging begins at this point, and the phage head is filled. The second cut, at the S end, is determined by the amount of DNA that can fit in the phage head. One headful exceeds the length of the phage DNA sequences, which explains point 2. This is called a headful packaging mechanism.

Mapping the location of Mu insertions in Mu lysogens indicates that the number of possible insertion sites is huge. If a culture is lysogenized by Mu and Mu-induced mutations are scored, it is found that the frequency of insertion in a particular gene is roughly proportional to the size of the gene. That is, if gene a is twice as large as gene b, there will be, on the average, twice as many a^- mutants as b^- mutants. Collections of insertions into a single gene, however, show that insertions into all sites are not equally probable. For example, among 75 Mu insertions in the *lacZ* gene, only 50 insertion sites were seen, and they tended to be in one region of the gene. Because the gene contains 3063 bp, one would expect that if the insertion was completely random, no insertion sites would be duplicated. No particular base sequences, however, were seen at the insertion sites, indicating that the specificity is not due to recognition of a specific base sequence.

Because transposons insert at many different sites and cause strong, polar mutations and the insertions can often be directly selected using a transposon-encoded antibiotic resistance, transposons are extremely useful for constructing mutant strains. Use of transposons for strain construction is described in Chapter 19.

TRANSPOSONS AND EVOLUTION

Apart from conferring antibiotic resistance, transposons do not seem, at first glance, to be of any value to a cell. This point may be made even more strongly for the IS elements. Thus, it is reasonable to ask why transposons have been maintained over evolutionary time. Generally a gene that serves no purpose to a cell will ultimately be mutated until it is unrecognizable, nonfunctional, and often deleted. For some time, it has been thought that transposons are an example of what has been termed "selfish" DNA—that is, a DNA segment that has persisted only because it has evolved to maintain itself. For example, a cell with a copy of a useless IS element will keep the transposon if transposition (which duplicates the element) occurs at a higher rate than mutation or loss of the element. This argument may indeed explain the persistence of many transposons. Experiments have suggested, however, that although a transposon does not provide a selective advantage to an individual cell, its presence in the long run confers an advantage to the cell population. The argument is based on reasoning that because evolution proceeds by accumulating mutations, a process that accelerates mutation may speed up evolution.

In Chapter 4, continuous growth of bacteria in a chemostat was described. Chemostat cultures are useful in studying mutation rates and evolution because both constant conditions and cell multiplication can be maintained indefinitely. Several laboratories have compared the multiplication rate of a cell that contains a transposon (Tn5 and Tn10 have been studied) with an isogenic cell that lacks the transposon. The generation times of these cultures determined in the standard way are not measurably different. In a chemostat, however, it is possible to prepare an initial mixture consisting of, for example, 10^5 wild-type *E. coli* cells and one *E. coli* cell with a Tn10 insertion. After prolonged growth (hundreds of generations), the ratio of transposon-containing cells to wild-type cells can be determined. Study of numerous chemostat cultures showed that some cultures had maintained the original 10^{-5} ratio, but others consisted mainly of cells containing Tn10. No antibiotic was present, so cells containing Tn10 did not have any selective advantage for this reason. Cultures in which the Tn10-containing cells had "won" were examined for transposition to determine whether a transposition event accounted for the more rapid growth. The number of copies of the Tet^r gene present in Tn10 had not increased, but the number of copies of the IS$10R$ terminus of Tn10 had usually increased by one. (Tn10 is a composite transposon with two slightly different forms of IS10—IS$10L$ and IS$10R$—at the left and right termini.) No increase in either Tn10 or IS$10R$ was observed in cultures in which the wild-type cells had won. Furthermore, biochemical analysis showed that the IS$10R$ insertion in the Tn10-containing cells was localized in a fairly small region of the bacterial chromosome. Thus, this insertion was associated with a slightly higher growth rate.

The interpretation of this and related experiments is that transposition produced beneficial mutations—that is, cells containing the transposons evolved faster. A similar effect has also been observed with cells containing mutator genes (see Chapter 10), that is, genes that confer on the cell an enhanced mutation rate.

It is clear that a significantly enhanced mutation rate produced by a transposon could also be deleterious. In bacteria, because transposition is often a replicative process, the number of elements would seem to be capable of increasing indefinitely. This would be harmful to the host cell and hence to the transposon because ultimately the host genome would be damaged beyond its ability to carry out necessary cellular functions. Thus, it seems likely that transposition would be limited in some way, and indeed transposition seems to be carefully regulated. Evidence for such a system comes from experiments in which a transposition is introduced into a cell that has no copies of the element. Transposition begins within several generations and occurs repeatedly. Soon the rate of transposition decreases, and ultimately the cell acquires a stable number of copies of the transposon; the particular number is characteristic of each element. The regulatory mechanism varies in different transposons.

The number of copies of a particular transposon in a bacterium is usually less than 10. In the higher eukaryotes, however, in which up to 30% of the DNA may consist of transposable elements, the number of copies of an individual element may be in the hundreds of thousands.

KEY TERMS

cointegrate	replicative transposition
insertion sequences	resolvase
nonreplicative transposition	transposon
phage Mu	transposase
precise excision	

QUESTIONS AND PROBLEMS

1. What two components are characteristic of transposable elements?

2. If you were determining the base sequence of a long DNA segment, what features of the sequence would make you confident that a transposable element was present?

3. A particular transposable element is observed at five different locations. Will the flanking target sequences be the same at each site?

4. A transposable element is observed at several different sites in different bacterial strains. Each element is flanked by a pair of target sequences. What characteristic is common to these target sequences?

5. What is a typical length of a target sequence?

6. What is a composite transposon?

7. Does a composite transposon necessarily have terminal inverted repeats?

8. Are all drug-resistance genes located on transposons?

9. Why does the presence of an IS element in a particular genetic region often cause polarity?

10. What is the evidence that transposition in bacteria includes a replicative step?

11. When transposition occurs in prokaryotes, two base sequences are duplicated. What are they?

12. When transposition occurs, is there any site selectivity with respect to the new site?

13. What term is used to describe a plasmid that results from transposon-mediated fusion of two plasmids, one of which contained a transposon? How many copies of the transposon are present in the fused molecule?

14. In the Tn3 system, what enzyme is responsible for formation of a cointegrate?

15. In the Tn3 system, what enzyme breaks down a cointegrate into two independent replicons? How many copies of the transposable element exist after action of this enzyme?

16. Name five genetic phenomena mediated by transposable elements.

17. In homologous recombination, two DNA molecules break and rejoin. If the event is physically reciprocal, the amount of DNA present is the same as that present before recombination; if the event is nonreciprocal, the total amount of DNA present decreases. What can be said about the relative amounts of DNA before and after transposition?

18. Transposition is a type of nonhomologous exchange, suggesting that matching of sequences is not part of the process. There seems to be some sequence effect, however, in that transposition occurs to some sites more than to others. The mechanism of this effect is unknown. Suggest what type(s) of things might account for this effect.

19. Explain how insertion of a transposon that lacks a transcription-termination signal can cause genetic polarity.

20. You have isolated a segment of DNA having about 500 bp. Several genetic experiments have suggested but have not proved that the segment contains a transposon. What physical or chemical experiment would you do to test the hypothesis?

21. An Amp^r plasmid whose replication is temperature-sensitive is introduced into an Amp^s cell by the CaCl$_2$ transformation method. After growth for several generations, 10^7 cells are plated on agar containing ampicillin at 42°C. Fifty colonies form.

    a. Give three mechanisms that could explain the presence of these colonies.
    b. Which mechanisms would not occur in a RecA$^-$ cell?

22. You have isolated two different insertion mutations in phage λ. How would you determine whether the two insertions are the same transposon?

23. A particular IS element has inserted into different positions in two λ phages. In one case, the insertion is at position 55; in the other, at position 75 (both measured on a scale of 100 from the left end of the phage). The IS element has a length equal to 1% of the size of one DNA. One of these phage variants shows polar effects on downstream genes; the other does not.

    a. Explain the difference in the polar effects.
    b. How could your explanation best be tested by electron microscopy?

24. Two bacterial genes (*chk* and *lat*) are near one another. When *chk$^-$* mutants are isolated (they arise spontaneously at a frequency of about 10^{-6} mutants per cell per generation), about 1% of these mutants are also *lat$^-$*. This is a surprisingly high frequency for spontaneous formation of double mutants.

    a. Could you guess what might be the cause of this phenomenon?
    b. Would a measurement of this frequency in a RecA$^-$ cell be informative? Explain.
    c. Would studying the effect of chemical mutagens be informative? Explain.

25. It is sometimes desirable to remove a transposon from a bacterium. This is possible because excision of transposons occurs spontaneously, although at a low frequency and by an unknown mechanism. If a transposon were inserted in the *lac* gene, one would only have to obtain a Lac$^+$ revertant colony to obtain cells that had lost the transposon. In some genes, revertants are not detected in such a straightforward way;

however, the following technique is successful for transposons carrying some antibiotic-resistance markers. Consider a culture of cells in which a transposon carrying the Tet^r gene is present in gene x rendering the cells X$^-$. Tetracycline is added to a growing culture, and a few minutes later penicillin is added. Penicillin kills only growing cells. The Tet^s cells are inhibited by the tetracycline. Tet^r cells grow in the presence of tetracycline and hence are killed by the penicillin. After many hours of growth, the cells are plated on agar lacking tetracycline. At a low but easily detected frequency, colonies form whose cells lack a functional Tet^r gene. Describe three mechanisms for generating these Tet^s cells and indicate which result in transposon loss.

26. When Mu infects a bacterium, DNA is inserted into the bacterial chromosome. Is all of the DNA inserted?

27. If a single burst experiment were done with Mu, would all progeny in a particular burst have the same bacterial sequences?

28. If Mu DNA is denatured and renatured and examined by electron microscopy, double-stranded molecules with unpaired ends result. Would you ever expect to see molecules without frayed ends?

29. Because Mu obligatorily inserts into the bacterial chromosome during its replication cycle and because insertion frequently occurs in a bacterial gene, thereby inactivating the gene, should one conclude that most infections do not lead to phage production?

REFERENCES

Bender, J., and N. Kleckner. 1986. Genetic evidence that Tn*10* transposes by a non-replicative mechanism. *Cell*, 45, 801.

*Berg, D., and M. Howe. 1989. *Mobile DNA*. American Society for Microbiology, Washington, D.C.

Campbell, A. 1981. Evolutionary significance of accessory DNA elements in bacteria. *Ann. Rev. Microbiol.*, 35, 55.

*Cohen, S. N., and J. A. Shapiro. 1980. Transposable genetic elements. *Scientific American*, February, p. 40.

Craigie, R., and K. Mizuuchi. 1985. Mechanism of transposition of bacteriophage Mu: structure of a transposition intermediate. *Cell*, 41, 867.

Kleckner, N. 1990. Regulating Tn*10* and IS *10* transposition. *Genetics*, 124, 449.

*Kleckner, N., J. Roth, and D. Botstein. 1977. Genetic engineering in vivo using translocatable drug-resistance elements. *J. Mol. Biol.*, 116, 125.

Kleckner, N., J. Bender, and S. Gottesman. 1991. Uses of transposons with emphasis on Tn*10*. *Methods Enzymol.*, 204, 139.

Nevers, P., and H. Saedler. 1977. Transposable genetic elements as agents of gene instability and chromosomal rearrangements. *Nature*, 268, 109.

Rogers, J. 1985. Origins of repeated DNA. *Nature*, 317, 765.

Shapiro, J. A. 1979. Molecular model for the transposition and replication of bacteriophage Mu and other transposable elements. *Proc. Natl. Acad. Sci.*, 76, 1933.

Symonds, N., A. Toussaint, P. van der Putte, and M. Howe. 1987. *Phage Mu*. Cold Spring Harbor Laboratory Press.

*Resources for additional information.

Bacterial Transformation

Bacterial transformation is a process in which a recipient cell acquires genes from free DNA molecules in the surrounding medium. In the laboratory, transformation is accomplished by isolating DNA from donor cells and then adding this DNA to a suspension of recipient cells; however, transformation also occurs in nature. Transformation is an important tool for genetic mapping for certain organisms. In addition, the ability to transfer plasmid or phage DNA into a recipient cell is important in genetic engineering (see Chapter 20).

DISCOVERY OF TRANSFORMATION

The discovery of bacterial transformation was one of the most important events in biology because it provided the foundation for modern microbial and molecular genetics. This discovery led to the demonstration that DNA is the genetic material.

Bacterial pneumonia in mammals is caused by certain strains of *Streptococcus pneumoniae*. Cells of these strains are surrounded by a polysaccharide capsule that protects the bacterium from the immune system of the infected animal and thereby enables the bacterium to cause disease. When a pathogenic (disease-causing) *S. pneumoniae* is grown on a nutrient agar, the enveloping capsule gives the bacterial colony a glistening, smooth (S) appearance. Some mutant strains of *S. pneumoniae* lack the enzyme activity required to synthesize the capsular polysaccharide, and these bacteria form colonies that have a rough (R) surface. R strains do not cause pneumonia because without their capsules the cells are inactivated by the host immune system.

Both R and S strains of *S. pneumoniae* breed true except for rare mutations in R strains that give rise to the S phenotype and rare mutations in S strains that give rise to the R phenotype. Because simple mutations can cause conversion between the forms, it was recognized quite early that the S and R phenotypes are determined by different allelic forms of a simple genetic unit.

In 1928, a seminal observation was made by Griffith: Mice injected with either living R pneumococci or with heat-killed S cells remained healthy, but mice injected with a mixture containing a small number of R bacteria *and* a large number of heat-killed S cells died of pneumonia. Bacteria isolated from blood samples of the dead mice produced pure S cultures having a capsule typical of the heat-killed S cells (Figure 13-1). Evidently dead S cells could in some way provide the living R bacteria with the ability to withstand the immunological

Figure 13-1. Production of disease-producing type S *S. pneumoniae* by infecting a mouse with a mixture of living type R cells, which alone do not cause disease, and heat-killed type S cells.

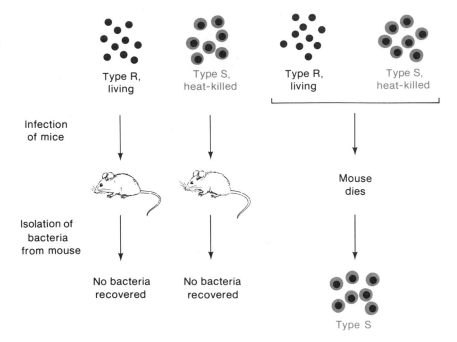

system of the mouse, multiply, and cause pneumonia. Furthermore, cultures derived from these changed or "transformed" bacteria retained the ability to cause pneumonia.

Later experiments showed that living S cells can also be produced (at low frequency) when R cells are grown in a liquid medium containing heat-killed S cells. Furthermore, extracts of S cells, from which intact cells and large pieces of cellular debris (mostly cell walls and membranes) had been removed by filtration, were as effective as intact S cells in causing specific transformation of R cells to the S phenotype. In 1944, Avery, MacLeod, and McCarty carried out a critical experiment that led to our current understanding of the phenomenon. They purified DNA from S cells and found that addition of minute amounts of this DNA to growing cultures of R cells consistently resulted in production of smooth colonies whose cells contained type-S capsular polysaccharide. Their DNA samples contained traces of protein and RNA, which complicated the interpretation of the data somewhat. The transforming activity, however, was not altered by treatment with enzymes that degrade proteins or by treatment with RNase but was completely destroyed by DNases (Figure 13-2). These experiments and several more sophisticated variants done over the following few years demonstrated that the substance responsible for genetic transformation was the DNA of the donor cells and hence that DNA is the genetic material.

BIOLOGY OF TRANSFORMATION

Transformation requires that the cells are in the proper physiological state and that an appropriate selection is available.

Detection of Transformation

Similar to most methods of gene transfer, transformation is a relatively rare event. Transformation is usually detected by selection for inheritance of a phenotype from the donor DNA by the recipient cells. For example, purified DNA obtained

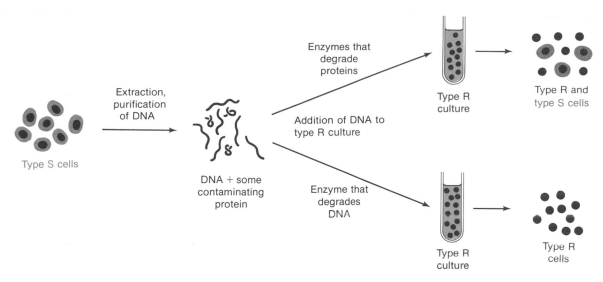

Figure 13-2. A diagram of the experiment that demonstrated that DNA is the active material in bacterial transformation.

from an erythromycin-resistant (Ery^r) culture of *S. pneumoniae* is mixed with cells from an Ery^s culture. After a period of incubation (allowing time for DNA uptake and time for phenotypic expression of the antibiotic resistance), the cells are plated on agar containing erythromycin. Formation of Ery^r colonies above the mutation frequency for Ery^s to Ery^r (about 10^{-8}) indicates that transformation has occurred. For *S. pneumoniae*, the maximum frequency of transformation for most antibiotic-resistance markers is 0.1 to 1%.

Competence

Transformation begins with uptake of a DNA fragment (the bacterial chromosome is usually broken extensively during isolation) from the surrounding medium by a recipient cell and terminates in recombinational exchange of part of the donor DNA with the homologous segment of the recipient chromosome. Probably most bacterial species are capable of the recombination step, but the ability of most bacteria to take up DNA efficiently is limited. Even in a species capable of transformation, DNA can penetrate only a small fraction of the cells in a growing population. Incubation of cells of these species under certain conditions, however, yields a population of cells in which the uptake of DNA is greatly enhanced, by a factor of 10^4 to 10^6. A culture of such cells is said to be "**competent**." The conditions required to produce competence and the fraction of the cells that are competent vary from species to species.

A great deal of effort has gone into understanding the mechanism of competence, but still the phenomenon is far from clear. Competence appears to result from changes in the cell wall of the bacteria and probably is associated with the synthesis of cell-wall material at a particular stage of the life cycle of the bacterium. In the course of developing competence, receptors of some kind are either formed or activated on the cell wall; these receptors are responsible for initial binding of the DNA. The number of active receptors varies widely from one organism to the next—for example, about 80 for *S. pneumoniae*, 50 for *Bacillus subtilis*, and 4 for *Haemophilus influenzae*. Competence usually arises at a specific stage of growth of a culture, typically late log phase, and highly depends on the growth medium and the degree of aeration of the culture. The particular medium and growth conditions that produce maximum competence vary from one bacterial species to the next. With *S. pneumoniae*, competence is generated just as the cells are entering stationary phase, but the competent state persists for only a few minutes (Figure 13-3). Competence initially arises in only a small fraction of the cells. Associated with early development of competence is the excretion by these cells of one or more proteins called "competence factors." These proteins convert the remainder of the cells in the population to the competent state, probably by some direct action on the receptors. Purified competence factors can also induce competence in most cultures in a way that does not depend strongly on the growth phase of the culture, which suggests that the medium, temperature, degree of aeration, and so on needed to produce a competent culture probably serve only to trigger synthesis or release of competence factors by a small number of cells. Competence factors are species specific and induce competence only in closely related species. The mechanism of action of competence factors is unknown.

The fraction of a culture that becomes competent and the duration of the competent state are also species dependent. For example, conditions have been found that produce 100% competence in *S. pneumoniae*, but competence persists for only a few minutes. In contrast, with *B. subtilis*, no more than 20% of the cells become competent, but the state lasts for several hours. Possibly the inability of most bacterial species to be transformed is only a reflection of lack of knowledge of the treatments necessary to induce competence.

The ability of a cell to take up DNA can also be elicited in a wide variety of bacterial strains by the $CaCl_2$ transformation technique described in Chapter 11. The state of the cells produced by this treatment, however, is quite different from that of naturally competent cells. Although almost any DNA molecule, of a wide range of sizes and either fragmented or intact, can be taken up by a cell treated with $CaCl_2$, stable genotypic change usually occurs only with intact replication systems, such as plasmids and phage DNA molecules. The mechanism of DNA uptake is different in competent cells versus $CaCl_2$-treated cells.

DNA Uptake

DNA uptake by competent cells has been examined in two ways. In one type of experiment, cells are exposed to radioactive DNA and then to a DNase capable of degrading DNA completely. These experiments indicate that there are at least two stages of interaction of DNA with competent cells. The first is a brief reversible-binding stage in which the DNA can be washed from the cells and in which the DNA is also completely sensitive to DNase degradation. In the second, or irreversible, stage, DNA cannot be washed from the cells, and it is

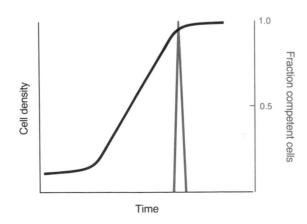

Figure 13-3. Time interval (orange) in which cells of *S. pneumoniae* are competent.

resistant to DNase. Irreversible interaction is an energy-requiring process that is associated with passage of the DNA through the cell membrane.

Although stable genotypic transformation is highly species specific, DNA uptake is normally not. This might be expected because all DNA molecules probably look the same to the outside of a bacterium. Because uptake of foreign DNA cannot be detected by genetic transformation, as it recombines poorly with host DNA, a competition assay has been used to show that a particular DNA molecule can be taken up by a competent cell. In one such experiment, different samples of penicillin-sensitive (Pen^s) S. pneumoniae were mixed with a fixed amount of DNA isolated from a Pen^r culture and with various amounts of calf thymus DNA. Increasing amounts of calf thymus DNA reduced the number of Pen^r transformants (Figure 13-4). The foreign DNA bound to the receptor sites and thereby competitively inhibited uptake of S. pneumoniae DNA. An exception to the rule of nonspecificity of DNA uptake is found in H. influenzae, which takes up only H. influenzae DNA. Apparently particular sequences of about 10 bp, which are rare in other DNAs, are needed for the initial binding to the cell surface. The mechanism of transformation of H. influenzae differs from that in most other bacteria in other ways also.

Denaturation of a DNA sample capable of transformation eliminates all transforming activity. Also, addition of competing but denatured calf thymus DNA does not inhibit transformation of S. pneumoniae by S. pneumoniae DNA (see Figure 13-4), so presumably single-stranded DNA does not bind to the receptors of competent cells. Studies with denatured radioactive S. pneumoniae DNA confirm this conclusion. Evidence that the denatured state is responsible for the lack of uptake comes from renaturation experiments. If the denatured DNA is renatured, both transforming activity and the ability to compete with normal transforming DNA are restored.

A simple genetic test shows that the state of DNA is altered during uptake. A Str^s culture of S. pneumoniae is exposed to DNA from a Str^r strain. After 15 minutes, the cells are washed free of excess unincorporated DNA and divided into two portions. After a period of time sufficient to allow expression of the Str gene, one part is plated on medium containing streptomycin, and a significant fraction of the cells prove to be transformed to the Str^r phenotype. DNA is isolated from the remainder of the 15-minute sample and tested for its ability to transform a fresh culture of competent Str^s cells, and transformation to streptomycin resistance occurs. Thus, transforming DNA is present in the 15-minute sample. In a related experiment, cells are plated, and DNA is isolated only 2 minutes after initial exposure to the DNA rather than after 15 minutes. In this case, plating yields the same number of Str^r colonies, indicating that washing the cells at this time (2 minutes) does not affect the transformation process. The DNA

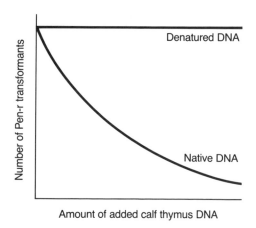

Figure 13-4. Reduction in transformation of Pen^s S. pneumoniae to the Pen^r phenotype by addition of various amounts of native calf thymus DNA to a fixed amount of S. pneumoniae transforming DNA. Denatured calf thymus DNA does not cause the reduction.

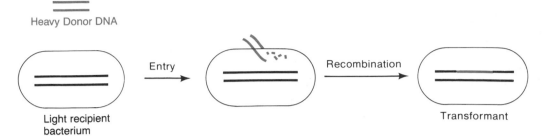

Figure 13-5. An experiment showing that in *S. pneumoniae* one strand of transforming DNA is destroyed and the complementary strand is inserted. This conclusion is based on the finding that transformants contain hybrid (heavy-light) segments but no fully heavy segments.

isolated from the cells, however, shows no transforming activity. Active transforming DNA is adsorbed to and enters competent cells; at first (2 minutes), it is in a nontransforming state, after which (before 15 minutes) it is converted to a transforming state. The period during which no transforming DNA can be isolated from potentially transformed cells (about the first 10 minutes) is called the eclipse. Physical studies with radioactive transforming *S. pneumoniae* DNA have shown what is happening. Associated with irreversible binding and DNA uptake is degradation of one strand of the transforming DNA by a nuclease contained in a multiprotein assembly of the cell membrane (Figure 13-5). Similar results have been obtained with *B. subtilis*.

A great many physical experiments indicate that only a single strand of DNA enters a competent *S. pneumoniae* cell. None of these experiments, however, show whether the partial degradation is essential for DNA uptake. As discussed in Chapter 1, a common approach to answer such a question is to attempt to isolate a degradation-deficient mutant and determine whether it can still take up DNA. Such a mutant has been isolated on a color-indicator plate, using a principle quite different from that used to identify sugar-utilization mutants—a color change that occurs when the dye methyl green binds to DNA. Colonies formed on agar containing both methyl green and free DNA are surrounded by a ring whose color differs from that of the bulk of the agar. This color difference results from the excretion of DNases by aging cells in mature colonies. The DNase degrades the DNA in the agar, causing the methyl green to change color locally to that of DNA-free dye. When a mutagenized *S. pneumoniae* culture was plated on DNA-methyl green agar, some colonies were found that lacked the colored ring. These mutants were called Noz⁻. Biochemical analysis of Noz⁻ cells showed that they lack a particular membrane nuclease, they do not degrade transforming DNA to the single-stranded form, and transforming DNA is not taken up by the cell. The properties of these mutants strongly suggest that degradation of one of the DNA strands is essential for DNA uptake in *S. pneumoniae*.

With *H. influenzae*, an eclipse period does not occur, and degradation does not accompany DNA uptake. In fact, many studies indicate that natural transformation in *H. influenzae* (a gram-negative bacterium) proceeds by a mechanism that is different from natural transformation in the gram-positive bacteria.

MOLECULAR MECHANISM OF TRANSFORMATION

In attempting to understand transformation in *S. pneumoniae*, two possible mechanisms were originally considered: (1) The incoming donor allele replaces the allele of the recipient; that is, DNA substitution occurs. (2) The DNA fragment containing the donor allele is added to the genome of the recipient, by insertion adjacent to the recipient allele, thereby forming a gene duplication (partial diploid). The first mechanism was demonstrated by a study of reciprocal transformation. First, an x^- strain was transformed with DNA from a strain carrying a dominant x^+ allele. Then the newly formed x^+ strain was exposed to DNA from an x^- strain, and it was found that x^- transformants arose at the same frequency as did x^+ transformants in the first experiment. If DNA addition occurred, the initial transformant would have had the genotype x^+x^-. If DNA from this transformant were used as donor DNA, a transforming DNA fragment would almost always carry both the x^+ and x^- alleles, so x^- cells would be exceedingly rare in the second transformation. Thus it was clear that DNA addition does not occur in transformation.

Another type of experiment using two closely linked markers made the point even more convincingly. In this experiment, the recipient bacterium (recipient 1) contained two closely linked loci m and n for antibiotic sensitivity, and the donor was resistant to both antibiotics, MN. Three genotypes, Mn, nM, and MN, were found with roughly the same frequency, which shows that M and N are linked (i.e., often carried on the same DNA molecule—discussed in a later section). In a second experiment, donor DNA with the genotype Mn was used to transform an mN recipient, and the M marker was selected. If the mechanism of transformation were donor addition, the genotypes of M recipients would be MmN and $MmNn$, both of which would have the N phenotype (Figure 13-6). Colonies with the MN and Mn phenotypes, however, were equally frequent, which agrees with the allelic substitution idea. That is, the transformation event that resulted in incorporation of the M marker was often associated with replacement of the N marker of the recipient and replacement by n, which was carried on the same DNA molecule as the M marker.

As stated earlier, in transformation with S. *pneumoniae*, uptake of the transforming DNA into the cell is accompanied by digestion of one of the input strands. Consequently, only a single strand is available for interaction with the recipient DNA. The single strand is, in an unknown way, protected from nuclease attack—possibly it is coated with the same protein (SSB protein) used to maintain single-stranded regions in a replication fork (see Chapter 8). Physical experiments in which high-density-labeled (heavy) DNA is used to transform a bacterium whose DNA has low density (light DNA) show that a single strand of the donor DNA, or a portion thereof, is linearly inserted into the recipient DNA (see Figure 13-5). Aided by a bacterial protein, equivalent to the *Escherichia coli* RecA protein, which facilitates DNA pairing in recombination (see Chapter 14), the incoming single-stranded fragment causes local unwinding of the recipient chromosome,

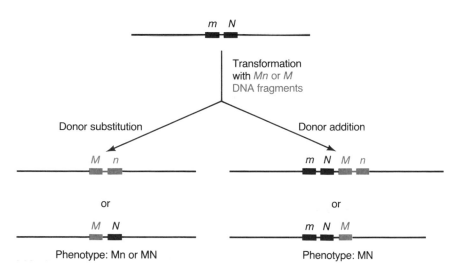

Figure 13-6. An experiment that shows that transformation occurs by donor substitution rather than donor addition. If the donor is Mn and the recipient is mN, transformants selected for the M phenotype are either Mn or MN. The addition model predicts that only MN will be recovered.

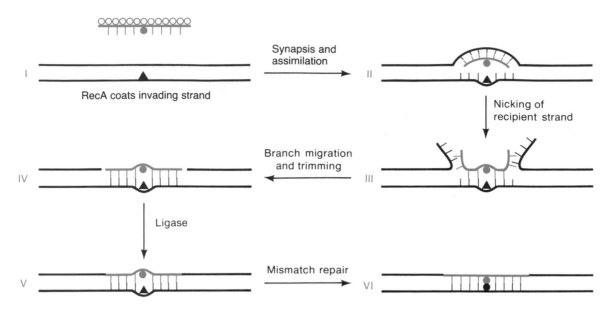

Figure 13-7.   Integration of single-stranded donor DNA by bacterial transformation.

presumably from the 5' end. By an unknown mechanism, the displaced single strand is cut. Unwinding of the recipient DNA continues at the end of the assimilated DNA, which allows the proportion of the invading strand that is base-paired to the recipient strand to increase (Figure 13-7). (This process is called branch migration. Trimming nucleases act simultaneously to remove free ends, which may be in donor or recipient DNA, and ultimately DNA ligase seals the nicks. The result is a heteroduplex region containing a mismatched base pair.) The outcome of the process—that is, whether or not the donor marker is recovered in the progeny cells—depends on whether or not mismatch repair occurs and, if so, whether or not the unpaired base in the donor or recipient strand is removed. For some markers, known as high-efficiency markers, either the mismatch is always corrected to match the donor genotype, or there is little or no repair. In the latter case, following replication of the chromosome and cell division, one cell has the donor genotype, and the other has the recipient genotype. Because the plating conditions used to detect recombinants are usually chosen to allow growth only of recombinants (for example, an antibiotic is added, or a necessary amino acid is absent), the colony that forms consists exclusively of recombinant bacteria. For the low-efficiency markers, mismatch repair usually removes the mismatched base from the donor, and the cell retains the recipient genotype. Proof that this is the correct explanation for the two types of markers comes from the isolation of *S. pneumoniae* mutants, called Hex⁻, that have a higher spontaneous mutation frequency. These mutants seem to have normal editing activity and methylate DNA at the normal rate and hence are thought to be defective in mismatch repair. When Hex⁻ mutants are transformed, all markers appear to be transformed as high-efficiency markers, probably because one of the two daughter cells after cell division always has the donor genotype.

MAPPING BY TRANSFORMATION

Transformation is the only technique available for gene mapping in some species. The technique is different from but analogous to that described in Chapter 1. For example, genetic mapping is usually accomplished by measuring recombination frequencies between markers. In transformation, however, several effects make frequency measurement an unreliable procedure. The main problem is that the probability of recovery of a donor marker in a recipient depends on the molecular weight of the donor DNA fragment and, more significantly, on the marker itself. In the previous section, we saw the distinction between a low-efficiency and a high-efficiency marker. Because this distinction is based on the probability of mismatch repair in a particular direction, which is probably determined by local base sequences (among other things), different alleles of the same gene are incorporated at different frequencies. Furthermore, reciprocal transformations do not yield the same transformation frequencies; that is, transformation of a donor Z allele to a recipient z strain will not, in general, occur at the same frequency as transformation of a donor z allele to a Z strain because of differences in mismatch repair. Clearly these effects cause the recombination frequency between a donor marker and a nearby resident marker to depend on features other than the distance between the markers, so recombination frequency cannot be used to generate a meaningful map.

Mapping by transformation experiments is based instead on the principle that two markers transform together if they are near enough to be carried on the same DNA fragment. When DNA is isolated from a donor bacterium, each chromosome is invariably broken into several hundred small fragments. With highly competent recipient cells and excess DNA, transformation of most genes occurs at a frequency of about one cell per 10^3. If two genes y and z are so widely separated in the donor chromosome that they are always carried on two different DNA fragments, at the same total DNA concentration the probability of simultaneous transformation (cotransformation) of a y^-z^- recipient to wild type is roughly the square of that number, or one y^+z^+ transformant per 10^6 recipient cells. If the two genes, however, are so near one another that they are often present on a single DNA fragment, the frequency of cotransformation is nearly the same as the frequency of single-gene transformation, or nearly one wild-type transformant per 10^3 recipients. Thus, cotransformation at a frequency substantially greater than the product of the two single-gene transformations implies physical proximity of two genes. Studies of the ability of various pairs of genes to be cotransformed yields gene order. For example, if genes y and z and genes z and x can be cotransformed, but genes y and x either cannot be cotransformed or cotransform at exceedingly low frequency, the gene order must be $y\ z\ x$ (Figure 13-8).

The use of cotransformation frequencies in mapping is not always straightforward. For example, when all or most cells in a population are not competent, the existence of a high transformation frequency is not sufficient to conclude that two markers are linked. This can be seen in the following example. Consider two markers Y and Z, each of which transforms at a frequency of one transformant per 100 recipient cells (frequency = 10^{-2}). If Y and Z are unlinked, the probability of cotransduction should be the product of the individual frequencies, or one cotransformant per 10^4 cells. Suppose in a particular experiment, however, that only 5% of the population were competent. In that case, the single transformants would arise not at a frequency of 10^{-2} but actually at one per five competent cells, or a real frequency of 0.2. In the absence of linkage, the expected number of double transformants would then be $(0.2)(0.2) = 0.04$, or one transformant for every 500 cells in the entire population. That is, the expected number of doubles arising at random would be only one-fifth the number of singles. If indeed doubles were found at a frequency of $^1/_{500} = 0.002$, which is high compared with the value of 10^{-4} that might be expected if all cells were competent, one would be tempted to conclude that the doubles arose by cotransformation. The fact that the number of cotransformants is one-fifth the number of single transformants would probably not make one suspicious because DNA molecules are fragmented at random, and not all molecules would be expected to carry both

Figure 13-8. Cotransformation of linked markers. Markers *a* and *b* are near enough to each other that they will often be present on the same fragment, as are markers *b* and *c*. Markers *a* and *c* are not. The gene order must therefore be *a b c*.

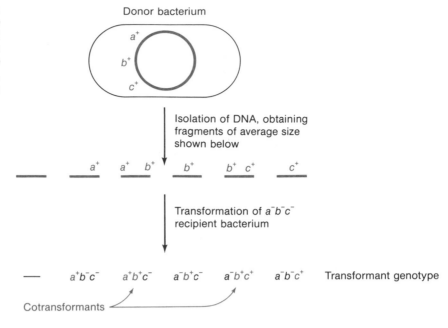

Note: No $a^+b^+c^+$ or $a^+b^-c^+$ cotransformants

markers. Thus, unless competence is high, some test is needed to assess whether or not a particular transformation value does indeed represent linkage. The most quantitative test available is a dilution test. For a population of cells, the overall transformation frequency of a single marker is clearly a function of DNA concentration: As fewer DNA molecules are added to a culture, there will be fewer transformants. Except for extremely high DNA concentrations at which the cells are saturated with DNA, the transformation frequency decreases as a direct linear function of DNA concentration. Thus, for transformation of the markers Y and Z, the rate of decrease of the frequency of transformation of Y, Z, or linked YZ with decreasing DNA concentration will be the same (Figure 13-9). If there are no DNA molecules that contain both markers, however, the production of YZ transformants depends on the cell interacting with both Y-containing and Z-containing DNA molecules. In this case, a 10-fold decrease in DNA concentration will lower the probability of the double interaction by a factor of 10^2, so the curve relating frequency and DNA will be steeper.

When carrying out transformational mapping, a tool unique to transforma-

Figure 13-9. A dilution test for cotransformation. If *a* and *b* are linked, the cotransformation frequency falls off with decreasing concentration at the same rate as the individual genes. If they are unlinked, the decrease is much greater.

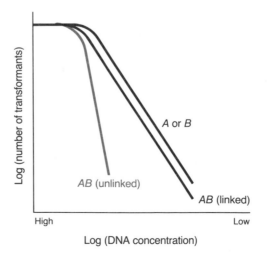

tion is available—the size of isolated DNA fragments can be controlled, so the probability of cotransformation of two genes can be related to the average molecular weight of the donor DNA. Such a relation enables a physical map to be constructed from transformation data. For example, by measuring the cotransformation frequency as a function of the molecular weight of the transforming DNA, the physical separation between the genes can be roughly determined. The frequency of transformation of any gene ultimately decreases because for small fragments the trimming seen in Figure 13-7 can destroy the gene in the donor DNA. This decrease, however, does not occur until the average molecular weight of the DNA is less than about 5×10^5. If the cotransduction frequency decreases as the size of the fragments decreases, it is likely that the markers are cotransformed.

For many bacterial species, transformation yields only a minimal genetic map that provides little information about gene organization. The problem is that many bacteria—for example, S. pneumoniae and H. influenzae—grow only in complex media, so the only genetic markers that are available are resistance to antibiotics and a few traits concerning polysaccharide metabolism. The situation has been different with B. subtilis, which grows well in minimal medium. Furthermore, nutritional mutants are easily obtained, and there are numerous phages, which also allows the use of phage-resistance markers. Several hundred markers are known for B. subtilis. Even with B. subtilis, however, a problem arises when using cotransformation to map. The problem follows from the fact that the bacterial chromosome always breaks during isolation, and the size of the donor DNA fragment limits the size of the region that can be examined. Thus, co transformational mapping invariably yields distinct linkage groups. It is possible with B. subtilis to transform with DNA having a mean molecular weight of about 25×10^6. Such a molecule can contain up to about 30 genes, but only some of these yield genetic markers (for example, most mutations may not be detectable simply by plating tests). Such a fragment is about 1% the size of the B. subtilis genome, so each fragment can yield information about only a few genes in a small region of the genome. Of course, fragmentation of the chromosome during DNA isolation occurs at random, so the region of the chromosome contained in one fragment overlaps the region contained in many other fragments. If available genetic markers were distributed uniformly over the chromosome, this overlap would result in a map that is a single linkage group. If there were regions of the chromosome, however, no more than 1 or 2% of the chromosome in length, that are devoid of markers, these regions would break the sequence of overlaps and thereby fragment the map. Indeed, cotransformational mapping of the B. subtilis chromosome yields about five distinct linkage groups. A classic physical experiment, however, provided the missing information for ordering these groups.

Consider a circular DNA molecule whose replication begins at a unique site and which replicates unidirectionally. Genetic markers are replicated in order A, B, C, . . . , Z from the replication origin to the replication terminus. After replication has begun, there will be twice as many A markers in the DNA than Z markers. As replication proceeds, the number of copies of each marker will double successively in the order in which they are replicated. Thus, in a synchronized culture in which all DNA molecules are in the same stage of replication, the successive times at which the number of copies of each marker doubles will tell the temporal sequence of replication of each marker. If DNA could be isolated from such a culture at various times and used as donor DNA in a transformation experiment, a simple measurement of transformation frequencies would yield the gene order (because transformation frequency is related to DNA concentration). For a variety of reasons, the primary one being that it is difficult to measure transformation frequencies with sufficient accuracy for the experiment, this technique is inadequate. The following variation on the experiment, however, works.

B. subtilis is a spore-forming bacterium. Spores are a stable inactive form of bacteria that form when certain bacterial species are subjected to extreme stress, such as starvation for nutrients. In forming a spore, replication of the bacterial chromosome is completed and not reinitiated. Spores do not divide and are stable for years. When spores are exposed transiently to high temperature and placed in a good growth medium, they germinate—that is, they are converted to a normal bacterial form and grow and multiply. Initiation of DNA replication occurs synchronously, cells divide at nearly the same time, and the cells remain in fairly good synchrony for about two generations. If spores are germinated in a growth medium containing D_2O (heavy water), the daughter DNA strands that are produced will contain deuterium (D) instead of hydrogen and will thus have a higher density than the parental strands. Thus, semiconservative replication can be followed by isolating DNA at various times and sedimenting the DNA to equilibrium in CsCl solutions (that is, the classic technique of Meselson and Stahl, described in Chapter 8). As each region of the DNA replicates, fragments obtained from that region will be shifted from the normal low density to the density of hybrid deuterium-hydrogen DNA. With this in mind, the following experiment was done: Spores were germinated in heavy medium as just described, and at various times DNA was isolated and centrifuged in CsCl. The hybrid (replicated) and light (unreplicated) fractions were isolated from each time sample and tested by transformation for the presence of each of a large number of markers (Figure 13-10). Replication of a marker was detected as a

Figure 13-10. An experiment showing the temporal order of replication of individual genes. (a) A unidirectionally replicating DNA molecule, showing genes *a–k*, that has replicated for a portion of the bacterial life cycle in medium containing D_2O (heavy medium). (b) The results of CsCl density-gradient centrifugation of DNA extracted from cells at three different times after the density shift. The individual letters show the genes carried on transforming DNA from two regions of the density gradient. Note that, as always, DNA is fragmented during isolation from bacteria.

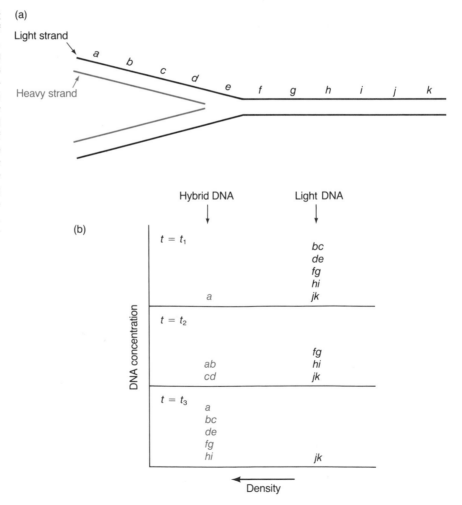

switch from finding the marker in the light DNA to finding it in the hybrid DNA. The switching time for each marker indicates when that marker replicates, and the sequence of times provides a physical map that locates each marker along the chromosome. Such a physical map can be combined with the numerous linkage maps derived from independent cotransformation experiments to join the various linkage groups into a single map.

Unfortunately, there is a major complication with this analysis: that *B. subtilis*, similar to *E. coli*, replicates its DNA bidirectionally (see Chapter 8). Thus, transfer of two markers from the light to the hybrid region at nearby times does not mean that they are nearby in the chromosome; they could be duplicated by replication forks moving in opposite directions, which would mean that the markers are widely separated. Thus, a marker *A* that replicated after one-fourth of a generation could be located in a 100-unit circular map at either 25 or 75, and marker *B* that replicated slightly later (say, three units later) could be at either 28 or 72. If one of the linkage segments, however, includes both *A* and *B*, they are both either at positions 25 and 28 or at positions 72 and 75. By examining enough markers, the mapping puzzle can be solved, and an excellent genetic map of *B. subtilis* has been obtained. This map has been supplemented by ordering genes by the generalized transduction procedures described in Chapter 18.

OTHER USES OF TRANSFORMATION

Transformation can be used for purposes other than mapping. One is to determine whether certain chemical agents act directly on DNA. When a bacterium or an animal is exposed to a particular chemical and mutagenesis occurs, one cannot be sure, just from the production of mutants, that the chemical acts directly on DNA. Alternatively, a metabolite of the chemical could be the active substance, or the chemical could induce a mutagenic pathway of some kind. If pure DNA is exposed to the chemical, however, and then used in a transformation experiment, the production of mutations would indicate a direct effect on DNA. Such an experiment would normally be done in a cotransformation experiment. Consider two closely linked genes, for example, Str and *lac*. If DNA from a donor strain with the genotype *lac⁻* Str^r was used to transform a *lac⁺* Str^s recipient and Str^r was the selected marker, some of the Str^r cells would also be *lac⁻*. If the donor strain were *lac⁺*, however, the *lac⁻* marker could not appear in the recipient. Thus, if DNA from a *lac⁺* Str^r donor was treated with a known mutagen and the mutagen acted directly on DNA, some *lac⁻* cells would be found among the Str^r transformants. Thus, the existence of *lac* Str^r transformants would indicate that the mutagen is directly active on DNA.

Transformation is also useful in identifying DNA. For example, in an experiment in which DNA isolated from a cell infected with a phage is fractionated by zonal sedimentation or electrophoresis, it is often necessary to locate the bacterial and phage DNA among various fractions. With *B. subtilis*, testing each fraction for transformation of bacterial markers can be used to identify bacterial DNA.

TRANSFORMATION IN NATURE

Transformation is usually carried out in the laboratory with purified DNA. The question, however, has frequently been raised whether transformation occurs in nature. Given the fact that evolution has proceeded in part by mutation and in part by recombination, it seems likely that transformation could have evolved in organisms that lack other mechanisms for exchanging DNA. What is necessary for transformation to occur in nature is the release of DNA from cells and the generation of competence. Many gram-positive bacteria, for example, *B. subtilis*, lyse and release their DNA when cells age or are subjected to starvation or various deleterious environments. Thus, there is a ready supply of DNA in the environment. More informative is the fact that transformation has been observed when genetically marked cultures of *B. subtilis* are mixed. Presumably some DNA is released in the cultures, and a small fraction of the cells are always competent, leading to occasional transformation. Furthermore, transformation has been observed when different marked bacteria are placed in the gut of a mouse. Transformation in nature is certainly an infrequent event. Given the time allotted for evolutionary change, however, it may be an important force in microbial evolution.

KEY TERMS

competence   linked markers
cotransformation   transformant

QUESTIONS AND PROBLEMS

1. What experimental result showed that transformation consists of a permanent genetic change?

2. What argument was used by Avery and coworkers to rule out protein or RNA as the genetic material?

3. Following publication of the transformation experiments of Avery, MacLeod, and McCarty, opponents of the DNA = gene theory, who believed that genes were made of proteins, argued that the transformation was caused by proteins that were contaminating the DNA sample.

    a. If transformation was indeed carried out by protein rather than DNA molecules and if the DNA preparation used contained at most 0.02% protein, how many protein molecules (each consisting of about 300 amino acids) would have been present in 1 ml of a DNA solution at a concentration of 10^{-7} mg/ml?

    b. If protein was the active agent in transformation, would the number calculated in part (a) account for the fact that in a typical transformation experiment 1000 transformants result from 0.0001 g of *S. pneumoniae* DNA?

4. Suppose you wish to prove to a skeptic that transformation is mediated by DNA and not by protein. You do not have available pure enzymes that degrade DNA or protein. You have worked hard, however, and have shown (i) that 50 different genetic traits can be transformed and (ii) that transformation is inefficient and highly dependent on DNA concentration. The only piece of nonmicrobiological laboratory equipment you possess is an ultracentrifuge, and with it you are able both to measure the molecular weight of DNA and to fractionate DNA according to molecular weight. Design a simple experiment to prove that the transforming principle is DNA. *Hint:* Think about linkage.

5. A critic of the interpretation of the transformation experiment might say that protein is the genetic substance and the protein can penetrate the cell only when the protein is bound to DNA. Thus, the loss of transforming activity that accompanies boiling of transforming DNA might be a result of dissociation of protein from the DNA. Assume that you have current knowledge about the denaturation of proteins and of DNA, about the effect of low and high pH on the chemical and physical properties of DNA and protein, and about the ionic strength dependence of the binding of protein to DNA. Design an experiment to prove that the critic is incorrect.

6. Molecular biology was in a formative stage when Avery and his colleagues performed the transformation experiment. Many kinds of instruments and techniques are now available that would have simplified their investigation. Suggest a way that CsCl density gradient centrifugation might have been used to determine whether DNA or protein is the genetic material.

7. A Lac⁻ *S. pneumoniae* is treated with DNA from a Lac⁺ strain and plated on a nonselective color-indicator medium in which Lac⁺ and Lac⁻ cells produce red and white colonies. What types of colonies will form on the medium?

8. Transformation sometimes occurs in nature by spontaneous lysis of a small number of cells and uptake of the released DNA by naturally competent cells. You are examining a number of bacterial species to see which ones can engage in intercellular exchange. A pair of strains of a bacteria that grow in Camembert cheese and have different genetic markers are mixed together. After a period of time, the bacteria are plated on selective media, and recombinant colonies form. You have reason to believe that transformation is occurring, but your colleague thinks that it is a new example of bacterial conjugation (such as an $F^+ \times F^-$ mating). What simple experiment might you do to distinguish these two alternatives?

9. A transformation experiment is carried out using donor DNA that is $A^+B^+C^+$ and a recipient that is $A^-B^-C^-$. A^+ transformants are selected and tested further. Of these, 64% are B^+ and none are C^+. Also, B^+ are selected, and 8% are also C^+. What is the gene order?

10. In a transformation experiment using DNA at a concentration of 10 g/ml, 3.5×10^6 A^+ transformants, of which 3×10^5 were also B^+, were detected. In another experiment, the DNA was instead at 1 μg/ml; the results were: A^+, 2.3×10^5; A^+B^+, $<10^3$. What can be said about the linkage of genes A and B?

11. Two different *B. subtilis* strains, carrying different nearby *lac* mutations, *lacZ1* and *lacZ2*, were transformed by a single DNA sample from a Lac⁺ cell. Lac⁺ recombinants were detected by plating on minimal medium containing lactose as the sole carbon source. With the *lacZ1* strain, the number of Lac⁺ transformants was 5×10^4/ml, for the *lacZ2* strain, the value was 2×10^3. Explain the difference between these values.

12. A competent culture of *S. pneumoniae* is mixed with [^{14}C]DNA having a radioactivity of 10^4 counts per minute (cpm). After 5 minutes, the cells are centrifuged. No acid-precipitable radioactivity can be found in the supernatant fluid. (Recall that only nucleic acid polymers are acid precipitable; nucleotides are not.) How much radioactivity do you expect to find associated with the cells?

13. A transformation experiment is done in which the donor strain is $A^+B^-C^+$ and the recipient is $A^-B^+C^-$. In one part of the experiment, A^+ cells are selected, and 250 are tested further. The associated unselected markers and the number of each genotype are: B^-C^+, 12; B^-C^-, 3; B^+C^+, 100; B^+C^-, 135. In the second part of the experiment, C^+ transformants are selected, and 250 of these are also tested. The unselected genotypes and their numbers are: A^+B^-, 17; A^+B^+, 85; A^-B^-, 13; A^-B^+, 142. What is the order of the three genes?

REFERENCES

Avery, O. T., C. M. MacLeod, and M. McCarty. 1944. Studies on the chemical nature of the substance inducing transformation of pneumococcal types. *J. Exp. Med.*, 79, 137.

Chassy, B., A. Mercenier, and L. Flickinger. 1988. Transformation of bacteria by electroporation. *Trends Biotechnol.*, 6, 303.

Cosloy, S. D., and M. Oishi. 1973. Genetic transformation in *E. coli* K-12. *Proc. Natl. Acad. Sci.*, 71, 84.

Dubnau, D. 1991. Genetic competence in *Bacillus subtilis*. *Microbiol. Rev.*, 55, 395.

*Hanahan, D. 1987. Mechanisms of DNA transformation. In J. Ingram, F. Neidhardt, K. Low, B. Magasanik, M. Schaecter, and H. Umbarger (eds.), Escherichia coli *and* Salmonella typhimurium: *Cellular and Molecular Biology.* American Society for Microbiology, Washington, D.C.

Hotchkiss, R. D., and J. Marmur. 1954. Double marker transformations as evidence of linked factors in desoxyribonucleate transforming agents. *Proc. Natl. Acad. Sci.*, 40, 55.

Kornberg, A., and T. Baker. 1992. *DNA Replication*, Second Edition. W.H. Freeman and Co., New York.

*Resources for additional information.

Lacks, S. 1962. Molecular fate of DNA in genetic transformation in *Pneumococcus*. *J. Mol Biol.*, 5, 119.

Mandel, M., and A. Higa. 1970. Calcium-dependent bacteriophage DNA infection. *J. Mol. Biol.*, 53, 159.

Mazodier, P., and J. Davies. 1991. Gene transfer between distantly related bacteria. *Ann. Rev. Genet.*, 25, 147.

*McCarty, M. 1985. *The Transforming Principle: Discovering that Genes Are Made of DNA*. Norton.

Shigekawa, K., and W. Dower. 1988. Electroporation of eukaryotes and prokaryotes: A general approach to the introduction of macromolecules into cells. *BioTechniques*, 6, 742.

Stewart, G., and C. Carlson. 1986. The biology of natural transformation. *Ann. Rev. Microbiol.*, 40, 211.

Bacterial Conjugation

In Chapter 11, plasmid transfer between donor and recipient cells was described. It was also mentioned briefly that some plasmids can integrate into the host chromosome; when they do so, their ability to transfer is retained, and in this way the chromosome itself becomes a giant transferable element. A well-studied example of chromosome transfer is mediated by the F plasmid. Properties of chromosome transfer and the recombination events that occur in the recipient cell after transfer are the topics of this chapter.

INSERTION OF F INTO THE *E. coli* CHROMOSOME

In addition to the genes carried by the F plasmid that mediate transfer of F to an F^- cell and regulate plasmid replication, F also contains three transposons, IS*1*, IS*2*, and $\gamma\delta$ (also called Tn*1000*). These elements do not participate in any of the systems responsible for replication, maintenance, or transfer, but they provide a mechanism for integration of F into the chromosome. When F integrates into the chromosome, the cell becomes an Hfr donor. Hfr is an acronym for high frequency of recombination, which refers to the fact that chromosomal genes are transferred from an Hfr cell to an F^- cell with much higher frequency than from an F^+ cell.

Integration of F is a reciprocal DNA exchange analogous to the recombination event that allows phage λ to integrate into the chromosome. DNA sequencing studies indicate than an integrated F sequence is always flanked by two copies (in direct repeat) of one of the IS elements found in the F plasmid. There are two ways by which Hfr cells might arise in a cell containing F:

1. *Homologous recombination.* If the recombinational event is physically reciprocal, homologous recombination between two identical IS elements, one in the chromosome and one in F, would yield flanking IS elements. Homologous recombination requires either the bacterial Rec system (described later) or a transposon gene product (Figure 14-1).
2. *Replicon fusion.* An Hfr cell could arise by formation of a cointegrate mediated by an IS element in F, resulting in the duplication of a target sequence in the chromosome.

Mechanism 1 requires a copy of the transposon on the chromosome, but mechanism 2 does not. Note that the consequences of these two mechanisms are the same—insertion of F into the chromosome flanked by two copies of the IS element in direct repeat. It is likely that Hfr strains arise by both mechanisms for the following reasons. Integration of F into the *Escherichia coli* chromosome has several characteristics that differ from λ integration:

1. The site where F integrates into the chromosome is not unique. Integration occurs predominantly in the IS3 element located near 94.5 on the physical map, but integration also occurs in the other IS elements (Figure 14-2).

2. Integration can occur at many sites in the chromosome. More than 20 major sites are known (Figure 14-3). The frequency of formation of a particular Hfr strain varies from one site to the next, indicating that integration occurs more readily at some sites than others. DNA sequencing studies have shown that at each of these sites there is a copy of IS2, IS3, or another IS in the chromosome, and the frequency of forming an Hfr at that site represents the probability of recombination with these elements. Integration at the major sites requires a functional RecA product.

3. F can integrate in both clockwise and counterclockwise orientations, depending on the orientation of the IS element in the chromosome. An inserted F mediates DNA transfer sequentially in the order of genes on the chromosome because F transfer is initiated by rolling circle replication from the transfer origin *oriT*, and transfer occurs in a single direction. Some Hfr strains transfer the chromosome in one direction, and others transfer genes in reverse order. There are a few instances in which F can integrate

Figure 14-1. Integration of F by a reciprocal exchange between an IS element in F and a homologous sequence in the bacterial chromosome.

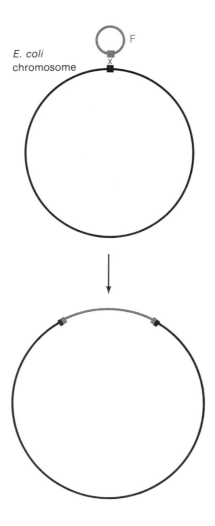

E. coli chromosome

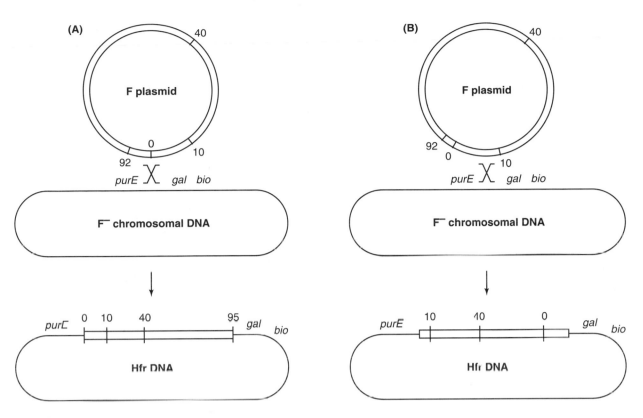

Figure 14-2. Formation of two different Hfr strains. The exchange in both molecules occurs at the points indicated by the cross-over. (A) An Hfr produced by a cross-over near 0 kb. (B) An Hfr produced by a cross-over at about 8 kb.

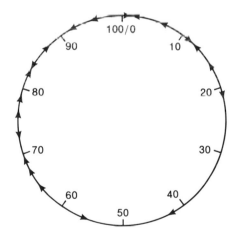

Figure 14-3. *E. coli* genetic map showing 20 different origins for Hfr transfer. The arrows show the direction of transfer.

in either orientation at a single site (note the sites at positions 12 and 83 in Figure 14-3). That is, if F integrates between genes *p* and *q*, one orientation would cause the transfer order to be *q r s . . . z a . . . n o p*, and the other orientation would yield the order *p o n . . . a z . . . s r q*.

Hfr TRANSFER

When F becomes integrated into the bacterial chromosome, the chromosome remains a single, circular DNA molecule, and F behaves as if it were part of this chromosome, increasing the size of the chromosome. Once integrated into the chromosome, F can transfer the entire chromosome from the Hfr cell to an F⁻ cell. The stages of transfer are much like those by which F is transferred to F⁻ cells: pairing of donor and recipient, rolling circle replication in the donor, and conversion of the transferred single-stranded DNA to double-stranded DNA by replication in the recipient. The replication and associated transfer of the Hfr DNA, which is controlled by F, is initiated in the Hfr chromosome at the same point in F at which replication and transfer begin within an F plasmid (*oriT*). Thus, a portion of F is the first DNA transferred, chromosomal genes are transferred next, and the remaining part of F is the last DNA to enter the recipient. In contrast to F or F' plasmids, however, in an Hfr the transferred DNA does not circularize and is not capable of further replication in the recipient.

Several differences between F transfer and Hfr transfer are notable (Figure 14-4):

1. It takes about 100 minutes for the entire *E. coli* chromosome to be transferred, in contrast with about 2 minutes for transfer of F. The difference in time is a result of the relative sizes of F and the chromosome.

2. During transfer of Hfr DNA to a recipient cell, the mating pair usually breaks apart before the entire chromosome is transferred, as a result of brownian motion (resulting from bombardment by solvent molecules). On the average, several hundred genes are transferred before the cells separate.

3. In a mating between Hfr and F⁻ cells, the F⁻ recipient usually remains F⁻ because cell separation almost always occurs before the final segment of F is transferred. The recipient does not gain the ability to engage in subsequent transfer unless it has received all of the F DNA from the donor—that is, unless chromosome transfer has been fully completed.

4. In Hfr transfer, although the transferred DNA fragment does not circularize and cannot replicate, the donor DNA may recombine with the chromosome of the recipient, generating recombinants in the F⁻ cell. For example, in a mating between Hfr *leu⁺* cells and F⁻ *leu⁻* cells, F⁻ *leu⁺* cells arise (Figure 14-5). Note that the genotype of the donor is unchanged because the transferred DNA strand is replaced by concurrent replication in the donor.

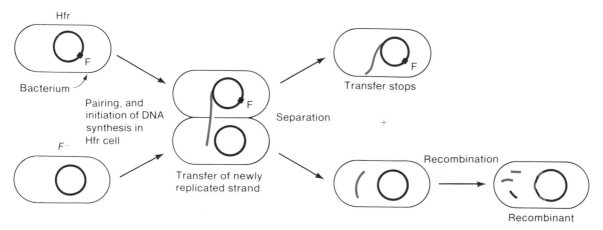

Figure 14-4.   A diagram showing mating between an Hfr donor bacterium and an F⁻ recipient bacterium. The two bacteria form a pair. Then DNA replication begins in the donor, adjacent to F, and a replica of F is transferred to the recipient. By random motion, the mating cells break apart. At a later time, a portion of the transferred DNA exchanges with the corresponding piece in the recipient. The intact recipient chromosome then replicates; the fragments do not replicate and are ultimately lost in the course of cell division. Although not shown, only a single strand is transferred, and this is converted to double-stranded DNA in the recipient.

The presence of the new chromosomal fragment in the recipient stimulates the cell's recombination system that causes genetic exchanges to occur, and a recombinant F^- cell often results. As in F and F' matings (see Chapter 11), to recognize the recombinants, some means is needed to distinguish the donor and recipient cells. As before, this is done by using cells that have genetic differences that can be recognized by growth of a colony on agar. For example, when mating Hfr leu^+ Str^s × F^- leu^- Str^r, plating on minimal medium lacking leucine and containing streptomycin selects against both the original F^- (leu^-) recipient cells and the Hfr (Str^s) donor cells and allows Leu^+ Str^r recombinants to form colonies. Recall that the transferred allele used to select against the parental recipient cells (Leu^+ in this case) is called a selected marker, and the allele used to select against the parental donor cells (Str^r in this case) is called the counterselective marker. The resulting recombinant cells are often called exconjugants.

Interrupted Mating and Time-of-Entry Mapping

Shortly after Hfr and F^- cultures are mixed, transfer of the chromosome from the Hfr begins. Transfer is not synchronous but is initiated over a period of several minutes. Each type of Hfr transfers genes in an order determined by the site of insertion and the orientation of F. Because there is a single transfer origin, the order of transfer will reflect the gene order in the chromosome. Transfer of the Hfr chromosome proceeds at a constant rate. Thus, the times at which particular genetic loci enter an F^- recipient are simply related to the positions of these loci on the chromosome, and a genetic map can be obtained from the time of entry of each gene (for an example, see Box 14-1).

Certain features of the transfer process in Hfr cells and information concerning the arrangement of bacterial genes in Hfr and F^- cells can be obtained by mechanically interrupting transfer during mating. The time at which a particular gene is transferred can be determined by purposely breaking the mating cells apart at various times and noting the earliest time at which breakage no longer prevents recombinants from appearing. (This can be done by violently agitating the suspension of mating cells in a kitchen blender.) This procedure is called the interrupted-mating technique. When this is done with Hfr × F^- matings, it is observed that the number of recombinants of any particular genotype increases with the time during which the cells are in contact (Table 14-1). The reason for the increase is slow transfer of the Hfr DNA.

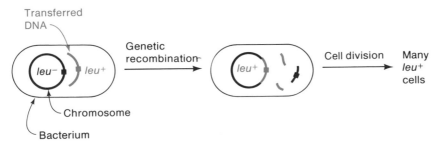

Figure 14-5. Conversion of a leu^- cell to a leu^+ cell by incorporation of a segment of the transferred DNA. The fragments remaining after recombination do not replicate and are gradually diluted as the recombinant cell divides. This drawing is schematic and does not attempt to distinguish between various proposed mechanisms of exchange; for example, possibly only a single strand of transferred DNA is inserted.

14-1. MAPPING AN Hfr INSERTION

Rationale: Hfr donor strains can transfer chromosomal DNA to F⁻ recipients. Genes that are behind and close to the origin of rolling circle replication (*oriT*) are transferred at high frequency and genes that are further from *oriT* are transferred at lower frequency. Once transferred to a recipient cell, wild-type copies of genes on the donor DNA can recombine with the recipient chromosome to repair mutations in the homologous genes. The frequency that a mutation is repaired is proportional to the frequency of transfer from the donor Hfr and thus the distance of the gene from *oriT*.

Example: A new Hfr was isolated. To determine where the Hfr maps, it was mated with five different auxotrophic recipient strains, selecting for prototrophic (i.e., nonauxotrophic) recombinants. The map positions of the auxotrophic mutations are shown below.

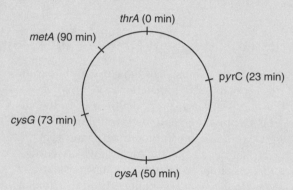

Based on the following results, where does the Hfr map? Which direction does the Hfr transfer the chromosome?

Recipient strain	Number of colonies on minimal glucose medium
thrA	7
pyrC	30
cysA	200
cysG	1000
metA	0

Answer: The new Hfr maps between *cysG* at 73 minutes and *metA* at 90 minutes, and the direction of transfer is counterclockwise.

Greater insight into the transfer process can be obtained by observing the results of a mating with several genetic markers. Consider the mating

$$\text{Hfr } a^+ \, b^+ \, c^+ \, d^+ \, e^+ \, \text{Str}^s \times \text{F}^- \, a^- \, b^- \, c^- \, d^- \, e^- \, \text{Str}^r$$

At various times after mixing the cells, samples are taken, agitated violently, and then plated on media containing streptomycin and one of four different combinations of the five substances A through E (for example, B, C, D, and E, but no A; or A, C, D, and E, but no B). Colonies that form on the medium lacking A are $a^+ \, \text{Str}^r$, those growing without B are $b^+ \, \text{Str}^r$, and so forth. All of these data can be plotted on a single graph to give a set of curves, as shown in Figure 14-6a. Four features of this set of curves are notable:

Table 14-1   Data showing the production of Leu⁺ Str^r recombinants in a cross between Hfr *leu⁺ str*^S and F⁻*leu⁻str*^r cells when mating is interrupted at various times

Minutes after mating	Number of Leu⁺ Str^r recombinants per 100 Hfr cells
0	0
3	0
6	6
9	15
12	24
15	33
18	42
21	43
24	43
27	43

Note: Extrapolation of the data to a value of zero recombinants indicates a time of entry of 4 min.

1. The number of recombinants in each curve increases with time of mating.
2. For each type of recombinant, there is a time before which no recombinants are detected.
3. Each curve has a linear region that can be extrapolated back to the time axis, defining a time of entry for each locus $a^+, b^+, \ldots, e^+$.
4. The number of recombinants of each type reaches a maximum, the plateau value, the value of which decreases with successive times of entry.

The time of entry phenomenon simply reflects the Hfr transfer of chromosomal genes. Transfer begins at a particular point in the Hfr chromosome, the transfer origin of F. Genes are transferred in linear order to the recipient. The time of entry of a gene is the time at which that gene first enters a recipient in the population. All donor cells do not start transferring DNA at the same time, so the number of recombinants increases with time; separation of a mating pair prevents further transfer and limits the number of recombinants seen at a particular time. Sometime after the time of entry, the transferred DNA undergoes genetic recombination in the recipient, forming a recombinant cell.

The times of entry of the genes used in the mating just described can be placed on a map (see Figure 14-6b). Mating the donor with a second recipient whose genotype is $b^- e^- f^- g^- h^-$ Str^r can then be used to locate the three genes f, g, and h (see Figure 14-6c). Because genes b and e are common to both maps, the two maps can be combined to form a more complete map (see Figure 14-6d).

Studies with other Hfr cell lines can also help to localize the map positions of genes (for an example, see Box 14-2). Because F can integrate at numerous sites in the chromosome and in two different orientations, different Hfr strains (for example, Figure 14-6c and e) yield different sets of curves, distinguishable by their origins and directions of transfer. Combining the maps obtained for different Hfr strains yields a composite map that is circular (see Figure 14-6f). The circularity of the map is a result of the circularity of the *E. coli* chromosome in the recipient cells and the multiple points of integration of the F plasmid in donors; if F could integrate at only one site, the map would be linear. Notice, however, that the map circularity does not prove that the Hfr chromosome is circular (this was proven by physical experiments, such as the autoradiogram shown in Figure 8-3).

Nearly 2000 genes have been mapped in *E. coli*—about half of all of the genes. Actually time-of-entry mapping is accurate only to about 1 minute of the map, and each minute contains about 30 to 50 genes. Thus, precise mapping is accomplished by transduction, as is described in Chapter 18. An abbreviated *E. coli* map, showing a small number of genes, is shown in Figure 14-7. Detailed genetic maps exist for *E. coli* and *Salmonella typhimurium*. The genetic maps of *E. coli* and *S. typhimurium* are similar. Furthermore, hundreds of mutations have been mapped within certain genes, such as the *hisG* gene and *putA* gene.

Rate of Chromosome Transfer

An important feature of F conjugation is that the time interval between the entry of different genes is independent of either the Hfr cell line or the F⁻ recipient. For example, the times of entry of the *thr, lac,* and *trp* genes with HfrH are 5, 13, and 32.5 minutes, yielding intervals of 8 minutes for *thr – lac* and 19.5 minutes for *lac – trp*. With HfrB7, which transfers in the opposite order, the times of entry for *trp, lac,* and *thr* are 3, 22.5, and 30.5 minutes. These values again yield the intervals of 19.5 minutes for *trp – lac* and 8 minutes for *thr – lac*. In fact, experiments of this sort first led to the suggestion that DNA transfer occurs at a constant rate. Physical experiments confirmed this point. In these experiments, Hfr *tsxʳ* cells, whose DNA was [¹⁴C]labeled, were mated with F⁻ *tsxˢ* cells. At various times after mixing the cells, a great excess of phage T6 was added, which caused lysis of the sensitive F⁻ cells within about 1 minute. The amount

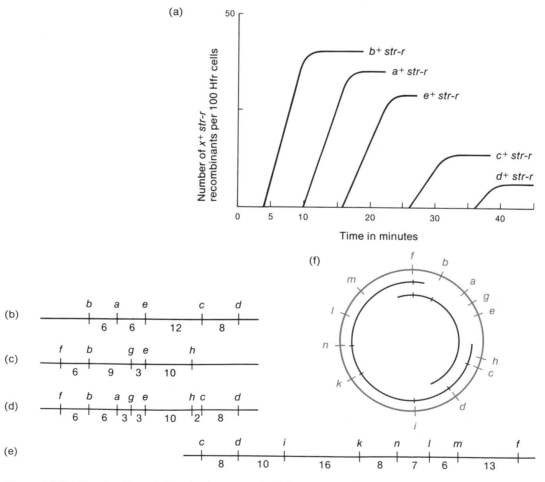

Figure 14-6. Construction of the circular map of *E. coli.* (a) Time-of-entry curves for a particular Hfr strain. (b)–(e) Four linear maps obtained from four different Hfr types. (f) The circular map derived from the data in panels (a)–(e).

14-2. Hfr MAPPING OF A CHROMOSOMAL MUTATION

Example: A new mutant was isolated that is unable to use acetate as a carbon source (*ace*). To determine where the mutation maps, it was mated with the four different *ace*⁺ Hfr donor strains shown below. (Arrowheads indicate the location and direction of transfer from each different Hfr.)

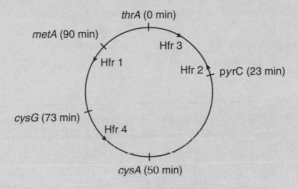

Given the following results, where does the *ace* mutation map?

Donor strain	Acc⁺ colonies
Hfr 1	1000
Hfr 2	5
Hfr 3	1000
Hfr 4	80

Answer: The *ace* mutation must lie between Hfr 1 (at about 85 minutes) and Hfr 3 (at about 5 minutes) on the chromosome.

of [¹⁴C]DNA released was measured and found to increase linearly with time after mixing. The absolute rate of transfer is very dependent on the culture conditions during mating. For example, decreasing the temperature during mating from 37° to 32° halves the rate of transfer.

Hfr Mapping

In *E. coli* and *S. typhimurium*, Hfr mapping is often used to localize a gene to a relatively small region of the chromosome (typically 5 to 10 minutes), then fine structure mapping of the gene is done by generalized transduction. Hfr mapping, however, was once commonly used for fine structure mapping in *E. coli* and *S. typhimurium*, and it is still extensively used for some bacteria that lack a good transducing phage. Thus it is worthwhile to understand how fine structure Hfr mapping is done.

This technique was first used to map the individual genes of the *lac* operon. Consider an Hfr that transfers in the order *pro – lac – purE*, in which a *purE* mutation introduces a nutritional requirement that can be satisfied by adenine. The gene order of the *lacZ* and *lacY* genes with respect to *purE* can be determined by comparing the results of cross (I) Hfr *purE*⁺ *lacZ*⁺ *lacY*⁺ Str^s × F⁻ *purE*⁻ *lacZ*⁻ *lacY*⁺ Str^r with the reciprocal cross (II) Hfr *purE*⁺ *lacZ*⁻ *lacY*⁺ Str^s × F⁻ *purE*⁻ *lacZ*⁺ *lacY*⁻ Str^r (Figure 14-8). In each case, Pur⁺ Str^r recombinants are selected by plating on minimal agar lacking adenine and containing streptomycin, and these recombinants are tested for the Lac⁺ phenotype by replica plating onto lactose color-indicator plates. Note that if the order is $purE - lacY - lacZ$, the number of $pur^+ lacY^+ lacZ^+$ Str^r recombinants would be smaller in cross I than in cross II because more exchanges would be required in cross I. With order II, $pur^+ lacY^+ lacZ^+$ Str^r recombinants would predominate in cross I. Experimental data show that the order is $purE - lacZ - lacY$.

Note that this technique of looking at relative recombination values yields the gene order but not the map distances, which must be obtained from the actual values of recombinant frequencies. This can be done in a straightforward way once the gene order has been determined, using either the recombination frequencies in the three-factor cross or frequencies obtained from separate two-factor crosses. Consider three mutations $lacZ1$, $lacZ2$, and $lacZ3$, which have the gene order with respect to $purE$ of $purE - lacZ1 - lacZ2 - lacZ3$. Two crosses were done:

(1) Hfr $pur^+ lacZ1^+ lacZ2$ Str^s × F$^-$ $purE^- lacZ1^- lacZ2^+$ Str^r
(2) Hfr $pur^+ lacZ2^+ lacZ3$ Str^s × F$^-$ $purE^- lacZ2^- lacZ3^+$ Str^r

Pur⁺ Str^r colonies were selected. These were further tested for the Lac⁺ phenotype. For the first cross, the fraction (number of Pur⁺ Lac⁺ Str^r)/(number of Pur⁺ Str^r) is a measure of the recombination frequency between $lacZ1$ and $lacZ2$. The same proportion for the second cross measures the recombination frequency between $lacZ2$ and $lacZ3$. If the values are 0.12 and 0.08, the genetic map is $lacZ1 - 0.12 - lacZ2 - 0.08 - lacZ3$. Doing this for numerous lac mutations—that is, determining the gene order from reciprocal crosses and evaluating genetic distances from the recombination frequencies—can yield an approximate genetic map.

Mapping by this procedure is limited by the reversion frequencies of individual mutations and the number of recombinant colonies that can conveniently be tested by replica plating. The smallest frequency measured in an early study of the lac operon was 3×10^{-5}.

Figure 14-7. A partial genetic map of *E. coli* showing the positions of 52 genes that are well mapped and commonly used in conjugation experiments. The units of the map are minutes. The map is 100 minutes long, which represents the time required to transfer an entire chromosome at 37°C. The threonine (*thr*) locus has been labeled 0 minutes. Note that fewer genes have been mapped in some regions than other regions of the chromosome—this does not indicate that such regions actually have few genes, simply that the regions have not yet been well characterized. Recent systematic DNA sequencing of about 12% of the *E. coli* chromosome indicates that the genome is very efficiently packed with information; there is about one gene per kilobase of DNA. Thus, *E. coli* has about 4000 genes, only half of which have been identified by mutants or by comparison with the sequences of known proteins.

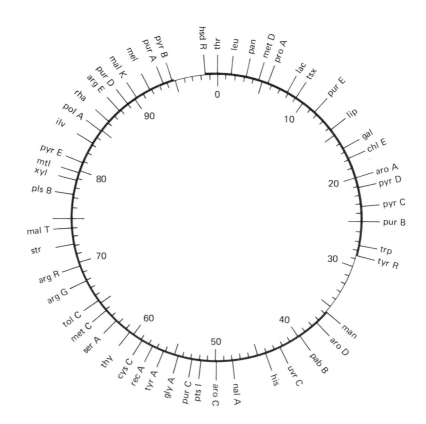

It should be noted that a linkage map generated in this way is in terms of map units, whereas a time-of-entry map has units of minutes. Relating map units to time units is not a straightforward process, and in fact the *E. coli* and *S. typhimurium* genetic maps are shown in minutes. In Chapter 18, we see how transduction can yield a fine structure map in minutes.

Mapping "Unselected" Recessive Markers

Consider a *met⁻* (methionine requiring) marker in an F⁻ cell. Determining the time of entry of this marker is straightforward because one can simply perform a time-of-entry experiment with an Hfr that is Met⁺. Plating the mating mixture on agar lacking methionine enables Met⁺ recombinants to grow. Suppose the *met⁻* mutation, however, was originally isolated in an Hfr strain. In that case, mapping is less direct because inability to grow on a plate lacking methionine is the trait brought to the recipient by the Hfr, and transfer of that trait would be recognized only by a small decrease in the number of colonies with time. For example, if only a few percent of the cells receive the mutation, entry of the mutation would decrease the number of recipients by only a few percent, which would be unrecognizable against the statistical fluctuation in number of colonies on different plates corresponding to the same time.

The *met⁻* mutation, however, can easily be mapped by linkage analysis, as is seen in the following example: Consider the mating Hfr *met⁻ leu⁺ lac⁺* Str^s × F⁻ *met⁺ leu⁻ lac* Str^r. Interrupted mating can be done and samples plated on two different media: (1) lactose color indicator agar, which contains all amino acids, and (2) medium containing methionine but no leucine. Both types of media contain streptomycin to select against the donor cells. The plates select for the following recombinants: (1) *lac⁺* Str^r and (2) *leu⁺* Str^r. These recombinants are tested for the *met⁻* mutation by replica plating colonies from plates 1 and 2 onto medium containing leucine but no methionine. Data are shown in Table 14-2. Note that at first all *leu⁺* colonies are *met⁺*, but at later times *leu⁺ met⁻* recombinants have formed. This result shows clearly that *leu* enters the recipient before the *met⁻* mutation because the Hfr marker is present before the marker in the recipient.

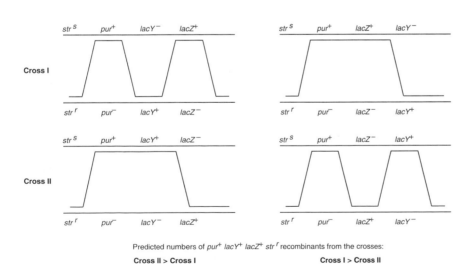

Predicted numbers of *pur⁺ lacY⁺ lacZ⁺ str*^{r} recombinants from the crosses:

Cross II > Cross I   **Cross I > Cross II**

Figure 14-8. Determination of the relative positions of the *lacZ* (denoted Z) and *lacY* (denoted Y) genes with respect to the *purE* gene, by comparing the results of reciprocal crosses.

Table 14-2   Data showing mapping of a *met⁻* mutation as an unselected marker in the cross Hfr *met⁻leu⁺lac⁺str^s* × F⁻*met⁺leu⁻lac⁻str^r*

Minutes after mixing	Leu⁺ Str^r recombinants		Lac⁺ Str^r recombinants	
	Number	% Met⁻	Number	% Met⁻
0	0	—	0	—
3	0	—	0	—
6	7	0	0	—
9	14	36	0	—
12	23	48	3	100
15	32	60	6	100
18	41	70	14	100
21	42	79	20	100
24	42	81	26	100
27	42	81	27	100

Note: The times of entries from these values are: *leu*, 4 min; *met*, 7 min; *lac*, 11 min.

Also, all *lac⁺* recombinants are *met⁻*, indicating that the *met⁻* marker was transferred before *lac*. The gene order is clearly *leu – met – lac* with this Hfr strain. Note that by plotting the fraction of *leu⁺* recombinants that are *met⁻* as a function of time, one obtains the time of entry of the *met⁻* mutation. In this experiment, the *met⁻* marker is not used in the initial selection; hence it is said to be an unselected marker.

Some markers can be mapped only as unselected markers. An example is a phage-resistance marker. Consider a cross between an Hfr that is resistant to phage T6 (Tsx^r) and a recipient that is Tsx^s. For example, in the cross Hfr *lac⁺ tsx^r* Str^s × F⁻ *lac⁻ tsx^s* Str^r, the Tsx^r and Tsx^s phenotypes can be tested by plating on a medium that has been previously spread with about 10^8 T6 phage: Tsx^r cells can form colonies on such a medium, but Tsx^s cells cannot. If the mating mixture were plated directly on plates containing streptomycin and T6, no colonies would form because (1) the Hfr cells will be inhibited by the streptomycin, (2) the Tsx^s recipients would be lysed by the T6, and (3) cells that have received the *tsx^r* allele would still possess normal (Tsx^s) T6 receptors on their cell walls and hence would also be killed. Note that the reciprocal cross with a Tsx^s Hfr and a Tsx^r recipient would also not work because all recipients would grow except for the few that had acquired the Tsx^s allele; hence recombination could be detected only by a small decrease in the total number of cells. The *tsx* gene, however, can be mapped easily if the allele is treated as an unselected marker. That is, *lac⁺* Str^r recombinant colonies can be selected and then tested for phage sensitivity. As in the mapping of the *met* marker, one would observe that many *lac⁺* cells would be Tsx^s if *lac* entered the recipient before *tsx*.

Hfr Collections

Collections of Hfrs inserted at known positions on the chromosome can be useful for genetic mapping. Collections of Hfrs have been constructed in *E. coli* and *Salmonella*. Such Hfr collections can be used to determine quickly the approximate map location of a new mutant. Each of the donor Hfrs is mated with the mutant recipient for a short time, then the exconjugants are plated on a selective medium to determine the number of wild-type recombinants. Only those Hfrs that are located near the chromosomal position of the mutant gene and in the correct orientation will transfer the corresponding wild-type gene at a high frequency. By using a collection of Hfrs that transfer an antibiotic-resistant transposon early after conjugation, it is possible to select for the exconjugants even if it is not possible to select directly against the mutant gene in the recipient.

Chromosome Transfer by F⁺ Cultures

F was originally detected by virtue of its ability to mediate transfer of chromosomal genes to a recipient cell. For the most part, this transfer is a result of the presence of rare Hfr cells in a predominantly F⁺ culture. This was first demonstrated by a fluctuation test of the type used to show the nature of spontaneous mutations (described in Chapter 10). In this test, the ability to transfer a particular gene (*thr*) by 50 small cultures was compared with the same ability of 50 aliquots of a large culture. It was found that the number of recombinants produced by mating each of the 50 aliquots of the large culture with an appropriate recipient culture ranged from 10 to 23. The mean value was 16, and the variance was 13. In contrast, with the 50 individual cultures, the number of recombinants ranged from 1 to 116; the mean was 15, but the variance was 351. As seen in Chapter 10, the large variance implies that Hfr cells arose at various times (as "mutations" of F⁺ cells) in the small cultures, producing clones of Hfr cells in each culture. (If every F⁺ cell was capable of a transient plasmid-chromosome association that could lead to the ability to transfer, Hfr "jackpot" cultures would not have been observed, and the variance among the 50 small cultures would have been no greater than the variance of the 50 aliquots of the large culture.)

Isolation of Hfr Strains

Thus, a typical F⁺ culture will contain rare Hfr cells. These Hfr cells can be isolated from the culture by a series of replica platings. For example, if about 10^7 F⁺ Leu⁺ Str^s cells are spread on a nutrient agar plate, the master plate, and allowed to grow until a confluent lawn of cells forms, within the cell lawn there will be a few microscopic clones derived from individual Hfr cells. A velvet pad is touched to the surface of the master plate and then replicated to a minimal agar plate containing streptomycin with an F⁻ Leu⁻ Str^r strain spread on the surface. Because DNA transfer is not immediately inhibited by streptomycin, the few Hfr cells transferred to the surface of the minimal plate engage in conjugation with the Leu⁻ Str^r recipient cells. Neither the Hfr cells nor the recipients can produce any visible growth on the plate because the Hfr cells are Str^s and the recipient cells are Leu⁻. Any Leu⁺ Str^r recombinants, however, will grow and produce small colonies. The positions of these colonies indicate regions on the original F⁺ plate where there are microcolonies of Hfr cells. These regions consist of some Hfr cells but will be contaminated with a great excess of F⁺ cells. The regions can be scraped from the plate, diluted, and spread on a new nutrient agar plate, one plate for each region, yielding new master plates. These plates are considerably enriched for Hfr cells compared with the original F⁺ plate. Each of these plates is replicated onto minimal streptomycin plates spread with the recipient cells as before, and regions in which recombinants arise are again noted. Cells are then scraped from the appropriate region of the confluent donor plates, spread on a new set of nutrient agar plates, and retested. After several cycles of replica plating, the ratio of Hfr cells to F⁺ cells becomes high enough that the master plate can be seeded with a few hundred cells so individual Hfr colonies can be found.

Isolation of F' Plasmids

An Hfr cell is produced when F integrates stably into the chromosome. At a very low frequency (about 10^{-7} per generation), F can also excise out of the chromosome. Excision is often imprecise, resulting in an excised circular plasmid, which contains genes that were adjacent to F in the chromosome (Figure 14-9). The excised DNA is an F' plasmid.

F' plasmids can be isolated from Hfr cultures by two straightforward techniques. One procedure is based on the fact that in an Hfr mating, the F segment is transferred in two stages. If cells are separated any time before 100 minutes after mixing donor and recipient cells, no recipient will receive a complete copy of the F plasmid. Furthermore, if mating is interrupted at 30 minutes, no markers that enter after 30 minutes will appear in recombinants. In contrast, when an F' forms by a rare aberrant excision, the plasmid contains a complete copy of F and the chromosomal segment that would normally be transferred last or first. Thus, in a brief mating, recipient cells that receive a terminal marker often carry an F'. A simple test is to mate the exconjugant with a suitable recipient; if the late marker is efficiently transferred shortly after mating, the recombinant carries an F'. The problem with this procedure is that if a small number of mating pairs are not separated by the technique used to interrupt mating, transfer can

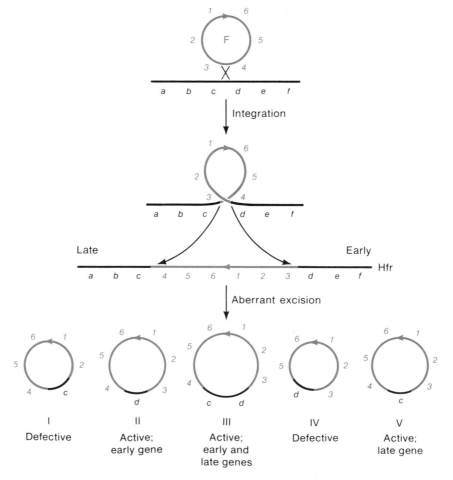

Figure 14-9.   (a) Formation of various F' plasmids by aberrant excision from a particular Hfr strain. Plasmids I and IV have lost F genes and hence are defective. If the plasmids are replication defective, they cannot be maintained and hence will not be detected. The usual means of detection of F' plasmids is by gene transfer, usually genes that are transferred late by an Hfr cell, at a time sufficiently early that the genes could not have been transferred by the Hfr cell. Thus, I and IV are normally not detected because the defects in these plasmids are defects in transfer; similarly, a type II plasmid will not be found because it contains only early genes.

continue on the plate, and a late marker will appear to have been transferred early. These recombinants, however, will not show up as F'-containing cells in the second mating. Nonetheless, the formation of an F' is such a rare event that the number of apparent colonies that arise owing to inefficient interruption of the mating is sometimes greater than the number of cells containing an F', requiring testing of a large number of colonies. This problem can be avoided by mating with a mutant that prevents homologous recombination, such as a *recA* mutant. In *recA* recipients, Hfr recombinants are not formed, but because inheritance of an F' does not require recombination, transfer of a marker on an F' is easily detected. Mating with a *recA* recipient also allows isolation of an F' plasmid containing an early marker.

Note in Figure 14-9 that a given Hfr strain can produce several F' plasmids because the positions of imprecise excision and hence the extent of the bacterial segment can vary. Use of Hfr strains with different transfer origins allows the isolation of F' plasmids that include any region of the *E. coli* chromosome. In fact, F' plasmids have been isolated that cover the entire *E. coli* genetic map.

Using transposons as a "portable region of homology" between the F' and the chromosome (see Chapter 12), it is possible to isolate Hfrs integrated any place in the chromosome where a transposon is located—that is, essentially anywhere on the chromosome. An example is shown in Box 14-3.

Chromosome Transfer Mediated by F' Plasmids

Chromosome transfer by F'-containing strains occurs at about 10^5 greater frequency than by F' strains. This phenomenon is called chromosome mobilization. Insight into the mechanism comes from the observation that this increased efficiency of chromosome transfer requires a $recA^+$ gene in the F' strain. Chromosome mobilization is a result of reciprocal recombination between the chromosomal segment of the F' and the homologous region of the chromosome itself. The result is insertion of the plasmid into the chromosome, effectively generating an Hfr cell. This cell differs from a typical Hfr in that the chromosomal genes of the F' are present in the recombinant donor in two copies: one that is transferred immediately after pair formation and the other transferred as the final markers (Figure 14-10). Further evidence for the recombination model comes from the fact that if the chromosome carries a large deletion that includes all of the chromosomal genes of the F', the plasmid cannot cause chromosome mobilization, and the transfer of chromosomal markers occurs at the same low frequency as from an F^+ cell.

By constructing an Hfr at a desired site using a transposon as a region of homology as shown in Box 14-3, it is possible to isolate an F' that carries genes immediately adjacent to the chromosomal transposon insertion. An example is shown in Box 14-4.

RECOMBINATION IN RECIPIENT CELLS

The final stage of bacterial conjugation is the incorporation of a transferred DNA fragment into the chromosome of the recipient to generate a recombinant cell.

Necessity for a Double Exchange

When a linear DNA fragment enters a recipient cell, it cannot be stably maintained through subsequent cell divisions because the fragment is unable to replicate. Three factors prevent replication: (1) the fragment generally lacks a replication origin; (2) except for a few phages, only circular DNA can replicate in bacteria; and (3) linear fragments are degraded by cellular nucleases. (The situation is quite different when a circular plasmid is transferred because the plasmid is an intact self-replicating unit.) Thus maintenance of the genes in the fragment requires recombination of the DNA into the chromosome. Because

14-3. USE OF TRANSPOSON Tn*10* FOR Hfr MAPPING

Example: An F' (Ts) Tn*10 lac*⁺ plasmid was integrated into a chromosomal Tn*10* insertion located near a new mutation called *proZ* in *S. typhimurium*. The Hfr was selected in two steps: First the F' was mated into the *proZ* recipient selecting for Lac⁺ at 30°C; then derivatives of the F' (Ts) Tn*10 lac*⁺/*proZ* strain that could grow on lactose at 42°C were selected. Because replication of the F' is temperature sensitive, the *lac* genes on the F' are rapidly lost at 42°C. Only cells with the F' integrated into the chromosome are Lac⁺ at 42°C.

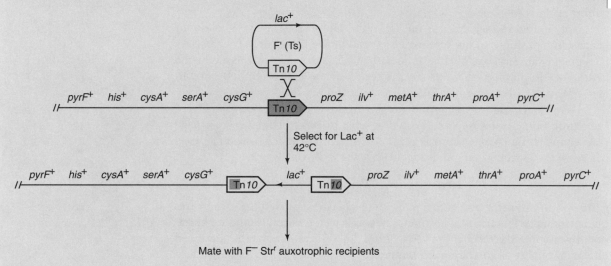

Because the Hfr is integrated into a Tn*10* insertion located near the *proZ* gene, the map location of the Hfr can be used to infer the position of the *proZ* gene. The Hfr was mapped by mating with auxotrophic recipients. Based on the results, shown below, where does the Tn*10* insertion map?

Auxotrophic Recipient*	Map Position (min)	Number of Recombinants† Tn*10*(A)	Tn*10*(B)
thrA Strʳ	0	++	–
proA Strʳ	7	+	–
pyrC Strʳ	23	+	+
pyrF Strʳ	33	+	+
his Strʳ	42	–	+
cysA Strʳ	50	–	++
serA Strʳ	63	–	++
cysG Strʳ	73	–	+++
ilv Strʳ	83	+++	–
metA Strʳ	90	+++	–

*Streptomycin resistance (Strʳ) in the recipients is due to a ribosomal mutation (*rpsL*).

†The Tn*10* insertions labeled (A) and (B) are in opposite orientations in the chromosome; hence the Hfr transfers chromosomal DNA in opposite directions from the two Hfrs. +++ indicates more than 1000 colonies, ++ indicates 200 to 1000 colonies, + indicates 50 to 200 colonies, and – indicates fewer than 50 colonies.

Answer: The Tn*10* insertion near *proZ* maps between 73 and 83 minutes on the *S. typhimurium* chromosome.

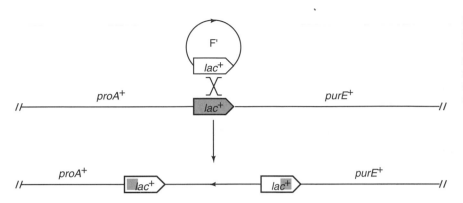

Figure 14-10. Formation of an Hfr by recombination between bacterial genes carried on an F' and the homologous chromosomal genes. Note that the genes are duplicated on the chromosome of the Hfr.

the transferred DNA is linear and the chromosome is circular, two exchanges are necessary: one on each side of the genes that are incorporated (see Figure 14-8). Multiple exchanges can also occur, but the number of exchanges must always be an even number.

The molecular mechanism of recombination has been studied for more than 20 years but is still not fully understood. Physical evidence suggests that only a single strand is integrated into the chromosome. Recombination is stimulated by the presence of a specific base sequence (called *chi*), which occur about once every 5000 bp in both the donor and recipient DNA. The frequency of recombination can be quite high in *E. coli*. For example, in crosses involving genetic markers near the leading end of the chromosome transferred by an Hfr, the frequency of recombination is very high—about 20% per minute of DNA transferred. That is, when two genetic markers are 1 minute apart, recombination between the markers is found in 20% of the recombinants. The frequency of recombination between a marker close to the leading end and a second marker that is 5 minutes away is sufficiently high that the markers appear to be unlinked. This high frequency of recombination observed close to the leading end of an Hfr is attributed to the entry of the RecBCD protein complex at the double-strand end (see later in this chapter). Consistent with this idea, the frequency of recombination is much lower—about 1–5% per minute of DNA transferred—for markers located more than 3 minutes from the leading end.

Anomalous Plateau Values

The value of a plateau region reflects both the efficiency of transfer of a marker and the probability of recombination. In general, the probability that conjugation will be disrupted before transfer is complete is the major factor in determining plateau values—because the probability of disruption is greater with longer times, the value continually decreases with time of entry (see Figure 14-6). Two types of markers, however, do not follow this pattern: those that enter very early and those near the counterselective marker (that is, the mutation in the recipient used to select against the donor cells).

Because two recombination events are needed for integration of a marker, the recombination frequency is affected by the distance of the marker from the ends of the fragment. Thus, if transfer is interrupted shortly after a marker enters a recipient, the probability of recombination will be low because there is only a short length of sequence homology for DNA exchange. In general, there is a sufficient length of DNA between a marker and the end of the fragment because DNA transfer continues after the marker enters the cell. Markers near the transfer origin present a special case because the distance between a very early marker and the beginning of the early segment of F may be quite small.

14-4. ISOLATION OF AN F' FACTOR

Example: A chromosomal *melB*::Tn*10* insertion was used to isolate an F' that carries the *S. typhimurium* *ace*⁺ genes. First an F'(Ts) Tn*10* *lac*⁺ was used to construct an Hfr located within the *melB*::Tn*10* insertion by selecting for growth on lactose at 42°C. Growth of the resulting Hfr at 30°C allowed the F' to excise from the chromosome and exist as a self-replicating plasmid. Occasionally the F' excised aberrantly, producing an F' that carries adjacent genes as shown below. The desired F' factors that carry the *ace*⁺ genes were identified by mating into an *ace*⁻ *recA*⁻ strain and selecting for Ace⁺.

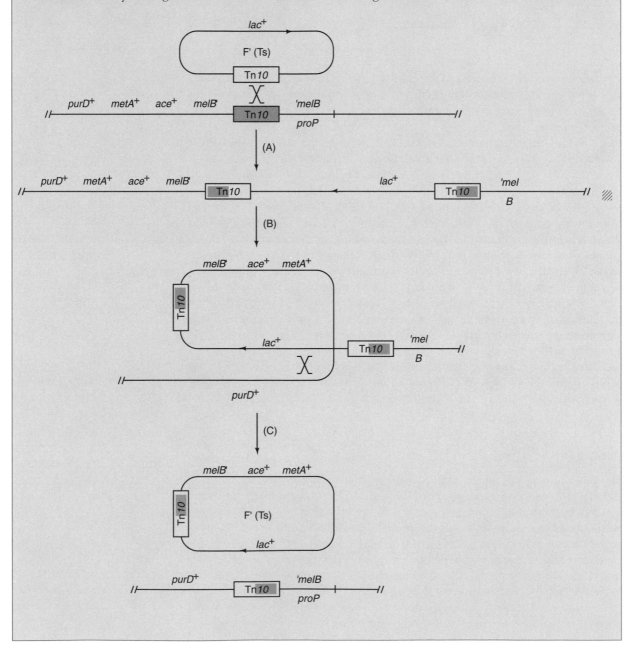

This is reflected in an anomalously low value of the plateau frequency for very early markers.

Figure 14-11 shows that later markers may also have anomalously low plateau values. This occurs whenever a marker is quite near the counterselective marker.

For example, if the Hfr is Str^s and the F$^-$ is Str^r, and Str^r is used as the counter-selection, a donor marker within a minute or so of the *str* gene (*malT* in the figure) will have a low probability of being recovered in a recombinant because an exchange must occur between the donor marker and the *str* gene. Otherwise the recipient would be Str^s and would not form a colony on the selective plate.

Efficiency of Transfer from an Hfr

Transfer of an F between *E. coli* cells occurs with high efficiency: Each donor cell can transfer a copy of the F plasmid within 20 minutes after mixing donor and recipient cultures. For example, if the ratio of F' *lac* Str^s cells to F$^-$ *lac* Str^r cells in a mating is 1:10, roughly 10% of the F$^-$ cells acquire F' *lac* in 20 minutes. In Hfr crosses, however, plateau values for early markers range from 20 to 50 recombinants per 100 Hfr cells, the maximum depending on the donor-recipient pair. One would, of course, not expect plateau values ever to reach 100% for several reasons: (1) The presence of a transferred DNA fragment does not mean that DNA exchange must occur, (2) recombination events within the fragment do not necessarily lead to integration of a particular genetic marker contained in the fragment, and (3) recombination may lead to base-pair mismatches that might be repaired in the recipient. Nonetheless, some information about transfer and recombination can be obtained from knowledge of the efficiency of transfer.

Transfer efficiency can be measured genetically by mating an Hfr strain that is lysogenic for phage λ with a recipient that is not a λ lysogen. Recall from Chapter 5 that lysogens are immune to infection by a phage that is the same as the prophage. This phenomenon, which is discussed in greater detail in Chapter 17, is a result of the synthesis of a repressor protein by the prophage. That is, if λ infects a λ lysogen, the incoming phage DNA is repressed by the λ repressor, which prevents both transcription and replication of the infecting phage DNA. When λ infects a nonlysogenic cell, however, no repressor is present, and a lytic cycle normally occurs. Mating of an Hfr cell that is lysogenic for λ with a nonlysogenic recipient is like infecting a nonlysogen with λ. Early in the mating, the transferred Hfr DNA behaves in a normal way, stimulating recombination in the recipient. Once the λ prophage enters a nonlysogenic recipient, however, which does not contain any λ repressor, transcription of the prophage begins, and a lytic

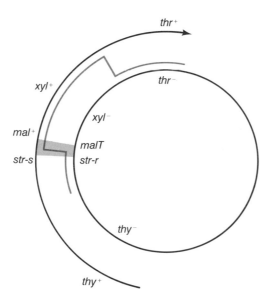

Figure 14-11. Recombination near a counterselective marker. The orange line indicates a possible Mal$^+$ Str^r recombinant. The first exchange can be anywhere clockwise from the *mal* marker, but the second exchange must be in the shaded region. A later marker, such as *thy*, may have a higher plateau region; even though fewer *thy*$^+$ markers are transferred, the possible exchange regions are fairly large.

cycle ensues. Thus, a nonlysogenic recipient receiving a prophage will lyse and hence can never make a colony. This phenomenon is called **zygotic induction**.

Zygotic induction can be demonstrated in two ways. In one test, an Hfr Str^s λ lysogen is mixed with nonlysogenic Str^r recipients, and the mixture is plated on a lawn of Str^r indicator cells in soft agar containing streptomycin. Cells receiving a prophage behave similar to a phage-infected cell: They release λ phage, which can form plaques in the lawn of indicator bacteria. The second procedure is based on the aberration produced by zygotic induction on a set of time-of-entry curves. Consider an Hfr λ lysogenic strain whose genes *a, b, c, d, e,* and *f* have times of entry 5, 10, 15, 20, 25, and 30 minutes when mated with a lysogenic recipient. The mating would result in six standard time-of-entry curves. If this strain was mated with a nonlysogenic recipient, however, and the prophage attachment site was between genes *b* and *c*, two changes in the set of curves would be observed: (1) No recombinants containing any of the genes *c, d, e,* or *f* would be formed, and (2) the number of recombinants containing gene *b* would be markedly reduced. Genes *c, d, e,* and *f* enter the recipient after the λ prophage, so all recipient cells that acquired these genes have already received λ and hence are destined to die. The second observation can be explained by the fact that many of the cells that have received gene *b* will also have received the λ prophage several minutes later and hence are killed. Thus, the main evidence for zygotic induction is the failure to transfer genes that enter after a particular time. This was how zygotic induction was first discovered.

Zygotic induction is a measure of DNA transfer because it is independent of subsequent genetic exchange processes. That is, any cell that receives the prophage becomes an infective center, so the number of infective centers indicates the number of cells to which DNA has been transferred. If the prophage is transferred within about 15 minutes after mixing the cultures, the number of infective centers equals the number of Hfr cells. Thus, transfer, or at least initiation of transfer, is 100% efficient. Comparison of this value with the number of recombinants for fairly early markers shows that at least half of the recipients engage in genetic recombination after receipt of a DNA fragment.

Rec Mutants

In 1965, Clark began a search for mutants defective for genetic recombination by looking for mutant F$^-$ cells that could not form recombinants after mating with an Hfr donor. Three genes were initially identified: *recA, recB,* and *recC*. These mutants were not defective in any stage of conjugation other than recombination because F' plasmids transferred at the usual rate and efficiency to each of the mutants. When mated with an Hfr selecting for recombination of an early marker, the recombination frequency was decreased by a factor of 10^6 or more with a *recA* recipient and by about 10^3 with either *recB* or *recC* recipients. (The residual recombination observed with *recB* and *recC* mutants is probably a result of alternative recombination functions provided by other genes.) The *rec* genes were roughly mapped by noting that a few recombinants appeared for late markers because once the particular *rec*$^+$ gene entered the recipient from the Hfr, recombinants could form. The recombination frequencies, however, were quite low because considerable degradation of transferred DNA occurred with the *rec* mutants, especially in *recA* mutants. Precise mapping of these genes was accomplished by transduction, a procedure to be described in Chapter 18.

The products of the *recB* and *recC* genes are subunits of an enzyme now known as the RecBCD protein. The RecA gene product is a multifunctional protein whose function in recombination is to bring two DNA molecules together.

OTHER PROPERTIES OF *rec* MUTANTS

Mutant cells lacking the ability to recombine have an additional phenotype—they are sensitive to DNA damage. For example, both *recA* and *recBC* mutants are killed by much smaller doses of ultraviolet light than are *rec*$^+$ cells (see Figure 9-10). The main reason for this sensitivity is that both the *recA* and *recBC* gene products are required for a major repair pathway, recombination repair (see Chapter 9). *recA* mutants are more UV sensitive than *recBC* mutants because the RecA protein is also required for SOS repair (see Chapter 9).

The DNA of *recA* mutants is unstable: It is continually degraded and resynthesized. The degradation is most striking after irradiation with ultraviolet light. After ultraviolet irradiation, more than half of the bacterial DNA is enzymatically degraded to short oligonucleotides. Because the degradation does not occur in a mutant lacking both *recA* and *recBCD*, it was inferred that the product of the *recBCD* genes is a nuclease responsible for the degradation. This rampant degradation of DNA is called the "reckless" phenotype.

Another feature of *recA* mutants is their slow growth compared with *rec*$^+$ strains. Rec$^+$ *E. coli* cells divide about every 25 minutes at 37°C in rich growth media. In contrast, a typical *recA* mutant may have a doubling time of 40 to 60 minutes. One reason *recA* mutants grow more slowly is probably due to reckless DNA degradation. The growth defect, however, is more extreme for *recBC*$^-$ mutants, which typically divide every 100 to 120 minutes. These growth defects are probably the result of many processes that are either inoperative or functioning inefficiently in *rec* mutants or possibly the result of newly turned-on aberrant processes. For example, *recBCD* mutants occasionally grow and divide without DNA replication, and a daughter cell completely lacking DNA results. Of course, such cells cannot grow further.

RecA PROTEIN AND ITS FUNCTION

Pairing of DNA molecules by RecA protein is essential to all modes of homologous recombination. The RecA protein has two major biochemical activities: (1) It binds to single-stranded DNA, and (2) it facilitates the self-proteolytic cleavage of certain proteins. Its DNA-binding activity is the feature that is relevant to recombination (the other property is a regulatory function). When acting as a DNA-binding protein, the RecA protein mediates nonspecific pairing of DNA molecules and homology-dependent strand invasion. Figure 14-12 shows several RecA-mediated DNA-DNA interactions that have been carried out with purified RecA protein and DNA molecules. The structures shown are stable and are held together by AT and GC base pairs between complementary base sequences. Study of the three interactions shown in Figure 14-12 (and others that are more complex) and of pairs of DNA molecules that will not interact has shown that stable pairing depends on two things:

1. One molecule must be single-stranded or have a single-stranded region.
2. At least one of the molecules must have a free end.

The first requirement can be provided by supercoiling of a DNA molecule that lacks a free end. (For example, the bacterial chromosome is supercoiled.)

The RecA-mediated interactions shown in Figure 14-12 are the end result of a sequence of three steps: presynaptic binding of RecA protein to single-stranded DNA, synapsis, and postsynaptic strand exchange (Figure 14-13). These stages were first elucidated in a study of the RecA-mediated pairing of a double-stranded circle and a linearized fragment of one of the complementary strands. Details of each of these steps are described next:

Figure 14-12. Three interactions mediated by the RecA protein; sc indicates that the circle is supercoiled.

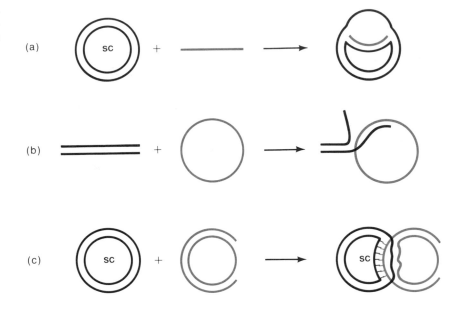

1. *Polymerization of RecA protein on single-stranded DNA.* If single-stranded DNA is mixed with RecA protein, a nucleoprotein filament forms in which the single-stranded DNA is coated with RecA protein.
2. *Synapsis.* In the presence of ATP, the nucleoprotein forms a complex with the double-stranded circle. The initial interaction is not between homologous regions. After the initial sequence-independent interaction, the two DNA molecules move relative to one another until homologous sequences come into contact; this is called homologous alignment. When homologous sequences are aligned, the two strands are not yet intertwined.
3. *Postsynaptic strand exchange.* Homologously aligned, but not intertwined, strands are bound together only weakly, and the structure is quite unstable. Once a homologously aligned region forms at the end of the single strand, RecA actively promotes the displacement of a strand from the double-stranded molecule and assimilation of the new strand. RecA acts as a helicase, unwinding the double-stranded DNA in advance of the forming heteroduplex.

Each of the interactions shown in Figure 14-12 (the three single-strand and double-strand interactions) can be explained by the multistep RecA-mediated process just described. Because it seems clear that both single-stranded DNA and a free end are required, most models of recombination include an early step in which one DNA strand is nicked and, in a variety of ways, unwound from the nick.

The role of RecA protein in pairing of single-stranded DNA to a double-stranded molecule is consistent with physical experiments (mentioned earlier) that indicate that only one strand of the transferred DNA combines with the chromosome of the recipient.

RecBCD PROTEIN COMPLEX

In addition to the RecA protein, the proteins encoded by the *recB*, *recC*, and *recD* genes are needed for recombination after conjugation and transduction. The RecBCD proteins form a multifunctional complex with exonuclease, endonuclease, helicase, and ATPase activities. The molecular mechanism of RecBCD-mediated recombination is not yet understood, but one possible model is shown in Figure 14-4.

The RecBCD protein complex enters at the end of a linear double-stranded DNA molecule (i.e., at a double-strand break). The RecBCD complex uses the energy of ATP to move along the DNA helix, hydrolyzing one molecule of ATP to ADP per bp it moves along the DNA. As it moves along the DNA, RecBCD complex progressively unwinds the strands (helicase activity), but the DNA is unwound faster than it is released, leaving a loop of unwound, single-stranded DNA associated with RecBCD. The RecBCD complex also degrades one of the unwound DNA strands (exonuclease activity). Upon reaching a *Chi* site, the RecBCD endonuclease nicks the single-stranded DNA within the loop, cutting the DNA very close to the *Chi* site. In addition, the *Chi* site somehow turns off the exonuclease activity of RecBCD complex (possibly by causing dissociation of the RecD subunit from the complex). The RecBC protein complex continues to move along the DNA and the helicase activity continues to unwind the DNA strands, resulting in a single-stranded DNA tail with a 3'-OH end. RecA protein binds to the single-stranded DNA and initiates pairing and strand exchange with a homologous double-stranded DNA molecule.

Other Recombination Pathways in *E. coli*

Recombination is decreased about 10^6-fold in *recA* mutants, suggesting that RecA protein is required for homologous recombination in *E. coli*. In contrast, in *RecBC* mutants, recombination is decreased only about 1000-fold. The residual recom-

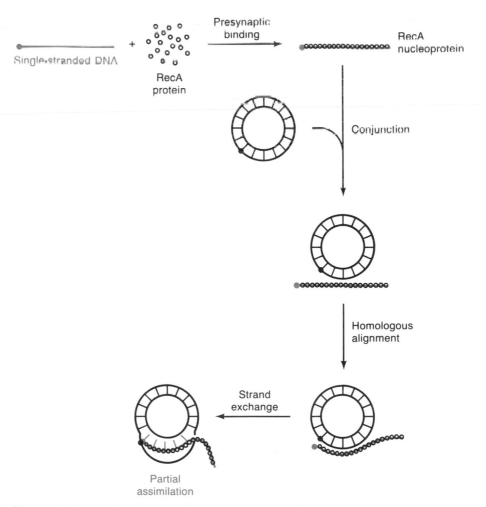

Figure 14-13. RecA-mediated pairing that leads to assimilation of a complementary strand.

bination probably results from products of two other recombination pathways—
the RecE and RecF pathway. These two alternative recombination pathways are
inefficient in wild-type cells, but their activity can be increased by mutations
(Figure 14-15).

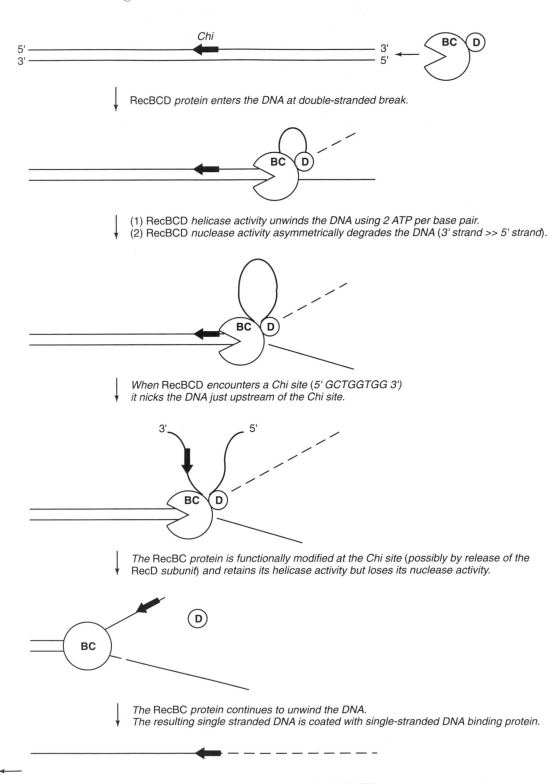

Figure 14-14.   A potential model for unwinding of double-stranded DNA by the RecBC protein.

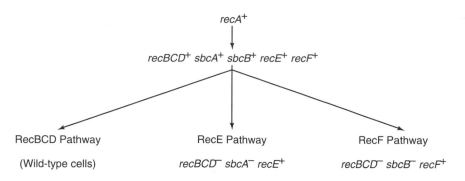

Figure 14-15. Recombination pathways in *E. coli*.

Similar to the *recBCD* gene product, the *recE* gene encodes a nuclease. The *recE* nuclease, called exonuclease VIII, however, is normally repressed by the *sbcA* gene product: *recBC⁻ sbcA⁺* strains do not express exonuclease VIII, so they are deficient for recombination. (*sbc* stands for suppressor of *recBC*.) In contrast, *recBC⁻ sbcA⁻* strains lack the repressor for *recE*, so they make high levels of exonuclease VIII, allowing recombination to occur efficiently. The *recE* and *sbcA* genes are located on a defective, cryptic prophage called Rac. The Rac prophage is related to λ, and the RecE pathway is mechanistically similar to the λ Red recombination pathway.

The second alternative pathway is called the RecF pathway. The recombination activity of the *recF* gene is normally inhibited by the *sbcB* gene product—exonuclease I. In *recBC⁻ sbcB⁻* strains, which lack exonuclease I, recombination occurs efficiently. The molecular basis for the inhibition of the *recF* activity by *sbcB* is not known, but one possible explanation is that exonuclease I degrades a recombination intermediate required by this pathway. Although the RecF pathway is not active on linear DNA fragments in wild-type cells, the RecF pathway seems to be a major pathway for plasmid-plasmid recombination in wild-type cells.

ROLE OF *rec* GENES IN PHAGE REPLICATION

The RecBCD nuclease seems to have deleterious effects on phages that engage in rolling circle replication in one stage of their life cycle (see Chapter 5). These phages make proteins that specifically inhibit the RecBCD nuclease at the appropriate stage in their life cycle. For example, an early λ gene called *gam* encodes a protein that binds directly to the RecBCD protein and inhibits its nuclease activity. Rolling circle replication does not occur in *gam* mutants; hence *gam* mutants have a dramatic decrease in burst size. Progeny phage are still made in *gam* mutants but by an alternative recombination pathway. This pathway requires a recombinational step mediated by a λ gene called *red*, which is responsible for most genetic recombination in λ. Thus, *red gam* double mutants are defective because neither the *red*-mediated recombination pathway nor the *gam*-mediated rolling circle pathway is active. A few progeny phage can still be made because the RecA system provides yet another alternative recombination pathway. Thus, λ *red⁻ gam⁻* phage produces no progeny when infecting a *recA⁻* cell. Phage production is restored in a *recA recBCD* host because the *gam* product is no longer needed to eliminate activity of the RecBCD protein, allowing rolling circle replication to occur.

CHROMOSOME TRANSFER IN OTHER BACTERIA

F can be transferred to several enterobacteria other than *E. coli*. For example, F can be mated into *S. typhimurium*, and Hfr strains of *S. typhimurium* have also been isolated. The *S. typhimurium* genetic map is almost identical to that of *E. coli* except for a large inverted region, about 20 minutes long, that includes the terminus. Hfr strains of *Citrobacter freundii* and *Erwinia chrysanthemi* have also been obtained. F is not stably maintained, however, in all enterobacteria, nor can it be integrated in all species in which the plasmid is stable. Thus, because of the value of conjugation for bacterial genetics, other systems have been developed for different species.

In Chapter 11, broad host range plasmids were described, which can be maintained in many bacterial genera. Two of these plasmids, RP4 and R68.45, have been particularly valuable for chromosome transfer in other species and have been used for mapping *Acinetobacter* and *Rhizobium* species. These plasmids, however, are somewhat difficult to use because they transfer at a frequency of about 10^{-8} per cell and do not form stable Hfr cell lines. Therefore genetically modified R'-like forms of the plasmids have been developed that transfer with greater frequency from a single origin. These modified plasmids share a homologous region between plasmid and chromosome, which enables F'-like, RecA-dependent chromosome mobilization to occur. Two methods have been used to create the necessary homology.

1. Recombinant DNA techniques (see Chapter 20) have been used to create an RP4 variant that contains a segment of chromosome in the plasmid.
2. A transposon is inserted into the plasmid, and the plasmid is transferred into a cell whose chromosome also contains the transposon inserted in a gene.

Chromosome mobilization using these R' plasmids does not occur in all bacterial species. For example, an RP4-Mu plasmid behaves like an F' and mobilizes the chromosomes of *E. coli* and *Klebsiella pneumoniae* but not of *Rhizobium leguminosarum* or *Proteus mirabilis*, even though the plasmid can be transferred to and from these bacteria.

So far, all bacteria discussed have been gram-negative. Conjugative plasmids have also been found in gram-positive bacteria (including *Streptococcus, Streptomyces, Clostridium,* and *Bacterioides*) and the Archaea, *Halobacterium*. Some plasmids transfer as efficiently as F does in *E. coli*, but in most cases, chromosome transfer occurs at a low efficiency, about 10^{-7} per donor. The mechanism of conjugation in gram-positive bacteria is totally different from that of *E. coli*. Contact is not made via pili. Instead cell-cell contact is mediated by diffusible proteins, called pheromones, which are released by recipient cells and cause clumping of large numbers of donor and recipient cells.

KEY TERMS

Chi site	RecBCD
donor	recipient
exonuclease	replicon fusion
Hfr mapping	*sbc*
homologous recombination	strand exchange
integration	synapsis
interrupted mating	time-of-entry mapping
mating pair	unselected markers
RecA	zygotic induction

1. How does an Hfr cell differ from an F^+ cell?

2. What is the function of a counterselective marker in an $Hfr \times F^-$ mating?

3. Distinguish F^+ and Hfr transfer with respect to the amount of genetic material (DNA) transferred and the intactness of the transferred unit.

4. Genes p, f, and q have times of entry of 7, 11, and 19 minutes. What is the gene order, and what are the map distances, in time units, with respect to the transfer origin?

5. How are F' plasmids produced?

6. An Hfr strain with genotype $met^- his^+ leu^+ trp^+$ and that transfers the met gene very late was mated with a $leu^- met^+ trp^- his^-(Ts)$ recipient. The his(Ts) mutation introduces a requirement for histidine at 42°C. After mating for several hours, the mixture was diluted and plated on minimal media with four different supplements. The plates were incubated at 42°C. The supplements in the plates and the number of colonies per plate are the following:

His + Trp	250
His + Leu	50
Leu + Trp	500
His	10

   a. What is the purpose of the met^- mutation in the Hfr strain in this experiment?
   b. Which genes entered first, second, and third?
   c. You now know the relative order of these three markers, but you know nothing about their exact location on the chromosome. What type of experiment would tell you where these markers are located on the chromosome?

7. An Hfr donor whose genotype is $a^+ b^+ c^+ Str^s$ was mated with a recipient whose genotype is $a^- b^- c \; Str^r$; the order of transfer was a-b-c. None of these genes are transferred early, the distance between a and b is the same as the distance between b and c, and none of the markers are near Str. Recombinants are selected as usual by plating on agar lacking particular nutrients and containing streptomycin. Which of the following are true? (Several answers are.) Explain.

   a. $a^+ Str^r$ colonies $> c^+ Str^r$ colonies.
   b. $b^+ Str^r$ colonies $< c^+ Str^r$ colonies.
   c. $a^+ b^+ Str^r$ colonies $< b^+ Str^r$ colonies.
   d. $a^+ b^+ Str^r$ colonies $= b^+ Str^r$ colonies.
   e. Most $a^+ c^+ Str^r$ colonies will also be b^+.
   f. Most $b^- c^+ Str^r$ colonies will also be a^-.
   g. $a^+ b^+ c \; Str^r$ colonies $< a^+ b^- c \; Str^r$ colonies.

8. Suppose you collect a large number of galactose-requiring (Gal^-) bacterial mutants and identify three closely linked genes (designated $galA$, $galB$, and $galC$) by complementation and rough mapping studies. You wish to order these genes and learn something about the genetic structure of the galactose region of DNA. To order the genes, you mate an Hfr having the genotype $bio^+ gal^+ Str^s$ with various F^- strains having the genotype $gal^- bio \; Str^r$. The Hfr transfers the bio locus later than the gal locus. You select $bio^+ Str^r$ recombinants and measure the fraction of these that have the genotype gal^+. For each of the mutations, the following fractions are observed: $galA$, 0.65; $galB^-$, 0.72; $galC^-$, 0.84. What is the gene order relative to the bio locus?

9. The order of four genes in an Hfr strain is $a \; b \; c \; d$. In a cross between an Hfr donor that has genotype $a^+ b^+ c^+ d^+ x \; Str^s$ and a female that has genotype $a^- b^- c^- d^- x^+ Str^r$, 90% of the $d^+ Str^r$ recombinants are x^-, and 100% of the $c^+ d^+ Str^r$ recombinants are x^-. The times of entry of a, b, c, and d are 5, 10, 15, and 20 minutes; the str gene enters at 55 minutes. Where is x located?

10. An Hfr donor of genotype $a^+ b^+ c^+ d^+ Str^r$ is mated with an F^- recipient having genotype $a^- b^- c^- d \; Str^r$. Genes a, b, c, and d are spaced equally. A time-of-entry experiment is carried out, and the data shown in the table below are obtained. What are the times of entry for each gene? Explain the low recombination frequency in the plateau region for $d^+ Str^r$ recombinants.

Time of mating, in min	Number of recombinants of indicated genotype per 100 Hfr			
	A^+str^r	b^+str^r	c^+str^r	d^+str^r
0	0.01	0.006	0.008	0.0001
10	5	0.1	0.01	0.0004
15	50	3	0.1	0.001
20	100	35	2	0.001
25	105	80	20	0.1
30	110	82	43	0.2
40	105	80	40	0.3
50	105	80	40	0.4
60	105	81	42	0.4
70	103	80	41	0.4

11. Suppose you have isolated two independent arginine-requiring (Arg⁻) mutant strains from a parent *E. coli* strain that already requires methionine (Met⁻) and is resistant to streptomycin. You mate the two mutants (1 and 2) with an Hfr strain whose genotype is $arg^+ met^+$ Str^s. Using the interrupted mating technique, you obtain the time-of-entry curves with the following characteristics: With mutant 1, the arg^+ Str^r recombinant curve extrapolates to 4 minutes, and the time of entry of *met* is 6 minutes. With mutant 2, the time of entry of *met* is again 6 minutes, but the data for the arg^+ Str^r recombinants extrapolate to 20 minutes. Explain the difference observed in the two matings.

12. After a brief mating between an Hfr whose genotype is $pro^+ pur^+ lac^+$ and a female whose genotype is F⁻ $pro^- pur^- lac$ Str^r, many $lac^+ pur^+ pro$ Str^r recombinants are found. A few $pro^+ lac^- pur^-$ Str^r recombinants also arise, and all of these are Hfr donors. Explain this result and state the location of F in the Hfr. (The three genes *pro*, *lac*, and *pur* are near one another.)

13. List two biochemical activities of the RecA protein.

14. What steps in the RecA-mediated synapsis process could occur between a double-stranded circle and a completely nonhomologous single strand?

15. In RecA-mediated synapsis, has base-pairing occurred in the stage of homologous alignment?

16. Name two properties of RecA that are important for recombination and state the stage at which each is important.

17. An Hfr strain transfers genes in alphabetical order. Would you expect to obtain F'y plasmids lacking gene z?

18. An Hfr cell transfers genes in the order $ghi \ldots def$. Which types of F' plasmids could be derived from this strain?

19. An Hfr strain transfers genes in alphabetical order. When using tetracycline sensitivity as a counterselective marker, the number of h^+ Tet^r colonies is 1000-fold lower than h^+ Str^r colonies found when using streptomycin sensitivity as a counterselective marker. Explain the difference.

20. An Hfr strain transfers genes in order abc. In an Hfr $a^+ b^+ c^+$ Str^s × F⁻ $a^- b^- c^-$ Str^r mating, will all b^+ Str^r recombinants have received the a^+ marker, and will all b^+ Str^r recombinants also be a^+?

21. An Hfr transfer genes in alphabetical order. A variant strain V13, known to be lysogenic for a phage XP1, is found that transfers only to gene e. That is, genes past e never seem to be transferred (at least, F-containing recombinants are never formed), and the frequency of transfer of genes $a^- d$ is less than for the normal Hfr strain. An F⁻ strain S132 has the property that transfer from V13 to S132 is normal; that is, time-of-entry and plateau values are the same as between a normal Hfr strain and a normal F⁻ strain. Suggest an explanation.

REFERENCES

Bachmann, B. 1990. Linkage map of *Escherichia coli* K-12, Edition 8. *Microbiol. Rev.*, 54, 130

Clark, A. J., and A. D. Margulies. 1965. Isolation and characterization of recombination-deficient mutants of *E. coli* K-12. *Proc. Natl. Acad. Sci.*, 53, 451.

Dixon, D., and S. Kowalczykowski. 1993. The recombination hotspot χ is a regulatory sequence that acts by attenuating the nuclease activity of the *E. coli* RecBCD enzyme. *Cell*, 73, 87.

Frost, L. S. 1992. Bacterial conjugation: everybody's doin' it. *Can. J. Microbiol.*, 38:1091.

Holloway, B., and K. B. Low. 1987. F-Prime and R-Prime Factors. In F. Neidhardt, J. Ingraham, K. B. Low, B. Magasanik, M. Schaechter, and H. E. Umbarger (eds.), Escherichia coli *and* Salmonella typhimurium: *Cellular and Molecular Biology*. American Society for Microbiology, Washington, D.C.

Jacob, F., and E. L. Wollman. 1961. *Sexuality and the Genetics of Bacteria*. Academic Press, New York.

Kucherlapati, R., and G. R. Smith. (eds) 1988. *Genetic Recombination*. American Society for Microbiology, Washington, D.C.

Low, K. B. 1987. Hfr Strains of *Escherichia coli* K-12. In F. Neidhardt, J. Ingraham, K. B. Low, B. Magasanik, M. Schaechter, and H. E. Umbarger (eds.), Escherichia coli *and* Salmonella typhimurium: *Cellular and Molecular Biology*. American Society for Microbiology, Washington, D.C.

Sanderson, K., and J. Roth. 1988. Linkage map of *Salmonella typhimurium*, Edition VII. *Microbiol. Rev.*, 52, 485.

Sarathy, P. V., and O. Siddiqi. 1973. DNA synthesis during bacterial mating. II. Is DNA replication in the Hfr obligatory for chromosome transfer? *J. Mol. Biol.*, 78, 443.

Singer, M., T. Baker, G. Schnitzler, S. Deischel, M. Goel, W. Dove, K. Jaacks, A. Grossman, J. Erickson, and C. Cross. 1989. A collection of strains containing genetically linked alternating antibiotic resistance elements for genetic mapping of *Escherichia coli*. *Microbiol. Rev.*, 53, 1.

*Smith, G. 1990. RecBCD enzyme. In F. Eckstein and D. Lilley (eds.), *Nucleic Acids and Molecular Biology*, vol. 1. Springer-Verlag.

*Smith, G. 1991. Conjugational recombination in *E. coli*: myths and mechanisms. *Cell*, 64, 19.

*Stahl, F. W. 1979. Specific sites in generalized recombination. *Ann. Rev. Genetics*, 13, 7.

*Stahl, F. W. 1987. Genetic recombination. *Scientific Am.*, 256, 91.

Weinstock, G. 1987. General recombination in *Escherichia coli*. In F. Neidhardt, J. Ingraham, K. B. Low, B. Magasanik, M. Schaechter, and H. E. Umbarger (eds.), Escherichia coli *and* Salmonella typhimurium: *Cellular and Molecular Biology*. American Society for Microbiology, Washington, D.C.

*Resources for additional information.

Genetics of Phage T4

Bacteriophages have played an important role in the development of molecular genetics. Studies on phage, in particular the *Escherichia coli* phages T4 and λ and the *Salmonella typhimurium* phage P22, have led to the discovery of many basic phenomena concerning replication, transcription, regulation, and recombination. Numerous examples are given throughout this book. In this chapter, we are concerned primarily with recombination processes, mapping, genome organization, and genetic procedures used to understand features of the life cycles of particular phages. The chapter begins with some general features of phages, supplementing the material presented in Chapter 5. Phage T4, however, comprises most of the chapter. For the biology of phages having single-stranded DNA or RNA, the reader should consult the references. You might find it useful to review Chapter 5 before studying this chapter.

PHAGE MUTANTS

Phage mutants have provided important insights about phage biology. In the 1950s and early 1960s, most available phage mutants were plaque-morphology mutants, in which the appearance of a plaque differs from that of the wild-type phage. A common variation is plaque size. For example, wild-type T4 produces a fairly small plaque, about 1 mm in diameter. The size results from a poorly understood phenomenon called "**lysis inhibition**." That is, when the multiplicity of infection is high, as is the case when a plaque approaches maturity, lysis is delayed or inhibited. Thus, late in the development of a wild-type T4 plaque, growth of uninfected bacteria catches up with phage multiplication, and nutrient is exhausted, which terminates enlargement of the plaque. T4 *rII* mutants do not undergo lysis inhibition and produce plaques several millimeters in diameter (Figure 15-1). Other T4 mutations lead to turbid plaques (*tu* mutants), lead to small plaques (mutations that reduce the enzymatic activity of lysozyme), or allow plaques to form on media containing inhibitory acridine compounds (*ac* mutants). Another important type of mutation are host range (*h*) mutations, which allow the phage to infect cells that are resistant to wild-type phage; for example, a host range mutation alters the phage tail allowing adsorption of T4 both to wild-type *E. coli* B and to the mutant B/4 (see Chapter 5). T4 and T4*h* can be distinguished by plating on a mixture of B and B/4; wild-type phage grows only on B and hence forms a turbid plaque (owing to unimpeded growth of B/4), whereas T4*h* lyses both bacteria and yields a normal plaque.

A great deal of genetic research has been possible with these mutants, but because plaque-morphology mutations are limited to a few genes, a fairly complete description of the phage genome was not possible with these alone. The isolation of conditional lethal mutations overcame this problem. For example, nonsense mutations allow phage growth (and hence plaque formation) only on sup^- bacterial strains, which contain a nonsense termination suppressor (see Chapter 10). The mutants fail to grow on wild-type (designated sup^+ or sup^0) bacteria but make normal or nearly normal plaques on sup^- bacteria. These mutations were isolated by brute-force screening techniques. A phage population was heavily mutagenized and plated on a sup^- host, and then the plaques were replica-plated onto a sup^0 lawn. Hundreds of thousands of plaques were tested, which led to the isolation of thousands of mutants.

The nonsense mutations just described prevent the phage mutants from plating on wild-type bacteria; hence they are located only in genes whose activity is essential to the phage. The concept of essential and nonessential genes requires some clarification. A gene is usually considered to be essential if a mutation in that gene prevents growth. For example, if a phage mutation prevents plaque formation, the gene is essential. This does not mean, however, that a "nonessential" gene does not have an important role. For example, although a typical burst size for an infected bacterium is 50 to 100, a burst size of four is usually sufficient to form a plaque: Mutations in many so-called nonessential genes reduce the burst size markedly, although not enough to prevent plaque formation.

A gene may be nonessential for four reasons: (1) There may be another functional copy of the gene. For example, an identical gene may be present in the bacterium, or another phage gene may have the same function as the mutant gene. This duplication may be of some value to the phage because it might increase the burst size by providing a higher concentration of an essential enzyme; alternatively, in nature there may be hosts that lack the gene, and the gene will then be essential for growth. (2) The gene is not required for phage growth but in some way increases either the rate of phage production or the burst size. In both cases (1 and 2), the gene confers an evolutionary advantage. (3) The gene is not needed in the laboratory, but it enables the phage to cope with special situations met in nature. (4) The gene is always unnecessary. This possibility is less likely because a truly useless gene would in general lack the selective advantage needed for the phage to retain it in the face of continued, long-term spontaneous mutagenesis.

Figure 15-1. A portion of a plate showing T4 r^+ (small plaques) and rII mutants (large plaques). (Courtesy of A. H. Doermann.)

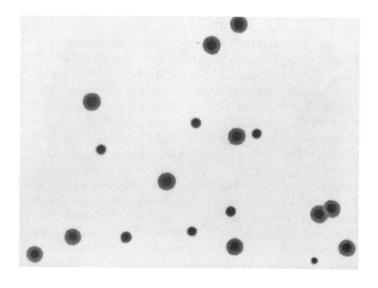

GENETIC RECOMBINATION IN PHAGES

Genetic recombination in phages was discovered by Hershey and Rotman in the late 1940s. They were studying phage T2, which is closely related to phage T4. They infected an *E. coli* culture with two different genetically marked T2 phages rh^+ and r^+h—each at a multiplicity of infection of about five. Roughly 98% of the progeny had the parental genotypes, but 2% consisted of about equal numbers of rh and r^+h^+ phage. On replating, the rh and r^+h^+ phage bred true, indicating that these genotypes were due to recombination. Furthermore, if *E. coli* was mixedly infected with the recombinants, genetic recombination occurred, again producing 98% rh and r^+h^+ and 2% rh^+ and r^+h phage (the original parental genotype). This discovery was a milestone in phage genetics, rendering phages suitable objects for genetic analysis. In this section, we examine several features of the recombination process in phage T4.

Effect of the Parental Ratio on Recombination Frequencies

For recombination between different phages to occur, a cell must be infected with several copies of the parental phages. Furthermore, to analyze the recombination frequency, the multiplicity of infection must be high enough that all bacteria receive both types of parental phage. Because individual phages, however, will be distributed among host cells according to the Poisson distribution (see Chapter 5), all bacteria will not receive the same number of both parents. This is an important consideration because the recombination frequency depends on the parental ratio.

To obtain reproducible recombination frequencies in different experiments, the ratio of parental phage must be kept constant. Ideally one would also like to obtain maximum recombination frequencies. A simple calculation shows the optimal ratio. Consider an infection by two phages for which the proportion of one phage is p and the proportion of the other phage is $1 - p$. Because the phage DNA molecules pair at random, the fraction of pairings between identical parents, which lead to no recombination, is $p^2 + (1-p)^2$, in which each term is the contribution of one of the parents. Recombination results only from pairings between different parents, and the fraction of such pairings is $1 - [p^2 + (1-p)^2] = 2p - 2p^2$. To determine the value of p that results in the maximal number of recombinants, we set the derivative of this expression, $2 - 4p$, equal to zero and solve for p. This shows that $p = 0.5$ gives a maximum: Recombination frequency is maximal when both parents are present in equal numbers. The Poisson distribution does not affect this conclusion because with a multiplicity of infection (MOI) = 5 of each parent, there will be as many cells infected with 4 of one parent and 1 of the other as 4 of the second parent and 1 of the first.

Reciprocity in Genetic Recombination

In the cross described in the introduction to this section, both recombinants were produced in equivalent numbers, a situation that is common in phage crosses. When this occurs, recombination is said to be reciprocal. In an attempt to determine whether this is the case for phage systems, single burst analysis (see Chapter 5) was carried out. In these experiments, cells infected by the rh^+ and r^+h parents were diluted shortly after infection and placed in individual culture tubes such that, on the average, one infected cell was present in every 10 tubes. After an incubation period sufficient for lysis to occur, each tube was plated, and the genetic composition of the plaques was determined for each plate. The observation was surprising: In individual bursts, recombinants did not occur in equal numbers. Some bursts contained parental phages plus only one of the recombinants, and others contained greatly unequal numbers of the two recombinants (of course, most bursts contained no recombinants). A clue to what was happening came from the observations that (1) the parental types also occurred in rather unequal numbers, and (2) summing all of the genotypes from all bursts yielded equal numbers of both recombinants. These observations indicated that some statistical process was occurring. Later biochemical studies of infected cells showed that only about half of the progeny phage DNA was packaged into phage heads and that DNA molecules were selected at random for packaging. Thus the apparent nonreciprocity could be a result of fluctuations caused by random selection of molecules from the pool of newly synthesized DNA. Statistical analyses of the composition of hundreds of single bursts were carried out to determine whether the existence of one recombinant type correlated with the presence of the other recombinant in the same burst, which would be expected if recombination was truly reciprocal. This proved to be the case, and recombination with phage T4 is believed to be a physically reciprocal process.

The original observation of apparent lack of reciprocity raises a general question of whether reciprocal recombinants should be expected as the outcome of a biochemical exchange process that is itself reciprocal. There is a difference between being genetically nonreciprocal and being physically nonreciprocal. This distinction is exemplified in Figure 15-2. In panel (a), an exchange occurs between markers a and b. No DNA is lost; the exchange is physically reciprocal, and both recombinant types are found; thus, the exchange is also genetically reciprocal. In panel (b), breaks occur on both sides of marker a; the products in the first row are physically reciprocal. However, a^+/a^- heteroduplexes are generated in the overlap region (second row). Physical experiments indicate that such overlaps can be large. If there were no mismatch repair, each DNA molecule would replicate, and two parental genotypes and the two recombinant genotypes would result; the exchange would appear to be reciprocal. If mismatch repair were to occur before replication and the negative allele were converted to the positive allele in both products of the first row of (b), the products would be one parental molecule and one recombinant molecule, as shown in the third row; genetically the exchange would be nonreciprocal. Actually there are many ways by which a physically reciprocal exchange can show genetic nonreciprocity; however, there are no simple models that can generate genetic reciprocity from an exchange that is physically nonreciprocal. Determining that an exchange is genetically reciprocal puts constraints on hypotheses about the physical event.

Recombination by Breakage and Rejoining of DNA Molecules

In 1931, studies with both *Drosophila* and maize demonstrated that recombination was associated with physical exchange of chromosomes. These experiments used mutants whose chromosomes carried physically recognizable features (knobs) that were either associated with or near the mutant site. Recombinant organisms were selected, and cells were examined microscopically. It was found that the chromosomes of recombinants possessed both morphological features, suggesting that the individual chromosomes had broken and reassembled during gamete formation, a physically reciprocal process. This idea persisted until the single-burst experiments described in the preceding section were done. In the time interval between the original observation and the ultimate explanation of the lack of apparent reciprocity resulting from statistical sampling, other suggestions were made for the mechanism of recombination. The discovery in the

1950s that all genetic information in phages resides in DNA placed the various models on a DNA level. Insight into the mechanism of recombination was obtained in 1962 by experiments with *E. coli* phage λ. These experiments used the density-labeling technique (see Chapter 2) to determine the contribution of parental phage DNA molecules to recombinant progeny and showed that in forming recombinants parental DNA molecules break and rejoin. Meselson and Weigle did the original experiments, but the results are somewhat complex to interpret (because of the simultaneous activity of several recombination systems), so a simplified version is described.

Two types of λ phage were prepared: A^+R^-, whose DNA contained a heavy isotope in both strands, and A^-R^+, whose DNA contained a light isotope in both strands. (*A* and *R* are genetic markers opposite the termini of the λ map.) In addition, the DNA of both phages carried mutations to eliminate all recombination except that determined by the bacterial recombination genes. So any recombination observed is specifically due to the bacterial recombination system. Mutations in the phage and in the bacteria also prevented the initiation of phage DNA synthesis so that density changes caused by physical exchange of parental DNA were not obscured by DNA replication. Bacteria were infected with both phages, and the infection was allowed to proceed until lysis occurred. Progeny phage, which contained only parental material, were centrifuged to equilibrium in a CsCl solution, and the density distribution of the genotypes

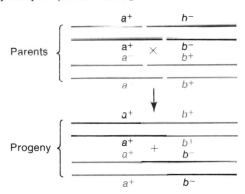

(a) Genetically and physically reciprocal exchange

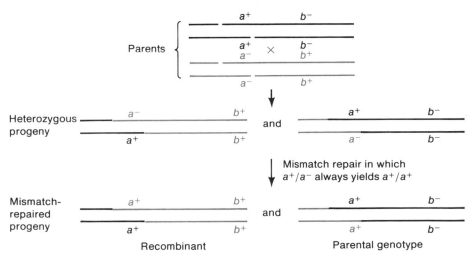

(b) Physically reciprocal but genetically nonreciprocal exchange

Figure 15-2.   Reciprocal and nonreciprocal exchanges.

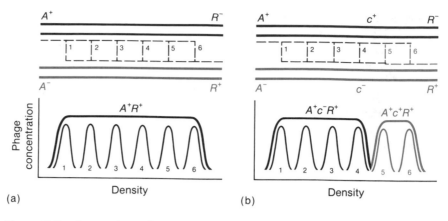

Figure 15-3. An experiment demonstrating breakage of DNA molecules and reunion to form recombinants. In the upper part of each panel, heavy lines represent high-density (black) and low-density (orange) *E. coli* phage λ carrying the indicated markers. After crossing the phage, using conditions that prevent DNA replication, as described in the text, progeny phage are centrifuged to equilibrium in CsCl. The thin lines in the distribution curve in the lower part of each panel show the expected density distribution for each of the numbered exchanges shown in the upper part of the panel. The heavy curves are the expected and observed distributions for all recombinant progeny taken together. (a) The uniform distribution obtained with two markers. (b) When three markers are used, the density distribution shows that $A^+ R^+$ phage are also c^+ only if breakage and reunion occurs to the right of the *c* marker, as expected.

of all phage particles was determined. Figure 15-3(a) shows that recombinant phage (A^+R^+ and A^-R^-) were found to range in density from fully heavy to fully light. That is, each recombinant particle contained material from both parents, indicating that physical exchange of material had occurred in forming the recombinants. When a central marker (in the *c* gene) was included (panel (*b*)), the density distribution showed that the c^+ marker from the A^+ parent appeared in A^+R^+ recombinants only if breakage and reunion occurred to the right of the marker, More complex experiments, in which DNA replication was permitted and in which other recombination systems were active, also gave evidence for breakage and reunion.

Effect of Deletions on Recombination Frequency

Measurement of recombination frequencies can be used to detect a large deletion. Consider two markers *a* and *b* that recombine with a frequency of 25%. A mutation *d* is found that from three-factor crosses maps between these markers. In two-factor crosses, the recombination frequencies for the intervals a^- and b^- are much less than 25%. One possible explanation is that *d* is a large deletion. If so, the recombination frequency between *a* and *b* if both parental phages contain *d* will be severely depressed because there is less material between *a* and *b*. Such an observation has often led to the discovery of large deletions.

GENETIC MAPPING OF PHAGE T4

The earliest genetic mapping experiments were carried out with *E. coli* phages T4 and λ. For example, a T4 phage with the mutations *r48* (large plaque) and *tu42* (plaque with a light turbid halo) was crossed with the wild-type T4:

tu42 r48 (turbid, large plaques) × *tu⁺ r⁺* (clear, small plaques)

The results are shown in Figure 15-4. Four plaque types appeared—the parental types, *tu42 r+* (turbid, small plaque recombinants) and *tu+ r42* (clear, large plaque recombinants)—as indicated. Note that both parental and recombinant types are easily identifiable. The recombination frequency is defined as:

$$\text{Recombination frequency} = \frac{\text{Number of recombinant phage}}{\text{total number of phage}} \times 100\%$$

Combining the data from numerous crosses generates a typical linkage map. Frequently different bacteria are used to detect recombinants, and all recombinant types are not seen. For example, in a cross between two nonsense termination mutants, one usually plates the lysate on a *sup⁻* host to score the total number of progeny phage and on a *sup⁰* host to count the number of wild-type recombinants. The double mutant recombinants are not seen. Because reciprocity in bulk lysates is the rule, however, the total number of recombinants is assumed to be twice the number of wild-type recombinants counted.

Genetic mapping of phage T4 led to an understanding of unexpected aspects of genome organization and phage production.

Genetic Map of T4 Is Circular

The T4 genome is quite a large DNA molecule (166 kb). In the phage life cycle, many progeny DNA molecules are present in the cell, so T4 DNA molecules engage in several (about five) rounds of recombination. This causes the recombination frequency per unit length of DNA to be fairly high. This high frequency of recombination means that markers do not have to be far apart (only a few percent of the genome length) before they appear to assort randomly. Thus, genetic mapping in which genetic distances are determined can be carried out only with markers that are quite close to one another.

In the early days of T4 mapping, when the number of available markers was quite limited, the genetic map appeared linear. The map shown in Figure 15-5 showed no unusual features until attempts were made to confirm map positions by three-factor crosses. Recall from Chapter 1 that a three-factor cross gives the gene order unambiguously because of the six possible recombinant classes;

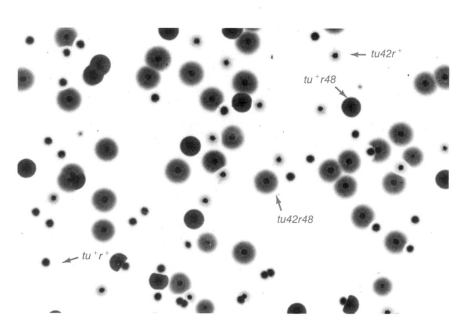

Figure 15-4. Progeny of a cross between *E. coli* T4 phage *tu+ r+* (a plaque is labeled at the lower left) and *tu42 r48* (center). Two types of recombinant plaques are found; representative plaques are labeled. (Courtesy of A. H. Doermann.)

the two rarest classes can be assumed to be a result of double exchanges. When the cross *r67 h42 ac⁺ × r⁺ h⁺ ac41* was carried out, the rarest recombinant classes were observed to be *r67 h⁺ ac⁺* and *r⁺ h42 ac41*, which indicates the map order *r67-h42-ac41*. This order conflicted with the map shown in the figure, which says that *r67* is closer to *ac* than to *h*. Other three-factor crosses also placed the ends of the linear map adjacent to one another. The resolution of this paradox was the proposal that the genetic map was circular, as shown at the bottom of the figure.

Possible Explanations for the T4 Circular Map

The most obvious explanation for a circular genetic map is that the DNA itself is circular. Although it is now possible to show whether or not T4 DNA is circular by electron microscopy, this technique was not available at the time that the circular map was discovered. A simple measurement of the viscosity of a T4 DNA sample exposed to a nuclease provided the necessary information. The viscosity of a solution of a macromolecule is affected primarily by the shape of the molecule at constant molecular weight: A long thin molecule yields solutions with a higher viscosity than a short molecule. Indeed, solutions of DNA have high viscosities because, for a given molecular weight, DNA molecules are much longer than most macromolecules. The crucial experiment consisted of placing a sample of T4 DNA molecules in a viscometer with a small amount of DNase and measuring the viscosity of the solution over time. The viscosity of a linear DNA sample decreases continually as strand breakage causes the molecules to become shorter. For a circular DNA molecule, however, the first break will linearize the circle, extending the molecule, and hence cause the viscosity to rise; subsequent breaks then decrease the viscosity. Thus, if T4 DNA were linear, the viscosity of DNase-treated T4 DNA should decrease continually with time; if it were circular, the viscosity should first rise and then decrease. Accurate viscosity measurements showed that there was no rise in viscosity and hence that the DNA is linear.

Although the DNA in the phage is linear, a simple explanation of the circular genetic map would be that the DNA circularizes after infection. Another possible explanation, however, is that the individual DNA molecules are terminally repetitive (redundant) in the sense that the gene order is *abc . . . xyzabc*; thus, a population of T4 DNA molecules is cyclically permuted in the sense that individual molecules have the gene orders *bcd . . . yzabcd, cde . . . zabcde, def . . . abcdef*, and so forth.

Figure 15-5. Conversion of the early linear map of phage T4 to a circular map by a three-factor cross using the markers shown in orange. This cross yielded the marker order *ac41 h42 r67*.

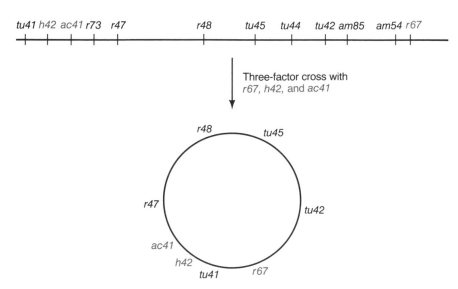

Phage Heterozygotes

A heterozygote contains two alleles of a gene—in most experiments, wild-type and mutant. In diploid organisms, this arrangement arises when one member of a homologous pair of chromosomes contains the wild-type allele and the other homologue contains the mutant allele. A phage contains only a single chromosome, so any heterozygotes must have another structure. Phage T4 heterozygotes were first observed in mixed infections with r^+ (small plaque) phage and rII (large plaque) phage. A small fraction of the progeny produce unusual mottled plaques (Figure 15-6). If phage are isolated from these plaques and replated on fresh bacteria, half of the resulting plaques are r^+ and half are rII. When replated again, both of these phage breed true. A simple explanation for heterozygotes of this type is that the phage that produce mottled plaques contain a heteroduplex of one strand with the r^+ allele and the complementary strand with the rII allele. Thus, the mottled plaque would not contain heterozygous particles but would consist of the progeny of an original heterozygote that produced both parental types when it replicated. Such heterozygotes would not be unexpected, and their existence was confirmed in two experiments, one with T4 and one with λ: (1) If *E. coli* is infected with both T4 r^+ and T4 rII in medium containing a partial inhibitor of DNA replication, the fraction of progeny that are r^+/rII heterozygotes increases—that is, the fraction of the plaques that are mottled increases. A reduced number of rounds of replication would decrease the number of overlap heterozygotes that would normally be converted to two parental types by DNA replication. (2) In the experiment shown in Figure 15-3, in which two λ phage were crossed, cI^+/cI^- heterozygotes (which produce a plaque with clear and turbid regions) were found but only at a density that would correspond to breakage in the central part of the λ DNA, where the cI gene is located.

Another feature of T4 heterozygotes indicated that a second type of heterozygote must also be present. Numerous deletions in T4 have been isolated. These deletions have little influence on most genetic processes other than bringing flanking genes nearer. The deletions, however, also increased the length of the heterozygous region. If phages bearing several closely linked markers are used in a mixed infection, a small fraction of progeny are found to be heterozygous for more than one marker. One explanation is that two markers can be present in an overlap region, which is certainly true. If the phages, however, also both carry a distant deletion (so distant that the deletion is not likely to affect the local recombination in any way), the number of heterozygotes for the two markers increases. Furthermore, when the deletion is present, heterozygotes containing more than two markers can be found that are undetectable if the deletion was absent. It is not obvious how a distant deletion could affect the length of an overlap

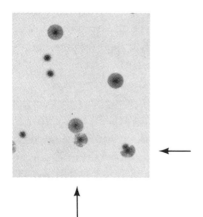

Figure 15-6. Results of a mixed infection with T4 r^+ (small plaques) and rII (large plaques) phage showing mottled plaques (arrows).

heterozygote. Furthermore, the number of these double heterozygotes is not increased by inhibiting DNA replication. Thus, it seems clear that T4 possesses two types of heterozygotes: those that increase in number when DNA replication is inhibited and those that increase in length when a deletion is present.

Streisinger and Stahl suggested that the second type of heterozygote is due to terminal redundancy. That is, they proposed that the ends of T4 DNA are direct repeats, for example, the gene order of the molecule would be $qr^+st \ldots abcd \ldots nopqrs$ (Figure 15-7). They explained the effect of a deletion by assuming that the amount of DNA contained in a phage head is fixed. Thus, if the deletion eliminated a gene or part of a gene, the terminally redundant region at ends of the phage DNA would be longer. Such a heterozygote would not produce a mottled plaque because progeny would also be terminal redundancy heterozygotes; to generate mottling, they proposed a replicating and packaging scheme by which a circularly permuted set of molecules would be generated. In a circularly permuted set, the markers in a terminal redundancy heterozygote separate from one another because the terminal redundancy of most unit-sized molecules does not include these markers. This scheme is described in the following section.

Packaging and Production of Cyclically Permuted, Terminally Redundant DNA Molecules

Replication of T4 DNA generates enormously long molecules (Figure 15-8) called concatemers. Packaging of T4 DNA into phage heads proceeds by cutting individual phage DNA units from these concatemers. The cuts are not made in unique base sequences in the DNA because, if they were, T4 DNA would not be cyclically permuted. Instead, the cuts are made at positions that are determined by the amount of DNA that can fit in a head. Presumably a free end of the DNA molecule enters the head, and this continues until the head is full; then the concatemer is cut. This is known as a headful packaging mechanism, and it explains how both terminal redundancy and cyclic permutation arise (Figure 15-9). The DNA content of a T4 particle is greater than the genome length. Thus, when cutting a headful from a concatemeric molecule, the final segment of DNA that is packaged is a duplicate of the DNA that is packaged first—that is, the packaged DNA is terminally redundant. The first segment of the second DNA molecule that is packaged is not the same as the first segment of the first phage. Furthermore, because the second phage must also be terminally redundant, a third phage-DNA molecule must begin with still another segment. Thus, the collection of DNA molecules in the phage produced by a single infected bacterium is a cyclically permuted set.

Proof of the explanations just given for the circular map came from physical experiments that directly demonstrated terminal redundancy of individual DNA molecules and cyclic permutation of the DNA population. Terminal redundancy can be shown by treating DNA with a DNase that removes bases only from the two 5' ends of the DNA (Figure 15-10). If the amount removed is greater than the terminally redundant segment, which from genetic arguments is about

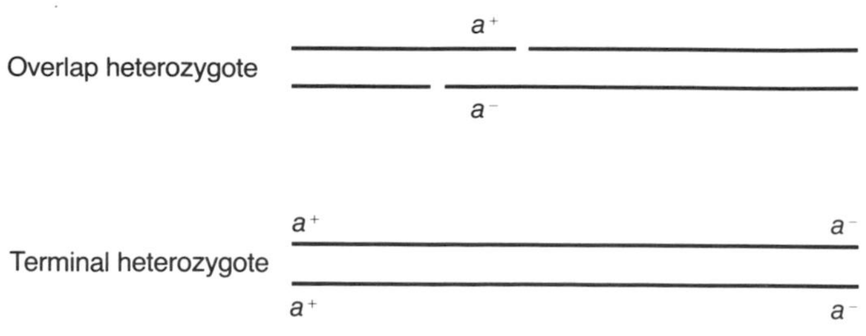

Figure 15-7. Two types of heterozygotes in T4. The overlap heterozygote is an immediate product of recombination, such as that in panel (b) of Figure 15-2. The a^+ and a^- alleles segregate when the DNA replicates. The terminal heterozygote occurs when the two alleles are present in the terminally redundant region. These alleles do not separate when the DNA replicates.

1% of the genome length, each molecule will be terminated by complementary single-stranded regions. For example, one terminus will have the sequence 3'-*abc* and the other the sequence *a'b'c'*-3', in which a prime denotes a complementary base. Hence if such treated DNA is exposed to renaturing conditions, the complementary termini will renature and a circular molecule will form. Such circular molecules were seen by electron microscopy, confirming the prediction of terminal redundancy. The length of the terminally redundant region was also determined by examining molecules in which the circle contained a small double-stranded segment flanked by two single-stranded regions. These arise in the following way: If the terminal single-stranded segments were 3'-*abcdef* and *xyz a'b'c'*-3', a double-stranded region consisting of renatured *abc* and *a'b'c'* will be

Figure 15-8. An electron micrograph of the replicating complex of T4 DNA. Note how it resembles the structure of *E. coli* DNA. (Courtesy of Joel Huberman.)

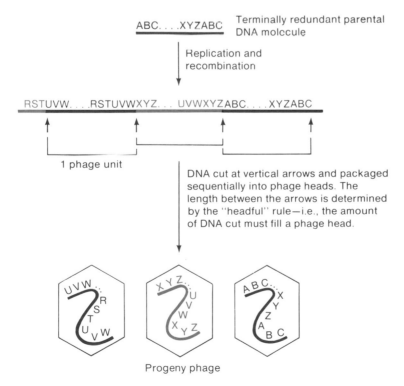

Progeny phage

Figure 15-9. Origin of cyclically permuted T4 DNA molecules. Alternate units are shown in different colors for clarity only.

flanked by single-stranded *xyz* and *def*. The length of the double-stranded segment is the length of the terminally redundant region.

Circular permutation was demonstrated by denaturing a T4 DNA sample and then allowing renaturation to occur. Because renaturation is a random process, pairing of unlike but complementary strands will occur and circular molecules bearing single-stranded termini will result (Figure 15-11). Electron micrographs of renatured DNA showed the predicted branched circles and thereby demonstrated that T4 DNA is cyclically permuted.

Further confirmation for the headful packaging model came from studies of T4 nonsense mutants with altered head proteins. When infecting sup^0 cells, these mutants produced aberrant heads that were sometimes larger and sometimes smaller (depending on the particular mutation) than the wild-type head. It was found that regardless of the head size, the aberrant heads were always filled to capacity. Some of the giant phage heads had as much as six times the normal amount of DNA.

Genetic Maps of Other Phages

Genetic maps have been obtained for many phages. Circularity is not a universal feature of these maps. For example, the map for phage T7 is linear, which corresponds to the linear DNA molecule contained in the phage. The DNA is terminally redundant but not cyclically permuted. T7 DNA also remains linear throughout the life cycle of the phage. In contrast, phage λ has a circular map (described later). Although the λ DNA molecule is linear in the phage head, it circularizes shortly after infection. The *S. typhimurium* phage P22 has a circular map, and similar to T4, its DNA is terminally redundant and cyclically permuted. Of the hundreds of phages whose DNA has been examined, the following types of linear DNA have been observed: terminally redundant and cyclically permuted, terminally redundant but not permuted, and neither redundant nor permuted. Furthermore, both linear and circular

Figure 15-10. A terminally redundant molecule and its identification by means of exonucleolytic digestion and circularization. A nonredundant DNA molecule cannot be circularized in this way.

(a) Terminally redundant DNA

5'		3'
A B C D E F G		W X Y Z A B C
A'B'C'D'E'F'G'		W'X'Y'Z'A'B'C' 5'
3'		

(b) After digestion with a 3' exonuclease

5'		
A B C D E F G		W X
F'G'		W'X'Y'Z'A'B'C' 5'

(c) After circularization of the molecule in (b)

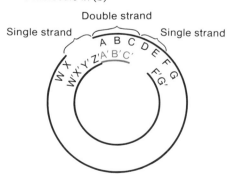

molecules have been isolated from various phages. Most DNA molecules, if they are linear, circularize after infection; T7 and closely related *E. coli* phages are rare exceptions.

FINE-STRUCTURE MAPPING OF THE T4 *rII* LOCUS

Until the mid-1940s, there were no examples of recombination between mutations within a gene. Such exchanges were not observable because of the lack of a system sensitive enough to detect low recombination frequencies. The first indication of intragenic recombination came from a study of the *lozenge* locus of *Drosophila*. The significance of the observation, however, was unclear because little was known about the substructure of the gene or, on a molecular basis, what a gene was. The situation changed dramatically in the mid-1950s— first, with the recognition that genes are segments of DNA and, second, with the work of Benzer, who carried out extraordinarily detailed mapping of the T4 *rII* locus. In his experiments, about 2400 independent mutations were mapped at 308 sites in the *rII* locus, with the goal of learning something about the internal organization of a gene.

Benzer's work exploited specific features of the T4 *rII* locus. Figure 15-1 showed the plaques formed by wild-type (T4 *r*⁺) phage and an *rII* mutant. The T4 *r*⁺ plaques are small with fuzzy edges, whereas the *rII* mutants produce larger plaques with sharp edges. Because recombination occurs with high frequency in T4, recombination between two *rII* mutations could be detected by the appearance of small *r*⁺ plaques—for example, certain *rII* markers recombine with

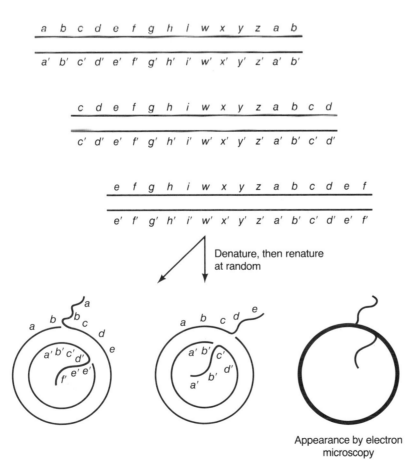

Denature, then renature
at random

Appearance by electron
microscopy

Figure 15-11. An electron microscopic test of cyclic permutation. A DNA sample is denatured until strands separate completely. Renaturation of strands with different termini produces double-stranded circular molecules (heavy line) with short single-stranded branches (thin lines).

frequencies as high as 8%. T4 *rII* mutants are easily isolated, as large-plaque mutants that arise at a frequency of about 10^{-5}. The mutation frequency can be enhanced considerably by treatment with various mutagens, so 20 plates with 500 plaques per plate will yield several mutants. Large-plaque mutants fall into three classes, *rI*, *rII*, and *rIII*, which map in different regions of the genome. The *rII* mutants were of special interest for fine-structure mapping because they have another property that Benzer realized would enable him to detect exceedingly low frequencies of recombination between different *rII* mutants—that is, *rII* mutants are conditional lethal mutants that fail to form plaques on *E. coli* strain K12 that is lysogenic for phage λ. The inability to plate on K12(λ) is a result of the activity of a λ gene, called *rex*, which is expressed in a lysogen. How the *rex* gene causes this inhibition has eluded geneticists now for more than 30 years, but understanding the mechanism underlying the plating deficiency is not necessary to take advantage of this system. There are five significant features of the *rII*-K12(λ) system:

1. The inhibition of T4 *rII* growth in K12(λ) is so complete that the rare plaques that do form are always revertants of some kind. The reversion frequency for many *rII* mutants is sufficiently low that recombination frequencies as low as 0.00001% could be detected. Actually the recombination frequencies in two-factor crosses are never that low, but this sensitivity makes possible multifactor crosses within the locus.

2. *E. coli* strain B supports the growth of both *rII* and r^+ phages, but the plaque morphology of the two types of phage are easily distinguished.

3. Crosses can be performed between different *rII* mutants by coinfecting *E. coli* B with the different phage. The number of recombinants can be measured simply by plating the progeny both on *E. coli* B (to detect all phage) and on K12(λ), on which only the r^+ recombinants grow. Because phage recombination is reciprocal, a double recombinant forms for every r^+ recombinant; thus the recombination frequency is twice the number of plaques formed on K12(λ) divided by the number of plaques formed on strain B.

4. Deletions in the *rII* locus are found at a reasonable frequency and can be identified as mutations for which no revertant plaques appear on plates containing *E. coli* K12(λ) and 10^8 mutant phage particles. In contrast, a typical *rII* point mutant will yield about 10 to 100 plaques.

5. Point mutants (that is, mutants that can revert) fall into two complementation groups, *rIIA* and *rIIB*. In a typical complementation test, K12(λ) bacteria are infected with two *rII* mutants. Infection by only one phage mutant yields no progeny. Mixed infection with some pairs of mutants, however, produces progeny in all the infected cells, and the progeny consist predominately of the two parental infecting phage types.

In the preliminary study of the *rII* system, about 50 *rII* mutants were sorted into complementation groups and mapped in the conventional way by performing two-factor crosses. The formidable goal of mapping the entire collection of about 2400 mutants, however, could not be achieved by directly carrying out two-factor or three-factor crosses because the number of required crosses would be more than a half million. Instead Benzer subdivided the *rII* locus into several segments and mapped each mutation to specific regions by deletion mapping. This procedure is based on a simple principle: A phage carrying a deletion cannot recombine with another phage having a mutation located in the region covered by the deletion to form wild-type recombinants because the wild-type allele is not present in either phage.

To locate mutations by deletion mapping, it was necessary to determine, at least approximately, the location and size of the deletion. This was accomplished first by crossing the deletions against one another. If two phage with overlapping *rII* deletions are crossed, r^+ recombinants cannot form because certain regions of the gene are missing in both phage. If the deletions do not overlap, however, wild-type recombinants can form. Thus, by crossing many pairs of deletion mutants, deletions could be ordered; the principle is shown in Figure 15-12. Of many *rII* deletions studied, seven large deletions were found that spanned the region in which all of the approximately 50 mapped *rII* point mutations were located. This yielded the relative order of the deletions. Then crossing a number of individual mutations against the deletions provided more detailed information about the limits of the deletions.

By using crosses between deletions and point mutations, Benzer was able to localize each of the 2400 mutations into one of seven major regions. He then used a set of smaller deletions, roughly four corresponding to each major deletion, to localize each mutation more precisely. In this way, the approximate location of each mutation could be determined by 11 crosses (seven with major deletions and four with small deletions), so with about 25,000 crosses, all mutations could be localized. To perform such a huge number of crosses (although much reduced from the more than 500,000 needed for mapping without deletions) would be a formidable task if each cross had to be done by a mixed infection of a growing bacterial culture, incubation until lysis occurred, and plating the lysate on both *E. coli* B and K12(λ). To reduce the labor, a spot test allowed 10 to 20 crosses to be carried out on a single plate. These spot tests were performed in the following way: A plate was prepared in which the soft agar contained about 10^8 K12(λ) cells, 10^6 B cells, and 10^7 phage carrying a deletion. If this plate were incubated, roughly 90% of the deletion phage would adsorb to the K12(λ), where they could not multiply. The remainder would infect the *E. coli* B and produce progeny, most of which would then adsorb to K12(λ). The K12(λ) always outnumber the B bacteria, so the small amount of lysis of the B cells could not produce a visible plaque. Thus these plates would produce a uniform lawn, appearing as if no phage had been present in the agar. To do a spot test, a drop containing about 10^7 particles of an *rII* mutant was placed in a small spot on the surface of the agar. After incubation of the plate, three types of spots were found:

1. *Complete clearing of the spot.* This observation is the result of complementation. If the mutant and the deletion are in different complementation groups, any K12(λ) infected with both phages will, by complementation, produce a burst containing both phage types. The two types of phage will continue to coinfect K12(λ) and ultimately lyse all bacteria in the region.

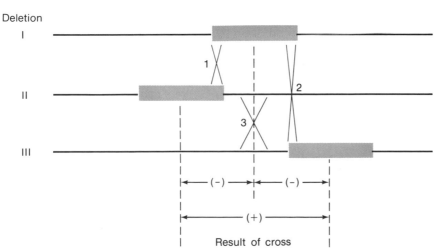

Figure 15-12. Mapping of deletions by the overlap method. The red boxes represent three deletions. In cross 1 (I × II) and cross 2 (I × III), no wild-type recombinants are produced (–) because the deletions overlap. Wild-type recombinants are produced—(+) in cross 3 (II × III)—because the deletions do not overlap. Indication of overlap by the lack of appearance of wild-type recombinants yields an unambiguous order of the deletions.

2. *A few plaques in the spot.* If both the deletion and the mutation are in the same complementation group, massive phage production (as just described) is not possible. If the deletion does not include the mutant site, however, recombination can occur in the B cells, producing a few r^+ phage that can then form a plaque on the excess K12(λ). Remember that recombination is a relatively rare event. (The recombination event cannot be detected in K12(λ) because the λ *rex* gene in the lysogen prevents growth of T4 phage.)

3. *No plaques.* If the deletion includes the mutant site, neither complementation nor recombination can occur, and no phage are produced.

This spot test enabled Benzer to determine quickly whether a mutant overlapped a particular deletion and, if it did not, whether it was in the *rIIA* or *rIIB* complementation group.

Another spot test was used to identify deletions and at the same time to localize the deletions. This is exemplified in Figure 15-13. In this case, two deletions are taken as standards: *r164* is in complementation group A, and *r196* is in group B. A mutant to be tested is mixed with *E. coli* B and K12(λ), as earlier, such that 10% of the bacteria are infected. The left panel of the figure, in which mutant 1 is tested, shows a background of plaques throughout the plate, indicating that this mutant is a point mutant (the plaques represent revertants). In addition, mutant 1 complements *r164* (complete clearing) and fails to recombine with *r196* (no plaques). Thus, mutant 1 is located in group B and must be within the *r196* deletion. The right panel shows that mutant 2 is a deletion because there is no background of plaques. Furthermore, mutant 2 recombines but does not complement *r164* (a few plaques) and complements *r196* (complete clearing). Thus, mutant 2 is a deletion in region A and does not overlap *r164*.

Once each mutation had been localized in a single region (determined by deletion mapping), Benzer used individual two-factor crosses to determine the frequency of recombination between the mutations and hence derived a detailed map of the *rII* locus. The following conclusions could be drawn from these data:

1. *The distribution of mutations was not random.* All of the 2400 mutants mapped into only 308 sites. Some sites were much more mutable than others. These sites are called hot spots (see also Chapter 10, Figure 10-6).

2. *Each complementation group represents a single gene.* All mutations in complementation group A mapped at one end of the map, and all mutations in complementation group B were at the other end of the map. This observation was the first real evidence that a complementation group represents a contiguous region of DNA, likely to encode a single polypeptide chain. Benzer introduced the term "cistron" to represent a region of the DNA that encoded a single polypeptide chain.

Figure 15-13. A spot test used to classify *rII* mutations. Dark regions indicate plaques (small spots) or cleared regions (large spot). Scattered plaques in the background indicate that the tested phage is a point mutation; absence of plaques implies a deletion. A large clear spot represents complementation. Small plaques in the spot result from recombination. See text for details.

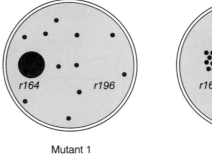

Mutant 1

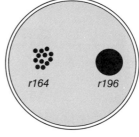

Mutant 2

This term was designed to replace the word "gene" but has never really done so. It remains, however, in the term polycistronic mRNA (see Chapter 6).

3. *Recombination can occur between adjacent nucleotide pairs*. The smallest recombination frequency observed between nearby markers was 0.02%. Because the total genetic map of T4 is about 1500 map units long, this represents $0.02/1500 = 1.33 \times 10^{-5}$ of the map. Assuming that recombination occurs with the same frequency in all regions of the DNA, from the size of the T4 DNA (1.66×10^5 nucleotide pairs), the smallest region in which recombination could occur is 2.3 nucleotide pairs. Because the assumption of uniform recombination at the nucleotide level is probably incorrect, and certainly mutations would not be recovered at all possible nucleotide positions, it seemed likely that recombination could occur between any pair of nucleotide pairs. In later years, physical measurements showed that the *rIIA* and *rIIB* genes consist of about 1800 and 850 nucleotide pairs. These numbers enabled map distances to be correlated roughly with physical distances. (The suggestion that genetic exchange could occur between mutations in adjacent nucleotide pairs was proved later by direct sequencing of the *E. coli trpE* protein made in recombinant cells.)

The fine-structure map of the *rII* region provided many important insights into molecular genetics. For example, it was first shown that the genetic code is a nonoverlapping triplet code by analysis of single-nucleotide insertion or deletion (frameshift) mutations in the *rIIB* cistron. Furthermore, the mechanism of action of numerous mutagens was worked out by studying the pattern of induction and reversion of mutations at particular sites in the *rII* region. The heterozygotes discussed earlier in this chapter, which led to an understanding both of the arrangement of genes in T4 DNA and the mechanism of packaging of DNA in the phage head, were *rII* heterozygotes. Finally, detailed studies of the mechanisms of recombination have been carried out by studying exchanges between *rII* mutations separated by various distances. The sensitivity of the *rII* system is great enough to detect multiple exchanges between as many as 10 markers; these experiments have yielded information about the clustering of exchanges and what may happen in or near a region of genetic exchange.

FEATURES OF THE T4 LIFE CYCLE

The T4 life cycle is typical for lytic phage. The timing of the life cycle is outlined as follows (with times shown in minutes at 37°C):

t = 0   Phage adsorbs to bacterial cell wall. Injection of phage DNA probably occurs within seconds of adsorption.

t = 1   Synthesis of host DNA, RNA, and protein is totally turned off.

t = 2   Synthesis of first mRNA begins.

t = 3   Degradation of bacterial DNA begins.

t = 5   Phage DNA synthesis is initiated.

t = 9   Synthesis of "late" mRNA begins.

r = 12 Completed heads and tails appear.

r = 15 First complete phage particle appears.

t = 22 Lysis of the bacteria and release of about 300 progeny phage occur.

The main feature to be noticed at this point is the orderly sequence of events. Figure 15-14 illustrates these events. In the sections that follow, a few of these stages are described further.

Taking Over the Cell

Shortly after infection, several events occur that enable the phage to turn off many bacterial functions necessary for continued bacterial growth. For example, the host RNA polymerase is modified in such a way that host promoters are poorly recognized. Second, host macromolecular synthesis is turned off. Finally, the first phage mRNA encodes DNases that rapidly degrade host DNA to nucleotides (this is not a common feature of phages).

Part of the takeover process is a series of events that allow transcription of T4 to occur in an orderly way. The complete pattern of early transcription of T4 DNA is complex. The basic pattern, however, which has also been observed in several other phages, is the following: Transcription of early mRNA starts at a single class of promoter by means of *E. coli* RNA polymerase (because that is the only polymerase in the cell). Then the polymerase is modified first by the addition of a small molecule, ADP-ribose, and later by the addition of phage-specified proteins, with the result that it no longer recognizes host promoters but instead initiates transcription at promoters for phage late mRNA promoters. Successive modifications of RNA polymerase provide one means of temporal control of the synthesis of many species of T4 mRNA.

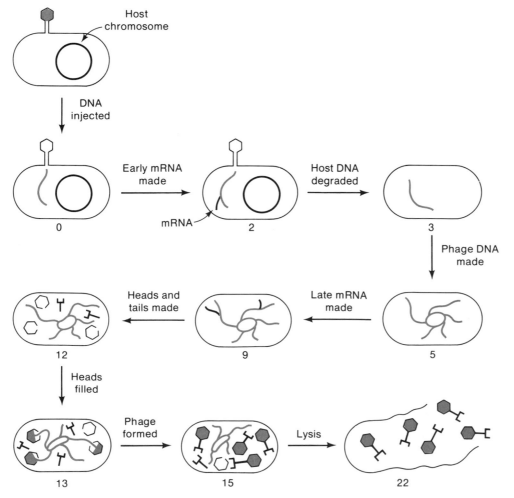

Figure 15-14. A schematic diagram of the life cycle of phage T4. The numbers represent time after injection in minutes at 37°C. For clarity, mRNA is drawn only at the time at which its synthesis begins.

Replication of T4 DNA

T4 DNA differs from typical DNA molecules in that it contains no cytosine (C). Instead a modified base, 5-hydroxymethylcytosine (HMC), pairs with guanine. Furthermore, the hydroxymethyl group has various glucoselike sugars covalently linked to it, so T4 DNA is covered with sugar chains (Figure 15-15). The presence of HMC and its glucosylation introduces particular requirements on the phage life cycle.

Five aspects of T4 DNA replication are especially interesting: (1) the source of nucleotides, (2) the synthesis of HMC, (3) the prevention of incorporation of cytosine, (4) glucosylation of T4 DNA, and (5) the enzymology of replication. The first four are discussed next. The enzymology is quite complex.

1. *Source of T4 DNA nucleotides: degradation of host DNA.* An early event in the T4 life cycle is the degradation of host DNA to deoxynucleoside monophosphates (dNMP). The responsible enzymes, which are active only on cytosine-containing DNA, cleave the host DNA to double-stranded fragments, which are then degraded to dNMP by a phage-encoded exonuclease. These mononucleotides are then used to resynthesize dATP, dTTP, dGTP, and dCTP by the usual *E. coli* enzymes, and this provides sufficient dNTP to synthesize 30 T4 DNA molecules. DNA precursors are synthesized de novo, but, to ensure an abundant supply of dNTP, five phage-encoded enzymes, which are virtually identical in activity to the *E. coli* enzymes, are also synthesized.

2. *Synthesis of HMC. E. coli* does not possess enzymes for forming HMC; therefore this is accomplished by two phage enzymes, which convert dCMP to dHDP. The *E. coli* enzyme, nucleoside phosphate kinase, which forms all nucleoside triphosphates in *E. coli*, then converts dHDP to dHTP, the immediate precursor of the HMC in the DNA.

3. *Prevention of incorporation of cytosine into T4 DNA.* T4 DNA polymerase cannot distinguish dCTP from dHTP, both of which can hydrogen bond to guanine. It is essential that no cytosine be incorporated into daughter T4 DNA strands because such cytosine-containing DNA would be a substrate for the T4 nucleases that degrade host DNA. For C to become part of daughter DNA molecules, dCMP must be converted to dCDP and then to dCTP. A phage enzyme, called dCTPase, degrades both dCDP and dCTP to dCMP.

   Another phage enzyme, dCMP deaminase, converts dCMP to dUMP, which then acquires a methyl group and becomes dTMP. This enzyme duplicates the activity of a similar *E. coli* enzyme (and hence is a product of one of the nonessential phage genes) but has an interesting economic function. The base composition of *E. coli* DNA is 50% (A+T) and T4 DNA is 66% (A+T). In *E. coli*, the ratio of dTTP to dCTP is about 1:1, in proportion to the T:C ratio in the DNA. The bacterial and phage dCMP deaminases, acting together, increase the amount of dTMP with respect to dCMP, so the ratio of dTTP to dHTP is 2:1, as is the T:HMC ratio in T4 DNA.

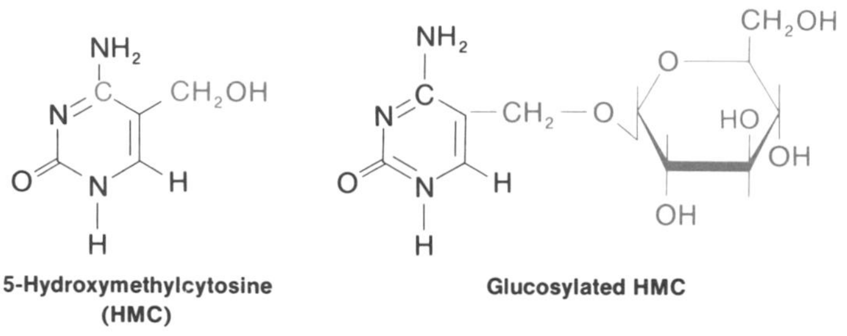

5-Hydroxymethylcytosine (HMC)

Glucosylated HMC

Figure 15-15. Nonglucosylated and glucosylated 5-hydroxymethylcytosine. If the CH_2OH (red) in HMC were replaced by hydrogen, the molecule would be cytosine.

Occasionally some C appears in progeny phage DNA. Presumably the T4 nucleases degrade this DNA shortly after synthesis; because the C is in only one strand of each daughter double helix, the DNA that is removed can be replaced by normal repair synthesis.

4. *Glucosylation of T4 DNA.* The presence of HMC in T4 DNA creates another problem for the phage because *E. coli* possesses an endonuclease that attacks certain sequences of nucleotides containing HMC (the normal role of this enzyme in *E. coli* is not known). To avoid this damage, the HMC residues in T4 DNA are glucosylated. This is accomplished by two phage enzymes that successively add two glucoses to HMC that is already in DNA. Thus, glucosylation protects the T4 DNA. The *E. coli*

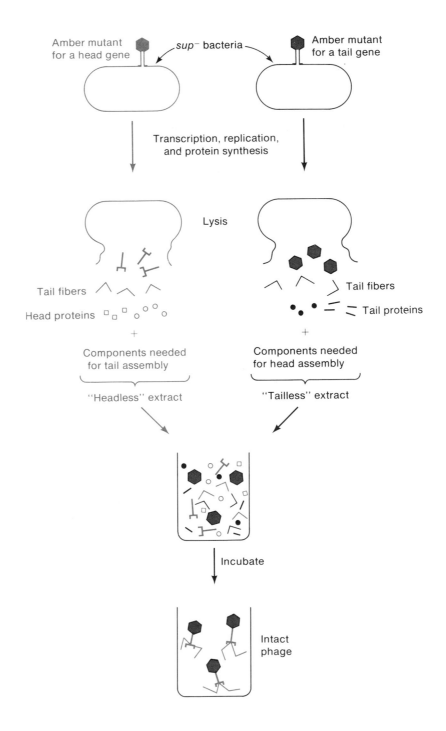

Figure 15-16. Production of intact T4 phage by in vitro complementation.

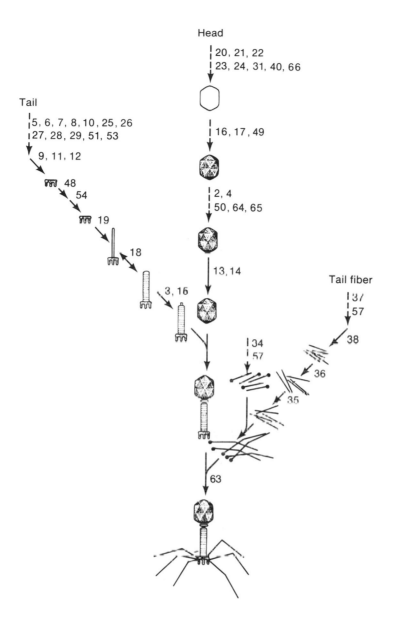

Figure 15-17. Morphogenetic pathway of T4 phage. The numbers designate T4 genes. (Courtesy of William Wood.)

endonuclease is inactive against glucosylated DNA; therefore glucosylation is a protective device. A simple genetic experiment shows that this is the only essential function of glucosylation. A T4 *agt⁻* mutant cannot carry out glucosylation, so its newly synthesized DNA is destroyed by the *E. coli* HMC nuclease. If an *E. coli* mutant (*rglB⁻*) that lacks this nuclease is used as a host, however, T4 *agt⁻* mutants grow normally even though nonglucosylated DNA is produced.

Production of T4 Phage Particles

Production of complete phage particles can be separated into two parts: assembly of heads, tails, and other structures and packaging of DNA of a sufficient length to provide a little more than one set of genes in the phage head. Packaging has been discussed in an earlier section.

Assembly of T4 (and many other phages) has been studied by two techniques, both of which require a large collection of phage mutants unable to make

Figure 15-18. Genetic map of phage T4 showing some, but not all, genes. The clustering of genes with related functions should be noted; although the tail-baseplate and tail-fiber genes form large clusters, other tail genes are distributed throughout the map. The solid and open regions indicate the locations of essential and non-essential genes. Control refers to genes needed to initiate various modes of transcription. The inner orange arrows indicate the direction and origins (but not the lengths) of various transcripts.

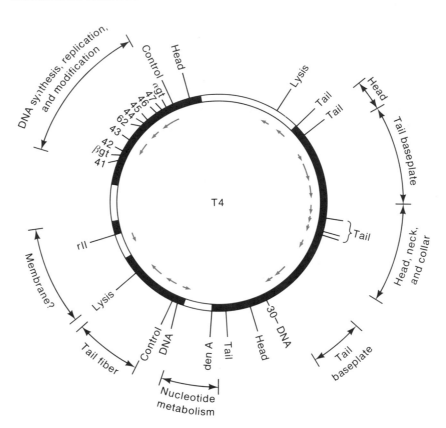

functional phage particles. In one method, different cultures of cells, each infected with a particular mutant, are lysed and examined by electron microscopy. This procedure shows that heads are made in the absence of tail synthesis and that tails are made by a mutant unable to synthesize heads. Thus head and tail assembly are independent processes. Electron microscopic examination of cell extracts infected with various head, tail, and tail fiber mutants has indicated the presence of partially assembled structures. Study of these structures and the components they contain has given information about the order of assembly of the various gene products. The second technique is a type of in vitro complementation assay. If two extracts of infected cells, one lacking heads and the other lacking tails, are mixed, functional phage particles assemble in vitro (Figures 15-16). The "headless" extract can also be fractionated, and a component can be isolated that allows a "tailless" extract to make tails. In this way, a protein in the tail assembly pathway can be isolated and identified.

These studies have shown that there are two types of components—structural proteins and morphogenetic enzymes. Some of the structural components assemble spontaneously to form phage structures, whereas others do so exceedingly slowly and hence need the help of enzymes. Genetic analysis shows that a few host-encoded factors are also needed for heady assembly; that is, *E. coli* mutants have been found that do not support the multiplication of T4. Infection of these mutants yields phage with aberrant heads. An abridged diagram of the assembly pathway is shown in Figure 15-17.

T4 GENE ORGANIZATION

More than 200 T4 genes have been identified. These genes account for about 90% of the DNA; thus, perhaps 20 genes remain to be found. T4 genes of known function can be divided into two classes: 82 metabolic genes and 53 particle-assembly genes. Of the 82 metabolic genes, only the 22 genes involved in DNA synthesis, transcription, and lysis are essential. The remaining 60 metabolic genes duplicate bacterial genes; particles in which these genes are mutated will grow, although occasionally they will have a smaller burst size. Of the 53 assembly genes, 34 code for structural proteins, and 19 code for the synthesis of enzymes and protein factors that are required catalytically for assembly. Thus, 17% of the DNA of phage T4 encodes essential metabolic functions, 39% is necessary for phage assembly, and 44% serves nonessential metabolic functions.

A genetic map of some of these genes is shown in Figure 15-18. A notable feature of the map is that genes having related functions are often adjacent and transcribed as part of polycistronic mRNA molecules. This is an efficient arrangement, allowing the synthesis of functionally related proteins to occur at nearly the same time and minimizing the number of regulatory elements required. Not all functionally related genes, however, are part of single transcription units, and some transcription units contain functionally discrete genes. The tendency to cluster related genes is common in many phage systems.

KEY TERMS

coinfection	hydroxymethylcytosine
complementation groups	lytic phage
cyclically permuted	plaque morphology
E. coli B	reciprocal recombination
E. coli K	rII
essential gene	Sup⁻
glucosylation	Sup°
headful packaging	terminal redundancy

QUESTIONS AND PROBLEMS

1. If an *E. coli* culture is simultaneously infected by phages T4 and T7, each at MOI = 5, only T4 phage will be produced. From what you know about T4 biology, propose a simple explanation.

2. What is the function of T4 dCTPase (deoxycytidine triphosphatase) in a T4 infection?

3. Describe the course of T4 phage DNA synthesis following infection of *E. coli* with a T4 mutant that cannot synthesize (a) cytidine hydroxymethylase or (b) α-glucosyl-transferase.

4. What is the basic principle used by T4 in regulating the transcription sequence of the mRNA molecules made before DNA replication begins?

5. Suppose you have a phage whose linear DNA is synthesized by the rolling circle mode and is packaged by "the headful rule" (that is, DNA is added to a head of fixed size until no more DNA can fit). The DNA, however, is normally neither terminally redundant nor cyclically permuted. You find a mutant strain of this phage, the DNA of which has a deletion in a nonessential gene. This phage is used to infect a bacterium, and many phage are produced. The DNA is isolated and is treated with an exonuclease, which removes a few bases from the 5'-P end. The treated DNA is then exposed to conditions that could circularize T4 DNA (if it were also pretreated with the exonuclease). This DNA is examined by electron microscopy. Will circles be found?

6. Most double-stranded DNA phages have several classes of mRNA that can be divided into two major groups, early mRNA and late mRNA. The genes carried on the early mRNA species vary from one phage to the next. Nonetheless, there are certain genes that are usually on early transcripts and some that are invariably on late transcripts. What are these genes?

7. What is the appearance of r^+ and *rII* phage when plated on a 1:1 mixture of *E. coli* B and K12(λ)?

8. In a cross between *rII A254* and *rII B82*, 26 plaques are found on a K12(λ) plate and 482 on a B plate, using phage at the same dilution. What is the recombination frequency between *rII A254* and *rII B82*?

9. Four *rII* deletions, A–D, are crossed against one another. Of the six crosses, three yielded recombinants: (1) A × B, (2) A × C, and (3) C × D. The remaining three—(4) A × D, (5) B × C, and (6) B × D—yielded no recombinants.

     a. What is the relation between these deletions?

     b. An *rII* mutation recombines with A, B, and C but not D. Where is this mutation located?

10. Most phages, of which T4 is certainly an example, all have life cycles in which a large fraction is devoted to late transcription. Why is the duration of time allotted to late transcription greater than the time for early transcription?

11. Three T4 *rII* deletions, A, B, and D, have the following properties: B × D yields r^+ recombinants, but A × B and A × D do not. What is the map order of the deletions?

12. Consider the deletions in Problem 11. A mutant yields r^+ recombinants with B and D but not with A. Locate the mutant with respect to the deletions.

13. A phage that produces large plaques maps in the *rII* region. If 10^7 particles are plated on K12(λ) cells, no plaques result. What do you know about the mutation?

14. A plate is prepared with 10^7 particles of a known *rII* A mutant, 10^8 *E. coli* K12(λ) cells, and 10^6 *E. coli* B cells. Two mutants are tested in spot tests. The background is plaque free. Mutant 1 gives a totally clear spot, and mutant 2 yields a few plaques. What do you know about the mutants?

15. Mutations *rII* A1 and *rII* B2 complement. Also, both mutations can recombine with the deletion *rII* X4 to yield r^+ recombinants. However, *rII* X4 fails to complement with either *rII* A1 or *rII* B2. Explain.

16. The genome of a phage that uses the headful packaging rule is represented as *ABCDEF . . . XYZAB*. It is terminally redundant and cyclically permuted, and individual molecules can be represented as *ABCDE . . . XYZAB, EFGHI . . . BCDEF*, and *JKLMN . . . GHIJK*. A deletion that removes genes *C* and *D* is isolated, and a phage lysate is prepared by infecting cells with this deletion. Which of the following sets of sequences should be observed among the phage progeny? (1) *ABEFG . . . XYZAB, EFGHI . . . ZABEF, YZABE . . . VWXYZ*; (2) *ABEFG . . . ZABEF, BEFGH . . . ABEFG, ZABEF . . . YZABE.*

17. Consider a phage with the strange property that productive infection occurs only if at least two phage adsorb to the bacterium. If you have 10^8 bacteria and add to this culture 3×10^8 phage, how many bacteria will be productively infected?

18. A new protein X appears in infected cells. Describe various experiments that you might perform to prove that the gene coding for X is phage encoded and is not encoded in host DNA.

19. How would you show whether a phage-encoded gene product is required throughout the infectious cycle or only at a unique time?

REFERENCES

Benzer, S. 1961. On the topography of the genetic fine structure of T4. *Proc. Natl. Acad. Sci. USA*, 47, 403.

Casjens, S. 1985. *Virus Assembly*. Jones and Bartlett, Boston.

Doermann, A. H. 1952. The intracellular growth of bacteriophages. I. Liberation of intracellular bacteriophage by premature lysis with another phage. *J. Gen. Physiol.*, 35, 645.

Doermann, A. H. 1983. Introduction to the early years of bacteriophage T4. In C. K. Matthews, et al. (eds.), *Bacteriophage T4*, p. 1. American Society for Microbiology, Washington, D.C.

Ellis, E. L., and M. Delbruck. 1939. The growth of bacteriophage. *J. Gen. Physiol.*, 22, 365.

Hershey, A. D. 1946. Spontaneous mutations in bacterial viruses. *Cold Spring Harb. Symp. Quant. Biol.*, 11, 67.

*Resources for additional information.

Hershey, A. D., and R. Rotman. 1949. Genetic recombination between host-range and plaque-type mutants of bacteriophage in single bacterial cells. *Genetics*, 34, 44.

Levinthal, C. 1954. Recombination in phage T2; its relation to heterozygosis and growth. *Genetics*, 39, 169.

MacHattie, L., et al. 1967. Terminal repetition in permuted bacteriophage DNA molecules. *J. Mol. Biol.*, 23, 355.

Matthews, C. K., et al. 1983. *Bacteriophage T4*, Second Edition. American Society for Microbiology, Washington, D.C.

Meselson, M., and J. Weigle. 1961. Chromosome breakage accompanying genetic recombination in bacteriophage. *Proc. Natl. Acad. Sci. USA*, 47, 857.

Streisinger, G., R. S. Edgar, and D. Denhardt. 1964. The chromosome structure in phage T4. I. The circularity of the linkage map. *Proc. Natl. Acad. Sci. USA*, 51, 775.

Streisinger, G., J. Emrich, and M. M. Stahl. 1967. Chromosome structure in phage T4. III. Terminal redundancy and length determination. *Proc. Natl. Acad. Sci. USA*, 57, 292.

Zinder, N. (ed.). 1975. *RNA Phages*. Cold Spring Harbor Laboratory Press, New York.

16

Lytic Growth of Phage λ

The *Escherichia coli* phage λ has two alternate life cycles—lytic and lysogenic growth. In this chapter, we consider only the former, which differs significantly from that seen for T4 in Chapter 15, the lysogenic cycle is described in detail in Chapter 17.

λ DNA AND ITS GENE ORGANIZATION

λ contains a linear double-stranded DNA molecule, consisting of 48,514 bp of known sequence. At each end of the DNA molecule, the 5' terminus extends 12 bases beyond the 3'-terminal nucleotide. The base sequences of these single-stranded terminal regions, which are known as cohesive ends, are complementary to one another (Figure 16-1). Thus, by forming base pairs between the cohesive ends, the linear λ molecule can circularize, yielding a circle with two single-strand breaks. Immediately after injection, λ DNA circularizes in this way, and *E. coli* DNA ligase seals the breaks to convert the DNA to a covalent circle.

Of 46 λ genes, 14 are nonessential for the lytic cycle, but only 7 are nonessential for both the lytic and the lysogenic cycles. Most λ proteins have been either purified or identified by gel electrophoresis. All regulatory sites, promoters, and termination sites are known.

The genetic map of λ is shown in Figure 16-2. A striking feature of the map is the clustering of genes according to function. For example, the head, tail, replication, and recombination genes form four distinct clusters. Even the genes needed for head-tail attachment lie between the head and tail genes. Many λ proteins—for example, regulatory proteins and those responsible for DNA synthesis—act at particular sites in the DNA. In general, these proteins are located adjacent to their sites of action (when there is a single site). For instance, the origin of DNA replication lies within the coding sequence for gene *O*, which encodes a protein for initiation of DNA replication, and the gene that generates the cohesive ends is located adjacent to one of the ends.

LYTIC LIFE CYCLE OF λ

The schedule of the lytic cycle of λ is fairly complex, probably because certain genes are used in both the lytic and the lysogenic cycles. The timing of the life cycle is outlined as follows (with time shown in minutes at 37°C):

t = 0   Phage adsorbs and DNA is injected.
t = 3   First ("pre-early") mRNA is synthesized.
t = 5   Two classes of early mRNA are synthesized.
t = 6   DNA replication begins.
t = 9   Synthesis of late mRNA begins.
t = 10   Structural proteins begin to be made.
t = 22   First phage particle is completed.
t = 45   Lysis and release of progeny phage.

Note that the cycle is 45 minutes long rather than the 22 to 25 minutes required for T4. The life cycles of most phages vary between 22 and 60 minutes at 37°C.

TRANSCRIPTION OF λ

With phage T4 and many other phages as well, timing of synthesis of the various mRNA molecules is accomplished primarily by mechanisms that determine the availability of promoters: the synthesis of a new RNA polymerase (phage T7) or the modification of the host polymerase (T4). In λ, the host RNA polymerase is also modified but not so that it recognizes unique phage promoters. Instead the modification enables RNA polymerase to ignore certain termination sites. Figure 16-3 shows a linear version of the λ genetic map, which includes the three regulatory genes *cro*, *N*, and *Q*; three promoters *pL*, *pR*, and *pR2*; the DNA-replication genes *O* and *P*; and five termination sites *tL1*, *tR1*, *tR2*, *tR3*, and *tR4* (Table 16-1). Seven mRNA molecules are also shown; the L (left) and R (right) transcripts are made in opposite directions from complementary DNA strands. The λ map is often drawn as a linear map, matching the linear DNA molecule in the phage head. In the standard orientation, gene *A* is the left end and gene *R* is at the right end.

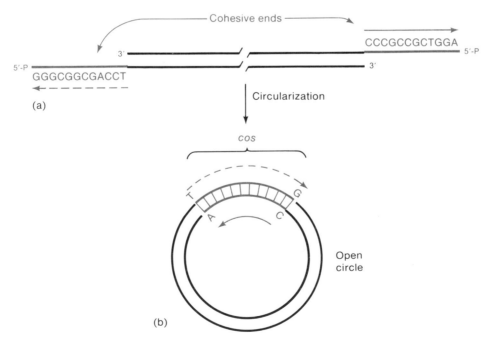

Figure 16-1. (a) A diagram of a λ DNA molecule showing the complementary single-stranded ends (cohesive ends). Note that 10 of the 12 bases are G or C. (b) Circularization by means of base pairing between the cohesive ends. The double-stranded region that is formed is designated *cos*.

An essential feature of the life cycle of phage T4 is rapid killing of the host and degradation of the host DNA. λ differs in this respect because in the lysogenic cycle the host must survive. Instead, even in the lytic cycle, λ multiplies while the host cell continues its normal function. Lytic growth of λ requires the sequential expression of the *O* and *P* genes, whose products are necessary for DNA synthesis followed by transcription of the genes encoding the structural proteins, and finally the packaging system and the lytic proteins. Both before and after transcription of genes *O* and *P*, however, other small transcripts are formed that encode the regulatory proteins responsible for turning transcription on and off at the appropriate times.

λ has two early promoters, *pL* and *pR*, from which synthesis of the RNA transcripts L1 and R1 are initiated. Transcription initially terminates at the sites *tL1* and *tR1*. L1 encodes only the N gene product, which is a major positive regulatory protein required for transcription of the *O* and *P* genes. Once synthesized, the N protein binds to the *nutL* (downstream of *pL*) and *nutR* (downstream of *pR*) sites on the mRNA. RNA polymerase interacts with N protein, which enables the polymerase to ignore the termination sites *tL1* and *tR1* forming the longer transcripts L2 and R2.

Once antitermination occurs, rightward transcription allows synthesis of the *O* and *P* gene products required for DNA-replication. The leftward transcript includes the *red* locus. This encodes two genes needed for genetic recombination, which plays an important role late in the life cycle. Because the Red proteins and the O and P proteins are catalytic, they do not have to be made continuously. The rightward transcript also encodes *cro*. When sufficient Cro protein is made, it acts as a repressor to turn off synthesis of all leftward mRNA by binding to the leftward operator *oL*. Rightward transcription terminates at a downstream termination site which is recognized by RNA polymerase even when it is modified by N protein.

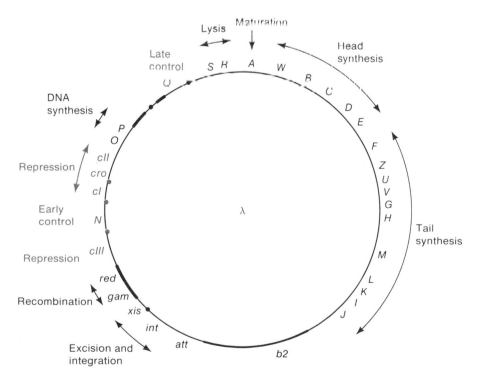

Figure 16-2. Genetic map of phage λ. Regulatory genes and functions are given in orange. All genes are not shown. Major regulatory sites are indicated by black solid circles. Regions nonessential for both the lytic and the lysogenic cycles are denoted by a heavy line.

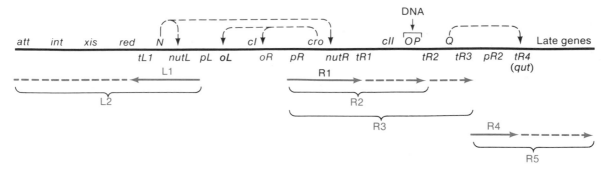

Figure 16-3. A genetic map of the regulatory genes of phage λ. Genes are listed above the line; sites are below the line. The mRNA molecules are orange. The dashed black arrows indicate the sites of action of the N, Cro, and Q proteins.

During this early-transcription period, rightward transcription provides enough mRNA that the concentrations of the O and P proteins reach values sufficient for efficient DNA replication. Somewhat later than the time when the Cro protein inhibits transcription from *pL*, the concentration of Cro increases to the point that it also binds to the rightward operator *oR* to block rightward mRNA synthesis. This ensures that wasteful synthesis of O and P proteins does not occur. Another positive regulator encoded by the *Q* gene is needed to turn on late mRNA synthesis.

Throughout this time, the tiny transcript R4 is synthesized continually from *pR2*, terminating at *tR4*. This transcript does not encode any known genes but is a leader for the late mRNA. Once R3 has been made, the *Q* gene product is expressed. Q protein binds to a DNA sequence called *qut* and prevents termination of RNA polymerase at *tR4*, allowing synthesis of the late mRNA. R4 is then extended to form transcript R5, the late mRNA, which encodes the head, tail, assembly, and lysis proteins.

The essential features of this highly efficient (although complex) regulatory system are as follows:

1. A λ-specific RNA polymerase is not made; the *E coli* RNA polymerase is used throughout the life cycle (as is true of T4) and is modified by accessory proteins to alter its specificity toward various DNA and RNA base sequences. The promoter specificity of RNA polymerase, however,

Table 16-1   Some sites and gene products in phage λ lytic growth

Site or gene product	Description
Site	
oL, oR	Left and right operators
pL, pR	Left and right promoters
tL (1,2)	Termination sites for leftward transcription
tR (1,2,3,4,5)	Termination sites for rightward transcription
Gene product	
Cro	Protein inhibitor of transcription from *pL* and *pR*
N	Antitermination protein acting at *tL1, tR1,* and *tR2*
O, P	Proteins required for DNA replication
Q	Antitermination protein acting at *tR4*
L	Messenger RNA synthesized in leftward direction
R	Messenger RNA synthesized in rightward direction
R4	Constitutively synthesized mRNA

is not altered; instead its ability to terminate transcription at certain termination sites is altered.

2. Inhibition of transcription occurs as a result of the repressor activity of Cro on the promoters *pL* and *pR*. This repression prevents wasteful synthesis of excess early gene products, which would compete with the synthesis of late gene products.

3. All the structural components are encoded in a single giant mRNA molecule, which is translated sequentially. Synthesis of the complete set of components takes many minutes, thereby delaying synthesis of intact heads and of a functional phage maturation system until the DNA replication system has provided many copies of λ DNA.

Note the series of delays—time required to transcribe particular early regions and time required to synthesize regulatory proteins. The result of these delays is that about 30 copies of λ DNA form before the maturation system is expressed, and 50 to 100 completed phage particles form before the onset of lysis.

Genetic Experiments That Gave Insight into λ Lytic Regulation

Most of the information about transcription in λ was derived from experiments in which newly synthesized RNA was labeled at various times after infection, and this labeled RNA was hybridized to various segments of λ DNA. In each case, the λ genes present in each segment of the DNA were known, so it was possible to determine the temporal sequence of transcription of all regions of the λ genome. Regulatory elements were identified by the fact that mutations in certain genes prevented synthesis of particular species of mRNA. For example, Q^- mutants were unable to synthesize the late mRNA, and N^- mutants failed to synthesize all mRNA, other than L1 and R1. A great deal of information, however, came from strictly genetic experiments, a few of which are described.

In the discussion that follows, several mutants are described, and we make statements about their inability to grow. One might reasonably ask how samples of these mutants are ever obtained. The answer is that, as is invariably the case with mutations in essential genes, the mutations are conditional. Most of these mutations are nonsense mutations, allowing the phage to be grown on suppressor-containing cells. Experiments in which the phage behave as mutants are done with strains lacking known suppressors (sup^0 strains).

The properties of gene N were first uncovered by genetic experiments. λ N amber mutants are unable to grow on a sup^0 host because the Q product is not synthesized, and hence no late mRNA is made. The *red* gene *exo*, which is downstream from N and encoded in L2, synthesizes an easily assayed exonuclease. N^- mutants do not make this exonuclease because L1 is never extended to form L2. A small deletion mutation, however, was isolated between N and *exo*; when coupled with an N^- mutation, this deletion enables the N^- phage to make the exonuclease. The results suggested that N acts at a downstream sequence, and the deletion removes this site. The N-sensitive termination site between genes P and Q was also identified by a genetic test. λ N^- mutants were plated on sup^0 bacteria to seek revertants. Some of these revertants mapped in N and restored N^- gene activity, as might be expected. One revertant, however, was a mutation that mapped between P and Q. This mutation, called *nin-5* (for N-*in*dependence) was a substitution of phage DNA by bacterial DNA; this substitution replaces the transcription-termination site and allows synthesis of Q protein without the antitermination activity of N protein.

In the preceding section, we mentioned that N protein interacts with **the *nut* site in the mRNA** to form a complex with RNA polymerase that can ignore certain terminator sequences. The interaction with N protein and RNA polymerase was also first recognized by a genetic observation. In an attempt to seek bacterial functions needed for λ phage production, bacteria were mutagenized, and mutants on which λ would not form plaques were sought. Many such mutants were found. Some of these were unable to adsorb λ, which was detectable by the ability of a λ *h* mutant to plate on these mutants, and these were not studied further. The remainder were called *gro* mutants. One class of these host mutants is called *nus* for N-utilization substance (originally called *groN*). In *nus⁻* hosts, λ⁺ behaves like λN⁻. That is, the N protein is made but fails to function in these bacteria. Such *nus* mutants map in several bacterial genes, including the gene encoding the β subunit of RNA polymerase and the *nusA* gene. Certain mutations in the λ*N* gene suppress the NusA phenotype, indicating that N protein and NusA protein interact. Biochemical evidence confirmed that the NusA protein binds to both N protein and RNA polymerase.

λ DNA REPLICATION AND PHAGE PRODUCTION

Most of the steps involved in λ DNA replication and the required proteins are known. This information has come from a variety of genetic and physical experiments and from studies with an in vitro replication system. Some of the observations are described in this section. We also see that replication and conversion of DNA to a form that can be packaged are coupled processes.

Genetic Experiments That Gave Insight into λ DNA Replication

In the initial major search for λ mutants, two classes of mutations that prevent λ DNA replication were found. These were mapped in the two adjacent genes, *O* and *P*, as already discussed. Further analysis provided evidence for a requirement for bacterial proteins—for example, λ could not grow on *E. coli* carrying mutations in known DNA replication genes, such as those encoding DNA polymerase III and the DnaJ protein. Some bacterial mutants were found in another way. For example, some of the Gro mutants mentioned in the preceding section proved to be a type called GroP; these supported the growth of λ carrying certain suppressed *P⁻* mutations. Similar experiments showed that the *groP* gene was identical to the *dnaB* gene, whose product is an enzyme essential for *E. coli* DNA synthesis.

Further information about the genetics of λ DNA replication came from a study of genes *O* and *P*, whose gene products are essential for DNA replication. In a search for revertants of *O⁻* mutations, most of the revertants mapped in gene *O*, as expected; however, some mapped in gene *P*. Furthermore, only revertants of certain *O⁻* mutations mapped in *P*. Similarly, one revertant of a particular *P⁻* mutation mapped in gene *O*. These experiments suggested that the *O* and *P* proteins interact. No GroO bacterial mutants were ever found, so O protein probably does not interact with any bacterial protein. Hence the genetic results suggest that O, P, and DnaB proteins form a complex with contact points between O and P proteins and between P and DnaB proteins.

λ replication begins at a unique origin called an *ori*. The λ origin *ori* was first identified by genetic experiments. A λ mutant was isolated that produced a very small plaque; it was called *ti12* for tiny. Infection by this mutant yielded only a few phage per cell, rather than a normal burst of about 50 to 100. In a mixed infection with wild-type λ, *ti12* appeared not to grow at all. For example, in a mixed infection with λ *ti12* and a genetically marked *ti⁺* phage, the infected cells produced about 50 *ti⁺* progeny phage per cell and no (less than 0.01) detectable *ti12* phage. This and several other experiments showed that *ti12* is unable to compete with another phage. Mapping of the mutation placed it adjacent to known *Q* mutations. The common clustering of λ genes according to function suggested that *ti12* affects DNA replication. However, *ti12* is clearly a site, because it could not be complemented by *ti⁺* in *trans*. Physical experiments using density-labeled *ti12* phage showed that it cannot initiate replication in a mixed infection. Thus, it was concluded that *ti12* is a mutation in *ori*. It is not an absolute-defective because, if so, it would never have been isolated (it would not be able to replicate at all). Rather it is a leaky mutation that fails to compete effectively for the replication-initiation system when a wild-type *ori* is present.

DNA Replication and Maturation: Coupled Processes

Following synthesis of the O and P products, replication of circular λ DNA begins. There are two modes of λ DNA replication: θ and rolling circle replication (Figure 16-4; see also Chapter 8). The θ replication increases the number of templates for transcription and further replication; the rolling circle replication provides the DNA for phage progeny. As the life cycle proceeds, θ replication stops, and rolling circle replication continues.

The DNA-cutting mechanism of λ differs from that used by T4 or Mu. The DNA found in a λ phage particle is linear and has single-stranded termini. These ends are joined when a circle forms, and the double-stranded region so formed is called a *cos* site (for cohesive site). Thus, every monomeric λ circle contains one *cos* site; however, a multimeric branch of a rolling circle contains many *cos* sites. The ends of the DNA molecule in the phage particle are always the single-stranded cohesive termini. The termini are formed by cleaving *cos* sites with a sequence-specific nuclease called terminase or Ter.

Figure 16-4 shows the three major species of intracellular λ DNA: circles, θ molecules, and rolling circles. Initially λ replicates by θ replication, but in time there is a gradual cessation of this mode and a transition to rolling circle replication. By the time heads and tails have been synthesized and Ter is active, rolling circles predominate. Note that because a rolling circle has two classes of *cos* sites—the one in the circle and those in the linear branch—some mechanism must exist for preventing cleavage of the one in the circle because if it were broken, replication would cease. This difficulty is circumvented by a site requirement of the Ter system: Efficient cleavage of a single *cos* site does not occur; if there are two *cos* sites and both are present on a single segment of DNA, cutting can occur. A λ DNA molecule can be cut from a linear branch by cleavage of two neighboring cos sites, but a pair of cuts in which one *cos* site is in the branch and the other is in the circle cannot be made. Ter-cutting, however, does not require that the DNA molecule be linear: A single λ unit can be cut from a dimeric circle that has been formed by genetic recombination. Physical experiments confirm

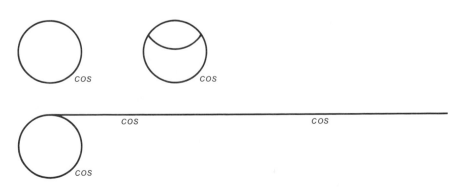

Figure 16-4. Three species of λ DNA present at the time maturation begins. The region containing the joined complementary, single-stranded termini (of the linear DNA molecule present in the phage head) is called *cos* (for cohesive site).

that a single cut in a monomeric circle does not occur efficiently.

The "two-*cos*-sites" rule explains how the first λ DNA unit is cut from a concatemeric branch of a rolling circle. This rule, however, would not allow excision of the second (adjacent) λ unit because this unit would be flanked by only one cos site and a free cohesive end. Hence only half of the DNA would be usable because only alternate segments of DNA would be packageable (Figure 16-5). The solution to this apparent lack of economy is that a free cohesive end and an adjacent *cos* site are also sufficient for DNA cutting to occur and allow sequential packaging. Thus the Ter-cutting rule may be restated as follows: Ter-cutting requires two *cos* sites or one *cos* site and a free cohesive end on a single DNA molecule. An experiment depicted in Figure 16-6 shows that the free end must be the end near the A gene of λ. (The A and R genes are not significant. We have used these genes to name the two ends of λ DNA to avoid possible ambiguity in the terms left and right.) In this experiment, three types of λ DNA molecules were prepared in vitro by joining the cohesive ends of either two intact DNA molecules (I) or one intact molecule plus one fragment containing either the A cohesive end (II) or the R cohesive end (III). (The fragments were prepared by breaking intact molecules near the center and then separating the fragments bearing a particular end from one another.) The molecules also contained a mutation in the P gene, so λ DNA replication was not possible. Three cultures of E. coli were separately transfected with these fragments. There was no DNA replication in the infection, but transcription occurred that resulted in synthesis of heads, tails, and the elements of the Ter system. Phage would be produced if the Ter system could cut λ units from these hybrid molecules. As shown in Figure 16-6, phage were obtained when cells were transfected with the dimer or with the molecule containing a free A end, but no phage were obtained when the fragment containing only a free R end was used. This and other experiments show that the packaging of λ DNA from the concatemeric branch of the rolling circle is polarized and proceeds from the A end to the R end.

Cutting at the *cos* sites and packaging of λ DNA are coupled: The Ter system is virtually inactive unless the Ter proteins are components of an empty λ head. Thus, when a bacterium is infected with a λ mutant unable to make an intact phage head (for example, an E^- mutant, which fails to make the major head protein), the phage DNA is not cleaved at *cos* sites.

The Ter system was first identified by genetic analysis of tandem dilysogens

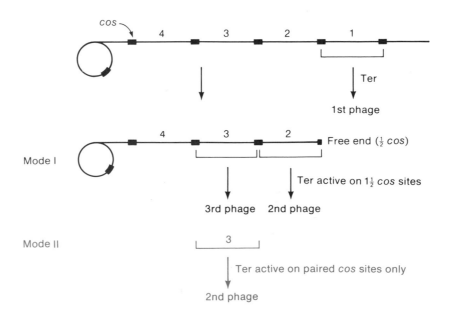

Figure 16-5. Two rules of packaging. In one mode (black), each λ unit is packaged. In the more limited mode (orange), alternate units are packaged. The more economical black mode is used by λ.

(cells with two adjacent prophages). (Properties of dilysogens are described in more detail in Chapter 17). Figure 16-7 shows a dilysogen that was constructed. The prophage on the left had the genotype $A^+ R^-$, and that on the right was $A^- R^+$. Both prophage were *int*⁻, which means that they lacked the ability to excise from the bacterial DNA by the normal route. In Chapter 5, it was pointed out that lysogens can be induced to produce phage; the details of the process were not given. When the dilysogen shown in Figure 16-7 was induced, phage were produced; however, all phage were $A^+ R^+$. When the prophage order was reversed, that is, an $A^- R^+$ prophage on the left and an $A^+ R^-$ prophage on the right, all phage produced were A^-R^-. Note that the phage produced always possessed the A allele to the right of the left *cos* site and the R allele to the left of the right *cos* site. These experiments demonstrated the existence of a genetic system that makes a cut between genes A and R, that is, within the *cos* sites.

Particle Assembly

Assembly of a completed λ phage particle requires both phage and bacterial genes. As in the study of T4 assembly, the pathway has been elucidated primarily by examining lysates of cells infected with various λ mutants. The basic observation is that there are four classes of phage mutations: those that (1) prevent head formation, (2) eliminate functional tails, (3) allow synthesis of heads and tails but not intact phage, and (4) prevent filling of the head with DNA. Isolation of *gro* mutants of *E. coli*, in which wild-type λ fails to produce heads, identified a role of a bacterial protein. The observation that phage production is restored to an infected *gro*⁻ cell by the presence of certain mutations in the λ E gene (which encodes the major head protein) indicates that the E protein and the bacterial Gro protein (GroE) interact. Complementation of *groE* mutations shows that there are two bacterial genes, *groES* and *groEL*. Certain *groES* mutations are reverted by mutations in the *groEL* gene, indicating that the GroEL and GroES proteins interact, and the GroEL and GroES proteins can be purified as a complex.

Figure 16-8 shows various stages of assembly that have been worked out, with details omitted. The process begins with an aggregation of many copies of

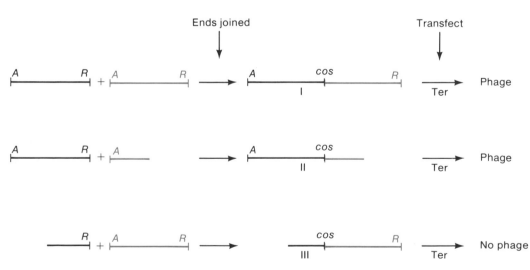

Figure 16-6. This experiment shows that one *cos* site and a free *A*-gene cohesive end are sufficient for activity of the Ter system. The cohesive ends are depicted by a vertical line. All molecules contained a mutation in the λ *P* gene to prevent DNA replication. Transfection refers to infection of a competent bacterium with phage λ DNA.

Figure 16-7. Phage produced by *int⁻* tandem dilysogens by terminase cutting at *cos* sites (dots). Both prophages are *int⁻*.

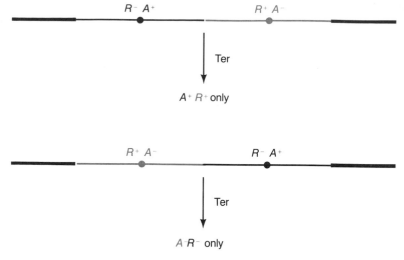

Figure 16-8. Diagram showing the pathway for assembly of phage λ. Most steps are fairly well understood except for the condensation of DNA into the phage head, the mechanism of which remains obscure.

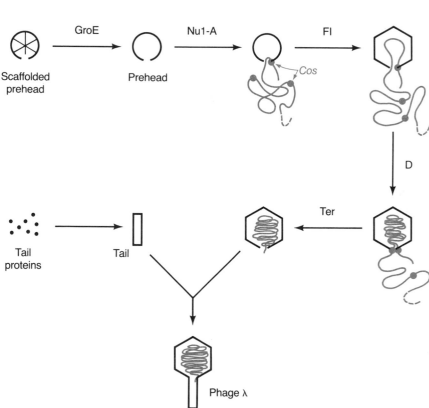

four-head proteins to form a scaffolded prehead, that is, a sphere with an internal supporting structure. This type of structure is the first assembly product in many phage systems. The GroESL complex then binds to the scaffold prehead, and the scaffolding is removed by the activity of bacterial proteases. (With some phages, the scaffolding dissociates spontaneously and can be reused.) Two phage-gene products, *Nul* and *A*, which also composed the Ter system, catalyze interaction of a short base sequence adjacent to one *cos* site with a point on the head (the neck) that will later become the head-tail attachment region. λ DNA then folds into the phage head; after a small amount of DNA enters, a conformational change in the λ E protein occurs, which causes an expansion of the spherical particle and formation of an icosahedron. The λ FI protein plays some role in this process. A small number of λ D-protein molecules are inserted into the head; at the same time, in an unknown way, λ DNA fills the head. Head-filling terminates when the next *cos* site is reached, at which point the Ter complex makes appropriate pairs of staggered cuts, thereby forming the single-stranded cohesive termini and releasing the unpackaged DNA from the filled head. A few bases at the end of one of the single-stranded termini of the λ DNA project from the head. In the meantime, tails have been assembled from several tail proteins. The tail is terminated by a head-tail connector protein, and the completed tail is bound to the head via both the short piece of single-stranded DNA and the λ head proteins in the neck. The free single-stranded terminus of the released DNA is available for binding to the neck of a second prehead, and so on. The GroESL protein complex required for the first step of λ head assembly is a member of a class of proteins called chaparonins that catalyze the folding of newly synthesized proteins. These chaparonin proteins are essential for growth of *E. coli*, and similiar proteins (called Hsp70s) are found throughout biology. GroESL was the first Hsp70 discovered and this was due to the role of the protein in phage assembly.

RECOMBINATION IN THE λ LIFE CYCLE

λ can engage in genetic recombination in a *recA*⁻ cell because λ encodes its own recombination system. A search for recombination genes began by seeking recombination-deficient mutants. These mutants were found by the following genetic test. A *recA*⁻ bacterial strain was prepared that contained a small segment of λ DNA that included gene *J*, which encodes the tail fiber protein. The *J* gene carried a host-range (*h*) type mutation enabling the phage to plate on a λ-resistant bacterium lacking the λ receptor in the outer membrane (*lamB*). We denote this strain *recA*⁻(λ *h*). This bacterium was infected with wild-type λ, and some phage were produced that carried the *h* marker from the bacterium and hence could plate on λʳ cells. Because the cell was *recA*⁻, the *h* marker must have been acquired by an exchange mediated by the λ recombination system. This observation was made with a spot test. Soft agar was seeded with a 1:1 mixture of the *recA*⁻ (λ *h*) cells and λʳ cells. A drop containing λ was spotted on the agar, and the plate was incubated. A completely clear spot developed because the wild-type λ lysed the *recA*⁻ (λ *h*) strain and the *h* recombinants lysed the λʳ cells. When mutagenized λ was used in this spot test, occasional turbid spots were found. These were formed by phage mutants that could not acquire the *h* marker and hence could only lyse one of the indicator cells, the *recA*⁻(λ*h*) strain. The mutant phage were called *red*, for recombination deficiency. Proof that the *red* locus is involved in genetic recombination came from combining a *red*⁻ mutation with one genetic marker in one phage and another marker in a second phage and showing that recombination could not occur between the markers. For example, in a mixed infection with λ*red*⁻ *A*⁻ and λ *red*⁻ *G*⁻, no *A*⁺*G*⁺ recombinants formed. Complementation tests between various *red*⁻ mutants showed that the *red* locus contains two genes called *bet* and *exo*. The *exo* gene encodes an exonuclease; the function of the *bet* gene is unknown. Many physical and biochemical experiments have been carried out with purified Bet and Exo proteins, but to date the mechanism of λ-mediated recombination is not fully understood.

An additional phenotype of Red⁻ mutants is their reduced burst size—roughly 20% that of a Red⁺ phage. An understanding of this phenomenon has come from physical studies in which intracellular DNA from infected cells was examined by electron microscopy. As the λ life cycle proceeds, θ replication halts, and rolling circle replication predominates. By the middle part of the life cycle, only a few rolling circles exist, which generate long concatemers from which λ DNA molecules are cut; in addition, there is a rather large amount of circular λ DNA. Because these have only a single *cos* site, they cannot be packaged and would be wasted were it not for the activity of the Red system: Red-mediated recombination converts most of these monomeric circles to dimeric circles, which have two *cos* sites, thus from each dimeric circle, a single λ DNA molecule can be cut out and packaged. Hence the Red system serves as a scavenger, avoiding wastage of the many circular DNA molecules that are generated early in the life cycle to serve as transcription templates.

In Chapter 14, it was pointed out that rolling circle replication is inhibited by the *E. coli* RecBCD protein and that λ has a gene, *gam*, whose product inactivates the RecBCD enzyme. A *gam⁻* mutant, similar to a *red⁻* mutant, has a decreased burst size because it cannot use the rolling circle pathway to produce packageable λ DNA (Table 16-2). A *red⁻ gam⁻* phage is even more defective because both pathways—the rolling circle pathway and the recombinant dimer pathway—are absent. A few phage, however, are still formed from dimers produced by the *E. coli* Rec system. As might be expected, λ *red⁻ gam⁻* cannot grow at all on a *recA⁻* cell because there is no pathway for the production of DNA molecules with two *cos* sites. Instead θ replication continues, and large numbers of nonpackageable λ circles are produced.

An interesting event occurs when λ *red⁻ gam⁻* infects a RecA⁺ cell. As just mentioned, the infection is quite defective, with a burst size of about four. Consequently, the plaques formed are small. With continual growth of this double mutant in a Rec⁺ culture, however, phage mutants that produce large plaques begin to accumulate. These mutant phage have acquired a third mutation, called *chi*; this mutation generates a DNA sequence that stimulates recombination mediated by the bacterial Rec system. Because of the increased Rec-mediated recombination, these *red⁻ gam⁻ chi* mutants produce a normal burst size, which accounts for the large plaque size. Chi mutants have attracted considerable attention among researchers who study the mechanism of Rec-mediated recombination because roughly one *chi* sequence exists in *E. coli* for every 4000 bp.

NONESSENTIAL GENES

When a phage is sedimented to equilibrium in CsCl solutions, it comes to rest at a density that is determined by the ratio of DNA (density 1.7 g/ml) and protein (density 1.3 g/ml). For λ, which is 50% DNA by weight, the density is roughly

Table 16-2   Plaque formation and maturation systems active when various λ mutants infect bacterial *rec* mutants

Phage mutant	Bacterial genotype	Plaque formation	Maturation system active	
			Rolling circle	Dimer formation
red⁺gam⁺	*recA⁺*	+	+	+
red⁺gam⁺	*recA⁻*	+	+	+
red⁻gam⁺	*recA⁺*	+ (small)	+	−
red⁻gam⁺	*recA⁻*	+ (small)	+	−
red⁺gam⁻	*recA⁺*	+ (small)	−	+
red⁺gam⁻	*recA⁻*	+ (small)	−	+
red⁻gam⁻	*recA⁺*	+ (very small)	−	Few (by Rec)
red⁻gam⁻	*recA⁻*	−	−	−
red⁻gam⁻	*recA⁻recBC⁻*	+ (small)	+	−
red⁻gam⁻chi	*recA⁺*	+	−	+ (by Rec)
red⁻gam⁻chi	*recA⁻*	−	−	−

1.5. If a centrifuge tube containing λ phage at equilibrium in CsCl, however, is fractionated by collecting successive drops, which are formed from successive layers of the tube, and each drop is plated, about 0.001% of the phage are present in a band at a slightly lower density, corresponding to a phage with 13% less DNA. Genetic analysis of these phage shows that they are deletions, extending from, but not including, the *J* gene to the attachment site *att*. (This deletion is shown in the lower part of the circular map in Figure 16-2.) The deleted region is called *b2* (b for buoyancy). Gel electrophoresis of the proteins synthesized by wild-type λ and the *b2* mutants indicates that seven proteins are missing in the deletion mutant. Clearly these proteins are not essential to λ, for if they were, the phage could not form plaques.

With phage T4, nonessential genes normally duplicate bacterial genes, and similar to the λ *red* gene, which is involved in one pathway to DNA maturation, mutations in some nonessential genes reduce the burst size. The *b2* mutants, however, have an apparently normal burst size. In fact, it has not been possible to detect any defect in the growth of a *b2* mutant. One could imagine that the *b2* genes might have a more subtle effect: For example, their absence might produce such a small decrease in the burst size that only on an evolutionary scale would λ *b+* (wild-type) outgrow λ *b2*. To test this, λ *b+* and λ *b2* were mixed in equal numbers and allowed to infect a bacterial culture. The phage progeny that were produced after several cycles of multiplication were taken, and another culture was infected with them. This process was continued through many cycles, and finally the ratio of *b+* and *b2* phage was measured (a simple plating test that is sensitive to the amount of DNA in a λ phage distinguishes the phage types). The expectation that there would be many more *b+* phage than *b2* phage was not realized. In fact, the *b2* phage outgrew the wild-type phage. A straightforward calculation shows that the difference in the multiplication rate is caused by the fact that it takes a slightly shorter time to replicate the deletion phage (because it has less DNA), and hence in one cycle of infection, a few more *b2* phage are made than wild-type phage. This phenomenon raises the following question: If λ *b+* is at a slight disadvantage compared with λ *b2*, how has the *b2* segment managed to survive evolutionary time? It seems likely that any deletion in a nonessential region would lead to ultimate replacement of the wild-type phage by the deletion phage.

It seems likely that the *b2* genes do have value in nature. In the laboratory, repeated propagation of λ samples, without occasional purification of a phage stock by preparing a new stock from a plaque, results in gradual accumulation of *b2* mutants. In contrast, λ has been isolated from the environment on various occasions and in several parts of the world (Europe, North America), and only λ *b+* has been isolated. This observation eliminates the possibility that most λ in nature are *b2* deletions and that it was just by chance that λ research began with a *b+* strain. This phenomenon, which is representative of numerous observations in the microbial world, is probably explained by the fact that laboratory conditions are never identical to natural conditions. In the laboratory, excellent growth media, controlled temperature and pH, and other idealized conditions are provided to growing microorganisms, for the purpose of obtaining reproducible data. In nature, however, conditions are far from ideal.

KEY TERMS

antitermination	*h* mutant
chaparone	head
cohesive ends	oL, oR, pL, pR
cos	prohead
cro	scaffold protein
exo, bet, gam	tail
GroEL and GroES	terminase

QUESTIONS AND PROBLEMS

1. What rules, which are different, determine the cutting sites in concatameric T4 and λ DNA?

2. Which transcripts of phage λ are made by an N^- mutant?

3. What functions are lacking in a λ Q^- mutant?

4. Certain λ specialized transducing particles carry the genes for synthesizing tryptophan, as shown here.

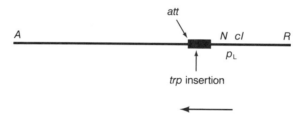

The tryptophan (*trp*) operon is transcribed in the direction indicated by the horizontal arrow: in the same direction as the phage mRNA that goes leftward from pL. There are several classes of aberrant phage that contain *trp* DNA (*trp*-transducing particles; see Chapter 18) that differ according to the size of the DNA molecule replaced by *E. coli trp* DNA. In each class, the size of the inserted bacterial DNA is roughly the same size as the phage DNA that is absent: The bacterial DNA contains the tryptophan synthetase gene, the insertion begins at *att* and moves to the right, and no essential phage genes are missing. These particles have the property that when they infect a bacterium lacking tryptophan synthetase, this enzyme is made.

   a. If the phage carries a point mutation in the λ N gene, no tryptophan synthetase is made. Why not? What part of the *trp* operon must be missing for this to be the case?

   b. Consider a *trp*-containing particle whose piece of bacterial DNA is so large that part of the N gene is replaced by bacterial DNA. Will any tryptophan enzymes be made? Explain briefly.

5. How many λ particles can be packaged from a dimer circle? A trimer circle?

6. If λ P^- infects *E. coli* at a low MOI, no phage are produced because there is no DNA replication. If MOI = 10 and either the bacterial or phage recombination systems are active, about one-third of the infected cells release one phage. Explain.

7. Suppose a λ phage has infected *E. coli* and is actively replicating its DNA. No DNA has been packaged because late mRNA has not yet been made. If a T4 phage is then added, will the λ phage still be made?

8. A large number of N^- phage are plated on a lawn of sup^0 bacteria. A few plaques arise, most of which are reversions in the N gene. One of the revertants, however, a small plaque-former, carries a point mutation that maps between genes P and Q. What do you think this mutation is? Why do you think the plaques are small?

9. The *E. coli* gene *dnaA* is needed to initiate a round of replication of the bacterial chromosome. Using a temperature-sensitive *dnaA* mutant (mutant above 40°C), suggest a simple plating test to determine whether the *dnaA⁻* gene product is used by λ. Do not suggest plating on the mutant bacteria because that will not work (why not?).

10. A mutant λ phage that makes a very tiny plaque on both sup^0 and sup^+ bacteria is isolated. It makes normal-sized plaques on both $groN\ sup^0$ and $groN\ sup^+$ bacteria. What kind of λ mutation would give this phenotype?

11. A λ mutant fails to form a plaque on sup^0, $supE$, and $supD$ bacteria but produces normal plaques on $supF$ cells. Why?

12. In a study of the size of the linear branch of the rolling circle in λ-infected cells, it is found that the branch increases in length until about 25 minutes after infection and then remains at a fairly constant length. What do you expect about the length of the branch if the cells are infected with a $\lambda\ E^-$ mutant?

13. $\lambda\ N^-$ mutants are unable to produce progeny phage when infecting sup^0 bacteria. Furthermore, the infected cells do not die. In fact, in such an infected culture and in the progeny of the infected cells, a large fraction of the cells contain $\lambda\ N^-$ DNA in plasmid form. Explain this phenomenon.

REFERENCES

*Arber, W. 1983. A beginner's guide to lambda biology. In *Lambda II* (R. W. Hendrix et al., eds.), p. 381. Cold Spring Harbor.

Friedman, D., and M. Gottesman. 1983. Lytic mode of lambda development. In *Lambda II* (R. W. Hendrix et al., eds.), p. 21. Cold Spring Harbor.

Furth, M., and S. Wickner. 1983. Lambda DNA replication. In *Lambda II* (R. W. Hendrix et al., eds.), p. 145. Cold Spring Harbor.

*Georgopoulos, C. 1992. The emergence of the chaperone machines. *Trends Biochem. Sci.*, 17, 295.

Hendrix, R. W., et al. (eds.). 1983. *Lambda II*. Cold Spring Harbor.

Hershey, A. D. (ed.). 1971. *The Bacteriophage Lambda*. Cold Spring Harbor

*Ptashne, M. 1992. *A Genetic Switch*, Second edition. Blackwell Scientific Publications.

*Resources for additional information.

Lysogeny

Infection of *Escherichia coli* by phage T4 or by λ in its lytic cycle results in lysis of the host cell and release of many progeny phage. In Chapter 5, an alternative life cycle, the lysogenic pathway, was described, in which phage are not produced. Instead a copy of phage DNA becomes quiescent, and the host cell survives and divides indefinitely. The host cell carrying quiescent phage DNA (a prophage) is called a lysogen. The prophage retains the potential to multiply: Many bacterial generations later, if the environmental conditions are right, the lysogenic state can be terminated and a lytic cycle started anew from the prophage. When this occurs, the host cell is killed, the prophage replicates, and progeny phage are formed and released. Except for the fact that the infection begins with an internal phage DNA molecule, this lytic pathway is no different from a lytic cycle initiated by infection with a phage particle. In this chapter, we examine lysogeny in some detail, referring primarily to λ. In particular, we emphasize the role that genetic analysis has played in understanding the lysogenic state and how it is achieved.

IMMUNITY AND REPRESSION

In Chapter 5, the appearance of plaques of temperate phage was described. Virulent phages, such as T4, form clear plaques because all bacteria in the region of the developing plaque are killed and lysed. Temperate phages, however, such as λ, form a plaque with a turbid center (see Figure 5-13). The turbidity is caused by cells that continue to grow and are not lysed by other phage in the plaque. A simple plating test demonstrates this resistance. A sterile wire is stabbed into the turbid center of a λ plaque and then streaked on a nutrient agar plate. The plate is incubated until colonies are formed. A drop of phage is spread in a line down the center of another plate, and the bacteria are plated in a line across the phage (Figure 17–5). Both sensitive and resistant cells grow on one side of the line, but the sensitive cells are killed after they cross the phage. The lysogens are "immune" to superinfecting phage, not simply phage-resistant bacterial mutants unable to adsorb phage. The cause of this resistance is examined in this section in some detail.

Causes of Immunity

The resistance of a λ lysogen to superinfection by λ is called immunity. Immunity is due to expression of a repressor in the lysogens. Phage λ contains a repressor-operator system (Figure 17-1). The repressor gene is called *cI*. The repressor protein binds to two operators, *oL* and *oR*, which are adjacent to and control two promoters, *pL* and *pR*. (The letters L and R mean leftward and rightward and refer to the direction of synthesis of two early mRNA molecules, when the genetic map is drawn as a linear map in a standard orientation; see Chapter 16, Figure 16-2 and 16-3).

In a lysogen, the *cI* repressor is synthesized continuously and in sufficient quantity that both operators, *oL* and *oR*, contain bound repressor molecules; thus *pL* and *pR* are unavailable to RNA polymerase, the transcription from the two early promoters is prevented. This block is sufficient to keep the prophage in an "off" state, enabling a lysogenic cell to grow indefinitely. When a nonlysogenic cell is infected by λ, the two operators of the incoming λ DNA molecule are unoccupied because phage repressor has not yet been made; thus transcription from *pL* and *pR* will occur. When a phage infects a lysogen, however, excess repressor molecules already present in the cytoplasm of the cell bind rapidly to the two operators in the infecting DNA molecule before RNA polymerase can bind to *pL* and *pR*. This operator-binding prevents the phage from proceeding into lytic development. The superinfecting DNA forms a covalent circle—phage gene products are not needed for circularization—but it cannot replicate because it lacks λ-specific replication-initiation proteins. The bacterium is unaffected by the presence of this DNA molecule and grows and divides normally, so the superinfecting DNA is progressively diluted out by bacterial multiplication.

Phage with mutations in the immunity system have been isolated. The two most important types are *cI* and *vir* mutants (Box 17-1):

1. A *cI⁻* mutant does not make a functional repressor. Hence *cI⁻* mutants make clear plaques because they always grow by the lytic pathway—that is, cI mutants are unable to form lysogens. These mutants are discussed in more detail shortly.
2. A *vir* (virulent) mutant carries mutations in *oL* or *oR* that prevent repressor binding. It also makes a clear plaque because it cannot establish repression; furthermore, it can superinfect and grow in a lysogen because it is insensitive to the repressor already present in the cell.

There are many temperate phages of *E. coli* other than λ. Two of these, which are closely related to λ (in that they have extensive regions of homology), are phages 21 and 434. Each of these phages has its own immune system—that is, its own repressor and repressor-specific operators. Thus, the 434 repressor cannot bind to a λ operator, and a λ repressor cannot bind to a 434 operator. Such a pair of phages is said to be "heteroimmune" with respect to one another. A temperate phage can form a plaque on a heteroimmune lysogen because the repressor made in the lysogen does not bind to the operator of the superinfecting phage. This is summarized in Table 17-1. The immunity region of the DNA, which includes the *cI* gene, operators, and promoters (and other elements to be described shortly), is given the genotypic symbol *imm*—for example, *immλ*, *imm21*, and *imm434*. Interesting hybrid phages have been created by crossing two heteroimmune phages and selecting a recombinant containing the immunity region of one phage and the remaining genes of the other heteroimmune phage. A prominent example is a λ-434 hybrid, which is

Figure 17-1. The repressor-operator system of λ showing the two early mRNA molecules. *cI* = Repressor gene; *p* = promoter; *O* = operator; L = left; R = right.

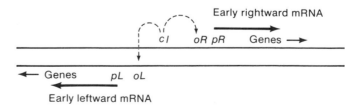

BOX 17-1. ANALOGOUS REGULATORY MUTANTS IN THE *lac* OPERON AND PHAGE LAMBDA

Two interactive groups of scientists working in the attic of the Institut Pasteur in Paris were responsible for elucidating the framework of both the *lac* operon and the phage λ regulatory systems. (In particular, François Jacob played major roles in elucidating both systems.) The formal logic of the two systems is the same if you consider lytic growth of λ as analogous to induction of the *lac* operon.

lac Mutant	λ Mutant	Explanation
lacI$^-$	λ*cI*$^-$	Constitutive mutants. No repressor is made, resulting in constitutive expression of *lacZYA* and lytic growth of λ. The wild-type allele is dominant in *trans* over the mutant allele.
lacO^c	λ*vir*	Constitutive mutants. Repressor cannot bind to mutant operator site, resulting in constitutive expression of *lac* and lytic growth of λ. *Cis* dominant.
lacI^S	λ*ind*$^-$	Noninducible mutants. Repressor is not inactivated by inducing conditions and thus always remains bound to the operator. The mutant allele is dominant in *trans* over the wild-type allele

genotypically designated *λimm434* (also occasionally *λ434hy*); this phage is shown in Figure 17-2.

Clear-Plaque Mutants of λ

Several types of λ mutants that produce clear plaques have been observed. The majority fall into three complementation groups, termed *cI*, *cII*, and *cIII*. Because turbidity results from immune cells growing in a plaque, these mutants must be defective in the immunity system. If a culture is infected with any one of these mutants, using conditions that would produce a high frequency of lysogenization with wild-type λ, most of the infected cells are killed. Rare (10^{-6}) survivors of *cI* infections are invariably λ-resistant; that is, a *cI* mutant always kills

Table 17-1 Ability of different phages to form plaques on homoimmune and heteroimmune lysogens

Superinfecting phage	Lysogen		
	B(λ)	B(434)	B(21)
λ	−	+	+
434	+	−	+
21	+	+	−

Note: The notation B(P) denotes a bacterium B lysogenic for phage P . + = forms a plaque; − = does not form a plaque.

a cell to which it adsorbs, if no repressor is present. Some cells, however, survive infections with *cII* mutants (10^{-5}) and *cIII* mutants (10^{-3}) and are stable lysogens. Another difference between these mutants can be seen in a mixed infection with any of the mutants and a wild-type λ, or between pairs of mutants, as shown in Table 17-2. The results indicate that *cI* mutants cannot produce a stable lysogen, but the rare lysogens formed by *cII* and *cIII* mutants behave as normal lysogens. These results have led to the conclusion that the *cI*-gene product is needed to maintain lysogeny, whereas the cII and cIII proteins are needed only to establish lysogeny. Thus is a mixed infection, the cII and cIII proteins can be supplied by a c^+ phage, and either the wild-type or the mutant phage can become a prophage with equal probability. Although λc^+ can prevent the *cI⁻* phage from killing the cell a single *cI⁻* prophage cannot be maintained. Biochemical experiments have shown that the *cI* gene encodes the repressor and the *cII* gene product is needed to activate transcription of the *cI* gene. The cIII protein is needed to stabilize the labile cIII protein. In a small fraction of cells infected with λ*cIII⁻*, the labile cII protein survives long enough for establishment of repression, which accounts for the fact that rare survivors of *cIII⁻* infections are lysogens. The biochemistry of this phenomenon is discussed in the next section.

The *vir* mutant described makes clear plaques. However, *vir* mutants, which require mutations in both operators, are not usually isolated as clear mutants on sensitive cells because the probability of finding a multiple mutant is quite low. These mutants are obtained by plating mutagenized λ on a lysogen and isolating phage from the very rare (10^{-8} or less) plaques that form.

λ *cI* Repressor Is Transcribed from Two Promoters

To establish lysogeny, repressor molecules must reach a certain concentration in the cell cytosol and must be synthesized more rapidly than the number of operator sites increases due to DNA replication of the phage—that is, the number of repressor molecules must be greater than or equal to the number of repressor

Figure 17-2. Formation of λ *imm434*. X indicates crossovers.

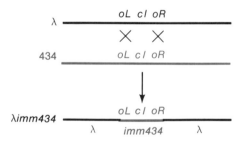

Table 17-2 Complementation of clear plaque mutants of λ for lysogeny

Phages	Lysogenization	Prophage in lysogen
cI⁻	No	—
cII⁻	Very rare	*cII⁻*
cIII⁻	Rare	*cIII⁻*
c^+ + *cI⁻*	Yes	c^+ only
c^+ + *cII⁻*	Yes	c^+ or *cII⁻*
c^+ + *cIII⁻*	Yes	c^+ or *cIII⁻*
cI⁻ + *cII⁻*	Yes	*cII⁻*
cI⁻ + *cIII⁻*	Yes	*cIII⁻*
cII⁻ + *cIII⁻*	Yes	*cII⁻* or *cIII⁻*

binding sites on the DNA. If the number of operator sites exceeds the number of repressor molecules, transcription will not be repressed, thus phage development and cell death will occur. Therefore, establishment of lysogeny requires an initial burst of repressor synthesis just after infection. In contrast, a lysogen contains only a single λ DNA molecule (the prophage), so less *cI* repressor is needed to maintain repression. Clearly transcription of the *cI* gene must occur in a lysogen, but transcription need not be strong.

In λ, high-level and low-level synthesis of the repressor is achieved by the existence of two promoters (Figure 17-3): (1) the "establishment promoter," *pRE*, which functions in an infected cell and which requires activation by the cII protein, and (2) the "maintenance" promoter, *pRM*, which functions in a lysogen and is regulated by the repressor itself. Thus, when lysogenization occurs (that is, when the phage has made enough cII protein to activate *pRE*), there is a burst of synthesis of "establishment" mRNA, and the repressor made from it activates the *pRM* promoter. As the amount of repressor increases, transcription of the *cII* gene is repressed, and no more cII protein is made. The cII protein is short-lived, and thus its concentration soon drops too low to maintain *pRE* in an active state, and synthesis of *pRE* mRNA terminates. Synthesis of *pRM* mRNA continues throughout successive generations, and translation of this mRNA maintains a concentration of the repressor sufficient to repress the prophage. As we have already stated, less repressor is needed in a lysogen; thus, for the sake of efficiency, less should be made when transcription is occurring from *pRM*. Actually in the fully "on" state, the two promoters are equally active in binding RNA polymerase; however, less repressor is made from the transcript initiated at *pRM* because the *pRE* transcript is efficiently translated while the *pRM* transcript is translated at a 10 fold lower efficiency. Thus the amount of repressor made from the two mRNA molecules is determined by the efficiency of translation. A summary of the events in the sequence of repressor synthesis is given in Table 17-3.

Thus the repressor is needed to turn on transcription from *pRM*. When the repressor concentration is high, however, transcription from *pRM* does not occur because the repressor also negatively regulates its own transcription. This regulatory mechanism enables λ to maintain a fairly constant repressor concentration, which is advantageous for two reasons: (1) In nature, bacteria have continually varying growth rates owing to fluctuations in the supply of nutrients. Regulation ensures that the repressor concentration does not diminish so much

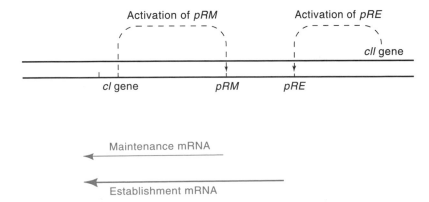

Figure 17-3. The *cI* gene of phage λ and adjacent regions, showing the two promoters for synthesis of *cI*-encoded mRNA. The heavy black lines are the two DNA strands. The thin and heavy orange arrows represent two mRNA molecules, both of which are transcribed from the lower DNA strand. The cII protein is translated from an mRNA molecule (not shown) transcribed rightward from the upper strand; this protein activates pRE, yielding the establishment mRNA (heavy red arrow). The cI protein activates pRM, yielding the maintenance mRNA.

that induction occurs spontaneously. (2) If unregulated, the repressor concentration might become so great that it would not be possible to induce the lysogen when needed.

Why should *pRM* be stimulated by the repressor? Because less repressor is made from *pRM* than from *pRE*, if λ only had a lysogenic pathway, synthesis of *pRM* mRNA could be constitutive. If a lytic cycle, however, is also to be possible, constitutive synthesis of *pRM* mRNA should be avoided—such synthesis might allow premature repression to occur, and this would prevent initiation of the lytic cycle.

To understand the mechanisms of these regulatory circuits, knowledge of the structure of the operators is required; this is presented in the next section.

The Operator: DNA Binding by Repressor and Cro

The operators *oL* and *oR* can be subdivided into six regions: *oL1*, *oL2*, *oL3*, *oR1*, *oR2*, and *oR3* (Figure 17-4). The base sequences of these regions are similar, and each region is capable of binding *cI* repressor, but the relative affinities of each isolated region for the repressor are not the same. The relative affinities of the operator sites for *cI* repressor are:

$$oL1 > oL2 > oL3 \quad \text{and} \quad oR1 > oR2 > oR3$$

Binding to *oL1* and *oL2* and to *oR1* and *oR2* is sequential and cooperative. Binding to *oR3* and *oL3* is not cooperative.

At low concentrations of repressor, *oL1* and *oR1* are occupied. Because the promoters *pL* and *pR* are adjacent to *oL1* and *oR1*, when repressor is bound to

Table 17-3   An outline of the sequence of events in cl repressor synthesis in the lysogenic pathway

1. Infection of cell
2. Transcription of rightward mRNA from *pR*
3. Activation of *pRE*
4. Transcription and translation of the cl repressor from *pRE* (high-level synthesis)
5. Activation of *pRM* by the cl repressor
6. Transcription and translation of the cl repressor from *pRM*
7. Turning off of *pR* by the cl protein
8. Degradation of the cll protein, so that *pRE* becomes inactive
9. Continued synthesis of the cl protein from *pRM* (low-level synthesis)

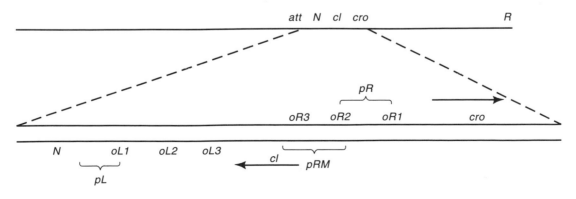

Figure 17-4.   A close look at the operator-promoter region of λ. The arrows denote the direction of transcription of the *cro* and *cl* genes.

these sites, RNA polymerase has reduced access to both promoters. The block is even more complete when $oL2$ and $oR2$ are also filled, as occurs in a lysogen. The promoter pRM is between $oR2$ and $oR3$, so if the repressor concentration is high, pRM is blocked; thus, when the concentration of cI repressor is high, it negatively regulates its own synthesis. Furthermore, RNA polymerase does not bind to pRM unless $oR1$ is occupied by repressor: Repressor bound to $oR1$ allows a protein-protein interaction with RNA polymerase, which significantly strengthens the binding of RNA polymerase to pRM.

Thus, if a cell infected by a λ phage is destined for lysogeny, the following events occur:

1. Transcription from pL and pR begins.
2. The cII protein is translated from the pR transcript (see Figure 16-3).
3. The cII protein enables RNA polymerase to transcribe from the pRE promoter. The cI repressor is made from the pRE transcript.
4. Repressor binds to $oL1$ and $oR1$, turning off transcription and hence synthesis of the cII protein. The cII protein is unstable, so transcription from pRE stops.
5. Occupation of $oR1$ and then $oR2$ causes pRM to be activated, so cI repressor continues to be synthesized from this promoter, although at a lower rate than from pRE.
6. The cI repressor continues to accumulate, so $oL2$ and $oR2$ become occupied, and repression of transcription from pL and pR becomes complete.
7. Transcription from pRM continues until the repressor concentration becomes so high that the protein binds to $oR3$. The activity of pRM is turned on and off to accommodate mild fluctuations in repressor concentration.

As we have described the system, there is no possibility for a lytic cycle. Entry into the lytic cycle requires the Cro protein, which is necessary for several features of the lytic cycle and to prevent synthesis of the cI repressor. The Cro protein is also a repressor that binds to the operators oL and oR. Its mode of action is based on its affinity for the operator sites; the affinity of Cro is opposite to that of the cI repressor:

$$oR3 > oR2 = oR1 \quad \text{and} \quad oL3 > oL2 > oL1$$

The sequence of events is as follows:

1. The cro gene is transcribed from pR (Figure 17-4).
2. After transcription of the cro gene, the cII gene, whose product is required to activate the pRE promoter, is transcribed.
3. Once the Cro protein is synthesized, it binds to $oR3$, preventing activation of pRM.
4. The concentration of the Cro protein increases, and $oR2$ and $oR1$ gradually become occupied by Cro. When these sites are filled, RNA polymerase loses access to pR, so synthesis of the Cro protein eventually stops.

Note that the Cro protein represses its own synthesis as well as repressing synthesis of the cII protein.

Recall that the lysogenic cycle requires leftward transcription from pRE and pL, and the lytic cycle requires extensive rightward transcription. With this in mind, we can summarize the competitive roles of the cI and Cro proteins. As shown in the discussion of Figure 17-4, cI and Cro engage in a mutually exclusive competition for the operator sites and thereby control the transcription events that involve oR (except for the earliest synthesis of N mRNA, oL is primarily involved in excision, as described later). Binding to $oR1$ and $oR2$ blocks pR, and binding to $oR3$ inhibits transcription from pRM. Cro and cI bind to these sites in a different order and hence constitute a bidirectional switch. When cI predominates, rightward transcription of *cro* from *pR* is off because *oR1* and *oR2* are occupied; leftward transcription of cI from *pRM* is autoregulated by cI (on, when there is little cI, and only *oR1* and *oR2* are occupied; off, when there is excess cI, and *oR3* is also occupied). When there is more Cro than cI, *oR3* is occupied first, so leftward transcription from *pRM* is always off; rightward transcription from *pR* continues unless the amount of Cro is quite high because binding of Cro to *oR3* is fairly weak. Thus, excess cI yields stable leftward transcription from *pRM* and the lysogenic pathway ensues, and a relative excess of Cro produces rightward transcription and the lytic pathway. What determines whether cI or Cro is in excess at the point in the life cycle when the choice must be made? Later we see that because the cII protein is responsible for beginning cI transcription, cII is the critical determinant of the decision between lysis and lysogeny.

LYSOGENY AND PROPHAGE INTEGRATION

Lysogeny predominates when cells in late log phase are infected with λ at a high multiplicity of infection (see Chapter 5). The result is establishment of immunity and integration of a prophage. In this section, we examine how lysogenization is detected and the genetic requirements for integration.

Tests for Lysogeny

When a culture is infected with λ under conditions that lead to efficient lysogeny, most of the cells survive and form colonies, How do you know if the colonies are lysogens or not? This is possible by a simple cross-streak test based on the immunity of lysogens to superinfection.

Figure 17-5 shows an example. A plate is prepared by streaking a thin line of wild-type λ phage across the agar surface. A second parallel streak is made with a heteroimmune phage such as λ*imm434*. A colony is then picked up with a sterile toothpick from a plate containing potential lysogens and drawn across the plate perpendicular to the phage streaks. The direction of streaking is such that the toothpick first passes through the line of λ and then through the line of *imm434* phage. The plate is then incubated until the bacteria in the streak show visible growth. If the colony is a nonlysogen, growth occurs only up to the first line of phage because at this point the bacteria are infected with the phage, these phage multiply in the infected cells, and their progeny lyse any remaining bacteria.

Figure 17-5. A test for lysogenization. The vertical pink bars are regions containing either *imm*1 or *imm434* phages. The horizontal bands represent growth of three types of cells that have been streaked from left to right.

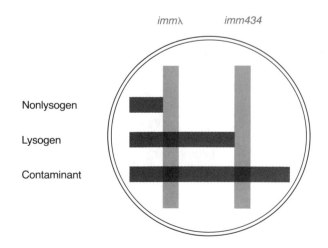

The pattern formed by a colony of a lysogen is quite different. In this case, the *immλ* phage cannot infect the immune cells; thus growth continues past this line of phage. When the cells reach the *imm434* streak, however, λ immunity cannot protect the lysogen from the *imm4343* phage, so cells are lysed downstream from that phage line. The reason for including the *imm434* phage in the test is to eliminate the possibility of scoring either a λ-resistant (adsorption-defective) mutant as a lysogen or a contaminating bacterium (something that fell onto the plate from the air, for example). Both the λ^r cell and the contaminant are resistant to λ; however, they are also resistant to *λimm434*, and hence the bacteria grow through both streaks. Thus, a λ lysogen in unambiguously identified by the pattern shown in Figure 17-5—resistance to *λimmλ* and lysis by *λimm434*.

Prophage Integration

When λ DNA integrates, it is inserted at a preferred position in the *E. coli* chromosome. This site is between the *gal* operon and the *bio* (biotin) operon; it is called the λ attachment site and designated *att*. Integration at a unique site was discovered by genetic mapping experiments:

1. Zygotic induction in a cross between a lysogenic Hfr strain and a nonlysogenic female (see Chapter 14) placed the prophage just after *gal* in the clockwise direction, when the *E. coli* map is drawn in a standard way.
2. Linkage analysis carried out by phage P1 transduction (discussed in Chapter 18) showed that in a lysogen, λ genes are linked both to the *gal* and *bio* loci.
3. The genetic linkage of *gal* and *bio* indicated that the map distance between *gal* and *bio* is considerably greater in a lysogen than in a nonlysogen.

A significant feature of the insertion mechanism was derived from the work of Campbell in the early 1960s, who obtained a λ prophage map by three-factor crosses with two λ genes and a *gal* marker. To do so, he used the specialized transducing phage *λgal*, described in Chapter 18. Mapping experiments with these particles showed that the *gal* genes are located within the central region of the λ map, indicating that the prophage map is a permutation of the map obtained by standard phage crosses (Figure 17-6). This permutation was explained by a mechanism of integration shown earlier in Figure 5-12. The essence of the mechanism, which is frequently called the Campbell model, is circularization of λ DNA followed by a single crossover between the phage and host DNA—precisely between the bacterial DNA attachment site and another attachment site that is located near the center of the linear phage DNA map. A phage protein, integrase (encoded by the *int* gene), recognizes the phase and bacterial attachment sites and catalyzes the physical breakage and rejoining of the phage and bacterial DNA.

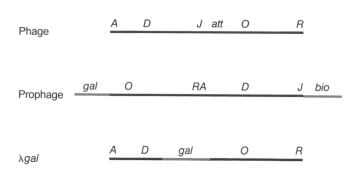

Figure 17-6. The order of genes on the DNA molecule in the phage head, in the prophage, and in *λgal*. λ genes are selected arbitrarily as reference points. Orange segments represent bacterial DNA.

The result is integration of the λ DNA into the bacterial DNA. Because the phage exchange site is not at the point where the ends of the DNA join to form a circle, the linear phage map is turned inside out, with the ends of the phage map in the middle of the prophage map.

Campbell's model was quickly accepted. It was clear, however, that the experiments that led to it did not rigorously prove that the prophage was inserted linearly into the bacterial DNA and did not exclude such unusual possibilities as the prophage being a branch to the *E. coli* chromosome. (The reason for the uncertainty was that *bio* transducing particles had not yet been discovered, so it was not possible to do a three-factor cross that included a *bio* marker.) A variety of experiments, however, both genetic and physical, proved the point. One significant genetic experiment made use of the fact that a segment of the *E. coli* chromosome was especially prone to the development of large deletions: the region covering the *trp* (tryptophan biosynthesis) and *tonB* (sensitivity to phage T1) genes. A λ variant was prepared by crossing λ with another temperate phage called φ80 that integrates at a site adjacent to *tonB trp*. Hybrid phages were obtained (much like the λ-434 hybrids used in immunity studies) that carried the attachment site of φ80 and much of the remainder of λ. This hybrid phage, λ-φ80, inserted at a site adjacent to the *tonB* locus, yielding the gene order λ-φ80 *tonB trp* (Figure 17-7). A lysogen was prepared, and cells were plated on agar that had previously been spread with phage T1. Only T1^r colonies formed. These were tested for the ability to grow on minimal agar lacking tryptophan to obtain *tonB trp* double mutants. These double mutants were found with fairly high frequency, suggesting that they were deletions that included both genes. To confirm that they were deletions, about 10^8 cells of each mutant were replated on agar lacking tryptophan. Mutants that showed no Trp$^+$ revertants were considered to be deletions. These were then tested for phage production to obtain cells with deletions that penetrated the λ-φ80 prophage. These deletion strains were then infected with a set of λ mutants that were defective in one of several genes that spanned the prophage map (e.g., λ A$^-$, D$^-$, Z$^-$, N$^-$, R$^-$). Production of a wild-type λ by recombination with prophage genes would indicate that a particular λ gene was present in the prophage. In this way, the prophage genes that remained in each deletion strain were determined. These phage-prophage crosses yielded a map that showed the order in which genes could be deleted by a deletion extending into the prophage from *tonB*. If a prophage were a branch on the *E. coli* chromosome, one would expect that either all prophage genes would be present or all would be absent. If a prophage were linearly inserted into the chromosome, deletions should remove contiguous genes starting from the *tonB* side of the prophage. As Figure 17-7 shows, the latter was observed.

A simple physical experiment confirmed that λ is linearly inserted into bac-

Figure 17-7. Various deletions entering the λ segment of a λ-φ80 hybrid prophage. The deletions remove the bacterial genes *tonB* and *trp*. The prophage is black, with the φ80 segment heavy. Bacterial DNA is orange.

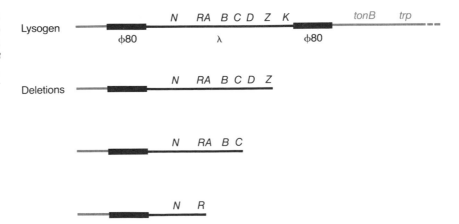

terial DNA. In this experiment, a bacterial strain was used in which the λ attachment site was deleted (by virtue of the large *gal-bio* deletion). This strain also contained a plasmid that carried the attachment site F'*gal attλ bio*. This strain was lysogenized, and several λ lysogens were obtained. To confirm that the prophage was in the plasmid, the strains were mated with nonlysogenic F⁻ cells; zygotic induction indicated a plasmid location. Plasmid DNA was then isolated from both the lysogens and the original nonlysogen and then tested by physical techniques for circularity and size. It was found that the F'(λ) plasmids were intact circles, similar to the nonlysogenic plasmid, and furthermore that the molecular weight was one λ unit greater than that of the nonlysogenic plasmid. Finally, heteroduplexes were prepared between F'*gal attλ bio* and F'*gal* (λ) *bio* and examined by electron microscopy (using the technique illustrated in Figures 2-9 and 2-10). Molecules consisting of a double-stranded circle the size of F' *gal attλ bio* with a single-stranded loop of λ length were seen, as expected of a plasmid with an inserted prophage.

Insertion-Defective Mutants

Stable lysogens require prophage insertion as well as repression because if λ does not insert into the chromosome, it will exist as an unstable circular plasmid, and λ-free cells will continually segregate out during cell division. In an effort to understand the insertion mechanism, λ mutants were sought that were defective in insertion. The basis of the isolation procedure was the expectation that an insertion-defective phage would still be able to establish repression. That is, it was predicted that infection of cells by an insertion-defective mutant, using conditions that normally lead to a strong lysogenic response, would result in survival of the cells, but the resulting cells would not be stable lysogens. Thus plaques formed by these mutants would still be turbid, but perhaps not quite as turbid as those produced by wild-type phage. The mutant hunt was carried out in the following way.

A λ sample was mutagenized and then plated. Many thousands of plaques were tested by stabbing a sterile needle first into the turbid center of a plaque and then onto the surface of a color-indicator (EMB) plate. The plate did not contain a carbohydrate as a carbon source, so cells on this medium should produce white colonies. Indeed, most of the colonies obtained from the plaques were white. A small fraction, however, were small purple colonies with ragged edges. Because each plaque contains both phage and bacteria and because a lysogen is immune to superinfection, stable lysogenic cells obtained from the plaque grow normally. If prophage integration had not occurred, however, the cells in the plaque would be a mixture of immune cells lacking an integrated prophage and sensitive cells that segregated out by division from an immune cell without an integrated prophage. Lysis of the sensitive cells would occur continually on the color-indicator plate because phage carried with the bacteria would infect these cells. This would cause the ragged edges and small size of the region of growth. Furthermore, phage-infected cells cause a decrease in the surrounding pH so the colony would be purple. Indeed some purple patches were found. Phage were then isolated from the plaques whose central cells produced purple growth, and phage lysates were prepared. These were tested for their ability to lysogenize in liquid medium. Streak tests of the sort depicted in Figure 17-5 showed that surviving colonies were lysed by both *immλ* and *imm434* phages, indicating that none of the survivors were stable lysogens.

Two types of phage mutants are unable to lysogenize—those lacking an active integration enzyme (or enzymes) and those with a mutant *att*. These classes were distinguished by the following complementation test. A culture was lysogenized by a mixture of equal multiplicities of infection of the λ mutant and λ*imm434*, the latter being a "helper" phage to provide the missing function. Conditions were used such that repression of both phages would be established. Survivors were tested for the presence of both the *imm*λ and the *imm434* phages. Most of the lysogenization-deficient mutants were able to lysogenize when complemented by the *imm434* phage, so their mutation resided in a protein. A small fraction could not be complemented, so their mutation presumably was in *att*; actually most of these proved to be deletions of all or part of the attachment site. The mutations that affected a protein were then tested against one another for complementation. All were found to be in a single complementation group, and hence in one gene. Mapping studies placed all mutations in a single region of the λ map, adjacent to *att*. The gene defined by these mutations was called *int*, for integrase.

Physical experiments confirmed the fact that *int⁻* mutants fail to integrate into the chromosome. In these experiments, cells in which the only *att*λ is in an F' plasmid were used. Such cells were infected with λ under conditions that lead to a lysogenic response, and the F' circle was observed to disappear and be replaced by a circle that was larger by one λ unit. When the cells were infected by an *int⁻* mutant, however, the F' was unaltered: It remained circular, and its size did not change.

Several important questions were raised by the discovery of the *int* gene: (1) Does integrase act at the phage and bacterial attachment sites, and how does the recombination occur? (2) Does integrase also participate in prophage excision? (3) Can integrase alone catalyze excision? These questions are answered in the next two sections.

Int-Promoted Recombination

In Chapter 16, recombination in mixed infection by genetically marked λ was described, and it was pointed out that this recombination is catalyzed by the λ *red* genes. In *rec*A⁻ hosts, however, recombination between some markers can occur even if the participating phages are *red⁻*. For example, with *red⁻* parents, no wild-type recombinants are observed in a cross between λ A⁻ K⁺ × λ A⁺ K⁻, but in the cross λ K⁺ P⁻ × λ K⁻ P⁺, wild-type recombinants arise with a frequency of about 8%. The difference between these crosses is the position of the markers with respect to *att*. A large number of crosses with different mutants showed that if all phages are *red⁻*, recombination occurs whenever the two markers are on opposite sides of *att* (e.g., *J* and *P*), but not if they are on the same side of *att* (like *A* and *J*), as shown in Figure 17-8. If the parents are *red⁻ int⁻*, no recombinants form with any pair of markers. The recombination observed between *red⁻*

Figure 17-8. Recombination between λ *red⁻* phages. Recombination occurs only between *att* sites (solid dots) by Int-promoted recombination.

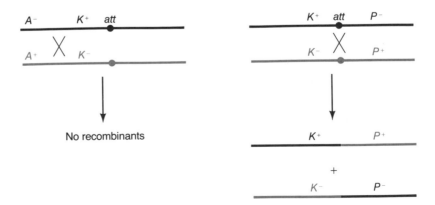

phage is mediated by integrase and is called Int-promoted recombination. Recombination occurs within the *att* site: If either of the phages contained a point mutation in *att*, no recombinants arose.

Physical experiments in which density-labeled *red⁻* phage were crossed (see Figure 15-3) showed the conservative nature of the exchange. That is, Int-promoted recombinants had a density expected for an exchange at *att* (Figure 17-9). (The symbols for the wild-type and recombinant *att* sites are shown in Figure 5-12.)

In Chapter 18, we describe unusual λ particles that arise by aberrant excision. If excisional cuts are made to the left of the *gal* gene and hence of the left prophage attachment site (BOP') and also left of the right prophage attachment site (POB'), a DNA molecule would result that carries *gal* and has the attachment site of BOP'. Similar aberrant excision with cuts to the right of attachment sites could generate particles containing *bio* and POB' (see Figure 18-6 for a drawing of these excision events). These particles are called *gal⁻* and *bio⁻* transducing particles. In a mixed infection with λ*gal* and λ*bio* particles, both of which are *red⁻*, Int-promoted recombination can generate both a normal phage (lacking bacterial genes and with the phage attachment site POP') and a phage (λ*gal bio*) carrying both the *gal* and *bio* genes and the bacterial attachment site BOB' (Figure 17-10).

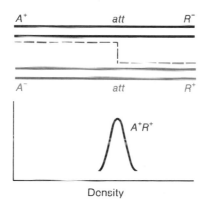

Figure 17-9. An experiment similar to that in Figure 15-3, which demonstrated breakage of DNA molecules and re-union to form recombinants. In the upper part, heavy lines represent high-density (black) and low-density (orange) *E. coli* phage λ carrying the indicated markers. The phage are *red⁻* and the bacterium is *recA⁻*, so only the *int* gene is active in recombination. After mixed infection with the two phage, using conditions that prevent DNA replication, progeny phage are centrifuged to equilibrium in CsCl. In contrast with the data shown in Figure 15-3, in this experiment, a single narrow band is produced because breakage and re-union occur only at *att*.

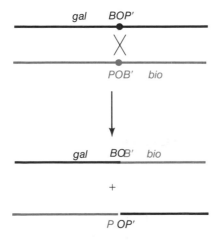

Figure 17-10. Int-promoted recombination between λ*gal* and λ*bio* phages (upper pair). The lower pair shows the two recombinants—λ*gal bio*, which carries BOB', and normal λ, which carries POP'.

Table 17-4 Recombination frequencies (percent) in Int-promoted crosses in which markers are on opposite sides of *att*.*

	λ (POP')	λgal (BOP')	λbio (POB')	λgal bio (BOB')
λ	9.7	28	15	10.3
λgal	–	4.8	10	0.07
λbio	–	–	n.t.†	n.t.
λgal bio	–	–	–	n.t.

*All phages are *red–* and the host is *recA–*.
†n.t. = not tested, because neither phage carries an *int* gene.

These four *red⁻* phages—λ *red*, λ *gal*, λ *bio*, and λ *gal bio*—were used in pairwise crosses to determine the frequency of Int-promoted recombination between different attachment sites. The rationale behind the experiment was the following: If the phage and bacterial attachment sites are identical, the prophage attachment sites would also be the same as these, and all four phage types would possess the same attachment site. In that case, the frequency of Int-promoted recombination should be the same in all crosses. If BOB' and POP' are different, however, you would expect each cross to yield a unique recombination frequency. Table 17-4 shows the results of these crosses: The recombination frequencies are different. Thus, the phage and bacterial attachment sites must also be different. Later experiments showed that the sequences of B, B', P, and P' are quite different, and the sequence of the central "O" region, often called the core sequence, is rich in AT pairs (Figure 17-11).

Integrase Reaction

Integrase has been purified, and the integration reaction can be carried out in vitro. Integrase has DNA-binding activity, binding strongly to POP'. It is a type I topoisomerase; that is, it can cause a break in one strand of a double helix, rotate one branch of the broken strand about the continuous strand, and then rejoin the ends. In addition to integrase, the reaction requires a host protein, integration host factor (IHF); this protein also binds to the attachment site and bends the DNA, allowing proper protein-protein and protein-DNA interactions for the recombination.

Integration of λ occurs by site-specific recombination between the *attP* (POP') and *attB* (BOB') sites (Figure 17-12). Int protein binds to the P site of *attP*, the B site of *attB* DNA, and several additional sites in the *attP* DNA to form a protein-DNA complex called an intasome. Int protein then nicks both the DNA on the 5' strand of the P site and the B site, forming a covalent phosphotyrosine bond between the 3' end of the nicked DNA and Int protein. The free 5'-OH group at the nicked DNA in the opposite *att* site attacks this phosphotyrosine bond, joining the DNA from the opposite strands and forming a Holiday junction. The DNA branch migrates across the core region to the right half of the inverted repeat, where the DNA is nicked on each 3' strand. The 3' strands are exchanged and the DNA is resealed, producing the recombinant *attL* (POB') and *attR* (BOP') sites (Figure 17-13).

Figure 17-11. Base sequence of the core of the λ *att* region.

```
- - - ┌──────────────┐ - - -
      │ GCTTTTTTATACTAA │
      │ CGAAAAAATATGATT │
- - - └──────────────┘ - - -
```

Coupling of Synthesis of Integrase and the *cI* Repressor

The synthesis of integrase is coupled to the synthesis of the *cI* repressor. This arrangement is efficient because integrase and the repressor are both needed in the lysogenic cycle and neither is needed in the lytic cycle. An outline of the regulatory pathway responsible for this coupling is shown in Figure 17-14. In the absence of a positive regulatory element (the product of the λ gene *cII*), the promoters for both the *int* and *cI* genes are not recognized by RNA polymerase. Shortly after infection, the *cII* protein is synthesized (see Chapter 16). If the concentration of the protein is high enough, the *cII* protein binds to sites near the promoters for the *cI* (*pRE*) and *int* (*pI*) and activates binding of RNA polymerase. RNA polymerase then transcribes both genes, and the gene products are made.

PROPHAGE EXCISION

A lysogen is generally quite stable and can replicate nearly indefinitely without release of phage. When the lysogenic bacterium becomes damaged, however, it is advantageous to the phage to initiate a lytic cycle. This process, which is called prophage induction, results from de-repression of phage genes and excision of the prophage.

The *xis* Gene

In an earlier section, it was pointed out that *int⁻* lysogens could be produced by mixed infection with an *int⁻* phage and a helper phage. These lysogens cannot be induced to produce phage, which indicates that integrase is also necessary for excision. This observation raises the question whether integrase alone carries out excision.

If integrase alone could catalyze excision, we may ask why it does not excise the prophage shortly after integration occurs. Indeed, excision would seem to be the preferred reaction because kinetically two attachment sites in the same DNA molecule (that is, the host chromosome) ought to interact with each other more rapidly than two attachment sites in different DNA molecules (phage and bacterial DNA). A simple explanation could be that the insertion reaction BOB' × POP' is efficient, but the excision reaction BOP' × POB' is not. True excision, however, does occur with 100% efficiency (that is, all cells in a lysogenic population produce phage when induced), so another explanation is necessary. Instead, in addition to Int another protein is required for excision. The first indication that another protein is required was the discovery of a mutation that prevents excision but not integration. This mutation was isolated in the following way, making use of a valuable mutation in the *cI* gene.

Of the many mutations known in the *cI* gene, one temperature-sensitive mutation, *cI857*, has been of particular value. This mutation results in the formation of a *cI* repressor that is inactive at 42°C but active at 30°C; thus phage carrying *cI857* make clear plaques at 42°C and turbid plaques at 30°C. Furthermore, transcription of the prophage in a lysogen can be initiated merely by raising the temperature of the growth medium to 42°C. Another useful feature of the mutation, which is quite unusual for a temperature-sensitive mutation, is that the temperature-induced alteration in protein structure is fully and immediately reversible; that is, if the temperature is lowered after heating, the *cI* protein renatures, and repression is reestablished. This property gives rise to a phenomenon called prophage curing. If a bacterium lysogenic for λ*cI857* is heated to 42°C and this high temperature is maintained, after 45 minutes (the length of the phage

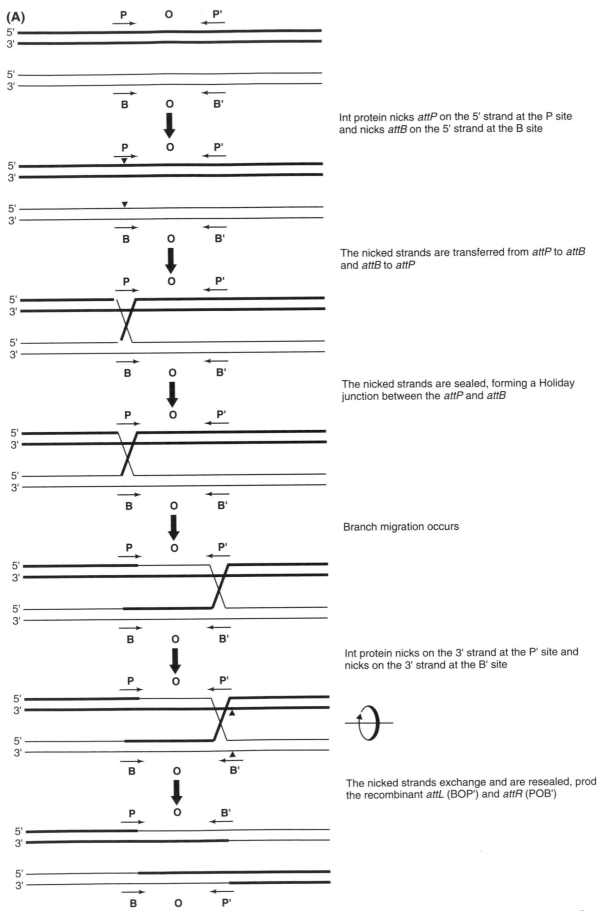

Figure 17-12. Integration of λ by site-specific recombination. (a) Molecular mechanisms of DNA exchange between *attP* (POP') and *attB* (BOB') (*continued*).

(B)

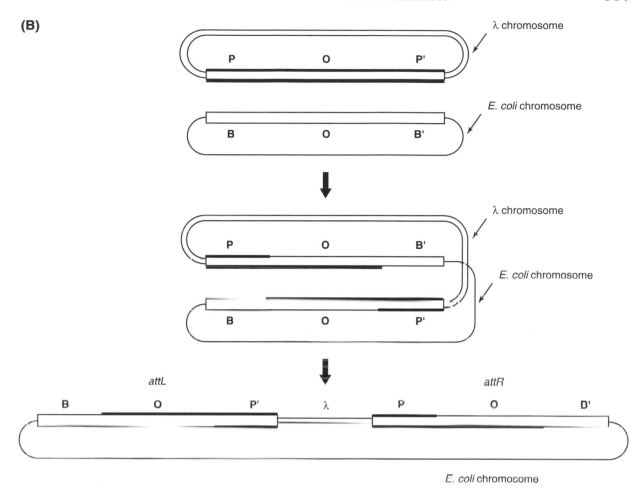

Figure 17-12. (b) Structure of the circular λ and *E. coli* chromosomes showing how the site-specific recombination results in integration of λ into the chromosome.

life cycle of an infecting λ), the cells lyse, and phage are produced. If the cells, however, are instead heated for 5 minutes at 42°C and then the temperature is dropped to 30°C, the cells survive. If these cells are allowed to grow for several generations and then plated, a large number of the colonies are nonlysogenic—they are cured of the prophage. Thus, 5 minutes at 42°C is sufficient to activate the excision system but not sufficient to establish a stable lytic cycle. The explanation is that restoration to 30°C reestablishes repression but cannot turn off the excision process if the mRNA encoding the excision system has been made.

Curing served as the basis for isolation of mutations in the excision system. A culture of a λ *cI857* lysogen was mutagenized and then plated at 30°C. After colonies formed, they were replica-plated onto a fresh plate and incubated for several generations at 30°C to initiate growth of the cells. The replica plates were then rapidly warmed to 42°C for 5 minutes and then cooled to 30°C. After growth for a few generations to allow segregation of cured cells, the plates were returned to 42°C and allowed to grow until colonies were visible. A lysogen cannot produce colonies at 42°C because of the presence of the *cI857* mutation, which leads to induction and lysis. Most of the cells transferred by replica plating, however, formed visible colonies because the cured cells produced by the temperature cycling were able to grow at 42°C. In a few cases, transferred cells did not grow: These blank spots represented colonies on the master plate that did not segregate cured cells. These colonies were removed from the master

Figure 17-13. The exchange that occurs in the integrase reaction showing the approximate positions of the strand breaks (arrows). The lowercase letters are bases in the flanking sequences B, B', P, or P'; the capital letters are in the O region.

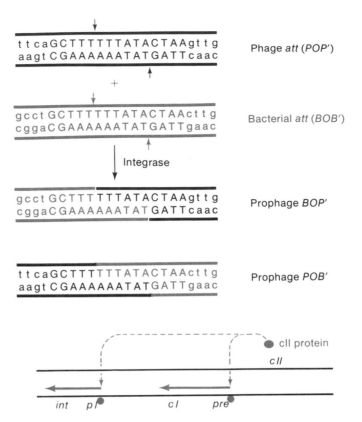

Figure 17-13. The exchange that occurs in the integrase reaction showing the approximate positions of the strand breaks (arrows). The lowercase letters are bases in the flanking sequences B, B', P, or P'; the capital letters are in the O region.

Figure 17-14. The role of the λ cII protein in stimulating synthesis of *cl* and *int* mRNA from pRE and pI. The mRNA molecules are drawn as orange arrows that indicate the direction of RNA synthesis. The arrows are located nearer the DNA strand that is copied. Dashed arrows point to the sites of binding of the cII protein (orange dots).

plate and tested further. First, it was shown that heating these cells did not lead to phage production. Infection of cultures derived from these mutants with wild-type λ*imm434*, however, which could provide all excision functions, and heating to 42°C, allowed *imm*λ progeny phage to develop from the prophage. Phage with the *imm*λ character were selected from these lysates and tested for the ability to lysogenize. Two types of mutants were present. Those that could not lysogenize were *int* mutations, whose recovery would be expected because integrase is clearly involved in excision. Some of these mutants lysogenized normally, but the lysogens could not be induced to produce progeny phage; that is, heating the lysogens resulted in cell death without phage production. The mutations carried by these phage were termed *xis* (pronounced "excise"). Complementation tests showed that all the *xis* mutations were in a single gene, and mapping experiments located the *xis* gene immediately adjacent to the *int* gene, in accord with the usual observation in λ of clustering of genes with related functions.

The *xis* gene product, excisionase, has been purified. It forms a complex with integrase, allowing integrase to bind more efficiently to the prophage attachment site BOP'; once bound to this site, integrase can make cuts in the core sequence and reform the BOB' and POP' sites. Thus the Xis-Int complex reverses the integration reaction, causing excision of the prophage (see Figure 5-12). The reactions between all attachment sites can be written:

$$BOB' + POP' \xrightarrow[\text{Int + Xis}]{\text{Int}} BOP' + POB'$$

The Xis-Int complex fails to catalyze the rightward reaction, so when excess excisionase is present, excision occurs.

The product of excision of a prophage from the *E. coli* chromosome is an intact chromosome and a circular λ molecule, which is also the arrangement present in an infected cell immediately after infection.

Avoidance of Excision During Lysogeny

In the lysogenic pathway, before *cI* repressor is made, transcription from *pL* occurs, so both the *int* and *xis* genes are transcribed. If both integrase and excisionase were present, a prophage that had just been inserted into the chromosome would immediately be excised, which clearly must be avoided. A simple mechanism prevents this excision (Figure 17-15).

As part of the progress along the lysogenic pathway, the activity of the cII protein turns on synthesis of the cI repressor from the *pRE* promoter and of integrase from the *pI* promoter. The *cI* repressor then acts on *pL* and turns off transcription of the *int* and *xis* genes from *pL*. The promoter *pI* is located *within* the *xis* gene downstream from the ribosomal binding site for synthesis of the *xis* product. Thus, this mRNA molecule provides integrase but not excisionase.

Prophage Induction

Although lysogens are generally quite stable, conditions that could cause death of the cell lead to de-repression (prophage induction) and phage production. The usual signal for de-repression is damage to the DNA. So far, all inducing agents, of which ultraviolet radiation has been studied most extensively, cause DNA damage that results in the activation of the SOS repair pathway (see Chapter 9). Furthermore, *recA⁻* lysogens are not inducible by these agents, which is consistent with a SOS effect because the RecA protein is an essential component of the SOS response. Evidence that a *recA⁻* mutation blocks only the damage-to-de-repression step comes from experiments in which *recA⁻* λ*cI857* lysogens are heated; with these strains, de-repression requires only thermal inactivation of the

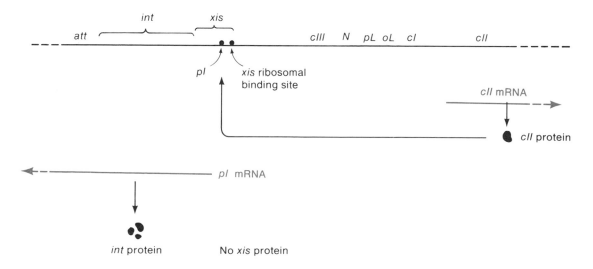

Figure 17-15. Prevention of synthesis of the Xis protein during lysogenization. Black lines are DNA molecules; orange lines are mRNA molecules. The cII protein activates pI, the *int* promoter, which is downstream from the ribosomal binding site of the Xis protein.

cI repressor. Thermal inactivation of the *cI* repressor induces lytic growth equally well in *recA⁻* and *recA⁺* lysogens of λ*cI857*.

DNA damage activates RecA protein to function as a protease facilitator RecA° (see Chapter 9). Following this activation, RecA° facilitates cleavage of the LexA protein, which is the repressor for all of the SOS genes, including recA itself, resulting in synthesis of the SOS gene products to high levels (see Chapter 9). When RecA° levels become high it facilitates the cleavage of the λ *cI* repressor (and the repressors of many other temperate phages), and this cleavage is responsible for induction. Cleavage inactivates the *cI* repressor, de-repression occurs, the early promoters *pL* and *pR* become available to RNA polymerase, and a lytic cycle ensues. Mutants of λ (called λ *ind⁻*) can be isolated that fail to be induced by UV light or other treatments that induce the SOS system. λ *ind⁻* mutants have mutations in the *cI* gene that block the autocleavage of the repressor. These mutants were isolated from a strain lysogenic for λ*cI857* and thus the λ *ind⁻* phage can be induced by temperate shift (the phage have two mutations in *cI*, *ind⁻* and *cI857*).

In the absence of a DNA-damaging agent, induction also occurs but at quite a low rate—about 10^{-4} cells per generation. This rare induction could result from spontaneous DNA damage, random fluctuations in *cI* repressor concentration, or occasional activation of the RecA to RecA°.

DECISION BETWEEN LYSIS AND LYSOGENY

The cII protein plays a central role in the choice between the lytic and lysogenic pathways. This protein has two important functions: (1) It activates the *pRE* promoter; and (2) it activates *pI*, the *int* promoter. For a lytic cycle to occur, either the synthesis or the activity of cII protein needs to be reduced to the point that *pRE* is not activated. If abundance or activity of cII protein initiates the decision between lysis or lysogeny, these features must be responsive to various signals. The critical property of the cII protein is its low stability: Its half life is about 2 minutes, owing to rapid degradation by cellular proteases. The cII protein levels are regulated by cIII protein, which increases the half-life of cII, by a protein-protein interaction. In a way that is unclear, cIII senses the multiplicity of infection (MOI), enhancing lysogeny when the MOI is high. The bacterial proteins HflA and HflB also affect the stability of cII protein by some effect on the protease activities, and these proteins respond in some way to the physiologic state of the cell, for example, starvation of the cells for required nutrients before infection can increase lysogenization 50- to 100-fold. The way the phage senses the cell physiology to modulate the lysis-lysogeny is not yet understood.

DILYSOGENS

Earlier we described an experiment in which a plasmid containing an inserted prophage was isolated and its DNA examined. In this experiment, most plasmids had increased in molecular weight by one λ unit; however, some plasmids were found to be two λ units larger, suggesting that the plasmid contained two prophages. This explanation was supported by another observation. When the culture was allowed to grow for many generations, the plasmid population became a mixture of those containing one λ unit and those containing two λ units. Furthermore, if the cell was *recA⁻*, the larger plasmid (two λ prophages) was stable. Thus it was hypothesized that these strains initially had a plasmid with two

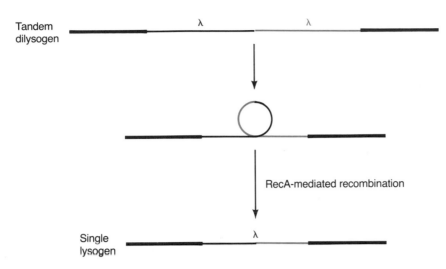

Figure 17-16. Conversion of a tandem dilysogen to a single lysogen by RecA-mediated recombination between homologous regions of λ. Heavy lines represent bacterial DNA.

λ prophages in tandem, and, in time, intramolecular reciprocal recombination occurred between homologous regions in the prophages, resulting in excision of one λ unit (Figure 17-16), a common observation when any DNA segment exists in tandem duplication.

OTHER MODES OF LYSOGENY

Most temperate phages form lysogens in the way described for λ—a prophage is inserted at a unique site in the host chromosome. Some phages have several chromosomal *att* sites, but this is rare. *E. coli* phage P1 is different from λ: Its prophage is not inserted into the chromosome at all. Following infection, P1 DNA circularizes and is repressed. In the lysogenic mode, it remains as a free supercoiled plasmid DNA molecule, roughly one or two per cell. Once per bacterial life cycle the P1 DNA replicates, and this replication is coupled to chromosomal replication. When the bacterium divides, each daughter cell receives a P1 plasmid. P1 has a Par function, similar to that which maintains plasmid number of cells containing low-copy-number plasmids (see Chapter 11). The mechanism of prophage maintenance is not as foolproof in phage P1 as in temperate phages that integrate DNA into a chromosome; for example, in each round of cell division, about 1 cell per 1000 fails to receive a copy of the P1 plasmid.

SOME PROPERTIES OF LYSOGENS

Lysogeny is a means of propagating a phage when a supply of sensitive bacteria has been exhausted. The prophage also has value to a bacterium because the bacterium survives the infection and is immune to homoimmune superinfection. Lysogeny may also have other effects on bacteria—often a lysogen has properties not present in the original bacterium. This process is called lysogenic conversion. One example is the resistance to phage T4 *rII* mutants conferred on *E. coli* by a λ prophage. A λ gene, *rex*, transcribed from the *pRM* promoter, blocks an early stage of infection by these mutants. Another example is found in *Corynebacterium diphtheriae*, the causative organism of diphtheria, which produces diphtheria toxin only when the bacterium is lysogenic for the phage β. Finally, a site of insertion may also be within a bacterial gene so prophage formation inactivates the gene.

USE OF λ TO SCREEN FOR CARCINOGENS

Apart from their intrinsic interest, phages have served molecular biology principally as a means of understanding fundamental molecular processes. Recently λ phage has also proved to be a valuable tool for screening for carcinogens. In Chapter 10, the Ames test was described in which carcinogens are detected by their ability to induce reversions of certain His⁻ mutations in *Salmonella typhimurium*. A similar test, the Devoret test, uses *E. coli* strains lysogenic for phage λ. Lysogens are spontaneously induced at a fairly low frequency, and in certain bacterial mutants the frequency is lower. Many carcinogens, however, cause DNA damage and are inducing agents. The induction can be detected quite simply. If one lysogenic cell is mixed with 10^8 bacteria and put on agar (exactly as might be done in detecting phage by plaque counting), the lysogen, which is not induced by this process, forms a microcolony (of at most 1000 cells) in the bacterial lawn. Occasionally late in the development of the colony, a cell is induced; this does not usually produce a plaque because depletion of nutrients in the agar limits cell growth, so the released phage cannot multiply. If an inducing agent (in this case, a carcinogen) is in the agar, however, the original lysogenic cell is induced to make phage at the time all of the bacteria in the agar begin to grow, so a plaque will form. If 10^5 lysogenic cells are put in the agar, the effectiveness of the inducing agent can be measured by the fraction of cells yielding plaques. This test, similar to the Ames test, uses liver extracts in the agar to activate carcinogens, successfully detecting the same carcinogens observed by the Ames test. The Devoret test, however, is often more sensitive, requiring lower concentrations of carcinogens; thus, it is capable of detecting some weak carcinogens that are not detected by the Ames test. Together these tests are being used to examine a large number of industrial chemicals and food additives.

The molecular basis of the Ames and Devoret tests is the same—the SOS repair system (see Chapter 9). Carcinogens damage DNA and the products of this damage bind to RecA, converting it to the protease facilitator, RecA°. In the Ames test, RecA° facilitates the proteolytic self-cleavage of the LexA and UmuD proteins, resulting in error-prone repair which increases the reversion frequency of the histidine mutation. In the Devoret test, RecA° facilitates the proteolytic autocleavage of λ repressor, resulting in phage induction and plaque formation.

KEY TERMS

attachment site (*att*)	lysogeny
	establishment
cI and Cro	maintenance
cII and cIII	oL, oR, pL, pR
clear plaque mutants	phage immunity
	pRE and pRM
excision (*xis*)	prophage induction
heteroimmune	site-specific recombination
integration (*int*)	temperate phage

QUESTIONS AND PROBLEMS

1. Which phages can plate on a λ lysogen: λ*imm*λ, λ*imm434*, λ*imm21*?

2. Which phages can plate on a lysogen of λ*imm434*: λ*imm*λ, λ*imm21*, λ*imm434*?

3. If two temperate phages are homoimmune, what phage elements must they have in common?

4. What is meant by a virulent mutant of a phage?

5. What plaque morphology is produced by a *cI*⁻ mutant and by a *vir* mutant?

6. What do you think is the difference, if any, between the appearance of a plaque of a *vir* phage on a homoimmune lysogen or a nonlysogen?

7. Phage 434 does not insert between the *gal* and *bio* genes because its attachment site is at another position. At what locus in the chromosome would you expect λ*imm434* to insert if only the promoter-operator region of phage 434 were present in the hybrid phage?

8. What symbol designates the bacterial, phage, and prophage *att* sites, and which is at the *gal* and at the *bio* end?

9. A phage has gene order *A B C att D E F*. What is the gene order in a prophage?

10. What proteins are needed for integration and for excision?

11. What gene products activate the *pRE* and *pRM* promoters for *cI* synthesis?

12. What attachment sites are carried by *gal*-containing and *bio*-containing specialized transducing phage?

13. Mutations in what genes or sites would prevent lysogenization?

14. An *int*⁻ mutant can integrate if the cell is coinfected with an *int*⁺ phage; the latter "helper" phage provides the missing function. Could a phage with an operator mutation (e.g., λ*vir*) be helped to integrate?

15. A λ*gal* phage can lysogenize a heteroimmune lysogen to form a tandem dilysogen. In theory, it could insert into the left (*gal*) or right (*bio*) prophage attachment sites. Genetic tests show that the gene order of the dilysogen is *gal* BOP' *R A gal* BOP' *R A* POB' *bio*, in which *A* and *R* are genes at the termini of the phage (given for reference only). Into which prophage *att* site did insertion occur?

16. You have isolated two independent mutant strains of the phage λ; both of these mutants form clear plaques. When either mutant alone infects *E. coli* at MOI = 10, no lysogenic survivors can be isolated. When *E. coli* is coinfected with 5 phage per cell of the first mutant plus 5 phage per cell of the second mutant, however, 10% of the cells survive the infection, and almost all of these survivors are lysogenic. When the lysogenic survivors are induced, 99% of the plaques produced are clear.

    a. What λ function or functions do you think are eliminated in these mutants?

    b. What is the most likely explanation for the increased frequency of lysogenization that occurs during coinfection by the two mutants?

17. Which of the following λ mutants cannot lysogenize at normal frequency: *N*⁻, *P*⁻, *Q*⁻?

18. Could an *A*⁺ *R*⁺ recombinant be produced in a cross between λ*A*⁻*R*⁺ *bio* and λ*A*⁺*R*⁻*bio*, if both the Rec and Red systems are inactive? Explain. The A and R genes are at opposite ends of the λ DNA molecule.

19. If a lysogen is irradiated with ultraviolet light and the cells are plated with an indicator bacterium, plaques form, roughly one per irradiated cell. If the cell is *recA*⁻, no plaques are formed. If the cell contains a temperature-sensitive *cI* repressor, phage are produced with both *recA*⁺ and *recA*⁻ bacteria. Explain this observation.

20. Infection of a λ lysogen with 10 λ phage particles does not result in phage development. This is the immune response. If the multiplicity of infection is 30, phage development occurs. In a dilysogen (two λ prophages), 60 phage particles per cell are needed to initiate a successful infection. Explain.

21. The cI857 temperature-sensitive repressor is reversibly denaturable in that if, after heating, a cell is returned to 34°C or less, the repressor renatures and becomes active.

    a. At 40°C, is the *cI* gene of a lysogen still transcribed?

    b. If the temperature is returned to 34°C, will the *cI* gene be transcribed?

    c. If a growing λ *cI857* lysogen is heated for 10 minutes and then cooled, phage development occurs on continued incubation in growth medium. If the lysogen is instead cooled after 5 minutes, there is no phage development and the bacteria survive. Explain.

22. In an *E. coli* tandem dilysogen, the left and right prophages have the genotypes *cI857 int⁻ P⁻* and *cI857 int⁻ A⁻*. The repressor of each is inactive at 42°C, and the prophages lack a functional integrase. The mutations *P⁻* and *A⁻* are, in this experiment, just genotypic markers located in the following order: *A F J att N P R*. If the lysogen is heated to 42°C, what will be the genotype or genotypes of the phage progeny?

23. Suppose you have a λ mutant that makes a clear plaque on *E. coli* strain A and a turbid plaque on strain B. How can this difference be explained?

24. If a λ lysogen is induced with ultraviolet light, a lysate results in which about one clear-plaque mutant per 10^4 phage is present. If a culture of a lysogen is grown without any prior induction and the culture fluid is analyzed for free phage that are spontaneously released, the fraction that form clear plaques is always greater than one per 10^4 phage.

    a. Explain the higher frequency of clear-plaque mutants among the spontaneously released phage.

    b. Among the spontaneously released phage, in which of the genes *cI, cII,* or *cIII* would you expect the mutations reside that yield a clear plaque?

25. A λ phage mutant called λ*b2* cannot lysogenize because it has a deletion starting just to the left of the O region of the attachment site; that is, its attachment site is del(OP'), in which delD represents a deletion. In a mixed infection with λ*bio* using conditions favoring the lysogenic response, dilysogens containing both λ*b2* and λ*bio* are found. Few single lysogens (these can be ignored in your thinking) and no (b2,b2) or (*bio,bio*) dilysogens are found. How have the (b2, *bio*) dilysogens formed?

26. Replication of λ DNA can be detected by hybridization of labeled λ mRNA with all of the DNA isolated from an infected cell or from an induced lysogen. As replication proceeds and the number of copies of λ DNA increases, more λ mRNA can be hybridized. Interpret the following observations.

    a. If an *xis⁻* lysogen is induced, the amount of hybridizable λ mRNA increases with time.

    b. Also, using *bio* mRNA and *gal* mRNA (both of which are easily obtainable) the amount of these mRNAs that can hybridize also increases with time.

    b. If an *int⁺* lysogen is induced, the amount of hybridizable *bio* and *gal* mRNA (as well as λ mRNA) increases, although fewer of these hybridize than in part (b).

27. A λ*gal* phage can integrate into the left prophage attachment of a heteroimmune lysogen. Integration, however, is not observed into the right prophage attachment site. Why not?

28. Suppose you had a λ lysogen and you wanted to add a mutation that would prevent adsorption of λ. How might you go about it? (Think about what you could plate the bacteria on.)

29. Suppose that you had a dilysogen that contained one wild-type λ phage and one λ*ind⁻* phage. Would this phage be induced by UV light?

REFERENCES

*Campbell, A. 1976. How viruses insert their DNA into the DNA of a host cell. *Scientific Am.*, December, p. 102.

Campbell, A. 1993. Thirty years ago in genetics: prophage insertion into bacterial chromosomes. *Genetics*, 133, 433.

Craig, N. L. 1988. The mechanism of conservative site-specific recombination. *Ann. Rev. Genet.*, 22, 77.

*Devoret, R. 1979. Bacterial test for potential carcinogens. *Scientific Am.*, August, p. 40.

*Resources for additional information.

Echols, H., and G. Guarneros. 1983. Control of integration and excision. In *Lambda II* (R. W. Hendrix et al., eds.), p. 75. Cold Spring Harbor.

Echols, H. 1980. Bacteriophage λ development. In *The Molecular Genetics of Development* (T. Leighten and W. Loomis, eds.), p. 1. Academic Press, New York.

*Friedman, D. 1988. IHF: A protein for all reasons. *Cell*, 55, 545.

Franklin, N. C., W. F. Dove, and C. Yanofsky. 1965. A linear insertion of a prophage into the chromosome of *E. coli* shown by deletion mapping. *Biochem. Biophys. Res. Commun.*, 18, 910.

Freifelder, D., and I. Kirschner. 1971. The formation of homoimmune double lysogens of phage 1 and the segregation of single lysogens from them. *Virology*, 44, 633.

Hendrix, R. W., et al. 1983. *Lambda II*. Cold Spring Harbor.

Hershey, A. D. (ed.). 1971. *The Bacteriophage Lambda*. Cold Spring Harbor.

Herskowitz, I., and Hagen, D. 1980. The lysis-lysogeny decision of phage λ: explicit programming and responsiveness. *Ann. Rev. Genet.*, 14, 399.

Jacob, F., and E. L. Wollman. 1953. Induction of phage development in lysogenic bacteria. *Cold Spring Harb. Symp. Quant. Biol.*, 18, 101.

Kaiser, A. D. 1957. Mutations in a temperate bacteriophage affecting its ability to lysogenize *E. coli*. *Virology*, 3, 42.

Landy. A. 1989. Dynamic, structural, and regulatory aspects of lambda site-specific recombination. *Ann. Rev. Biochem.*, 58, 913.

Lederberg, E. 1951. Lysogenicity in *E. coli* K-12. *Genetics*, 36, 560.

Lieb, M. 1953. Establishment of lysogenicity in *E. coli*. *J. Bacteriol.*, 65, 642.

Mizuuchi, M., and K. Mizuuchi. 1980. Integrative recombination of bacteriophage λ: extent of the DNA sequence involved in the attachment site function. *Proc. Natl. Acad. Sci.*, 77, 3220.

Numrych, T., R. Gumport, and J. Gardner. 1992. Characterization of the bacteriophage lambda excisionase (Xis) protein: the C-terminus is required for Xis-integrase cooperativity but not for DNA binding. *EMBO J.*, 11, 3797.

Oppenheim, A., and A. B. Oppenheim. 1978. Regulation of the *int* gene of bacteriophage λ: activation by the *cII* and *cIII* gene products and the roles of the *pI* and *pL* promoters. *Mol. Gen. Genet.*, 165, 39.

*Ptashne, M. 1992. *A Genetic Switch*, Second edition. Blackwell Scientific Publications.

Roberts, J. W., C. W. Roberts, and D. W. Mount. 1977. Inactivation and proteolytic cleavage of phage lambda repressor *in vitro* in an ATP-dependent reaction. *Proc. Natl. Acad. Sci.*, 74, 2283.

Roberts, J., and R. Devoret. 1983. Lysogenic induction. In *Lambda II* (R. W. Hendrix et al., eds.), p. 123. Cold spring Harbor.

Rothman, J. L. 1965. Transductional studies on the relation between prophage and host chromosome. *J. Mol. Biol.*, 12, 892.

Weisberg, R. A., and A. Landy. 1983. Site-specific recombination in phage lambda. In *Lambda II* (R. W. Hendrix et al., eds.), p. 211. Cold Spring Harbor.

Transduction

Transduction is a phenomenon in which bacterial DNA is transferred from one bacterial cell to another by a phage particle. Phage particles that contain bacterial DNA are called transducing particles. There are two types of transducing phage—generalized and specialized. Generalized transducing phage produce some particles that contain only DNA obtained from the host bacterium rather than phage DNA, and the bacterial DNA fragment can be derived from any part of the bacterial chromosome. Specialized transducing phage produce particles containing both phage and bacterial genes linked in a single DNA molecule, and the bacterial genes are obtained from a particular region of the bacterial chromosome. In this chapter, we describe generalized transduction by the *Salmonella typhimurium* phage P22 and specialized transduction by the *Escherichia coli* phage λ.

GENERALIZED TRANSDUCTION

Both the *E. coli* phage P1 and the *S. typhimurium* phage P22 are capable of generalized transduction. P1 packages about twice as much DNA (approximately 2 minutes or 100 kb) as P22 (approximately 1 minute or 50 kb), but certain P22 mutants are much more efficient transducing phage than P1. The mechanisms of generalized transduction by P22 is better understood, so it is described here in detail.

P22 is a temperate phage that infects *Salmonella* by binding to the O-antigen, part of the lipopolysaccharide on the outer membrane. After infection, P22 circularizes by recombination between terminal redundancies at each end of the phage DNA. During lytic growth, the circular genome of P22 initially undergoes several rounds of θ replication then changes to rolling circle replication. Rolling circle replication produces long concatemers of double-stranded P22 DNA. These concatemers are packaged into phage heads by a "headful" mechanism: Packaging is initiated by a P22 nuclease that cuts P22 DNA initially at a specific sequence called a "Pac site," the head is filled with about 44 kb of DNA, then the nuclease cuts the DNA and moves on to package a second headful of DNA into another phage head. Typically about five headfuls of DNA are packaged into phage heads from each concatemer. Because the P22 genome is only 42 kb, this yields the terminal redundancy at the ends of P22. Thus linear double-stranded DNA is packaged into new phage particles. When the cell lyses, it releases 50 to 100 new phage.

There are sequences on the *S. typhimurium* chromosome that are homologous to the P22 Pac site, called "pseudo-*pac*" sites. When P22 infects a cell, occasionally the P22 nuclease cuts one of these chromosomal sites and packages 44-kb fragments of chromosomal DNA into P22 phage heads. These phage heads contain only bacterial DNA, no phage DNA. The P22 particles carrying bacterial DNA (transducing particles) can inject this DNA into a new host, and the DNA can then recombine into the chromosome by homologous recombination. Because P22 can transfer DNA fragments from all regions of the chromosome, this process is called generalized transduction.

Normally the P22 nuclease is quite specific, so only about 1% of the phage heads actually carry bacterial DNA. In addition, all genes are not transduced at the same frequency because cleavage of the chromosome is not truly random. A derivative of P22, called P22 HT, is useful for generalized transduction. This phage has a high transducing (HT) frequency owing to a mutation that decreases the nuclease specificity of the *pac* sequence. About 50% of the P22 HT phage heads carry random transducing fragments of chromosomal DNA. Thus a P22 transducing particle may contain a fragment derived from any region of the host DNA, and a sufficiently large population of P22 phage will contain particles, at least one of which carries a given host gene. Because each transducing particle carries about 1% of the chromosome and about 50% of the phage are transducing particles, on the average, any particular gene is present in roughly one transducing particle per 200 P22 HT phage.

Phage P1 can be used for generalized transduction in *E. coli*, but fewer than 1% of the P1 phage heads carry transducing fragments.

What happens when DNA is injected into a bacterium from a generalized transducing particle? For example, consider a P1 transducing particle from wild-type *E. coli* containing a *leu*⁺ gene. If a *leu*⁺ transducing particle adsorbs to a bacterium whose genotype is *leu*⁻ and injects its DNA into the bacterium, the cell will survive because the phage head contains only bacterial genes and no phage genes. A double-recombination event exchanging the *leu*⁺ allele on the transduced DNA for the *leu*⁻ allele in the host chromosome will convert the genotype of the host cell from *leu*⁻ to *leu*⁺ (Figure 18-1). The Leu⁺ transductants can be selected on minimal medium without leucine as long as the *leu*⁻ allele does not have a high reversion frequency.

Figure 18-1. Transduction. The phage P1 infects a *leu*⁺ donor, yielding predominately viable P1 phage with an occasional one carrying bacterial DNA instead of phage DNA. If the phage population infects a bacterial culture, the viable phage will produce progeny phage, and the transducing particle will yield a transductant. Notice that the recombination step requires two crossovers. For simplicity, double-stranded DNA is drawn as a single line.

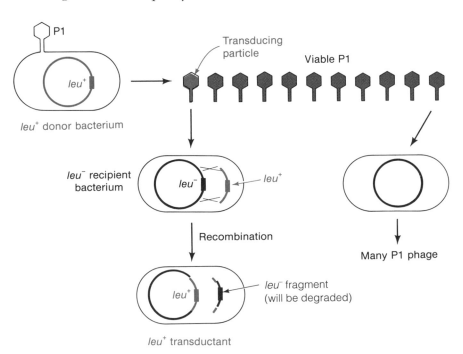

The donor cell can be either Rec⁺ or Rec⁻, but integration of the transducing fragment into the chromosome of the recipient requires homologous recombination. Thus, transduction requires that the recipient cell be Rec⁺; no transductants are observed with *recA* or *recBCD* recipients. The donor cell can be either Rec⁺ or Rec⁻.

What happens when a transducing fragment is brought into a *rec* mutant? The gene present on the transducing fragment can be transcribed and produce a functional gene product, allowing limited growth of a mutant cell on a selective plate. Because the DNA fragment cannot replicate, however, each time the host cell divides only one of the daughter cells receives the transduced fragment. The other cell may contain enough of the gene product in its cytoplasm to grow and divide a few times, but no more can be made. The cell with the fragment divides again, producing one fragment-free daughter cell that can divide a few times and one daughter cell with the original fragment. The result of this process is that the colony does not divide by exponential multiplication of the initial cell, so only tiny microcolonies called "abortive transductants" develop.

What is the evidence that generalized transducing particles contain only bacterial DNA? This was demonstrated using the DNA density-labeling technique (see Chapter 2) as shown in Figure 18-2. Host bacteria were grown for many generations in growth medium containing ¹⁵N and then infected with ¹⁴N-labeled P22 in medium containing ¹⁴N. The resulting phage progeny were centrifuged to equilibrium in concentrated CsCl, which separates the particles by density. The centrifuge tube was fractionated by puncturing the tube bottom and collecting successive drops, and each fraction was tested for the presence of both viable phage and transducing particles. The viable phage had the density of particles containing ¹⁴N-labeled DNA, but the transducing particles were found at the density of particles having ¹⁵N-labeled DNA, indicating that they contained little, if any, phage DNA.

COTRANSDUCTION AND LINKAGE

The small fragment of bacterial DNA contained in a P22 transducing particle carries about 50 genes, so transduction provides a valuable tool for linkage analysis of short regions of the bacterial genome. Consider a population of P22 prepared from a *leu⁺ gal⁺ bio⁺* bacterium. The phage lysate contains particles able to transfer any of these alleles to another bacterium; that is, a *leu⁺* particle can transduce a *leu⁻* bacterium to *leu⁺*, or a *gal⁺* particle can transduce a *gal⁻* bacterium to *gal⁺*. Thus, when a *leu⁻ gal⁻* culture is infected with phage, both *leu⁺*

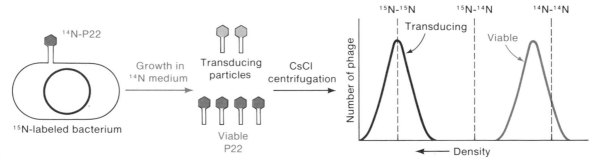

Figure 18-2. Demonstration that P22 transducing particles contain only bacterial DNA. Phage DNA are shown in orange, and bacterial DNA are shown in black.

gal⁻ and *leu⁻ gal⁺* bacteria (transductants) are produced. If the multiplicity of infection (MOI) is less than 1, however, *leu⁺ gal⁺* colonies rarely arise (Figure 18-3) because the *leu* and *gal* genes are too far part to be carried on the same DNA fragment, and few bacteria are infected with both a *leu⁺* particle and a *gal⁺* particle.

The situation is quite different with a recipient bacterium whose genotype is *gal⁻ bio⁻* because the *gal* and *bio* genes are separated by only about 0.5 minute. A P22 phage can carry about 1 minute of chromosomal DNA, so both sets of genes will often be present in a single DNA fragment carried in a transducing particle (see Figure 18-3). Not all *gal⁺* transducing particles, however, are *bio⁺*, and not all *bio⁺* particles are *gal⁺* because recombination of the transducing DNA sometimes occurs between the *gal* and *bio* genes. The probability that both markers will be in a single particle and hence the probability of simultaneous transduction of both markers (cotransduction) depends on how close the genes are to each other. Cotransduction of the *gal* and *bio* genes can be detected by plating infected cells on the appropriate growth medium. If *bio⁺* transductants are selected (by spreading the infected cells on a glucose-containing medium lacking biotin), both *gal⁺ bio⁺* and *gal⁻ bio⁺* colonies are produced. If these colonies are tested for the gal marker by replica plating onto galactose-biotin plates, about 20% will be *gal⁺ bio⁺* and 80% will be *gal⁻ bio⁺*; similarly, if *gal⁺* transductants are selected by growth on galactose-biotin plates and retested on a medium lacking biotin, about 20% (the *gal⁺ bio⁺* colonies) will grow. Thus, sometimes the *gal* and *bio* genes will be cotransduced, and sometimes only one of the two genes will be transduced.

Figure 18-3. Demonstration of linkage of the *gal* and *bio* genes by cotransduction.

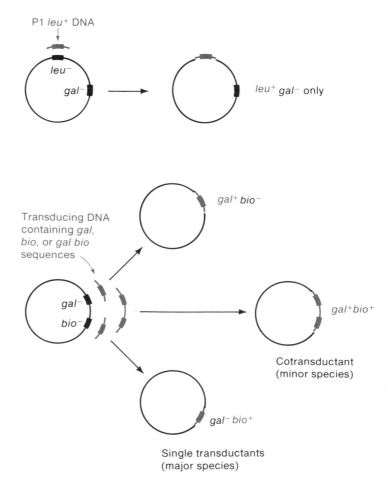

MAPPING BY COTRANSDUCTION

Cotransduction can be used to determine a genetic map of the relative distances between genetic markers and the order of the markers. The use of cotransduction to determine gene order is based on the fact that the closer genes are to each other, the greater the probability they will be cotransduced: That is, the frequency of cotransduction is inversely proportional to the distance between two genes (see Figure 13-8). When genes are close to each other, the results are ambiguous because of experimental uncertainty. As an example, consider three genes a, b, and c, for which the frequencies of cotransduction determined by two-factor crosses are: a-b, 90%; a-c, 33%; and b-c, 32%, each value having an error of about $\pm2\%$. The values indicate that b is closer to a than to c because 90% is greater than 32%. The order of the genes, however, cannot be deduced from these data alone because the difference between 32 and 33% is not significant, so although the order a-c-b can be excluded, the orders a-b-c and b-a-c are both consistent with the data.

Three-factor crosses can be used to determine the order of closely linked markers, such as the a and b genes just described, with respect to a third gene or the order of mutant sites within a single gene. This method is based on the principle that the probability of appearance of an event that requires four crossovers is much less than the probability of an event that requires only two crossovers. In this method, the results of reciprocal genetic crosses are compared. These crosses are shown for three hypothetical genes in Figure 18-4. In the first cross, the donor (P22) genotype is $a^+ b^- c^+$, and the recipient genotype is $a^- b^+ c^-$. In the second cross, + and – alleles are interchanged, so the donor genotype is $a^- b^+ c^-$, and the recipient genotype is $a^+ b^- c^+$. In each cross, the $a^+ b^+ c^+$ genotype is selected. Figure 18-4 shows that for order I, production of $a^+ b^+ c^+$ in cross 1 requires four crossovers, whereas in cross 2, only two crossovers are needed. Thus for order I, the expected frequency from cross 1 is much lower than that from cross 2; in contrast, for the alternative order II, the expected frequencies are roughly comparable for the two crosses, the differences depending only on the spacing between the markers selected. If for cross

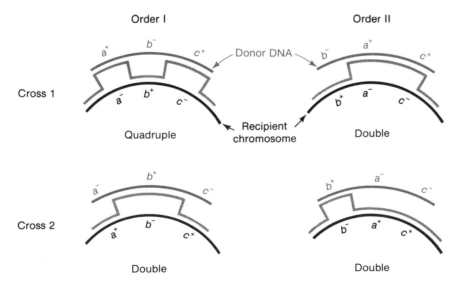

Figure 18-4. Use of reciprocal crosses to determine the order of markers *a* and *b*, relative to *c*, by transduction. The chromosome of the recipient is circular. The gray lines show the product of crossing over. The principle underlying the technique is that a double crossover has a greater probability of occurrence than a quadruple crossover.

Order I

Order II

Cross 1

Donor DNA

Quadruple

Double

Recipient chromosome

Cross 2

Double

Double

Selection for $a^+b^+c^+$ in all crosses

Expected results:

$$(a^+b^+c^+)_1 \ll (a^+b^+c^+)_2 \qquad\qquad (a^+b^+c^+)_1 \approx (a^+b^+c^+)_2$$

1 the frequency of $a^+ b^+ c^+$ is observed to be much less than that of cross 2, then the order must be $a\ b\ c$.

Consider transduction from a donor strain whose genotype is $A^+ B^- C^-$ to a recipient that is $A^- B^- C^+$. One phenotype (A^+) is selected by plating on medium lacking substance A, and 100 A^+ colonies are tested further for the B and C genotypes by replica plating. The data are shown in Table 18-1. The cotransduction frequencies in Table 18-1 show that C is nearer to A than B is to A, which is consistent with either of the orders A C B or C A B. These orders can be distinguished by diagramming the recombination events needed to generate the various recombinant genotypes. This is shown in Figure 18-5. The rule to analyze the data is the usual one: The *rarest* recombinant class requires the *maximum* number of exchanges. Table 18-1 shows that the rarest class is $A^+ B^+ C^+$. Figure 18-5 shows that this recombinant is produced by a quadruple exchange with order 1 and a double exchange with order 2. Thus, order 1 is the more likely one.

Such genetic crosses allow determination of gene order in a straightforward way but not genetic distance. The main problem in relating recombination frequencies to genetic distances is that even when two genes are close enough that they can be carried by a single transducing particle, many particles that carry one of the markers will not carry the other marker, and infection by such a particle cannot lead to transduction of the marker that it lacks. Thus, to convert frequencies to map distances can be complex. The genetic maps of *E. coli* and *S. typhimurium* are divided into 100 minutes (where 1 minute is an arbitrary unit approximately equal to 50 kb, not a unit of percent recombination as is common in eukaryotic genetics). Wu developed a mathematical formula that relates cotransduction frequencies to map distance:

$$C = (1 - d/L)^3$$

where C is the cotransduction frequency expressed as a decimal, d is the distance in minutes between a selected marker and an unselected marker, and L is the size of the transducing fragment in minutes. For P22, L is approximately 1 minute. Solving this equation for d gives:

$$d = L - L{-}C^{1/3}$$

Therefore for P22, the distance between two markers in minutes is:

$$d = 1 - (1)\,(C)^{1/3}$$

Hence for the genes A, B, and C in Table 18-1, the map is:

$$A \leftarrow 0.38 \rightarrow B \leftarrow 0.12 \rightarrow C$$

Table 18-1   Some data from a transduction experiment

Donor genotype: $A^+B^+C^-$
Recipient genotype: $A^-B^-C^+$

100 A^+ colonies selected for further analysis:

Genotypes observed	Number of colonies
$A^+B^+C^+$	5
$A^+B^+C^-$	19
$A^+B^-C^+$	49
$A^+B^-C^-$	27
	100

Cotransduction frequencies:

A^+B^+	(5 + 19)/100 = 0.24
A^+C^-	(19 + 27)/100 = 0.46

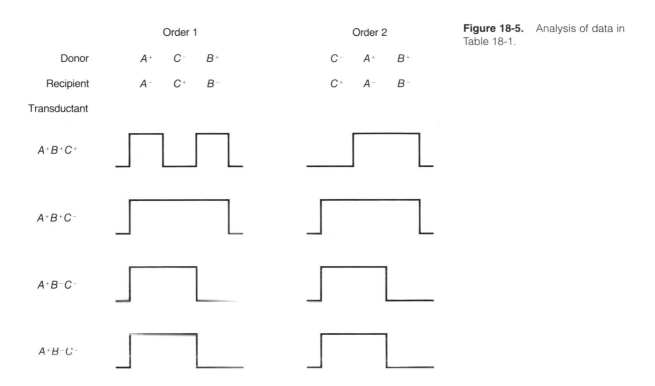

Figure 18-5. Analysis of data in Table 18-1.

SPECIALIZED TRANSDUCTION

In the generalized transducing systems discussed so far, transducing particles form during a lytic cycle by packaging host DNA fragments into phage heads. Because any region of the chromosome can be packaged, all possible host sequences are represented in a heterogeneous population of generalized transducing particles (assuming the population is large enough).

A different type of transducing particles can be produced during excision of an integrated prophage: specialized transducing particles. These particles differ from generalized transducing particles in four ways: (1) Specialized transducing particles contain both host and phage DNA linked in one continuous molecule; (2) only regions of the host DNA that flank the prophage are found in these particles; (3) a single transducing particle can serve as a template for production of a homogeneous population of identical transducing particles; and (4) they are produced only by induction of a lysogen.

Formation of Specialized Transducing Particles from a λ Lysogen

Specialized transducing particles form by the mechanism shown in Figure 18-6. When λ prophage is induced, an orderly sequence of events usually ensues in which the prophage DNA is precisely excised from the host DNA. In the case of phage λ, this is accomplished by site-specific recombination enzymes (encoded by the *int* and *xis* genes) acting on the left and right prophage attachment sites. At a very low frequency (about 1 cell per 10^6 to 10^7 cells), an excision error is made; two incorrect cuts are made, one within the prophage and the other in the bacterial DNA. The pair of abnormal cuts will not always yield a length of DNA that can fit in a λ phage head—it may be too large or too small. If the spacing between the cuts produces a molecule between 79 and 106% of the length of a normal λ phage-DNA molecule, however, packaging can occur. Because the prophage is located between the *E. coli gal* and *bio* genes and because the cut in the host DNA can be either to the right or the left of the prophage, transducing particles can arise that carry the *bio* genes (if the aberrant cut is to the right of λ) or the *gal* genes (if the aberrant cut is to the left of λ).

Formation of the λ*gal* and λ*bio* transducing particles causes loss of λ genes. λ*gal* particles lack the tail genes and sometimes head genes, both of which are located at the right end of the prophage; the λ*bio* particle lacks genes from the left end of the prophage (*int, xis, red*). The number of missing phage genes of course depends on the position of the cuts that generated the particle and thus correlates with the amount of bacterial DNA in the particle. The phage genes that are deleted are located at the prophage ends, but because of the permutation of the gene order in the prophage and the phage particle, the deleted phage genes are always from the central region of the phage DNA, as shown in Figure 18-6. The head and the tail genes are essential, so λ*gal* transducing particles are unable to form plaques: They are defective, and this is denoted λd*gal*. The genes missing in λ*bio* transducing particles are not essential for lytic growth, so λ*bio* phage are usually plaque forming. In some λ*bio* phages, the deletion extends into or past the *N* gene, making these phage defective; these are denoted λd*bio*. λ*gal* transducing particles can be isolated from lysogens that have a large deletion between *gal* and *att*L. In these strains, the *gal* locus is close enough to the prophage that *gal* can substitute for the nonessential *b2* region, so the resulting phage are not defective.

If λd*gal* transducing particles lack phage genes required to grow lytically, how are they formed? Notice that the transducing particle is produced by aberrant excision from a normal prophage. The prophage contains all of the essential genes, and hence the necessary gene products (head and tail proteins) are still present in the chromosome. Initially producing the defective transducing particles is not a problem, but there is a deficiency of essential gene products when the transducing particle infects a new cell. As just mentioned, the size of the bacterial DNA substituted for phage DNA in a specialized transducing particle can vary, depending on the location of the aberrant exchange. The location of the exchange can be determined by crosses between a particular λ*gal* phage and several normal λ phage, each carrying a different mutation with a known location. If recombination between the two mutant λ cannot produce wild-type λ, the mutation must be in the region of the substitution in that phage (Figure 18-7a). Similarly, mutations can be mapped by performing crosses with λ*gal* particles having substitutions of known extent (Figure 18-7b). The precise physical loca-

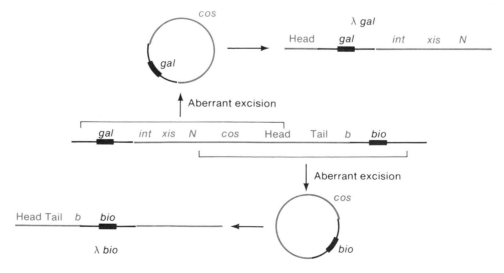

Figure 18-6.   Aberrant excision leading to production of λ*gal* and λ*bio* phages.

tion of the substitution can also be determined by heteroduplex analysis. DNA of a normal phage and a transducing particle is isolated, mixed, denatured and renatured (see Chapter 2), and examined by electron microscopy. Hybrid molecules can then be seen in which homologous single strands have joined to form double-stranded DNA, but nonhomologous single-stranded regions remain in unrenatured bubbles (Figure 18-8). Thus the end points of the substitution can be localized as the branch points of the bubble.

Specialized Transduction of a Nonlysogen

A specialized transducing particle can transduce a mutant bacterium in several ways. The mechanisms are the same for the λ*gal* and λ*bio* particles, so λ*gal* is used as an example. Consider a Gal cell that is infected with a lysate resulting from induction of a Gal⁺ culture lysogenic for λ. Under conditions that lead to

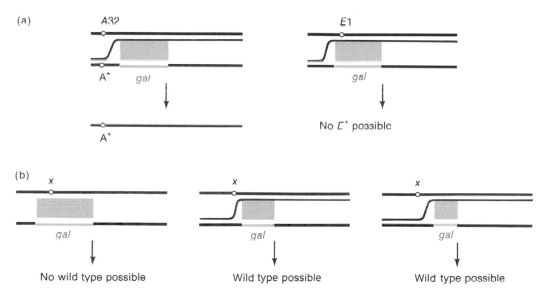

Figure 18-7. (a) Locating the left terminus of a *gal* substitution in a λ*gal* transducing particle. A cross is carried out between λ*gal* and normal λ marked either with the *A32* or *E1* mutations, whose locations are known. In the cross at the left, only the relevant recombinant is shown. (b) Locating a mutation x by crosses with three different λ*gal* particles, selecting for x⁺ phage. The shaded orange area indicates the nonhomologous regions in which crossing over does not occur. Note that the right end of the *gal* substitution is always the *att* site.

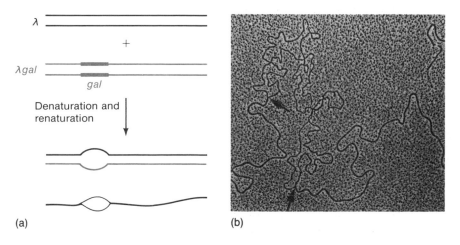

(a)   (b)

Figure 18-8. (a) A diagram showing the heteroduplexing technique for determining the size and location of a substitution. (b) An electron micrograph of λ and λ*gal* DNA. The arrows show the termini of the denaturation bubble. Single-stranded and double-stranded DNA have nearly the same width with the technique used to produce this micrograph. (Courtesy of Norman Davidson.)

the establishment of lysogeny (high MOI and late log-phase cells), this lysate will consist almost entirely of normal λ phage, but about 10^{-8} of the phage will be λdgal^+. If these infected cells are plated on galactose plates so only Gal$^+$ cells can grow, colonies are found at a very low frequency (about 0.001% of the infected cells). These colonies are of two types (Figure 18-9).

Type I consists of nonlysogenic cells, all of which are Gal$^+$. If the colony is resuspended and individual bacteria are allowed to form colonies, all the colonies are Gal$^+$. These contain stable gal^+ bacteria that have arisen as a result of two crossovers that replace the chromosomal gal^- gene with the gal^+ gene from the phage. Thus, the type I cells, which contain only one gal^+ gene, are haploids.

Type II cells contain a λdgal^+ prophage. If a type II colony is dispersed and individual cells are plated, about 1% of the colonies are Gal$^-$ and have lost the prophage. Type II cells have arisen by a single crossover within the gal genes. These cells now contain two copies of the gal genes, one gal^+ and one gal^-. These transductants are called heterogenotes, or partial diploids. Type II cells are incapable of being induced to produce progeny λdgal particles because the prophage lacks the tail genes. Transcription and synthesis of many phage-specific proteins, however, can occur.

When there are two copies of the gal genes in a cell, intramolecular genetic recombination can occur between the homologous DNA. There is an equal chance this will yield gal^+ or gal^- cells during continued growth of a type II gal^+ transductant, as shown in Figure 18-9. This results in the loss of both the λ prophage and one of the two copies of the gal gene.

Heterogenotes can also be produced in another way. If cells are infected at an MOI such that many cells are infected by both a λgal and a normal λ (which are present in excess in the initial lysate), a λgal-λ dilysogen can form. These arise by sequential insertions of the two phage. First, the normal λ inserts by the usual POP × BOB reaction. Once the prophage is established, the λgal, which has the attachment site, BOP can insert into the BOP at the gal end of the prophage (this is an efficient Int-promoted exchange). Thus the gene order in the dilysogen is

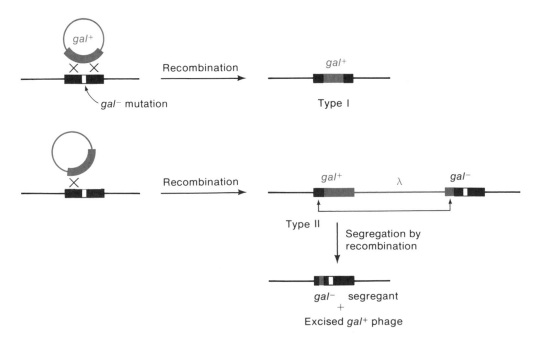

Figure 18-9.   Production of type I and type II λgal transductants and segregation of gal^- cells from a type II cell.

gal⁻ (chromosome)-λ*gal⁺* λ, with the *gal⁺* gene in the center of the dilysogen (try drawing this out).

Specialized Transduction of a Lysogen

A λ lysogen can also be transduced to produce a heterogenote (Figure 18-10). This occurs at a high frequency by a single crossover between the λ*gal* and the λ prophage or the chromosomal *gal* genes, similar to the recombination that produced type II cells. The resulting transductants are heterogenotes both for the *gal* operon and the prophage. They are more unstable than the type II transductants described earlier because the probability of intramolecular recombination leading to production of a *gal⁻* segregant is greater: It is not limited to the small *gal* duplication but can occur in both the duplicated *gal* genes and prophage DNA.

High-Frequency-Transducing Lysates

The λd*gal*-λ⁺ dilysogens just described are useful because when induced, the head and tail genes from λ⁺ can provide the structural components necessary for packaging λd*gal*. Thus when such a dilysogen is induced, both normal and λd*gal* particles are produced by the mechanism shown in Figure 18-11. Roughly equal numbers of the two types of particles are produced.

The dilysogens just described yield lysates half of whose phage are transducing particles. Such a lysate is called a high-frequency-transducing (HFT) lysate. A lysate formed from a single lysogen in which aberrant excision occurs only infrequently is called a low-frequency-transducing (LFT) lysate.

SPECIALIZED TRANSDUCING PHAGE AS A CLONING VEHICLE

Specialized transducing particles have a wide variety of uses. One of these is as a cloning vehicle. For example, a classic experiment used a φ80 specialized transducing phage to clone the *supF* gene from *E. coli*. Analysis of the *supF* gene in

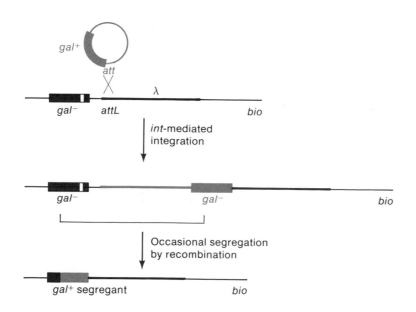

Figure 18-10. Transduction by prophage integration. Segregation by recombination in the *gal* genes occasionally occurs; this segregation can also occur at the *att* sites if some integrase is made. The infecting phage and the prophage must be heteroimmune for the insertion to occur.

Figure 18-11. Production of λ and λ*gal* from a tandem (λ, λ*gal*) dilysogen.

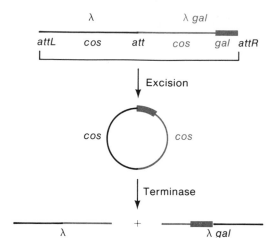

this phage provided the first evidence for the generation of a suppressor tRNA by a base change in the anticodon of a normal tRNA. Specialized transducing phage have also served as hybridization probes for identifying specific mRNA molecules. For example, early studies of the kinetics of transcription of the *lac* operon were carried out by radioactive labeling of RNA after induction of the operon and measuring the amount of *lac* mRNA by determining the amount of radioactive mRNA that hybridized to denatured φ80*lac* DNA. These types of cloning experiments are now usually done with specific cloning vectors (see Chapter 20), but the use of specialized transducing phage provided the initial paradigm for later cloning experiments.

KEY TERMS

concatemers	imprecise excision
cotransduction	λd*gal*
defective transducing phage	partial diploids
generalized transduction	prophage
high-frequency transducing lysates	selected marker
	specialized transduction

QUESTIONS AND PROBLEMS

1. If 10^8 phage λ infect 10^8 *gal*⁺ bacteria and all the phage adsorb, what fraction of the infected cells will yield *gal*⁺ transducing phage?

2. What are the differences between a phage that produces only specialized transducing particles, one that produces only generalized transducing particles, and one that can produce both?

3. Why are λ specialized-transducing particles generated only by induction rather than by lytic infection?

4. Some temperate phages that integrate their DNA do not seem to form specialized transducing particles. Give several possible reasons for this observation.

5. Specialized transducing particles of P22 carrying the *proA* and *proB* genes can be generated. These genes are immediately adjacent to the prophage attachment site on the bacterial chromosome. The *proA* and *proB* specialized transducing particles have the following behavior: (a) Lysates of the specialized transducing particles can go through the lytic cycle but only in mixed infections. (b) In single infections, the particles transduce by substitution; in mixed infections, however, the particles transduce by lysogenization. What is the molecular basis of each of the observations?

6. Under what circumstances could a lysate of phage P1 containing transducing particles carry phage λ?

7. In a classic experiment, *E. coli* cells, which are genetically unable to synthesize thymine, were fed 5-bromouracil to make their DNA "heavy" and then were infected with phage P1 and incubated in a medium containing (a) thymine, (b) 5-bromouracil, or (c) thymine plus ^{32}P until lysis occurred. Analysis of phage progeny by density-gradient centrifugation showed that transducing particles from the first two media had similar densities and that those from the third were not radioactive. From these observations, what do you conclude about the phage genetic material in transducing particles and the origin of transduced segments?

8. A bacterium with genotype $A^+ B^- C^-$ was transduced by a particle carrying the linked genes $A^- B^+ C^+$, and $A^+ B^+ C^+$ recombinants were selected. The reciprocal cross was also performed. The number of $A^+ B^+ C^+$ recombinants was the same in both the first cross and the reciprocal cross. What information does this give you about the order of the three genes?

9. P1 transduction analysis was carried out to determine the order of mutant sites in the *trpA* gene of *E. coli*, using a closely linked mutation, called *ant*, as an unselected marker. In each of the crosses listed here, Trp$^+$ recombinants were selected and then tested for the *ant*$^+$ or *ant*$^-$ allele. The donor genotype is given first, and the percent of Trp$^+$ recombinants that is *ant*$^+$ is given in parentheses.

   a. Cross 1: $ant^+ trpA34 \times ant^- trpA223$ (18% ant^+)
   b. Cross 2: $ant^+ trpA46 \times ant^- trpA223$ (52% ant^+)
   c. Cross 3: $ant^+ trpA223 \times ant^- trpA34$ (50% ant^+)
   d. Cross 4: $ant^+ trpA223 \times ant^- trpA46$ (20% ant^+)
   e. What is the order of the three *trpA* markers with respect to the *ant* gene?

10. Phage P1 is grown on $trpC^+ pyrF^- trpA^-$ cells and used to transduce $trpC^- pyrF^+ trpA^+$ recipient. Trp$^+$ transductants were selected and tested for the presence of the other markers. The following data were obtained: $trpC^+ pyrF^- trpA^- = 548$ colonies, $trpC^+ pyrF^+ trpA^- = 570$ colonies; $trpC^+ pyrF^- trpA^+ = 3$ colonies; $trpC^+ pyrF^+ trpA^+ = 90$ colonies.

   a. What is the order of the three genes?
   b. What are the cotransduction frequencies for *trpC* and *pyrF*, *trpA* and *pyrF*, and *trpC* and *trpA*?
   c. What is the map distance in minutes between the three markers?
   d. Why are the map distances not additive?

REFERENCES

Campbell, A. 1977. Defective bacteriophages and incomplete prophages. *Comprehensive Virol.*, 8, 259.

Ebel-Tsipis, J., D. Botstein, and M. S. Fox. 1972. Generalized transduction by phage P22 in *S. typhimurium*. I. Molecular origin of transducing DNA. *J. Mol. Biol.*, 71, 433.

Ebel-Tsipis, J., M. S. Fox, and D. Botstein. 1972. Generalized transduction by bacteriophage P22 in *S. typhimurium*. II. Mechanism of integration of transducing DNA. *J. Mol. Biol.*, 71, 449.

Echols, H., and D. Court. 1971. The role of helper phage in *gal* transduction. In *The Bacteriophage Lambda* (A. D. Hershey, ed.), p. 701. Cold Spring Harbor.

Gottesman, S., and J. R. Beckwith. 1969. Directed transposition of the arabinose operon: a technique for the isolation of specialized transducing bacteriophages for any *E. coli* gene. *J. Mol. Biol.*, 44, 117.

Margolin, P. 1987. Generalized transduction. In Escherichia coli *and* Salmonella typhimurium: *Cellular and Molecular Biology* (F. Neidhardt, J. Ingraham, K. B. Low, B. Magasanik, M. Schaechter, and H. E. Umbarger, eds.), American Society for Microbiology, Washington, D. C.

*Masters, M. 1985. Generalized transduction. In *Genetics of Bacteria* (J. Scaife, D. Leach, and A. Galizzi, eds.), p. 197. Academic Press, New York.

*Resources for additonal information.

Morse, M. L., E. M. Lederberg, and J. Lederberg. 1956. Transduction in *Escherichia coli* K12. *Genetics*, 41, 142.

Poteete, A. 1988. Bacteriophage P22. In *The Bacteriophages* (R. Calender, ed.), p. 647. Plenum, New York.

*Roth, J. 1970. Genetic techniques in studies of bacterial metabolism. *Methods Enzymol.*, 17, 1.

Schmieger, H. 1972. Phage P22 with increased or decreased transduction abilities. *Mol. Gen. Genet.*, 119, 75.

Shimada, K., R. A. Weisberg, and M. E. Gottesman. 1972. Prophage λ at unusual chromosomal locations. *J. Mol. Biol.*, 63, 483.

Smith-Keary, P. 1991. *Molecular Genetics: A Workbook*. Guilford.

Susskind, M., and D. Botstein. 1978. Molecular genetics of bacteriophage P22. *Microbiol. Rev.*, 42, 385.

Wall, J. D., and P. D. Harriman. 1974. Phage P1 mutants with altered transducing abilities for *E. coli*. *Virology*, 59, 532.

Wu, T. 1966. A model for three-point analysis of random general transduction. *Genetics*, 54, 405.

Zinder, N. D. 1953. Infective heredity in bacteria. *Cold Spring Harb. Symp. Quant. Biol.*, 18, 261.

Zinder, N. D., and J. Lederberg. 1952. Genetic exchange in *Salmonella*. *J. Bacteriol.*, 64, 679.

19

Strain Construction

A great deal of the actual laboratory work in microbial genetics involves constructing bacterial strains and phages containing particular genes or combinations of genes. A certain amount of this construction is done by mutagenesis, but often it is necessary to construct strains with combinations of existing mutations. Many of the techniques of strain construction used by geneticists involve standard procedures, but more often they involve a set of specialized, clever tricks. In this chapter, we give some examples of how strains can be constructed to give the reader a flavor of the ingenuity required in microbial genetics. The techniques are given in two parts: construction of bacterial strains and of phage strains. To facilitate understanding these procedures, an *Escherichia coli* map showing the mutations discussed in this chapter is provided in Figure 19-1.

CONSTRUCTION OF BACTERIAL STRAINS

Novel bacterial strains are constructed for two main reasons: to provide multiply mutant bacteria for mapping purposes and to create bacteria that have special properties to study specific genetic and biochemical processes. To accomplish these goals, the geneticist has available the techniques of mutagenesis, recombination, transposition, and so on, but these must often be coupled with powerful selection techniques that make it possible to find a rare bacterium with the desired genotype. Many of these general techniques were described earlier in this book. In this section, a variety of techniques, some of which illustrate general procedures, are used.

Isolation of Sugar-Utilization Mutants

Bacteria growing on rich (broth) plates containing the colorless dye triphenyltetrazolium chloride (usually called **tetrazolium**) reduce the dye to an intensely red compound. The reaction is inhibited at low pH. Thus, Lac⁻ bacteria growing on lactose-tetrazolium plates yield red colonies, whereas Lac⁺ bacteria, which produce acid by metabolizing lactose, produce white colonies. Tetrazolium plates are especially valuable in searches for sugar-utilization mutants because of the ease with which one red colony can be seen against a background of a large number of white colonies. In contrast, on EMB agar, one would seek a white colony against a background of purple colonies on a purple plate, which is much harder

Figure 19-1. An *E. coli* map, showing genes used in examples of strain construction in this chapter.

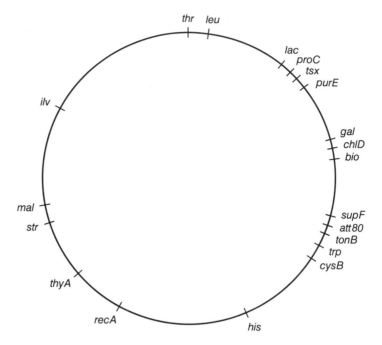

to see. On tetrazolium plates, one Lac⁻ colony can usually be detected among about 5000 colonies on one plate. Thus by using mutagens that raise the mutation frequency to about 10^{-5}, 100 plates would yield only 5 to 10 mutants.

If a multiple mutant, such as a Lac⁻ Gal⁻ Ara⁻ triple mutant, were desired, it could be isolated by a series of selections on tetrazolium plates. First, a Lac⁻ mutant could be isolated from a mutagenized wild-type culture by screening for red colonies on lactose-tetrazolium plates. A culture of this mutant could then be mutagenized and plated on galactose-tetrazolium plates to screen for a Gal⁻ mutant (which would be Lac⁻ Gal⁻). A third round of mutagenesis and plating on arabinose-tetrazolium plates would yield the triple mutant. The entire procedure would take less than 2 weeks, and most of the time would be spent waiting for colonies to grow.

Temperature-sensitive *lac* mutants, which are Lac⁻ at 42°C and Lac⁺ at 30°C, could be isolated by a slight modification of the procedure. In a primary selection, a mutagenized culture would be plated on lactose-tetrazolium plates, which are incubated at 42°C. Lac⁻ colonies would be selected and replated on lactose-tetrazolium plates at 30°C. Colonies that are red on a 42°C plate (Lac⁻) and white on a 30°C plate (Lac⁺) have the Lac⁻ temperature-sensitive phenotype.

Isolation of Thymine-Requiring Mutants

Mutants that require a supply of thymine in the growth medium have been valuable in the study of bacterial DNA synthesis for three reasons: (1) Their DNA can be made radioactive by growth in medium containing radioactive thymine (which is incorporated poorly by Thy⁺ cells); (2) a high-density label can be introduced by growing the mutants in medium with 5-bromouracil, which substitutes for and is much denser than thymine; and (3) by starving for thymine, DNA synthesis can be inhibited. Penicillin selection, which enriches for most auxotrophic mutants, fails with Thy⁻ mutants because Thy⁻ cells lose the ability to divide, and hence to form colonies, after about 15 minutes in a medium lacking thymine (this is called thymineless death). However, *thy⁻* mutants can be isolated by growth in medium containing either of the substances trimethoprim or aminopterin, which interfere with biochemical reactions that use the vitamin folic acid as a cofactor. For complex biochemical reasons, Thy⁻ mutants grow more rapidly than Thy⁺ cells in a medium containing trimethoprim and thymine. After about 100 generations of growth, which requires repeated subculturing a number of times in this medium, cultures consist primarily of mutants in the *thyA* gene.

Unusual Selections for Auxotrophic Deletion Mutants

In a discussion of the proof of λ prophage insertion in Chapter 17, it was pointed out that deletions in the *tonB-trp* region can be isolated by selecting cells that are resistant to phage T1 and then screening them for a Trp⁻ mutant. Because a double mutant would be exceedingly rare, most of these mutations turn out to be deletions. Similar techniques apply to other regions of the genome. For example, cells resistant to phage T6 also often require purines and yield deletions in this region of the *E. coli* chromosome.

Deletions of the histidine (*his*) genes can be isolated in a novel way. The temperate phage P2 differs from most other phages in that it can integrate at several different chromosomal sites. One of these is near the *his* operon. Frequently P2 is spontaneously excised from this site, leaving a cured cell (lacking a prophage). With fairly high frequency, the excision event is an aberrant one in which a large portion of the chromosome is also removed. This process produces a large deletion that invariably includes the *his* operon but may include many other nearby genes. These His⁻ mutants can be enriched by penicillin selection and isolated by replica plating. Another technique, however, is available. If an auxotrophic mutant is plated on minimal medium containing a growth-limiting amount of the nutrient, the colony will be quite small and can be isolated on the basis of its size. This technique is not applicable to most mutations because the mutation frequency is too low; that is, for a typical auxotroph, thousands of plates might be needed to find one tiny colony. This number could be reduced by vastly increasing the number of colonies per plate, but colony size is always quite variable on minimal plates containing more than a few hundred colonies. Deletions arising from P2 excision, however, occur fairly frequently, so the number of His⁻ mutants in a culture is high (about 10^{-3} in an overnight culture). Thus, this simple screening technique works well for isolating *his* deletions.

Strain Construction Using Existing Strains

One problem with isolating new strains by mutagenesis is that, after several rounds of mutagenesis, many unknown mutations will exist on the chromosome, and some of these unknown mutations may have important physiological phenotypes. Therefore new strains are most often created by transferring mutations from one strain to another. The two most common methods are bacterial conjugation and transduction. Conjugation is used when exchange of large segments of the chromosome is necessary. (For many bacteria that lack useful transducing phage, conjugation is the primary method of moving mutations into new strains.) Transduction is used when small DNA segments need to be transferred. Because the amount of DNA transferred is limited, transduction also is more likely to maintain isogenicity. For example, suppose two variants of one strain are needed, one carrying a particular *pro*⁻ mutation and another with a particular *trp*⁻ mutation. Both variants could be constructed by transferring the mutations from appropriate Hfr strains. Because there is no convenient way to control the amount of DNA that is exchanged in the recipient, the new strains could differ with respect to silent or leaky mutations present in different regions of the Hfr. With P1 transduction, however, a maximum of 2% of the chromosomal DNA can be transferred to the recipient, and an even smaller amount is usually retained after recombination has occurred; thus it is likely that the two desired variants will remain almost isogenic except for the desire *pro⁻* and *trp⁻* mutations.

In making strains by Hfr transfer, it is convenient, although not necessary, to use an Hfr whose origin is near the desired gene. For example, if the time of entry of the desired pro⁻ marker is less than 10 minutes, one can use a 1:1 ratio of Hfr to recipient, bypass any selective procedure, and merely collect several hundred recipient cells and test for the marker by replica plating. (A counterselection is required to prevent growth of the Hfr's.) More often, markers are selected or acquired by their linkage to a selective marker. For example, if a particular *pro⁻* allele is needed, a mating can be carried out between an Hfr *pur⁺ pro⁻* Str^s and an F⁻ *pur⁻ pro⁺* Str^r strain, plating on medium containing proline and lacking adenine to select for the Pur⁺ phenotype. Pur⁺ Str^r colonies are picked and tested by replica plating for the closely linked *pro⁻* mutation (Figure 19-2).

Similar selections can be carried out by generalized transduction. For example, a Lac⁻ cell can be made Lac⁺ merely by infecting it with P1 that has been grown on Lac⁺ cells and selecting for Lac⁺ transductants on lactose minimal medium. For cotransduction by phage P1, it is necessary to choose another marker that is within 2 minutes of the desired mutation in the time-of-entry map. For example, how could you construct a *leu⁻* strain that carries the suppressor *supF*? The *leu* gene is at about 1 minute on the *E. coli* map, and the *supF* gene is located at about 26 minutes in a cluster containing *tonB* (T1 resistance) and *cysB* (cysteine synthesis). Given a strain collection that includes the three strains *cysB⁻*, *leu⁻*, and *supF*, the *leu⁻ supF* mutant could be constructed as follows (Figure 19-3):

1. Make a T1-resistant (*tonB⁻*) *cysB⁻* strain by plating the *cysB* mutant on medium containing phage T1.
2. Grow P1 on the resulting *tonB⁻ cysB⁻* mutant and infect the *leu⁻* mutant. Select for transduction to *tonB⁻* by plating on medium containing phage T1. Replica-plate to minimal medium with leucine but no cysteine, and pick a *leu⁻ tonB⁻ cysB⁻* strain.
3. Grow P1 on the *supF* strain, which is also *cysB⁺*, and transduce the *leu⁻ tonB⁻ cysB⁻* strain to *cysB⁺*. Because *supF* is close to *cysB⁺*, it will be cotransduced in some of the *cysB⁺* transductants. The *supF* colonies can be found by testing for the growth of a conditional-mutant phage that can form a plaque on *supF* but not on *sup⁻* bacteria. Because *leu* is a long distance from *cysB*, the transductants will remain *leu⁻*.

A technical trick is necessary whenever P1 transduction is done. Because only about one particle in 10^4 to 10^5 P1 phage carries bacterial DNA, a large number of active P1 phage are present on a transduction plate. Thus it is necessary to prevent these phage from infecting and lysing or lysogenizing the transductants. The protection is provided by sequestering the Ca^{++} ion required for

Figure 19-2. Selection of a *proC⁻* strain by mating an Hfr *purE⁺ proC⁻* Str^s strain with an F⁻ *purE⁻ proC⁺* Str^r strain. Cells are plated on medium containing proline and streptomycin and lacking a purine.

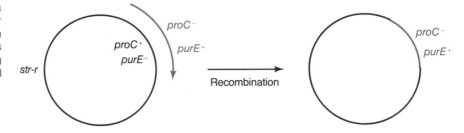

adsorption by P1: Initially P1 is allowed to adsorb to the bacteria in medium containing Ca^{++}, using a low multiplicity of infection, then the infected cells are spread on selective plates that contain citrate (which chelates the Ca^{++} ion). Thus, the progeny phage are not able to adsorb to the transductants.

Isolating Transposon Insertion Mutations in Genes

Transposons are widely used to generate mutations in particular genes. The mutations are especially valuable when complete absence of gene function is desired: Because the gene is interrupted, the mutation is essentially never leaky. Transposon mutants can be isolated by transposition from a "suicide" delivery system—that is, a phage or plasmid that is unable to replicate in or kill the recipient cells. For example, one way of transposition of Tn*10* in *E. coli* uses a λ derivative that carries a Tn*10* insertion in the *cI* gene and has nonsense mutations in genes required for lytic growth ($O^- P^-$): sup^0 cells are infected with the $\lambda O^- P^- cI::Tn10$ phage, and tetracycline-resistant colonies are selected. In each of the resulting Tet^r colonies, Tn*10* must have transposed into chromosomal genes because the $O^- P^-$ mutations prevent lytic growth of the λ in the sup^0 host, and the *cI* mutation prevents lysogeny by the λ phage. Because there are roughly 3000 nonessential genes in *E. coli*, several mutants with Tn*10* insertions in any particular gene can usually be found by simply replica plating about 10,000 colonies (50 plates with about 200 colonies per plate).

Using Transposon Insertions Near Genes

Often it is not possible to move genes into new strains by directly selecting for the mutation or for a nearby marker. In such cases, a transposon insertion near the gene can be used for cotransduction, selection for antibiotic resistance encoded by the transposon. For example, often *recA* mutants are needed to study complementation, recombination, SOS repair, or prophage induction. A *recA$^-$* mutant can be constructed by conjugation (Figure 19-4) or cotransduction. A

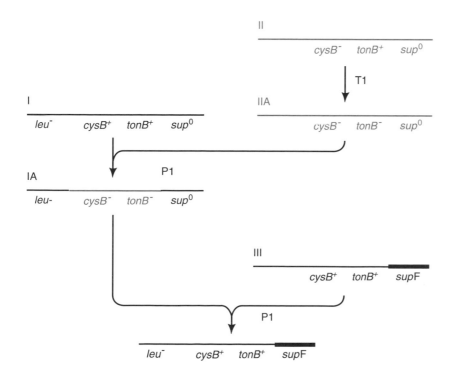

Figure 19-3. A scheme for constructing a *leu$^-$ supF* cell by successive changes, starting with strains I, II, and III. The markers *cysB, tonB,* and *sup^0* are within 1 minute of one another. The *leu* marker is actually 25 minutes away.

recA mutation cannot be directly selected, so it is moved from one strain to another by cotransduction with the *srl* gene, which maps near the *recA* gene. First a *srl*::Tn*10* mutation is transduced into the recipient selecting for tetracycline resistance encoded by Tn*10*. The resulting strain is unable to use sorbitol as a carbon source (*srl*⁻). Then the *srl*::Tn*10* mutant is transduced with phage grown on a *recA*⁻ mutation which cotransduces about 50% with Srl⁺ (Figure 19-5). The *recA*⁻ cotransductants can be identified by testing for ultraviolet sensitivity (recall that *recA*⁻ cells are very ultraviolet-sensitive; see Chapter 9).

Constructing New F's

F'*lac* inserts into the *E. coli* chromosome at fairly high frequency by Rec-mediated recombination between homologous *lac* regions on the plasmid and on the chromosome. If the chromosome contains a deletion for the entire *lac* operon, however, such integration is not possible. Integration still occurs at other sites, although at a 100-fold lower frequency, either by transposition of the IS on the F or by recombination between an IS on the F and a homologous IS on the chromosome. The alternative insertion sites are scattered throughout the chromosome. Integration can be detected if an F' is used that carries a temperature-sensitive mutation in the F replication function (see Chapter 14). Thus the F'(Ts) *lac*⁺ is transferred to a *lac*-deletion mutant, then the cells are plated on minimal lactose medium at 42°C to prevent replication of the F'(Ts) and to select for colonies that are Lac⁺. The resulting strains are Hfrs, so the position of integration can be located by determining the times of entry of various genes by mating with a suitable recipient.

The integration of F into the chromosome can occur in either the clockwise or the counterclockwise direction. The orientation can be ascertained from the time-of-entry curves.

Figure 19-4. A scheme for transferring the *recA*⁻ marker from one strain (I) to another (II). (a) Strain I is converted to HfrIA, which transfers *thy* and *recA* as early markers. (b) HfrIA is mated with strain II, to yield the desired *recA*⁻ strain (IIA).

(a) Construction of Hfr *recA*⁻. Mate Hfr *ilv*⁺ with F⁻ *ilv*⁻ *recA*⁻, and select Ilv⁺.

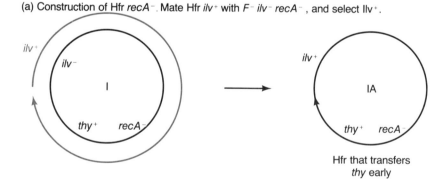

(b) Construction of *recA*⁻ recipient.

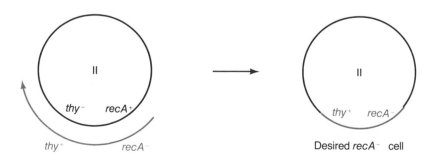

F'(Ts) *lac* can recombine with other plasmids to generate temperature-sensitive F' plasmids carrying genes other than *lac*. As long as a cell can be obtained in which there is a deletion covering the chromosomal gene carried on the F', that gene can be moved to other chromosomal sites.

Localized Mutagenesis

Some types of interesting mutations are quite rare. To isolate rare mutants might require heavy mutagenesis of the cells. Such heavy mutagenesis may produce multiple mutations in other genes. Thus the phenotype could result from the effects of several mutations. This problem can be avoided by mutagenizing only a small region of the bacterial chromosome. This is possible by mutagenizing a transducing lysate, a trick developed by Hong and Ames. For example, to find rare mutations that affect the active site of proline permease, a P22 lysate was grown on a *Salmonella* strain with a Tn*10* insertion near the proline permease gene (*putP*). This lysate was mutagenized with hydroxylamine, then a *putP*⁺ recipient was transduced, selecting for tetracycline resistance encoded by Tn*10*. The resulting transductants were replica plated onto medium with toxic proline analogs to screen for mutants with unique phenotypes. Such rare mutants, which were never found by general mutagenesis of the chromosome, were found at a frequency of 10^{-2} to 10^{-3} by this localized mutagenesis. An example is shown in Box 19-1.

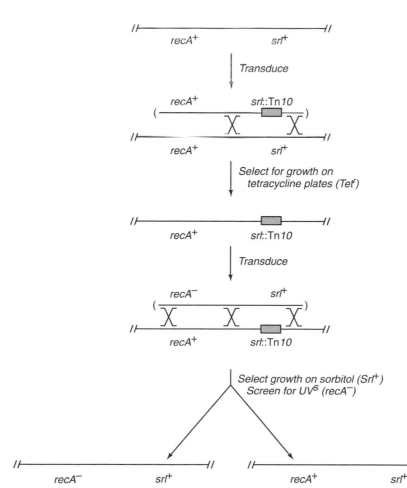

Figure 19-5. Use of closely linked transposon insertions for strain construction. An example showing the construction of a *recA* mutant using an *srl*::Tn*10* insertion that is 50% cotransducible with the *recA* gene.

BOX 19-1. LOCALIZED MUTAGENESIS WITH HYDROXYLAMINE

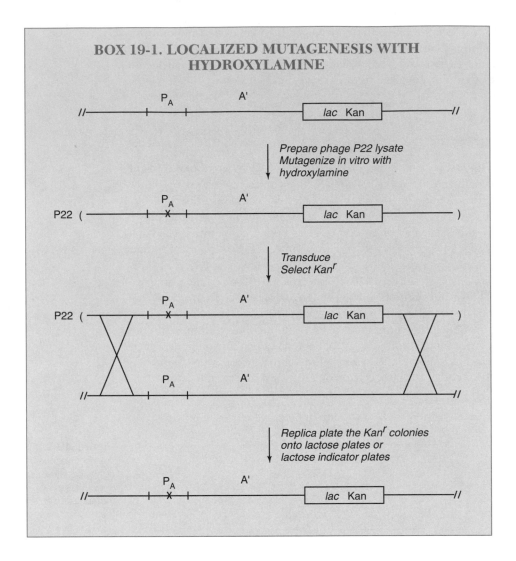

OPERON AND GENE FUSIONS

To study the regulation of gene expression, it is necessary to be able to assay for products of the genes in an operon. For many genes, the assays are either very tedious or inaccurate. In contrast, the assay for β-galactosidase is simple and precise. By constructing a fusion that places the *lac*Z gene under control of the promoter of an operon of interest, expression of the operon can be studied by assaying β-galactosidase activity.

Two types of such fusions are commonly used, operon and gene fusions (see Chapter 7). Operon fusions with *lac*Z place the transcription of the *lac*Z under the control of another promoter, but the *lac*Z gene retains its own translational start sites. Thus, operon fusions make a single mRNA that encodes a truncated polypeptide from the first part of the interrupted gene followed by the *lac*Z protein. The proteins made by an operon fusion are discrete. In contrast, a *lac*Z gene fusion places transcription and translation of the *lac*Z gene under control of another gene. Thus, gene fusions make a single mRNA that encodes a single hybrid polypeptide with the N-terminus of the interrupted gene product and β-galactosidase on the C-terminus.

One example of a gene fusion was described in Chapter 7—the fusion of the *lac* and *pur* operons was formed by a deletion between these genes. This fusion, however, was not particularly valuable because it linked *pur* genes to the *lac* promoter. Nonetheless, this fusion illustrates a general method: the introduction of a deletion between the promoter of one operon and a gene of another. Clearly a deletion cannot be put between the *lac* operon and any gene; a gene that is quite distant would require a gigantic deletion that would remove a large number of essential genes. Thus fusions are commonly constructed by moving the *lac* operon.

Using the preceding technique, we saw how to move the *lac* genes to the *tonB* locus. This relocation puts *lac* adjacent to the attachment site for phage φ80. Preparation of a φ80 lysogen and a search for transducing particles made by inducing the lysogen has yielded a φ80*lac* particle. This phage can be used to isolate a fusion between the *lac* and the *trp* operons that will put the *lacZ* gene under control of the *trp* promoter. Figure 19-6 shows a portion of the genetic map near the *trp* operon. Downstream from this operon is the *tonB* locus and the prophage attachment site for phage φ80. Using a Lac⁻ strain, φ80*lac* can be inserted into the attachment site. This yields the gene order *trp tonB lac*. In this strain, the *trp* and *lac* promoters still control their respective genes. These cells are then plated on agar seeded with T1. Surviving colonies are tested for the ability to grow on minimal plates lacking tryptophan. Those that cannot grow on these plates are T1^r and Trp⁻ and hence are likely to have deletions that extend from the *trp* operon through the *tonB* locus. These deletions are tested further for the ability to grow on lactose as a carbon source only *when tryptophan is absent.* Because tryptophan represses the *trp* operon, fusion strains that have removed the *lac* promoter will not be able to use lactose in the presence of tryptophan. Mutations in different genes within the *trp* operon can be distinguished by their ability to grow when various intermediates of the tryptophan biosynthetic pathway are provided. By such tests, the extent of the deletion into the *trp* operon can be determined.

Use of Mu*d* Phage to Isolate Operon and Gene Fusions

The fusion procedure described previously depends on the ability to isolate deletions in specific chromosomal regions. A variety of transposon based tools have been developed that eliminate this requirement. Some of the most useful vectors for constructing operon and gene fusions are derivatives of phage Mu. Mu is a phage that can transpose to essentially random sites in the chromosome: When it inserts within a gene, it disrupts the gene resulting in a null mutation (see Chapter 12). Casadaban constructed derivatives of Mu phage (called Mu*d*) that carry an antibiotic-resistance gene (such as ampicillin resistance or kanamycin resistance) and place a promoter-less *lac* operon just inside one end of Mu (Figure 19-7a,b). One type of Mu*d* phage contains translational start sites for *lacZ*, so it forms operon fusions. Another type of Mu*d* phage lacks the translational

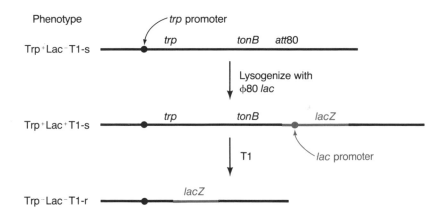

Figure 19-6. Construction of a fusion strain that couples the *trp* promoter and the *lacZ* gene.

start sites for *lacZ*, so it forms gene fusions resulting in a hybrid protein. (This nomenclature is often confused by nongeneticists: Sometimes operon fusions are mistakenly called gene fusions, and gene fusions are mistakenly called protein fusions. The hybrid proteins, however, result from a fusion between two genes, so the correct terminology is gene fusions.)

A variety of Mu*d* vectors are now commonly used, and the methods used for transposition of the different vectors vary. In each case, however, isolating fusions in a desired gene is quite similar. Transposition of Mu*d* is induced such that each cell in the population has a single Mu*d* insertion at a random site on the chromosome. Cells with Mu*d* insertions can be selected as antibiotic-resistant colonies. The antibiotic-resistant colonies can then be replica plated to screen for the desired mutants. Mutations in a particular gene are found at about 1 in 3000 Mu*d* insertion mutants.

Mu*d* insertions can occur in either orientation within a gene (Figure 19-7c, d). Hence about half of the Mu*d* operon fusion will be inserted in the proper orientation for transcription of β-galactosidase and form blue colonies on plates containing X-gal. (The indicator X-gal is a colorless compound that does not induce *lac* expression but produces a blue dye if cleaved by β-galactosidase.) Similar to operon fusions, about half of the Mu*d* gene fusions are inserted in the proper orientation for transcription of β-galactosidase. Gene fusions, however, must also be inserted in the correct reading frame for proper translation of β-galactosidase, so only one in six of the insertions will form blue colonies on plates containing X-gal.

Operon and gene fusions have a variety of uses. A few are described here:

1. *Regulation of gene expression.* Operon and gene fusions can be used to determine if regulation of gene expression occurs at the transcriptional or translational level. If the gene is regulated at the transcriptional level, β-galactosidase expression from both operon and gene fusions to a gene will be induced or repressed to a similar extent. In contrast, if the gene is regulated at the translational level, β-galactosidase expression from an operon fusion to the gene will not be induced or repressed, but β-galactosidase expression from a gene fusion to the gene will be.

2. *Isolation of regulatory mutants.* The many analogs and indicator plates useful for selecting or screening for *lac* regulatory mutants can also be

Figure 19-7. Mu*d* vectors for constructing genetic fusions. Transcription is indicated by straight arrows and translation is indicated by squiggly arrows. (a) A Mu*d* operon fusion. The '*trp*' sequence provides the necessary ribosome binding site for the *lac* operon. (b) a Mu*d* gene fusion. Note that the *lac* operon is missing the necessary translational start sites, so translation of *lac* must be initiated from the sites within the disrupted gene. (c and d) Mu*d* can insert in either orientation within a gene. Only one orientation (c) yields a fusion with the disrupted gene.

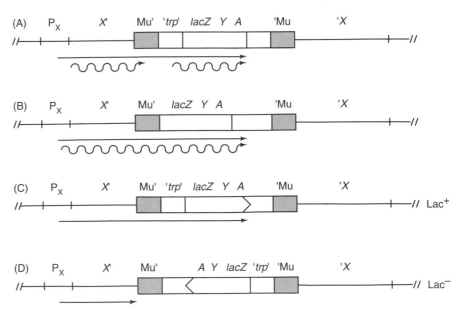

used with *lac* operon and gene fusions as well. For example, under repressing conditions, many Mu*d* operon fusions do not make sufficient β-galactosidase and lactose permease to allow growth on lactose as a sole carbon source. Thus regulatory mutants that result in increased transcription of the mutant gene can be isolated by simply plating the operon fusion mutants on minimal-lactose plates to select for Lac$^+$ mutants.

3. *Determining the direction of transcription.* Operon and gene fusions can also be used to determine the direction of transcription of a gene relative to other genes on the chromosome. One approach involves using the Mu*d* insertion as a region of homology with F' (Ts) lac as described in Chapter 7.

CONSTRUCTION OF PHAGE MUTANTS

As in the construction of bacterial mutants, a variety of techniques are used to construct phage mutants. Some are rather general, and others make use of specific properties of certain mutations, as in the following examples.

Production of Single Mutants

The production of a single mutant often involves little other than selection following mutagenesis. In some cases, it is possible to guess the properties of the mutants; for example, λ *red*$^-$, *int*$^-$, and *ris* mutations (see Chapters 16 and 17). In other cases, with plaque-morphology mutations, no selection is needed; that is, λ clear-plaque mutations and T4 *r* mutations can be isolated by simply inspecting the plates. Isolation of some mutants, however, requires brute force. For example, most of the nonsense mutations in λ were isolated by mutagenizing a phage suspension, plating it out on *sup*$^-$ bacteria, and then testing tens of thousands of plaques for suppressor-sensitive mutations by stabbing a lawn of *sup*^{0} bacteria. About 50 plaques can be tested on a single stab plate, so many plates were needed. The mutant phage were then classified by complementation.

Isolation and Phage with Multiple Mutations

Often phage with multiple mutations are required. For example, the λ vectors described previously that are used for Tn*10* transposition contain a clear plaque mutation and mutations in replication genes. How do you isolate phage with multiple mutations? The easiest type of double mutant phage to construct contains two mutations of different types, such as a plaque-morphology mutation and a suppressor-sensitive mutation. For example, a T4 *45*(Am) *rII* (most of the T4 genes are designated by numbers) can be isolated by coinfecting bacteria with a T4 *rII* mutant and a T4 *45* mutant and plating the progeny phage on a mixture of *sup*^{0} and *sup*$^-$ bacteria. T4 *45*$^+$ can infect both types of cells producing a clear plaque, whereas *45*(Am) mutants grow only on the *sup*$^-$ cells and hence form a turbid plaque. The *r*$^+$ and *rII* mutations are identified by plaque size, so the T4 *45*(Am) *rII* double mutant can be recognized as a large turbid plaque.

A more complex approach is needed to isolate a λ*cI*$^-$ *O*(Am) mutant:

1. Cross λ *cI*$^-$ *O*$^+$ with λ *cI*$^+$ *O*(Am), and plate on *sup*$^-$ bacteria. The progeny phage will include λ*cI*$^+$ *O*(Am), λ*cI*$^-$ *O*$^+$ (parents) and λ*cI*$^-$ *O*(Am), λ*cI*$^+$ *O*$^+$ (recombinants).

2. Select a few hundred clear (cI^-) plaques; these are either cI^- O^+ or cI^- O(Am). Test these clear plaques for the O^- allele by stabbing into lawns of sup^- and sup^0 bacteria. The desired cI^- O(Am) double mutants produce clear plaques on the sup^- bacteria and no plaques on the sup^0 bacteria.

A genetic trick can be used to isolate λcI^- P^- phage. Recall that many λP^- phage plate on GroP$^-$ cells (a class of *dnaB* mutation), whereas λP^+ do not. Thus, plating the progeny of the cross λcI^- P^+ × λcI^+ P^- on a GroP$^-$ host yields only the λcI^+ P^- parents, which produce turbid plaques, and the desired λcI^- P^- mutant, which produces clear plaques.

Construction of a Phage with Two Suppressor-Sensitive Mutations

In constructing a phage with two suppressor-sensitive mutations, it is rarely possible to select for the mutations directly. For example, consider the construction of λD(Am) R(Am). (You cannot simply cross λD(Am) and λR(Am) phages and select for the double mutant because only a wild-type recombinant can be selected by growth on sup^0 cells. Testing progeny for inability to grow on sup^0 bacteria will not distinguish a double mutant from either of the parents because neither the parents nor the double mutant will grow.) The solution is to use linked markers. For example, a cross could be made between λ cI^- D(Am) and λ c^+ R(Am) *red*$^-$ (Figure 19-8). Plating on *recA*$^-$ cells and picking turbid plaques allows isolation of recombinants with an exchange between the *red* and c markers. Unless there is a double crossover, the phage will be D(Am) c^+ R(Am). It is necessary to check that a double crossover has not occurred. This can be done by back crossing the recombinants with the parental phage. If the recombinants contain the D(Am) and R(Am) mutations, they will be unable to repair the inability of either parent to grow on sup^0 cells.

Isolation of λ Deletions

Deletion mutations can be useful, but there are no mutagenesis procedures that specifically generate deletions. Parkinson took advantage of a physical property of phage λ to develop a specific selection for deletion mutants. λ phage are unstable in the absence of divalent cations, particularly Mg^{++}. If λ phage are placed in a solution of ethylenediaminetetra-acetic acid (EDTA), a chelating agent that binds divalent cations and effectively removes them from solution, the phage burst open, releasing their DNA (and hence lose the ability to form plaques). The divalent cations serve two purposes: (1) They neutralize the charge repulsion between the negatively charged phosphates in the DNA, allowing the DNA to fold into a compact structure. If the charge is not neutralized, the DNA expands, which increases the internal pressure in the head. (2) They stabilize the phage head. When λ is placed in EDTA, a small fraction (10^{-6}) of the phage survive and remain able to form plaques.

Figure 19-8. A cross that can yield λ D^- Q^-. Plating on *recA*$^-$ bacteria and selecting turbid (c^+) plaques demand the exchange shown in orange. If only a single exchange occurs, the result is λ D^- c^+ Q^-.

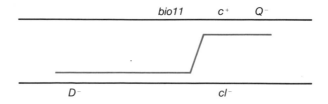

The survivors have deletions in nonessential regions of the phage DNA (usually in the *b2* region; see Figure 15-2), presumably because with less DNA in the phage head, there is less internal pressure. By using different concentrations of EDTA, it is possible to isolate deletions of different lengths: The higher the EDTA concentration, the larger the deletion. The maximum deletion reduces size of the λ DNA about 21%. (Smaller DNA molecules are not packaged properly.)

Construction of Partial Prophage Deletions

In Chapter 17, we described studies that used a bacterial strain containing only a part of a prophage. Several methods are available to prepare such strains. For example, consider a lysogen of λ *cI857* (Ts) *int⁻*, which has been made by use of a λ*int⁺* helper phage. If this lysogen is heated to 42°C, the cells will be killed because the cI repressor will be inactivated, and although the phage cannot excise (since it is *int⁻*), the functions that kill the host cell will still be produced. About 1 cell in 10^8, however, survive this treatment, and if the experiment is done on galactose color-indicator plates, some of these survivors will be Gal⁻. These cells have a deletion that removes the *gal* genes and extends into the prophage past the killing functions (that is, past gene *R*).

A related method makes use of the *chlD* gene, which is located in the region *gal-chlD-attλ-bio*. Wild-type *E. coli* cannot grow in the presence of the chlorate ion because the *chlD⁺* gene converts it to a toxic compound. Plating a λ lysogen on medium containing chlorate and selecting survivors that are also Gal⁻ yields deletions that remove all or part of the *gal* locus, through the *chlD* gene and generally into the λ prophage. When a nonlysogen is exposed to chlorate, deletions that remove the entire *gal-bio* region are common. This is a useful technique to delete *attλ* from *E. coli*.

Isolation of a Phage Carrying a Suppressor

Lysogenic phages may be used to isolate specialized transducing particles (see Chapter 17). The specialized transducing particles can be used to move a nearby gene into other cells. (In essence, the nearby gene is "cloned" on the phage particle.) In early studies of suppression, Beckwith isolated a phage that carried the *supF* gene. This ultimately led to an understanding of nonsense suppressors because it made it possible to determine the base sequence of the suppressor tRNA.

The *supF* gene is near the attachment site for phage φ80. Because specialized transducing particles are rare, a selection was needed to find the specialized transducing phage. A φ80 strain carrying a nonsense mutation (in a phage gene) suppressible by *supF* was used to lysogenize a *supF* strain. Then the lysogen was induced to obtain a lysate that contained some transducing particles. The lysate was plated on *sup⁰* bacteria. The original φ80 mutant could not form plaques because of the nonsense mutation (Table 19-1). About 1 phage per 10^6, however,

Table 19-1 Plaque-forming ability of a φ80 mutant and its *sup-3*-transducing variant on two bacterial strains

Bacteria	Plaque-forming ability		
	Phage φ80	Phage φ80(Am)	Page φ80 (Am)*supF*
sup⁰	Yes	No	Yes
supF	Yes	Yes	Yes

Note: Am refers to the presence of a chain termination mutation in a phage gene.

formed plaques; these were specialized transducing phages that carried the *supF* suppressor (φ80 *supF*).

The φ80 *supF* phage was used in many bacterial strain constructions to introduce a suppressor into *sup*⁰ strains. Merely by lysogenizing with the transducing phage, the strain acquired the suppressor phenotype.

Specialized transducing particles carrying other genes can also be isolated. Typically *att*λ or *att*80 is placed near the gene of interest, and then specialized transducing particles are isolated by the techniques already described.

KEY TERMS

genotype	screen
indicator medium	selection
localized mutagenesis	selective medium
phenotype	strains

QUESTIONS AND PROBLEMS

1. Starting with F⁻ *trp*⁻ *lac*⁻ Str^r, make a *trp*⁻ *gal*⁻ *lac*⁺ Str^r strain. You may make use of any of the following strains: F⁻ *gal*⁻ *lac*⁻, F⁻ *gal*⁻ *lac*⁺, Hfr *gal*⁻ *lac*⁺, and Hfr *gal*⁺ *lac*⁻, in which the Hfr's transfer clockwise from an origin at 99 minutes.

2. You have an Hfr Δ*lac* strain (Δ represents a deletion) that transfers the deletion a few minutes after mating. Your strain collection also contains numerous Hfr *lac*⁺ strains. Use these strains to replace a *lac*⁻ point mutation in an F⁻ strain by a *lac* deletion.

3. By using λ*cI857* (a λ with a temperature-sensitive *cI* repressor) as a tool, convert a *gal*⁺ strain to one that is deleted for *att*λ.

4. The *uvrB* gene is near the *bio* genes. Starting with a prototroph, make a *uvrB* deletion. No mating or transduction need be done. Any phage may be used as well as any tricks you can think of.

5. A highly useful *lac*⁻ mutation has been isolated in an Hfr. You would like to transfer this mutation to a particular F⁻ strain. Mating and selection for the Lac⁻ phenotype will not work very well because the *lac* gene is transferred fairly late by this Hfr; thus the fraction of recipients that will be Lac⁻ will be too small to detect on a color-indicator plate. Suggest a procedure that will work. Use phage T6 and any kind of plate.

6. Starting with three different T4 phages, each carrying one of the mutations *ac41*, *rII*, and *h*, make a triple mutant. Describe the bacteria and the plates that would be used.

7. Prepare a *uvrA*⁻ *uvrB*⁻ *leu*⁻ triple mutant, starting with *uvrA*⁻ and *leu*⁻ strains. The *uvrA* gene is at 91.6 minutes, and *uvrB* is in the *gal attl bio chlD uvrB* cluster at 18 to 19 minutes. A *malB* mutant (which makes cells resistant to λ) will be useful: *malB* is within 0.1 minutes of the *uvrA* gene. Be sure to include a test for the double mutant.

8. You have an F⁻ *leu*⁻ Str^r strain and wish to make it *leu*⁻ Str^s. Assuming you have a large collection of bacteria, how might you do it?

9. A *dna*(Ts) mutant is unable to synthesize DNA at high temperature but continues to make protein for a time equivalent to two to three generations. In the absence of DNA synthesis (that is, at a nonpermissive temperature), cell division is inhibited in *dna*(Ts) mutants. If the time at high temperature does not exceed a few generations, lowering the temperature allows DNA synthesis to be restored, and after a generation at a low (permissive) temperature, cell division resumes and proceeds until the proper ratio of DNA to protein is achieved. Suggest a procedure by which a *dna*(Ts) mutant might be isolated. No special plates are needed, but one or more of the following pieces of common laboratory equipment (e.g., a centrifuge, filter, or pH meter) are helpful.

10. Describe how you would isolate a Leu⁻ strain using Tn*10*. You may use λ *cI857 P*(Am) *O*(Am) Tn*10*.

11. Select a bacterial strain in which the promoter for the arabinose operon is linked to the *lacZ* gene.

12. Construct λ *cI857 red⁻ C*(Am) starting with λ *cI857 C*(Am), λ *bio-11 c⁺*, and λ *c⁺ red⁻*.

13. Starting with λ *bio-11 R⁻* and λ *cI857 A⁻*, isolate λ *cI857 A⁻ R⁻*. The gene order is *A -bio-11 -cI -R*, and the recombination frequency between *A* and *R* is about 10%.

14. Starting with λ *bio-11 c⁺ A⁻* and λ *cI857 R⁻*, make λ *cI857 A⁻ R⁻*. Note the difference between this construction and that in Problem 13. A critical point is the fairly high recombination frequency between *A* and *R* (about 10%). (*Hint*: This is somewhat of a brute-force selection.)

REFERENCES

Beckwith, J. 1963. Restoration of operon activity by suppressors. *Biochim. Biophys. Acta*, 76, 162.

Beckwith, J. 1981. A genetic approach to characterizing complex promoters in *E. coli. Cell*, 23, 307.

Beckwith, J. 1991. Strategies for finding mutants. *Methods Enzymol.*, 204, 3.

Casadaban, M. J., et al. 1977. Construction and use of gene fusions directed by bacteriophage Mu insertions. In *DNA Insertion Elements, Plasmids, and Episomes* (A. I. Bukhari et al., ed.), p. 531. Cold Spring Harbor.

Davis, R. W., D. Botstein, and J. R. Roth. 1980. *A Manual for Genetic Engineering. Advanced Bacterial Genetics*. Cold Spring Harbor Laboratory.

Dila, D., and S. Maloy. 1987. Proline transport in *Salmonella typhimurium: putP* permease mutants with altered substrate specificity. *J. Bacteriol.*, 168, 590.

Franklin, N. 1978. Genetic fusions for operon analysis. *Ann. Rev. Genet.*, 12, 193.

Gottesman, S., and J. R. Beckwith. 1969. Directed transposition of the arabinose operon: a technique for the isolation of specialized transducing bacteriophages for any *E. coli* gene. *J. Mol. Biol.*, 44, 117.

Groisman, E., and M. J. Casadaban. 1986. Mini-Mu bacteriophage with plasmid replicons for in vivo cloning and *lac* gene fusion. *J. Bacteriol.*, 168, 357.

*Maloy, S. 1989. *Experimental Techniques in Bacterial Genetics*. Jones and Bartlett, Boston.

*Manoil, C., J. Makalanos, and J. Beckwith. 1990. Alkaline phosphatase fusions: sensors of subcellular location. *J. Bacteriol.*, 172, 515.

+Miller, J. H. 1992. *A Short Course in Bacterial Genetics*. Cold Spring Harbor Laboratory.

Parkinson, J. S., and R. Huskey. 1971. Deletion mutants of bacteriophage lambda. Isolation and initial characterization. *J. Mol. Biol.*, 56, 369.

Roth, J. 1970. Genetic techniques in studies of bacterial metabolism. *Methods Enzymol.*, 17, 1.

Silhavy, T., and J. Beckwith. 1985. Uses of gene fusions for the study of biological problems. *Microbiol. Rev.*, 49, 398.

Silhavy, T., L. Enquist, and M. Berman. 1984. *Experiments with Gene Fusions*. Cold Spring Harbor Laboratory.

Vinopal, R. 1987. Selectable phenotypes. In Escherichia coli *and* Salmonella typhimurium: *Cellular and Molecular Biology*. (F. Neidhardt, J. Ingraham, K. B. Low, B. Magasanik, M. Schaechter, and H. E. Umbarger, eds.). American Society for Microbiology, Washington, D. C.

*Resources for additional information.

THE NEW MICROBIAL
GENETICS

Genetic Engineering

In addition to providing insight into basic biological questions, genetics can also be used to manipulate biological systems for scientific or economic reasons. Traditionally this genetic manipulation required mutagenesis, gene transfer, and genetic recombination followed by selection for desired characteristics. When using such techniques, geneticists have been forced to work with the random nature of mutagenic and recombination events, which required selective procedures, often quite complex, to find rare mutants with the desired genotype. Since the early 1970s, new techniques have been developed that allow the genotype of an organism to be modified in a directed and predetermined way. In this approach, called **recombinant DNA technology**, **genetic engineering**, or **gene cloning**, purified DNA fragments are isolated and recombined by in vitro manipulations. The basic technique is quite simple: Two DNA molecules are isolated and cut into fragments by one or more specialized enzymes, the fragments are joined together in a specific desired combination, then the recombinant DNA molecule is returned to a cell. The recombinant DNA technology has greatly enhanced our ability to manipulate genes and has revolutionized the study of gene structure and regulation.

JOINING DNA MOLECULES

The basic procedure of the recombinant DNA technique consists of two stages: (1) joining a desired DNA segment to a DNA molecule that is able to replicate and (2) introducing the recombinant DNA into cells where it can replicate to produce many copies of the desired DNA fragment (Figure 20-1). When this is done, the genes in the donor segment are said to be **cloned**, and the carrier molecule is the cloning **vector**.

Vectors

To be useful, a vector must have three properties: (1) It must be able to replicate, (2) there must be some way to introduce the vector DNA into a cell, and (3) there must be a simple selection or screen for the vector.

The three most common types of vectors used are plasmids, phage λ, phage M13, and viruses. The DNA of each of these vectors has a replication origin, carries genes that are identifiable by simple plating or biochemical tests, and can be introduced into appropriate host cells. There are a large number of different vectors that are used for different purposes. A few common vectors and their properties are listed in Table 20-1.

Restriction Enzymes

Restriction enzymes were discovered as an outgrowth of the study of host restriction and modification of phages, described for phage λ in Chapter 5. An attempt to account for the site specificity of the DNA cleavage led to the isolation of novel nucleases called **restriction endonucleases** or **restriction enzymes**. These are endonucleases that recognize a specific base sequence in a DNA molecule and make two cuts, *one in each strand* of the double helix, generating 3'-OH and 5'-P termini. More than 1000 different restriction enzymes have been purified from hundreds of different microorganisms. Most these enzymes recognize sequences that are nucleotide palindromes. That is, the DNA sequence of the recognition site has the structure:

$$
\begin{array}{c}
\text{A B C } | \text{ C}' \text{ B}' \text{ A}' \\
\text{A}' \text{ B}' \text{ C}' | \text{ C B A}
\end{array}
\quad \text{or} \quad
\begin{array}{c}
\text{A B } \text{X} \text{ B}' \text{ A}' \\
\text{A}' \text{ B}' \text{X}' \text{ B A}
\end{array}
\quad \text{or} \quad
\begin{array}{c}
\text{A B } | \text{ B}' \text{ A}' \\
\text{A}' \text{ B}' | \text{ B A}
\end{array}
$$

Where the capital letters represent bases, a prime indicates a complementary base, X is any base, and the vertical line is the axis of symmetry. A few recognition sites with more than six bases are known, but none have been observed that contain fewer than four bases.

Figure 20-1. An example of cloning. A fragment of frog DNA is joined to a cleaved plasmid vector. The hybrid plasmid is transformed into a bacterial host. The cloned frog DNA is carried to all progeny bacteria by replication of the plasmid.

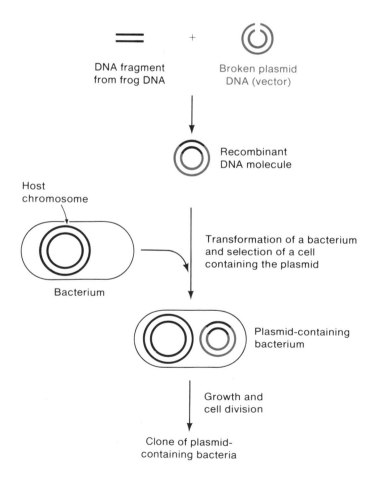

The specific recognition sequence for each enzyme differs in base sequence and size. Size of the recognition sequence is an important determinant of how many times the enzyme will cut a particular piece of DNA. Thus, the longer the recognition sequence, the lower the probability that the sequence will be found in a particular DNA molecule. Because there are 4 bases and any one of the bases could be present at a particular position in a DNA sequence, the probability that a particular base will occur at a particular position in a DNA sequence is 1/4. Therefore the probability of finding a specific sequence of nucleotides is $(1/4)^n$, where n is the number of nucleotides. Some restriction enzymes recognize restriction sites of only 4 bp in length. These enzymes are called **frequent cutters** because there is a high probability that this site will occur in a DNA molecule. For example, if all 4 bases occur with equal frequency, the probability of the site occurring is $P = (1/4)^4 = 1/256$. In other words, on the average you would expect to find this site once in every stretch of 256 bp. Restriction enzymes with recognition sequences of 6 bp are widely used in cloning because they cut less frequently and therefore produce a high proportion of DNA segments larger than 1 kb in size but small enough to be manageable for cloning. For example, a given 6 bp site is expected to occur once in every 4096 bp. Enzymes that have 8 to 10 bp recognition sequences have been discovered. These enzymes cut infrequently and produce DNA fragments hundreds of kilobases in size. They are used to map whole bacterial chromosomes and eukaryotic genomes by pulsed field gel electrophoresis (Chapter 2).

There are several major types of restriction enzymes: type I enzymes, which recognize a specific sequence but make cuts at random in the adjacent DNA; type II enzymes, which make cuts only *within* the recognition site, and type IIS enzymes, which make cuts at a set distance (less than 20 bp) outside of the recognition site. Type II enzymes are the most useful for genetic engineering since ends have a known sequence, making joining of the DNA fragments easier. All restriction enzymes of this class make two single-strand breaks, one break in each strand. There are two distinct arrangements of these breaks: (1) both breaks at the center of symmetry (generating **flush** or **blunt ends**) or (2) breaks that are symmetrically placed around the center of symmetry (generating **cohesive** or **"sticky" ends**). Some examples and their consequences are shown in Figure 20-2. Table 20-2 lists the sequences and cleavage sites for 12 useful restriction enzymes, nine of which generate cohesive sites and three of which yield flush ends.

Because a particular restriction enzyme recognizes a unique sequence, *the*

Table 20-1 Some common bacterial cloning vectors

Vector	Properties useful as a cloning vector
Plasmids	
pSC101	Low copy number vector, Tet^r, compatible with ColE1-type plasmids
pBR322	Medium copy number vector, Tet^r Amp^r
pUC18	High copy number vector, Amp^r, contains a multiple cloning site and Lac α-complementing fragment so insertions can be detected by "blue to white" screening
pUC19	Identical to pUC18 except the multiple cloning site is in the opposite orientation, allows cloning of a DNA fragment in both orientations
Phage	
M13mp18	Single-stranded phage vector, Amp^r, double-stranded RF contains multiple cloning site and Lac α-complementing fragment so insertions can be detected by "blue to white" screening
M13mp19	Identical to M13mp18 except the multiple cloning site is in the opposite orientation, allows cloning of a DNA fragment in both orientations
λgt	Phage λ cloning vector, allows clones to be maintained in a single copy on a λ lysogen

number of cuts made in a DNA fragment is limited. For example, a restriction enzyme that recognizes a 6 bp site will cut a typical bacterial DNA molecule, which contains roughly 3×10^6 bp, into a few hundred to a few thousand fragments. Smaller DNA molecules such as phage or plasmid DNA molecules may have fewer than 10 of the sites (frequently one or two and often none). Because of this specificity, *a particular restriction enzyme generates a unique set of fragments from a particular DNA molecule.* Another enzyme will generate a different set of fragments from the same DNA molecule. (Some enzymes—called **isoschizomers**—have the same site specificity.) Figure 20-3a shows the sites where phage λ DNA is cut by the enzymes *Eco*RI and *Bam*HI. The family of fragments generated by a single enzyme is usually detected by agarose gel electrophoresis of the digested DNA (Figure 20-3b). The fragments migrate at a rate that is a function of molecular weight (see Chapter 3), and their molecular weights can be determined by reference to fragments of known molecular weight that are run concurrently.

If a host contains a type I restriction enzyme, it is difficult to introduce unmodified, foreign DNA into the cell. To avoid this problem, most strains used for cloning foreign DNA contain mutations that inactivate any restriction enzymes. The *hsdR* gene encodes the endonuclease function, and the *hsdM* gene encodes a protein required for both the endonuclease and the modification functions (r^-m^-). Hence *hsdR* mutants are restriction$^-$ modification$^+$ (r^-m^+), and *hsdM* mutants are restriction$^-$ modification$^-$ (r^-m^-). *hsdR* mutants are most useful as hosts for cloning foreign genes because these mutants cannot restrict foreign DNA, but once the DNA is in the cell, it will be modified by the *hsdM* gene product.

Cloning a Restriction Fragment in a Plasmid

A particular restriction enzyme recognizes a specific base sequence. Furthermore, most restriction enzymes that produce cohesive ends generate complementary single-stranded termini at each end of the fragment. Hence the fragments generated by cutting one DNA molecule with a particular restriction enzyme have the same set of single-stranded ends as the fragments produced by cutting a different DNA molecule with the same enzyme. (Except when one of the DNA molecules is a linear DNA molecule, the ends of linear fragments would be different because they are not produced by cutting with the restriction enzyme—for example, phage λ cut with *Eco*RI). Therefore fragments from the DNA molecules of two different organisms (for example, a bacterium and a frog) can be joined by renaturation of the complementary single-stranded termini. If

Figure 20-2. Two types of cuts made by restriction enzymes. The arrows indicate the cleavage sites. The dashed line is the center of symmetry of the sequence.

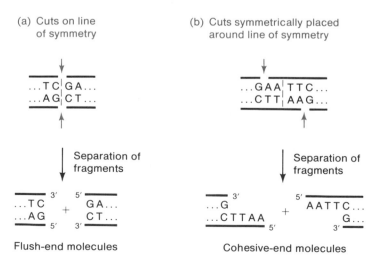

(a) Cuts on line of symmetry

...TC|GA...
...AG|CT...

Separation of fragments

...TC $^{3'}$ $^{5'}$ GA...
...AG   CT...
    $_{5'}$ $_{3'}$

Flush-end molecules

(b) Cuts symmetrically placed around line of symmetry

...GAA|TTC...
...CTT|AAG...

Separation of fragments

...G $^{3'}$ $^{5'}$ AATTC...
...CTTAA   G...
   $_{5'}$ $_{3'}$

Cohesive-end molecules

the hybrid DNA is sealed with DNA ligase (**ligated**) after base pairing, the fragments are covalently joined.

    If a DNA molecule is ligated to a plasmid, it will be replicated with the plasmid DNA. For example, Figure 20-4 shows a plasmid DNA molecule that has only one cleavage site for a particular restriction enzyme, which is also used to cleave frog DNA. If frog fragments are mixed with linearized plasmid DNA and the fragments are ligated, a circular plasmid DNA molecule containing frog DNA can be generated. Once the hybrid plasmid is moved into a bacterium by transformation, the inserted DNA fragment, which replicates as part of the plasmid, is **cloned**.

Joining of DNA Fragments by Addition of Homopolymers

The field of recombinant DNA research began in 1972 (just before the properties of restriction enzymes were understood) with the development of a general method for joining any two DNA molecules. The initial approach used the

Table 20-2   Some restriction endonucleases and their cleavage sites

Microorganism	Name of enzyme	Target sequence and cleavage sites
Generates cohesive ends		
E. coli	EcoRI	G↓A A | T T C C T T | A A↑G
Bacillus amyloliquefaciens H	BamHI	G↓G A | T C C C C T | A G↑G
B. globigii	BglII	A↓G A | T C T T C T | A G↑A
Haemophilus aegyptius	HaeII	Pu G C | G C↓Py Py↑C G | C G Pu
Haemophilus influenza	HindIII	A↓A G | C T T T T C | G A↑A
Providencia stuartii	PstI	C T G | C A↓G G↑A C | G T C
Streptococcus albus G	SalI	G↓T C | G A C C A G | C T↑G
Thermus aquaticus	TaqI	T↓C | G A A G | C↑T
Generates blunt ends		
Brevibacterium albidum	BalI	T G G↓C C A A C C | G G T↑
Haemophilus aegyptius	HaeI	(A) G G↓C C (T) (T) C C | G G (A)↑
Serratia marcescens	SmaI	C C C↓G G G G G G | C C C↑

Note: The vertical dashed line indicates the axis of dyad symmetry in each sequence. Arrows indicate the sites of cutting. The enzyme *Taq*I yields cohesive ends consisting of two nucleotides, whereas the cohesive ends produced by the other enzymes contain four nucleotides. The enzyme *Hae*I recognizes the sequence GGCC whether the adjacent base pair is A·T or T·A, as long as dyad symmetry is retained. Pu and Py refer to any purine and pyrimidine, respectively.

enzyme **terminal nucleotidyl transferase**, an unusual DNA polymerase obtained from animal tissue, which adds nucleotides to the 3'-OH group of an extended single-stranded segment of a DNA chain. The reaction differs from that of ordinary polymerases in that *it does not need a template strand*. To generate the extended single strand with a 3'-OH terminus, one need only treat the DNA molecule with a 5'-specific exonuclease to remove a few terminal nucleotides. In

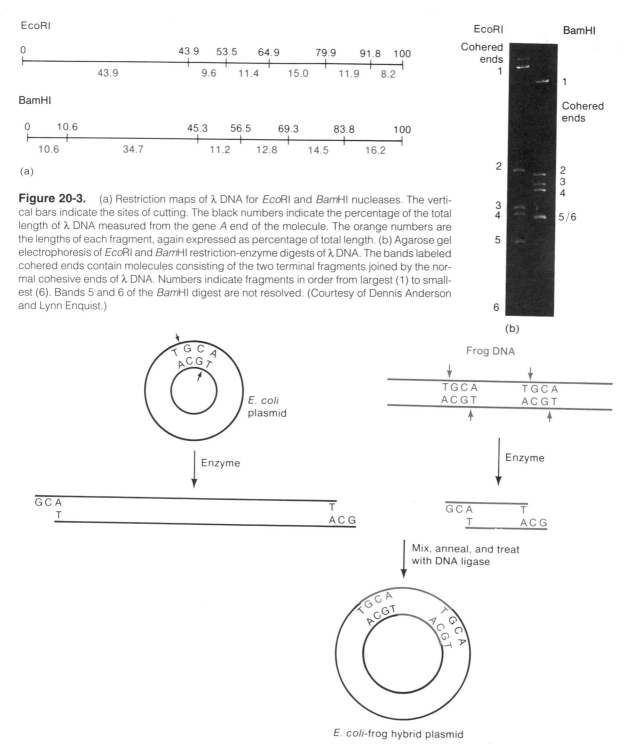

Figure 20-3. (a) Restriction maps of λ DNA for *Eco*RI and *Bam*HI nucleases. The vertical bars indicate the sites of cutting. The black numbers indicate the percentage of the total length of λ DNA measured from the gene *A* end of the molecule. The orange numbers are the lengths of each fragment, again expressed as percentage of total length. (b) Agarose gel electrophoresis of *Eco*RI and *Bam*HI restriction-enzyme digests of λ DNA. The bands labeled cohered ends contain molecules consisting of the two terminal fragments joined by the normal cohesive ends of λ DNA. Numbers indicate fragments in order from largest (1) to smallest (6). Bands 5 and 6 of the *Bam*HI digest are not resolved. (Courtesy of Dennis Anderson and Lynn Enquist.)

Figure 20-4. Construction of a hybrid DNA molecule from fragments derived from different organisms by using restriction enzymes. Such interspecies hybrids are often called chimeras. Short arrows indicate cleavage sites.

a reaction mixture consisting of the exonuclease-treated DNA, dATP, and terminal nucleotidyltransferase, poly(dA), tails will form at both the 3'-OH termini (Figure 20-5). If instead dTTP were provided, the DNA molecule would have poly(dT) tails. Two molecules can be joined if poly(dA) tails are put on one DNA molecule and poly(dT) tails on the second molecule, by allowing the poly(dA) tails to anneal with the poly(dT) tails, as shown in Figure 20-5. Completion of the joined molecule is accomplished by filling in any gaps with DNA polymerase I and sealing the remaining nicks with DNA ligase.

This method, which is called **homopolymer tailing**, is useful for joining DNA molecules that lack complementary ends. Such molecules were initially prepared by mechanical breakage of large DNA molecules as described subsequently.

Isolating Random Genomic DNA Fragments

Consider a gene G that is to be cloned in a plasmid. This is accomplished by inserting a DNA fragment containing G into the plasmid DNA. The simplest method is to treat both the plasmid DNA and the DNA containing G with the same restriction enzyme and anneal the resulting fragments. It may be, however, that every known restriction enzyme that cuts the plasmid also cuts within gene G and thereby inactivates the gene. How can the gene be cloned?

One approach is to fragment the donor DNA mechanically by **hydrodynamic shear** forces. If a DNA solution is passed rapidly through a small orifice (such as a hypodermic needle) or is stirred vigorously, the DNA molecules are broken into fragments of 3 to 30 kb. Because typical genes from prokaryotes are no larger than 1 to 2 kb, in most of the fragmented molecules the gene will remain intact. The fragments resulting from shearing usually have blunt ends. One problem with this approach is that it is not possible to recover easily the DNA fragments formed.

A second approach is to digest the genomic DNA partially with a restriction enzyme that cuts the DNA frequently. Typically restriction enzymes that recognize a 4 bp site are used. For example, *Salmonella typhimurium* DNA

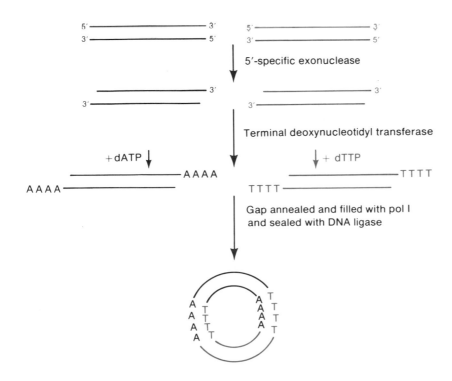

Figure 20-5. The joining of two molecules with complementary homopolymer tails.

can be partially digested with the restriction enzyme *Sau*3A to yield an average fragment size of 10 to 15 kb. If completely digested with *Sau*3A, the resulting DNA fragments would have an average of 256 bp. In the partial digest, any given site is cut in only some of the fragments. Thus the resulting random population of DNA fragments is likely to include an intact copy of all the *S. typhimurium* genes (Figure 20-6).

Blunt-End Ligation

Sometimes restriction enzyme sites that produce cohesive ends are not present in two DNA molecules at the desired places. In such cases, cloning the DNA fragment may require joining DNA molecules cut with different enzymes. One approach to this problem is to convert the single-stranded termini to blunt ends by "filling in" the ends with DNA polymerase or by removing the overhanging ends with the 3' - 5' nuclease of a DNA polymerase or a single-stranded nuclease. The blunt-ended DNA fragments can then be joined by T4 DNA ligase. *Escherichia coli* phage T4 produces a DNA ligase in large quantities in an infected cell. This enzyme is a typical ligase in that it seals nicks in double-stranded DNA having 3'-OH and 5'-P termini, but in contrast to most DNA ligases, it has the additional property of joining two DNA molecules with blunt ends.

Generally when the two blunt-ended fragments are joined, the recombinant site cannot be cut with either of the two enzymes used originally (Figure 20-7).

Linkers

Another way of joining a DNA fragment with single-stranded termini to a DNA fragment with blunt ends is to add a short DNA segment (a **linker**) (Figure 20-8) containing a restriction site to both ends of the blunt-ended molecule. This procedure is useful because at a later time the fragment can be recovered from the vector by treatment with a restriction enzyme that cuts the site in the linker.

INSERTION OF A DNA MOLECULE INTO A VECTOR

If a random collection of DNA fragments (for example, obtained by partial digestion of genomic DNA with a restriction enzyme or by shearing) is allowed to anneal with a vector, a large number of different hybrid molecules are obtained. This procedure is called "shotgun" cloning. If a particular gene is to be cloned, the vector possessing that gene must be identified from the collection of all vectors possessing foreign DNA. Identifying the desired clone requires a suitable selection or screen. Although selections or screens exist for many genes, there are certainly many cases in which the clone of interest is either so rare or so difficult to detect that it is necessary to purify the fragment containing the gene of interest before cloning it. For example, if the size of the desired restriction fragment is known, it can be separated by gel electrophoresis, cut out of the gel and purified, then directly ligated into a plasmid vector.

Isolation of cDNA and Using Reverse Transcriptase

Expression of a cloned gene is commonly used as an assay that the intact, functional gene has been cloned. For example, when the *lacZ* gene is properly cloned in a plasmid, β-galactosidase is expressed. Eukaryotic genes have different reg-

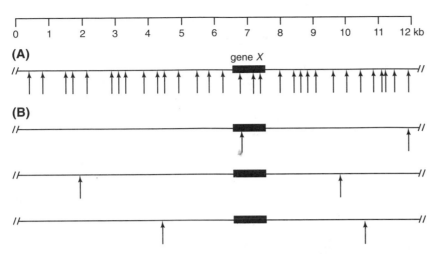

Figure 20-6. Isolation of random DNA fragments by partial digestion with a restriction enzyme that recognizes a 4 bp site that occurs frequently. (a) Complete digestion with *Sau*3A produces a large number of DNA fragments with an average size of 256 bp. Note that *Sau*3A cuts multiple times within gene *X*. (b) Partial digestion with *Sau*3A such that it cuts only once every 5 kb on the average. Note that some fragments will be cut within gene *X*, but many fragments will have only cuts outside of gene *X*, allowing the intact gene *X* to be directly cloned.

ulatory sites and often contain introns that disrupt the reading frame. Therefore in general, eukaryotic genes are not properly transcribed and translated into functional proteins when cloned in a bacterium. This problem can often be circumvented by cloning the DNA complementary to the mature mRNA—this DNA is called cDNA.

For example, assume that you wish to clone the rabbit gene for β-globin (a subunit of hemoglobin) in a bacterial plasmid. The recombinant plasmid could

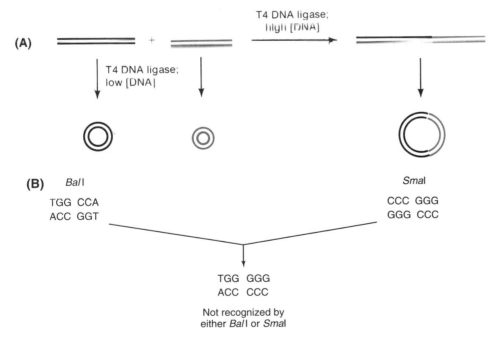

Figure 20-7. (a) Recombinant DNA molecules produced by joining blunt-ended DNA fragments. At low concentrations of DNA, intramolecular circularization is favored, whereas at higher concentrations of DNA, joining of two different DNA molecules is favored. (b) Often recombinant DNA molecules produced by joining joint-ended DNA fragments cannot be cut with either of the two enzymes used to produce the cut parent DNA molecules.

be formed by treating both rabbit cellular DNA and the *E. coli* plasmid DNA with the same restriction enzyme, mixing the fragments, annealing, ligating, and finally infecting *E. coli* with the DNA mixture containing the hybrid plasmid. The next step is to identify the bacterial colony containing this plasmid. It is not possible for a bacterium containing the β-globin gene to synthesize β-globin because the gene contains introns (see Chapter 6), and *E. coli* lacks the enzymes needed to process the introns properly to produce the mature eukaryotic mRNA. Other tests could be done to find the desired clone, such as hybridization of bacterial DNA with purified β-globin mRNA, but because only about one plasmid-containing cell in 10^5 contains the β-globin gene, this method is tedious. Therefore a more direct approach is useful.

Some types of animal cells, such as the reticulocyte, which produces β-globin, make a large amount of a few specific proteins. In these cells, the specific processed mRNA molecules constitute a large fraction of the total mRNA synthesized in the cell. Thus, it is possible to isolate mRNA from these cells that consists almost exclusively of a single mRNA species—in this case, β-globin mRNA. The purified mRNA can be used to clone the gene.

An enzyme present in certain animal viruses (called retroviruses), **reverse transcriptase**, is used in vitro to make the complementary DNA from an RNA template. A double-stranded DNA molecule obtained by reverse transcription of an mRNA is called **complementary DNA (cDNA)**. If the introns were removed from the RNA template before the RNA was isolated, the corresponding cDNA will contain an uninterrupted coding sequence. The cDNA generally has blunt ends.

DETECTION OF RECOMBINANT MOLECULES

Cloning DNA fragments in plasmid vectors is relatively straightforward, but detecting the recombinants is sometimes more difficult. If a vector is cleaved by a restriction enzyme and annealed with a random collection of restriction fragments, many types of molecules result: The vector can religate without an in-

Figure 20-8. Formation of recombinant DNA through use of a linker. The short arrows indicate the sites of cutting by the *Eco*RI enzyme.

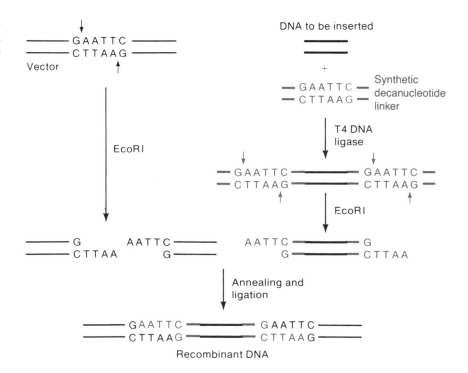

serted fragment, the vector may have one or more inserts, or the fragments may ligate together without the vector. Fragments without vector sequences cannot stably transform cells because such molecules lack a replication origin and replication genes. Some means, however, is needed to ensure that the transformed plasmids carry an inserted DNA fragment. Furthermore, even if the plasmids carry an inserted fragment, there is no guarantee that a plasmid containing donor DNA has the DNA segment of interest. Thus, some test is needed for detecting the desired plasmid.

Methods to Insure That a Plasmid Vector Contains a DNA Insertion

The plasmid pBR322 is a widely used vector. It is small (4361 bp) and has two different antibiotic-resistance markers: resistance to tetracycline (Tet^r) and to ampicillin (Amp^r). Thus, cells transformed with pBR322 are easily selected by growth on plates containing one of these antibiotics. Also, pBR322 has seven unique restriction-enzyme cleavage sites at which foreign DNA can be inserted.

Three procedures are commonly used to eliminate plasmids that lack foreign DNA. One procedure for detecting plasmids with insertions is called **insertional inactivation** (Figures 20-9 and 20-10). For example, the *Bam*III and

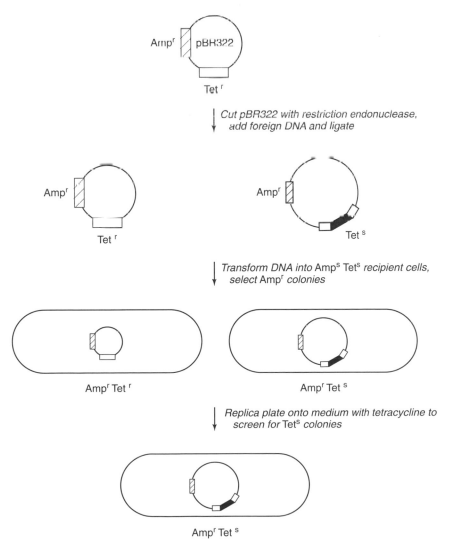

Figure 20-9. Method for isolating a particular restriction fragment.

*Sal*I sites on pBR322 are within the *tet* gene. Thus, insertion at either of these sites yields a plasmid that is Amp^r Tet^s because the *tet* gene is inactivated. If appropriate restriction fragments are ligated to pBR322 digested with the same restriction enzyme, then transformed into wild-type *E. coli* cells (Amp^s Tet^s) and plated on medium containing ampicillin, all the Amp^r colonies must possess the plasmid. These colonies can then be screened for sensitivity to tetracycline. All the Amp^r colonies will also be Tet^r unless the *tet* gene of pBR322 has been inactivated by insertion of foreign DNA. Thus, cells with an insertion cloned into the pBR322 DNA will be Amp^r Tet^s.

A second procedure is to treat a cleaved plasmid with **alkaline phosphatase**, an enzyme that removes 5'-P termini, leaving 5'-OH groups. Because DNA ligase requires a 5'-P, the plasmid cannot be ligated unless another DNA fragment with a 5'-P is inserted (Figure 20-11).

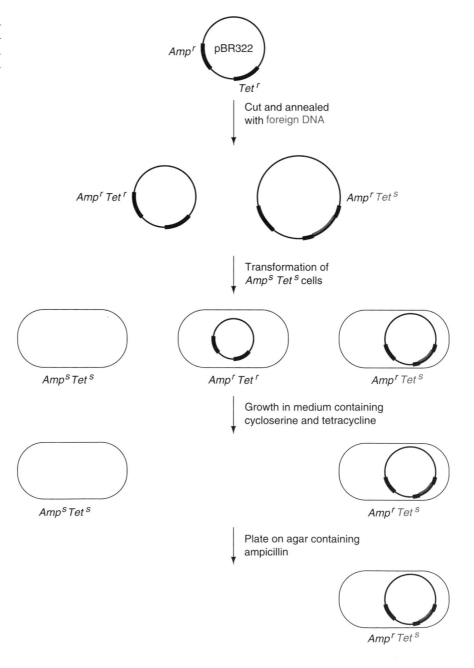

Figure 20-10. Cycloserine enrichment for clones with an insertion that inactivates the Tet^r gene. Cycloserine kills only growing cells.

A third procedure uses a special type of plasmid called a cosmid. **Cosmids** are plasmids that carry the phage λ *cos*-site. These vectors combine the features of a plasmid and phage λ to increase the probability of selecting a recombinant plasmid carrying foreign DNA. A typical cosmid is a circular ColE1 plasmid (see Chapter 11) carrying both the gene for resistance to the antibiotic rifampicin (Rif^r) and the *cos* site of phage λ (see Chapter 16). The cosmid has two cleavage sites for the *Hin*dIII restriction enzyme, which separate the *rif* gene from the *cos* site and the ColE1 region (Figure 20-12). In Chapter 16, the *cos* site–cutting (Ter) system responsible for packaging phage λ DNA into the phage head was described. The Ter system can act on a DNA molecule only if the following two conditions are satisfied: (1) The DNA molecule must contain two *cos* sites, and (2) the *cos* sites must be separated by no less than 38 kb and no more than 54 kb. The use of cosmids as vectors involves packaging λ DNA in vitro. The cosmid shown in Figure 20-12 cannot serve as a substrate for in vitro packaging because the DNA has only one *cos* site. If the cosmid is treated with the *Hin*dIII restriction enzyme, the resulting linear fragments still cannot be packaged. Panel (a) of the figure shows that if the molecules are annealed with one another, among the many resulting molecules, one is a dimer (or higher multimer) containing two *cos* sites. The Ter system, however, still cannot package this molecule because the *cos* sites are separated by only a few kilobases. Panel (b) shows that if a mixture of cleaved cosmid DNA and *Hin*dIII restriction fragments obtained from the DNA of another organism (e.g., camel DNA) is allowed to reanneal, the recombinants will include linear dimers with two *cos* sites separated by a sufficient amount of foreign DNA that their distance is suitable for the λ packaging system. Thus, in vitro packaging yields a collection of transducing particles containing a linear fragment of recombinant cosmid DNA, terminated at each end by the normal cohesive ends of λDNA. Note that packaging occurs only if the

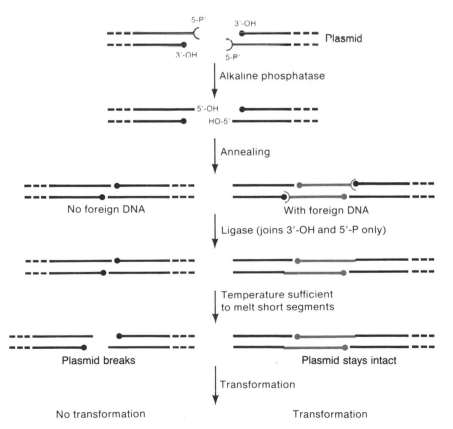

Figure 20-11. Method to prevent re-formation of a cleaved plasmid. The plasmid DNA, but not the foreign DNA, is treated with alkaline phosphatase, an enzyme that converts terminal 5'-P groups to 5'-OH groups.

cosmid contains foreign DNA (although it would be possible to package a multimer containing many copies of the *rif* gene in tandem). Panel (c) shows the result of infecting Rif^s *E. coli* with the cosmid-carrying transducing particles. (No phage development occurs because the particles lack the necessary phage genes.) The cosmid DNA is injected and circularizes via the λ cohesive termini. Once

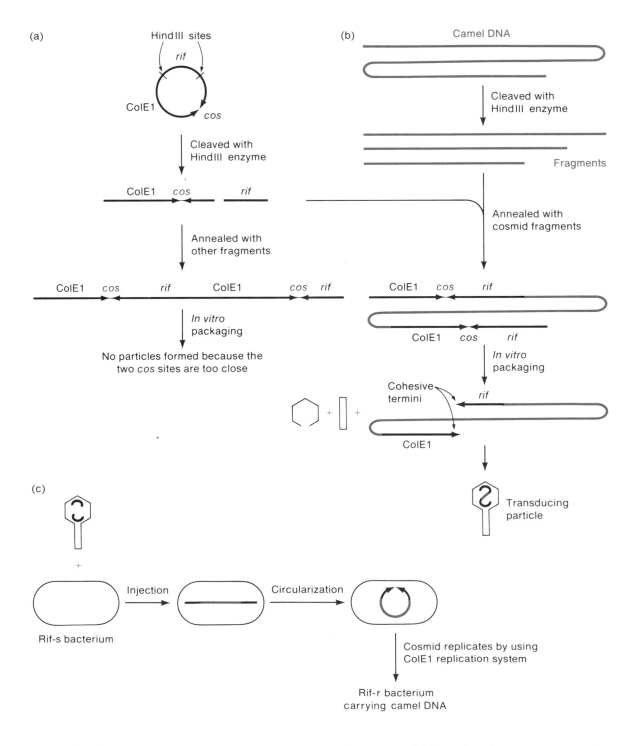

Figure 20-12. Use of a cosmid as a vector. (a) Cleavage of the cosmid. (b) Formation of a transducing particle. (c) Transduction into a new host.

the cosmid DNA has circularized and has been ligated, it can replicate, using the ColE1 replication system. Growth of the infected cells on plates containing rif-ampicin selects cells carrying recombinant cosmid clones with the *rif* gene, the ColE1 region, and the foreign DNA.

Methods to Insure That a Phage Vector Contains Foreign DNA

There are two procedures commonly used for cloning genes in *E. coli* phage λ. Both are based on the fact that λ has several centrally located genes that are not needed for lytic growth (*b2, int, xis, red, gam, cIII*). These genes are in the so-called nonessential region (see Chapter 16).

A derivative of phage λ called λgt4 has two *Eco*RI restriction sites near the termini of the nonessential region (wild-type λ has five *Eco*RI sites, but three of these sites were removed by genetic manipulation). To use this vector, the DNA is cleaved with *Eco*RI and the three fragments separated by gel electrophoresis. The terminal fragments are isolated, and the central fragment is discarded (Figure 20-13). The terminal fragments (called λ arms) can be joined via the complementary single-stranded termini, but the resulting DNA molecule is only 72% as long as that of wild-type λ DNA and thus is noninfective because the minimum length of DNA that can be packaged in a λ phage head is 77% of the full size. If foreign DNA is inserted, however, the resulting hybrid DNA becomes infective (that is, produces progeny phage) because the increased length of the DNA allows it to be packaged. Thus, if an *E. coli* culture is infected with DNA annealed from a mixture of the two terminal fragments and foreign DNA, any phage that is produced will contain foreign DNA.

Another λ derivative contains a section of the *lac* operon that includes the *lacZ* gene, the *lacP* promoter, and the *lacO* operator. There is a restriction site in *lacO*, so insertion of foreign DNA into the phage at that site yields a phage that expresses β-galactosidase constitutively (see Chapter 7). The indicator X-gal is a colorless compound that does not induce *lac* expression but produces a blue dye if cleaved by β-galactosidase. Thus, when a mixture of *lacO⁺* and *lacOᶜ* (constitutive) cells are plated on plates containing X-gal, the *lacOᶜ* colonies will be blue because only these cells synthesize β-galactosidase. Similarly, if *lacO⁺* and *lacOᶜ* phages are plated on

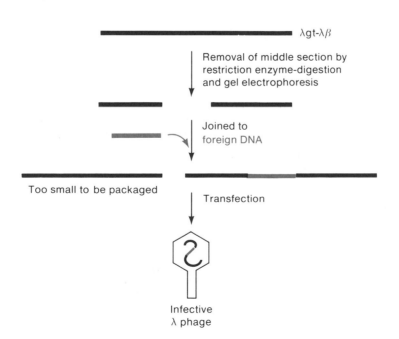

Figure 20-13. Packaging of recombinant DNA in a λ phage.

a lawn of *lacO*⁺ cells on agar containing X-gal, *lacO*⁺ phage will produce colorless plaques, but *lacO*ᶜ phage (which contain inserted foreign DNA) express β-galactosidase during lytic growth of the phage, producing blue plaques.

Identifying Recombinant Plasmids by Size

A plasmid containing a fragment of foreign DNA is often larger than the original uncleaved plasmid. This increase in size can be used to detect plasmids with an insert by agarose gel electrophoresis. Single colonies of plasmid-containing cells can be lysed and directly examined by electrophoresis: There is enough plasmid DNA in a single colony to be visible as a single band moving far ahead of the chromosomal DNA. The plasmid DNA moves a distance related to its molecular weight—the larger the DNA, the smaller the distance moved in a given time interval. Hence colonies containing larger plasmid DNA molecules, which contain donor DNA, move more slowly than those lacking donor DNA.

Identifying Clones by Complementation

The simplest procedure for detecting a particular cloned gene is complementation. For example, a gene from the histidine biosynthetic pathway of the Archaea *Methanobacteria voltae* was identified by cloning random fragments of *M. voltae* into a plasmid, transforming the DNA into an *E. coli his*⁻ mutant and selecting for growth on plates with no histidine. This method works only for genes that are able to complement: It is not generally successful when eukaryotic DNA is cloned in bacteria because usually either the eukaryotic DNA is not properly expressed or there is no corresponding bacterial gene. In such cases, the two procedures described in the following sections may be useful.

Identifying Clones by Hybridization

The **colony** or **in situ hybridization assay** allows detection of the presence of a particular gene (Figure 20-14). If an appropriate hybridization probe is available, it can be used to detect clones with the desired insertion. For example, if the DNA sequence of a homologous gene from another organism is known or if a partial amino acid sequence of the protein is known, a small oligonucleotide of the desired sequence can made, radiolabeled, and used as a hybridization probe. Colonies transformed with the recombinant plasmids are replicated onto a nitrocellulose filter. A portion of each colony remains on the agar, allowing it to be isolated after the hybridization. The filter is treated with NaOH, which lyses the cells and releases denatured DNA, and then the paper is dried, which fixes the denatured DNA to the paper. The filter is flooded with ³²P-labeled probe, which is complementary to the desired gene, then the filter is incubated under

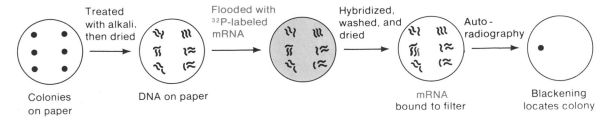

Figure 20-14.   Colony hybridization. The reference plate, from which the colonies on paper were obtained, is not shown.

conditions that favor renaturation. The filter is then washed to remove unbound [^{32}P]probe; bound radioactivity remains on the paper only if [^{32}P]probe has hybridized. The filter is dried and placed on autoradiographic film. Colonies containing clones that hybridize to the probe can be identified as a dark spot on the autoradiogram. The corresponding colony can be picked from the original plate.

A similar assay can be done with phage vectors. Phage are plated on a lawn of bacteria, and after plaques develop, a nitrocellulose filter is laid on the plate. Some phage from each plaque stick to the filter. The filter is then treated with alkali, as described earlier, to fix and denature the DNA, and the plaque whose phage contain the gene of interest are identified by hybridization. The desired plaque is then isolated from the reference plate.

Identifying Clones by Immunological Techniques

If the protein product of the gene of interest is made, two immunological techniques allow the protein-producing colony to be identified. If the protein is excreted, a **radioactive antibody test** can be used (Figure 20-15). Autoradiography shows which colony synthesized the gene product.

An **immunoprecipitation test** can also be used to identify a protein-producing colony by directly adding a specific antibody to the medium. If the protein is excreted, an antibody-antigen precipitate called **precipitin** forms a visible white ring around the colony producing the protein (Figure 20-16). Two slight modifications of this procedure allow one to detect a protein that is not excreted. In one modification, a host cell is used that is lysogenic for λcI857. A replica of the plate containing colonies to be tested is made by transferring the

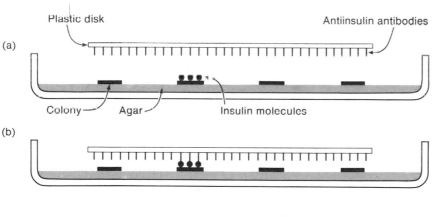
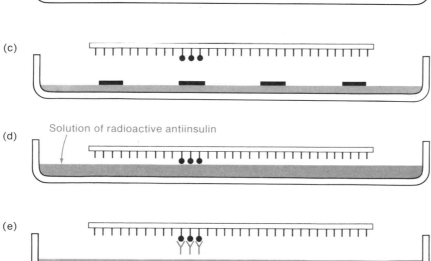

Figure 20-15. Radioactive antibody test. (a, b) A plastic disc coated with anti-insulin antibody is touched to the surface of colonies on a petri dish. (c) Insulin molecules from an insulin-producing colony bind to antibodies. (d) The disk is then dipped into a solution of radioactive anti-insulin. (e) Radioactivity adheres to the disk at positions of insulin-producing colonies.

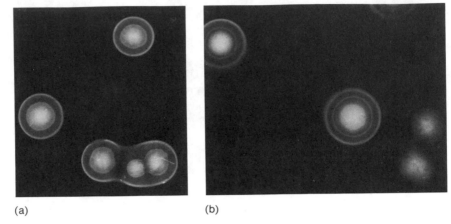

(a)   (b)

colonies onto agar containing antibody. When the colonies on the antibody plate become visible, the temperature is raised to 42°C to inactivate the temperature-sensitive *cI857* repressor. This induces lysis of many of the cells, thereby releasing the cell contents. The protein of interest then reacts with the antibody and forms a precipitin ring at the site of the colony. In the second modification, soft agar containing antibody and the enzyme lysozyme is poured on the colonies and allowed to harden. The lysozyme lyses the cells on the surface of the colony, releasing the intracellular proteins, resulting in formation of a precipitin ring around colonies that produce the appropriate protein.

SINGLE-STRANDED DNA CLONING VECTORS

It is sometimes useful to clone a particular single strand of DNA. This is possible using a single-stranded phage vector such as M13. M13 is a phage containing single-stranded circular DNA. Its replication cycle consists of two steps (Figure 20-17). First, a double-stranded circular intermediate (called replicating form [RF]) is made. Then the RF is used as a template for synthesis of the phage single-stranded DNA (the strand contained in the phage particle is called the + strand). If a foreign gene is cloned into the RF and cells are transformed with the recombinant RF, the resulting phage will contain only one of the complementary strands (Figure 20-18).

Cloning into M13 vectors makes use of a phenomenon called α-complementation (Figure 20-19). The active form of the *E. coli lacZ* gene product, β-galactosidase, is a tetramer consisting of four identical subunits of 1021 amino acids each. A small deletion (called ΔM15) of amino acids 11 through 41 prevents formation of the tetramer. A small N-terminal fragment of β-galactosidase (called the α-peptide) consisting of the first 92 amino acids, however, can interact with the defective ΔM15 polypeptide to form functional tetramers.

M13 cloning vectors are genetically engineered derivatives of the phage with the following features (Figure 20-20):

1. The phage carries a *Hinc*II fragment of the *lac* operon that includes the *lac* promoter and operator and the short N-terminal region of the *lacZ* gene required for α-complementation. M13 is unusual in that it does not kill its host cell. Progeny phage are instead extruded through the cell wall. Multiplication of infected cells, however, is slow, so turbid "plaques" result from the localized slow growth of the infected cells. Thus, when the M13 mutant carrying the *Hinc*II fragment infects a Lac⁻ host cell carrying the *lac* ZΔM15 deletion, cells within the plaque will be phenotypically Lac⁺. If IPTG (see Chapter 7) and X-gal are added to the plates,

the *lac* genes are induced, and the Lac$^+$ cells will appear blue. Thus, the M13 cloning vector forms blue plaques on this medium.

2. The *Hinc*II fragment has been engineered to contain a short sequence with cleavage sites for several restriction enzymes. For example, the M13 vector mp18 contains sites for *Eco*RI, *Sst*I, *Kpn*I, *Sma*I, *Bam*HI, *Xba*I, *Sal*I, *Pst*I, *Sph*I, and *Hind*III, in that order. Such sites are called **multiple cloning sites**. They increase the versatility of the vector by allowing the use of numerous restriction enzymes. Insertion of DNA into any of these sites destroys the α-complementing ability of the *Hinc*II fragment; hence M13 hybrid phage form colorless plaques (sometimes called "white" plaques). This "blue-white" screening procedure is also used with plasmid vectors, for example the pUC vectors (see Table 20-1).

Typically DNA is cloned into M13 vectors as follows: The double-stranded

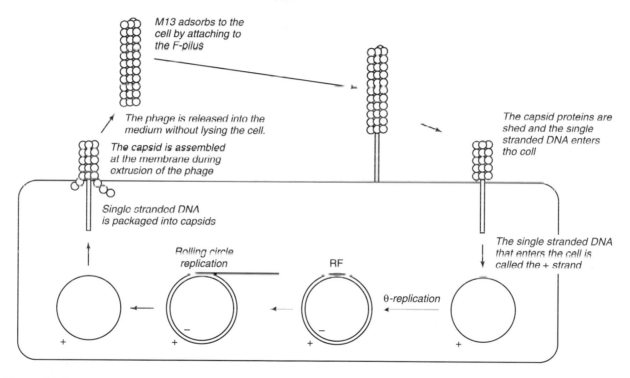

Figure 20-17. The life cycle of phage M13.

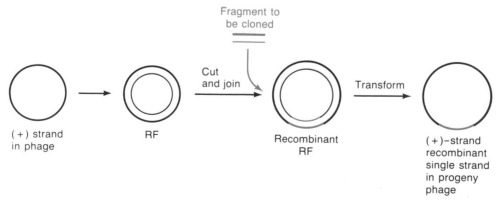

Figure 20-18. Cloning of a single strand in M13 by insertion of a double-stranded fragment into the RF.

Figure 20-19. α-Complementation of the *lacZ* gene.

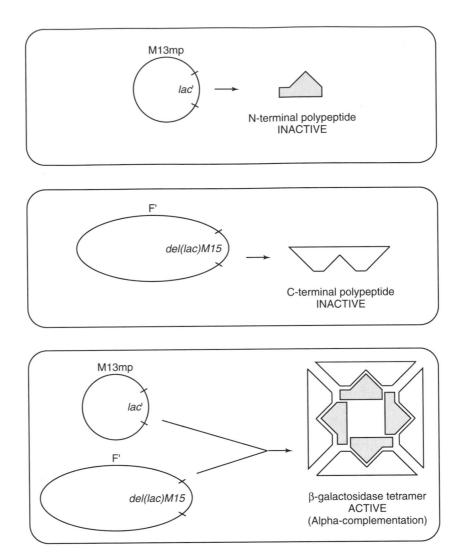

RF DNA purified from cells infected with the phage is cleaved with two restriction enzymes that cut within the multiple cloning site. The DNA to be cloned is also cut with the same enzymes. Because the resulting restriction fragments have different sticky ends, they can anneal only in one orientation. The DNA is ligated then transformed into a *lacZΔM15* host. The white plaques that contain phage with potential insertions are picked. Phage in one plaque are then propagated, and DNA is isolated from phage particles. Because the phage DNA is single-stranded, it contains only one of the complementary strands of the cloned fragment. To be able to isolate both individual strands, another M13 derivative was engineered to contain the multiple cloning site in the opposite orientation. Cloning of the same restriction fragment separately in both vectors yields two different recombinant phage, each of which yields one of the complementary strands of the cloned DNA (Figure 20-21.)

GENE BANKS

Often it is necessary to isolate different genes or DNA segments repeatedly from a single organism. It is quite time-consuming to go through the complete cloning procedure each time a new DNA segment is needed. Thus collections have been established of hybrid plasmid-containing bacteria that are of

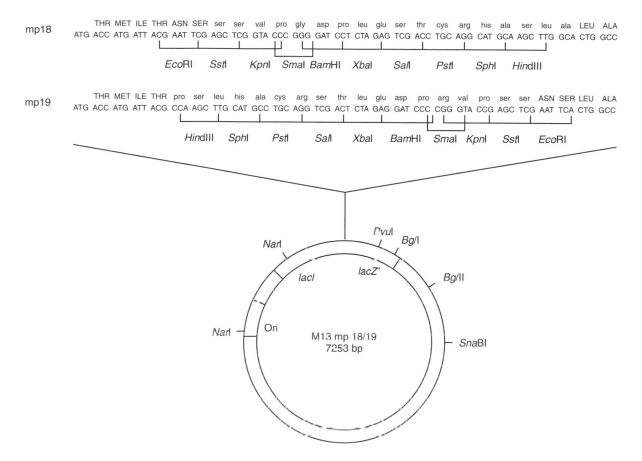

Figure 20-20. Examples of M13 vectors used for cloning. Only restriction enzymes that cut the vectors at a single site are shown on the restriction maps. The two vectors M13 mp18 and mp19 are identical except their multiple cloning sites are in opposite orientations to facilitate cloning DNA fragments in both directions. The multiple cloning sites of mp18 and mp19 are shown at the top of the figure. The amino acids in the α-peptide are shown with the amino acids encoded by the polylinker indicated in lowercase letters. The following restriction enzymes do not cut mp18 or mp19: *AatII, ApaI, AsuII, BclI, BssHIII, BstEII, BstXI, EcoRV, HpaI, MluI, MstI, NcoI, NheI, NotI, NruI, NsiI, SstIII, ScaI, SfiI, SinI, SpeI, StuI, XhoI, XmaI.*

sufficient size that each segment of the cellular DNA is represented in the collection. Such collections are called **gene banks** or **gene libraries**. To be useful, it is important to insure that a gene bank fully represents the entire genome.

Clarke and Carbon constructed one of the first gene banks, a collection of *E. coli* colonies containing a ColE1 plasmid in which yeast DNA had been cloned. To form the gene bank, purified ColE1 DNA was cleaved with *Eco*RI, which makes a single cut in this DNA molecule. The yeast DNA was not cleaved with a restriction enzyme; doing that would certainly have prevented many genes from being in the library as intact units because some genes would be cleaved by *Eco*RI. Instead yeast DNA was broken at random by hydrodynamic shear forces. To join the ColE1 and yeast DNA molecules, poly(dT) tails were added to the 3'-OH termini of the cleaved ColE1 DNA, and poly(dA) tails were added to the yeast DNA fragments. The molecules were joined by homopolymer tailing, and an *E. coli* strain was transformed with the ligated DNA. Cells containing the ColE1 plasmid are resistant to colicin E1, so plasmid-containing cells were selected on plates containing colicin E1. The colonies were then transferred to small wells filled with storage medium.

How many colonies do you need to feel confident that the gene bank represents the entire genome? The number of colonies required to make a

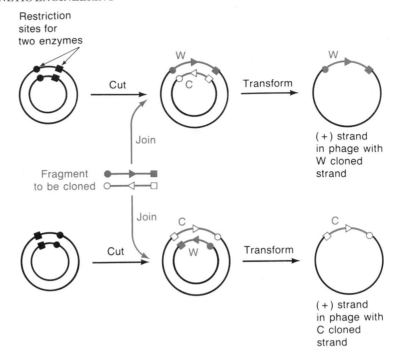

Figure 20-21. Separate cloning of complementary strands of a fragment by insertion in opposite orientations in M13 RF.

complete gene bank can be determined from the following calculation. Consider a DNA sample containing fragments of such a size that each fragment represents a fraction f of the genomic DNA. The probability P that a particular sequence is present in a collection of N colonies is:

$$P = 1 - (1 - f)^N$$

or

$$N = \frac{\ln(1 - P)}{\ln(1 - f)}$$

Assume that you want probability of 0.99 that every sequence is represented in the library—that is, $P = 0.99$. If the donor DNA is haploid yeast (molecular weight of DNA is about 10^{10}) and the average molecular weight of the sheared DNA is 8×10^6, $f = 8 \times 10^6/10^{10} = 0.0008$ and $N = [\ln(1 - 0.99)]/[\ln(1 - 0.0008)] = 5754$. Thus, if the gene bank contains about 5800 colonies, there is a 99% probability that any yeast gene will be present in at least one colony. Furthermore, if only a few colonies are selected from the library at random, it is likely that their yeast DNA sequences will not overlap. To construct a similar gene bank from cells with a large DNA content (such as *Drosophila*), the required number of colonies is about 300,000; this can be reduced roughly n-fold by increasing the size of the constituent fragments by a factor of n.

A clone containing a particular DNA segment can then be isolated from the gene bank by one of the screening procedures described earlier in this chapter.

ORDERED GENE BANKS

An **ordered gene bank** consists of a set of clones that carry known regions of an entire chromosome. Such an ordered gene bank of *E. coli* DNA cloned into a phage λ vector was constructed by Kohara (hence it is often called the "Kohara collection"). An ordered gene bank can be useful for mapping and cloning new genes. For example, if a new gene has been identified that maps near *put* at 22 minutes on the *E. coli* chromosome, it will be present on a λ clone from the Kohara collection carrying this region. The gene can be easily subcloned from the appropriate λ clone using known restriction sites. Alternatively, if a gene is cloned on a plasmid but not yet mapped, the map position of the gene can be determined by DNA-DNA hybridization. DNA from the plasmid clone is labeled and then hybridized with a filter containing spots of DNA from each individual λ clone in the ordered set. Only the spot that contains DNA from the homologous λ clone will hybridize with DNA from the plasmid clone, so the plasmid clone must map at the region of the chromosome corresponding to that particular λ clone.

Although an ordered gene bank is useful, such collections are not available for many organisms simply because it is a tremendous amount of work to determine the correct order of the clones in a gene bank.

KEY TERMS

alkaline phosphatase	isoschizomers
alpha-complementation	ligation
blunt ends	linkers
cDNA	multiple cloning sites
cohesive ends	ordered gene banks
complementation	palindrome
cosmids	phage M13
gene banks or libraries	recombinant DNA
gene cloning	restriction enzymes
genetic engineering	type I
hybridization	type II
hydrodynamic shear	reverse transcriptase
insertional inactivation	vector

QUESTIONS AND PROBLEMS

1. Restriction enzymes generate three types of termini. What are they?

2. What sequence features do restriction sites have?

3. Will the sequences 5'GGCC and 3'GGCC be cut by the same restriction enzyme?

4. Could the sequences 5'AGCGCT and 5'GGCGCC be cut by the same restriction enzyme?

5. What is the source of restriction enzymes?

6. Can a restriction enzyme cut at two sites that have different base sequences?

7. What property must two different restriction enzymes possess if they yield identical patterns of cleavage?

8. What unique property is possessed by the enzyme terminal nucleotidyl transferase compared with other DNA polymerases?

9. What are three methods that can be used to join fragments with blunt ends?

10. What are two methods that will ensure that a fragment will be joined to a vector in a unique direction?

11. If a plasmid vector cleaved at a single site is mixed with a restriction fragment, annealed, and ligated, some circular molecules are found that are smaller than the plasmid vector. What are these molecules?

12. What is the advantage of using a plasmid with two antibiotic-resistance genes as a cloning vehicle?

13. What are two methods that can be used to ensure that a particular fragment has been cloned?

14. What is one advantage of using a cosmid vector?

15. Restriction enzymes A and B are used to cut the DNA of phages T5 and T7. A particular T5 fragment is mixed with a particular T7 fragment, and a circle forms that can be sealed by DNA ligase. What can you conclude about these enzymes?

16. Phage X82 DNA is cleaved into six fragments by the enzyme *Bgl*I. A mutant is isolated whose plaques look quite different from the wild-type plaque. DNA isolated from the mutant is cleaved into only five fragments. Account for this change. (In general, explain why often a restriction enzyme may be found that yields n fragments with wild-type phage and $n - 1$ fragments with a mutant).

17. The A^+ allele of a cell is easily selected by growth on a medium lacking substance A. Repeated attempts, however, to clone the A gene by digestion of cellular DNA with the enzyme *Eco*RI (and ligating into a vector that has been cut with *Eco*RI) are unsuccessful. If *Hae*III is used, clones are easily found. Explain.

18. Suppose you had a plasmid with an Amp^r gene that contains an *Eco*RI site and a *lacZ* gene that contains an *Sal*I site. If you were trying to clone a particular gene, would it be easier if you used *Eco*RI or *Sal*I to cleave the DNA? How would you detect the recombinant plasmid?

19. A Kan^r Tet^r plasmid is treated with *Bgl*I, which cleaves the Kan^r (kanamycin) gene. The DNA is annealed with a *Bgl*I digest of *Neurospora* DNA and then used to transform *E. coli*.

    a. What antibiotic would you put in the growth medium to ensure that a colony has the plasmid?

    b. What antibiotic resistance phenotypes will be found among the colonies?

    c. Which phenotype do the *Neurospora* DNA have?

20. A *lac*$^+$ Tet^r plasmid is cleaved in the *lac* gene by a restriction enzyme. The restriction site for this enzyme contains four bases, and the cuts generate a two-base, single-stranded end. The particular site in the *lac* gene is in the first amino acid of the chain, changes of which rarely generate a mutant. The single-stranded ends are converted to blunt ends with DNA Pol I, and then the ends are joined by blunt-end ligation. A *lac*$^-$ Tet^s bacterium is transformed with the DNA, and Tet^r bacteria are selected. What will the Lac phenotype of the colonies be?

21. Consider a restriction enzyme that makes x cleavage sites within the *lac* operon, generating fragments containing (1) the promoter and operator, (2) the *lacZ* gene, and (3) the *lacY* gene. This collection of fragments is ligated with a vector and transformed into a *lacZ*$^-$*lacY*$^-$ host. Among the Lac$^+$ transformants, some clones make more β-galactosidase than permease, and others make more permease than β-galactosidase. Give two different explanations for these clones. (Think about the fact that annealing is a random process.)

22. Sometimes a cloned bacterial gene is not expressed because it has been separated from its promoter. The gene can be expressed if it can be cloned in a plasmid containing a promoter to which the gene can be linked. The gene must be oriented correctly, however, because the coding sequence must be linked to the strand containing the promoter. Unfortunately, when mixing the fragments with a cleaved plasmid, joining will occur in both possible orientations. Consider a plasmid containing the *E. coli lac* operon (complete with the *lac* promoter), with a *Bam*HI site separating the promoter and the genes. The *lac* sequence also contains restriction sites for several other enzymes. Explain how any one of these enzymes might be used in the cloning procedure to ensure that the gene of interest is linked to the *lac* promoter.

23. How many clones are needed to establish a gene library for monkey DNA (6×10^9 bp per cell) if fragments whose sizes average 2×10^4 bp are used and if you want to ensure that 99% of the monkey genes will be in the library?

24. Explain the following phenomenon. Restriction enzyme 1 acts at all sites acted on by restriction enzyme 2, yet enzyme 2 cannot act at any of the sites for enzyme 1.

25. A restriction digest of a particular DNA produced by *Eco*RI always yields five fragments. In one experiment, an extra fragment, which is quite faint and migrates more slowly than all of the other bands, is seen. How might you explain this fragment?

26. A DNA molecule is cleaved by a restriction enzyme and analyzed by gel electrophoresis. Only one sharp band is seen. Explain.

27. When DNA isolated from phage J2 is treated with the enzyme *Sal*I, eight fragments are produced, whose sizes are 1.3, 2.8, 3.6, 5.3, 7.4, 7.6, 8.1, and 11.4 kb. If J2 DNA is isolated from infected cells, however, only seven fragments are found, whose sizes are 1.3, 2.8, 7.4, 7.6, 8.1, 8.9, and 11.4. What is the likely form of the intracellular DNA?

28. If a plasmid is cleaved in five positions with a restriction enzyme that produces an extended 3' terminus, will all fragments be able to circularize?

REFERENCES

Ausubel, F., R. Brent, R. E. Kingston, D. D. Moore, J. G. Seidman, and J. A. Smith. 1987. *Current Protocols in Molecular Biology*. John Wiley and Sons.

Beers, R. F., and E. G. Bassett (eds.). 1977. *Recombinant Molecules: Impact on Science and Society*. Raven.

Berger, S., and A. Kimmel. 1987. *Guide to Molecular Cloning Techniques. Meth. Enzymol.*, vol. 152. Academic Press, New York.

Chauthaiwale, V., A. Therwath, and V. Deshpande. 1992. Bacteriophage lambda as a cloning vector. *Microbiol. Rev.*, 56, 577.

Cohen, S. N., et al. 1973. Construction of biologically functional bacterial plasmids *in vitro*. *Proc. Natl. Acad. Sci. USA*, 70, 3240.

Davis, R. W., D. Botstein, and J. R. Roth. 1980. *A Manual for Genetic Engineering. Advanced Bacterial Genetics*. Cold Spring Harbor.

Grunstein, M., and D. S. Hogness. 1975. Colony hybridization: a method for the isolation of cloned DNAs that contain a specific gene. *Proc. Natl. Acad. Sci.*, 72, 3961.

Lobban, P., and A. D. Kaiser. 1973. Enzymatic end-to-end joining of DNA molecules. *J. Mol. Biol.*, 78, 453.

Maniatis, T., et al. 1978. The isolation of structural genes from libraries of eucaryotic DNA. *Cell*, 15, 687.

McPherson, M. 1991. *Directed Mutagenesis: A Practical Approach*. IRL Press, New York.

Mertz, J., and R. Davis. 1972. Cleavage of DNA: RI restriction enzyme generates cohesive ends. *Proc. Natl. Acad. Sci.*, 69, 3370.

Murray, N. 1983. Phage lambda and molecular cloning. In *Lambda II* (R. W. Hendrix et al., eds.), p. 395. Cold Spring Harbor.

Murray, N. 1983. Lambda vectors. In *Lambda II* (R. W. Hendrix et al., eds.), p. 677. Cold Spring Harbor.

Russel, M. 1991. Filamentous phage assembly. *Mol. Microbiol.*, 5, 1607.

*Sambrook, J., E. F. Fritsch, and T. Maniatis. 1989. *Molecular Cloning: A Laboratory Manual*. Cold Spring Harbor Laboratory.

*Shortle, D. 1987. Genetic strategies for analyzing proteins. In *Protein Engineering* (D. Oxender and C. F. Fox, ed.). Alan R. Liss, Inc.

Smith, D. H. 1979. Nucleotide sequence specificity of restriction enzymes. *Science*, 205, 455.

*Watson, J., M. Gilman, J. Witkowski, and M. Zoller. 1992. *Recombinant DNA*, Second edition. Scientific American Books.

*Resources for additional information.

Applications
of Genetic Engineering

G enetic engineering has many practical applications. Restriction enzymes can be used to generate maps of genomic DNA to identify mutant sites and to produce mutations at particular sites. The majority of the applications, however, are cloning, mutating, and overexpressing genes, for example: (1) isolating and characterizing a particular gene, part of a gene, or region of a genome; (2) producing particular RNA and protein molecules in large quantities; (3) improving the production of biochemicals (such as enzymes and drugs) and commercially important organic chemicals; or (4) producing varieties of plants having particular desirable characteristics, such as requiring less water or resistance to disease.

RESTRICTION MAPPING

It is often useful to determine the location of restriction sites in a particular DNA molecule. A map showing the positions of restriction sites is called a **restriction map**. There are several ways to obtain such a map. One method, applicable to DNA molecules with both linear and circular forms, is shown in Figure 21-1. The number of fragments produced by cutting a circular DNA molecule is equal to the number of sites cut: When a circular DNA molecule is cut once, one linear fragment is produced; if the circular DNA is cut twice, two linear fragments are produced; and so on. In contrast, the number of fragments produced by cutting a linear DNA molecule is one more than the number of sites cut: When a linear DNA molecule is cut once, two linear fragments are produced, and so forth. The fragments are separated by gel electrophoresis, and their sizes are determined by comparison with standards of known size.

The most common way to obtain a restriction map is by doing **double digests** of the purified DNA using restriction enzymes. The procedure is to take three aliquots of purified DNA; treat aliquot #1 with the first enzyme, aliquot #2 with the second enzyme, and aliquot #3 with both enzymes (a **double digest**); and then compare the three resulting sets of fragments by gel electrophoresis. An example of a double digest of a circular plasmid is shown in Box 21-1. A diagram showing the sets of fragments that might be generated from a double digest of linear DNA such as phage λ is shown in Figure 21-2. The value of this procedure is that it yields restriction maps for both enzymes simultaneously. If the number of cuts is small (for example, six), this procedure may be sufficient for unambiguous ordering of the fragments.

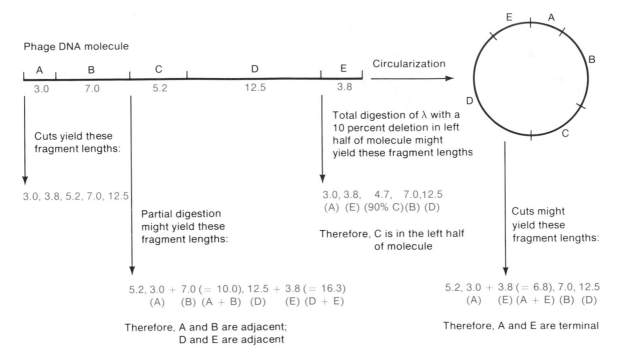

Figure 21-1.   A possible scheme for determining the positions of restriction cuts in a phage λ DNA molecule, which has a linear and a circular form.

Sometimes all the DNA molecules in the solution will not have been cut, and faint bands formed by the uncut fragments can be seen. This is called a **partial digest**. The molecular weight of the partially digested bands are the sum of the molecular weights of two fragments that have been completely digested, which indicates that the two fragments are adjacent in the original uncleaved molecule. Another approach for determining the restriction map of linear DNA fragments involves labeling the termini with ^{32}P before cleavage. Labeling can be done using the enzyme polynucleotide kinase, an enzymatic reaction in which only the 5' terminus of a DNA strand is labeled. The fragment is then partially digested with a restriction enzyme such that on the average each DNA molecule is cut only once. The ^{32}P-labeled fragments are then separated by gel electrophoresis, and the size of the fragments is determined by autoradiography (Figure 21-3).

The position of some fragments can also be determined by examining insertion or deletion mutants. If the insertion is not cut by the restriction enzyme, the fragment with the insertion will appear larger. If the insertion is cut by the restriction enzyme, the fragment will disappear and be replaced by two different bands. Likewise, if a deletion does not eliminate a restriction site, a smaller fragment will appear in the digest of the deletion mutant. If a deletion instead removes a restriction site, two bands will be absent, and a new band will appear; the size of the new band equals the sum of the sizes of the missing fragments minus the size of the deletion.

SITE-DIRECTED MUTAGENESIS

Random mutagenesis is necessary for the initial characterization of a gene. The resulting mutations can identify the phenotype of null mutants and with appropriate selections or screens, may identify specific residues involved in the structure and function of the gene product. Once a gene has been identified and cloned

BOX 21-1. AN EXAMPLE OF RESTRICTION MAPPING BY DOUBLE DIGESTS

Example: Purified plasmid DNA is digested with three restriction enzymes: *Eco*RI, *Hin*dIII, and *Pst*I. The results are shown in the following table:

	Enzyme	Fragment sizes (kb)
a	*Eco*RI	20
b	*Hin*dIII	20
c	*Eco*RI + *Hin*dIII	13.5 + 6.5
d	*Pst*I	10 + 7 + 3
e	*Eco*RI + *Pst*I	9 + 7 + 3 + 1
f	*Hin*dIII + *Pst*I	10 + 4.5 + 3 + 2.5

Both *Eco*RI (a) and *Hin*dIII (b) cut the plasmid only once.
The *Eco*RI site is 6.5 kb from the *Hin*dIII site (c). Thus, the following restriction map could be drawn:

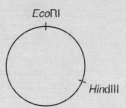

*Pst*I cuts the plasmid at three sites.
*Eco*RI cuts the 10 kb *Pst*I fragment into 9-kb + 1-kb fragments.
Based on these data, four restriction maps are possible:

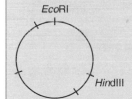 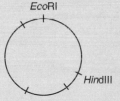 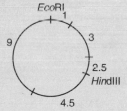 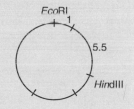

*Hin*dIII cuts the 9-kb *Pst*I fragment into 4.5-kb and 2.5-kb fragments.
Since *Hin*dIII does not cut the 10-kb *Pst*I fragment, the first two maps are ruled out.
Since the *Hin*dIII site is 6.5 kb from the *Eco*RI site, only the third map would give fragments of the sizes obtained.

and the DNA sequence determined, it is possible to mutagenize specific base sequences to determine the effect of particular changes on the gene product. A few approaches for constructing specific mutations are outlined in this section. In most of the procedures, the gene to be altered is carried on a plasmid. The plasmid is isolated, manipulated in vitro, and then returned to a bacterium for replication and gene expression.

Constructing Deletions

Several approaches can be used to make a deletions at a particular position. The simplest method consists of treating a plasmid with a restriction endonuclease to make two cuts surrounding a site to be deleted. The fragments are isolated, and the ends of the desired fragments are rejoined. The gene of interest now contains a deletion. With a detailed restriction map that identifies sites for many

Figure 21-2. An example of using double digests to determine the restriction map of a circular plasmid.

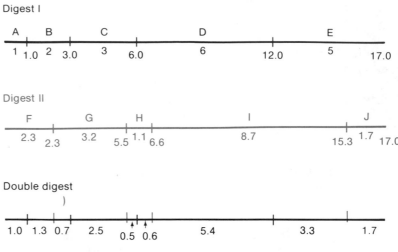

Figure 21-3. Using end-labeled fragments to determine the restriction map of linear DNA.

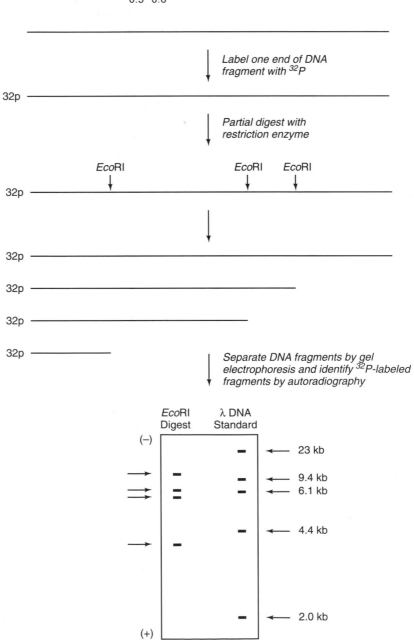

restriction enzymes, it is often possible to delete a particular region of DNA. The disadvantage of this technique is that deletions are usually quite large and are limited by the availability of appropriate restriction sites.

Other approaches use nucleases to yield deletions with random end points. One method involves cleaving the double-stranded plasmid DNA with a restriction enzyme that cuts the DNA at a single site and then using the exonuclease *Bal*31 to degrade the DNA in both directions. *Bal*31 yields deletions that extend in both directions from the restriction site. Another method involves cleaving double-stranded plasmid DNA with two restriction enzymes, each of which cuts the DNA at a single site—one of the sites with a 3'-overhang. The DNA is then digested with exonuclease III (ExoIII). ExoIII does not degrade DNA from a 3'-overhang; thus it generates deletions extending only in one direction. The size of the deletions generated by both *Bal*31 and ExoIII depends on the concentration of the exonuclease used and the time of the reaction. Thus, a set of deletions of different lengths can be obtained by treating the DNA for different amounts of time before stopping the reaction. The resulting DNA fragments can be converted to blunt ends and ligated with T4 DNA ligase.

An alternative approach can be used to isolate short deletions at a specific restriction site. The plasmid can be cut with an enzyme that makes a single staggered cut in the gene of interest. The short cohesive termini generated by the restriction enzyme are removed by a nuclease that specifically degrades single-stranded DNA (for example, S1 or mung bean nuclease), leaving blunt ends. The plasmid can then be resealed by blunt-end ligation with T4 DNA ligase, leaving deletions of 2 to 4 bp.

Constructing Point Mutations at Random Sites in a Particular Region of DNA

A plasmid clone can be mutagenized in vitro by exposing it to a chemical mutagen such as hydroxylamine or nitrous acid. Cells are transformed with the mutagenized DNA, and after two rounds of DNA replication, the gene contains one or more point mutations in the desired region.

Constructing Point Mutations at a Particular Base Pair

Sometimes it is useful to mutagenize a specific base pair in a DNA sequence to test its role directly—for example, a base pair believed to encode an amino acid at the active site of a protein or a base pair believed to be part of the DNA binding site of a regulatory protein. In such situations, it is possible to change one (or more) specific base pairs directly to a specified different base pair by "site-directed mutagenesis." There are several methods for site-directed mutagenesis, but one of the most common is the *dut ung* method (Figure 21-4).

POLYMERASE CHAIN REACTION

Polymerase chain reaction (**PCR**) is a technique that allows a specific DNA or RNA sequence to be amplified over 10^6-fold. Using PCR, it is possible to obtain sufficient amounts of a specific DNA sequence from a single colony to clone the DNA fragment or determine its DNA sequence.

PCR uses oligonucleotide primers that hybridize to the DNA on each side of the region to be amplified. The oligonucleotides are usually synthesized in vitro. To construct appropriate oligonucleotide primers, you must know the sequence of regions that flank the DNA to be amplified. For example, you might know the DNA sequence of other genes on either side of the region, you might be able to infer the DNA sequence from the amino acid sequence, you might know the DNA sequence of a homologous gene from another organism, or you might know the DNA sequence of the wild-type gene and want to know the DNA sequence change in a mutant.

PCR amplification of DNA involves three temperature-dependent steps

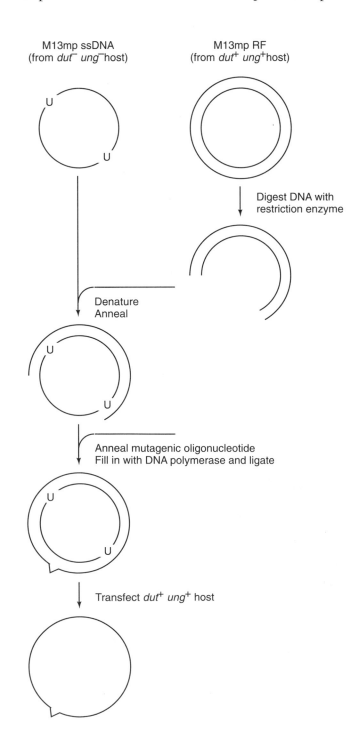

Figure 21-4. The *dut ung* method for site-specific muta-genesis. An M13 single-stranded template (+ strand) is grown on a *dut⁻* (dUTPase⁻) *ung⁻* (uracil N-glycosylase⁻) mutant to allow in-corporation of uracil into the DNA in place of thymine. A heterodu-plex is constructed by hybridizing the (+) strand to the (−) strand of a restriction fragment from M13 RF grown in a *dut⁺ ung⁺* host. A mu-tant oligonucleotide is hybridized to the single-stranded gap and double-stranded plasmid is syn-thesized using a DNA polymerase and then sealed with DNA ligase. When this plasmid is transformed into a *dut⁺ ung⁺* host, the uracil residues in the (+) strand are re-moved, making the (+) strand un-able to replicate. Hence, only the mutant (−) strand is replicated.

(Figure 21-5). First, the double-stranded template DNA is denatured to single-stranded DNA. Second, primers complementary to the 3' ends of opposite strands of the DNA are hybridized with the single-stranded template DNA. Third, DNA polymerase extends each primer, duplicating the segment of DNA between the primers. DNA synthesized during the first cycle has the 5' end of the primers and a variable 3' end. When these strands are denatured, the parental strand will

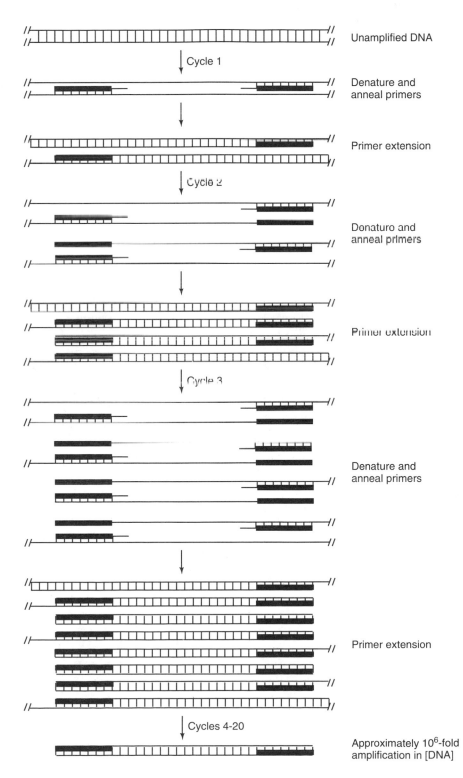

Figure 21-5.   Amplification of a DNA fragment by PCR.

Unamplified DNA

Cycle 1

Denature and anneal primers

Primer extension

Cycle 2

Denature and anneal primers

Primer extension

Cycle 3

Denature and anneal primers

Primer extension

Cycles 4-20

Approximately 10^6-fold amplification in [DNA]

Figure 21-6. The amount of DNA produced with each cycle of PCR.

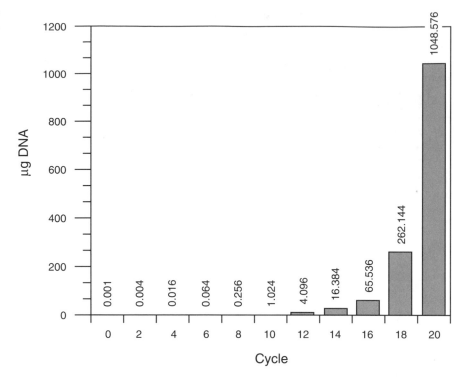

rehybridize to the primer, so the product with a variable 3' end will continue to be synthesized during subsequent cycles of PCR. These products, however, accumulate only arithmetically. The second cycle of denaturation, annealing, and primer extension produces discrete products with the 5' end of one primer and the 3' end of the other primer. Each strand of this discrete product is complementary to one of the two primers and thus acts as a template in subsequent cycles. Therefore these discrete products accumulate exponentially with each round of DNA amplification, resulting in 2^n-fold amplification of the DNA in n cycles of PCR (Figure 21-6). The cycles of heating and cooling are done automatically in an instrument called a thermocycler. Each step takes no more than a few minutes.

Originally PCR used *Escherichia coli* DNA Pol I, but because this enzyme rapidly loses activity at high temperatures, more enzyme had to be added after each cycle of heating. Now PCR uses DNA polymerases isolated from thermophilic bacteria and Archaea that can withstand the cycles of high temperature. *Taq* DNA polymerase (originally isolated from the bacterium, *Thermus aquaticus*) is most commonly used for PCR. *Taq* polymerase is a highly processive enzyme with an optimal temperature of 75°C to 80°C.

PRODUCTION OF PROTEINS FROM CLONED GENES

One of the common goals of genetic engineering is to produce a large quantity of a particular protein that is otherwise difficult to obtain (for example, proteins for which there are normally only a few molecules per cell). In principle, the method is straightforward: The gene is cloned onto a plasmid downstream of a strong promoter. With a high-copy-number plasmid or an actively replicating phage, synthesis of a gene product may often reach a concentration of about 1 to 5% of the cellular protein. In practice, the goal of producing large quantities of a desired protein is not always this simple because of a variety of theoretical and technical problems.

Expression of a Cloned Gene

Expression of a cloned gene requires that the gene is properly transcribed and translated. Many proteins are expressed at low levels in vivo and are detrimental to the growth of the cell when overexpressed at high levels, so it is often necessary to express a cloned gene under the control of a regulated promoter. If a restriction site separates the coding region of a gene from its promoter, the gene can be inserted into a plasmid vector under the control of another promoter. Vectors have been constructed that contain a ribosome binding site and a well-characterized, regulatable promoter in front of useful insertion sites. Such vectors are called "**expression vectors**."

For example, a common approach to control the expression of DNA-binding proteins has been to clone the structural gene for the DNA-binding protein downstream of the "***tac* promoter**." The *tac* promoter is a strong, hybrid promoter with the –35 region of the *trp* promoter, the –10 region of the *lac* promoter, and the *lac* operator. Expression of the *tac* promoter can be regulated by the Lac repressor protein made from the *lacI* gene. Usually very little Lac repressor is made from the *lac*+ gene — not enough repressor to turn off expression of a cloned gene on a multicopy plasmid. Therefore, the *tac* promoter is usually regulated by the *lacIq* gene, which contains a mutation in the *lacI* promoter that increases the activity of the constitutive *lacI* promoter. Strains with the *lacIq* mutation express the Lac repressor at high levels, causing strong repression of the *tac* promoter unless the inducer, isopropyl-β,D-thiogalactoside (IPTG), is added. When IPTG binds to Lac repressor, it causes an allosteric change, which prevents the repressor from binding the *lac* operator (Chapter 7). Thus, when regulated by the Lac repressor, the level of expression from the *tac* promoter increases in response to increasing concentrations of IPTG. By varying the IPTG concentration, the amount of a protein expressed from a cloned gene regulated by the *tac* promoter can be varied over several orders of magnitude.

For proper transcription, the restriction fragment must be inserted in the correct orientation downstream of the promoter and ribosomal binding site. If both ends of the fragment have the same restriction cleavage site, insertion into a vector can occur in two possible orientations. Insertion in both orientations occurs with equal probability; therefore one must isolate a recombinant vector that has the gene in the correct orientation (Figure 21-7). Simply providing a functional promoter and a ribosomal binding site in the correct orientation, however, does not guarantee that the gene will be correctly expressed.

The problem of gene expression is more difficult when cloning eukaryotic genes for several reasons:

1. Eukaryotic promoters usually are not recognized by bacterial RNA polymerase.
2. The mRNA transcribed from eukaryotic genes often lacks ribosomal binding sites recognized by bacterial ribosomes.
3. The mRNA may contain introns that must be excised (RNA processing).
4. The protein product of eukaryotic genes may need to be modified (posttranslational processing). For example, proinsulin requires particular proteolytic cleavages before it becomes active insulin.
5. Eukaryotic proteins often fold incorrectly in bacteria and are degraded by bacterial proteases.

The first two problems can usually be solved by cloning the gene downstream of a bacterial promoter and ribosomal binding site as described previously. The intron problem can be solved by cloning the cDNA. Correct posttranslational processing of many eukaryotic proteins is not possible, although in some cases the processing can be done in vitro after the protein is purified.

Figure 21-7. Insertion of a cloned gene in two orientations with respect to a promoter *p* in the vector.

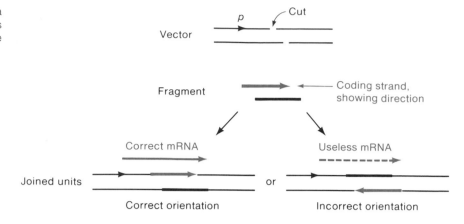

Degradation of many eukaryotic proteins can sometimes be decreased by using mutants defective for certain proteases (although not completely eliminated because there are many different proteases in *E. coli*). Nevertheless, in some cases, it is necessary to clone eukaryotic genes in a eukaryotic host such as yeast to express the functional protein correctly.

SOME APPLICATIONS OF GENETIC ENGINEERING

Genetic engineering of bacteria has other practical uses as well. A few examples are described in the following sections.

Construction of Industrially Important Bacteria

Bacteria with novel phenotypes can be produced by genetic engineering, sometimes by combining the features of several other bacteria. For example, several genes from different bacteria have been inserted into a single plasmid that has then been placed in a marine bacterium, yielding an organism capable of metabolizing petroleum; this organism has been used to clean up oil spills in the ocean. Furthermore, many biotechnology companies are at work designing bacteria that can synthesize industrially important chemicals. Bacteria have been designed that are able to compost waste more efficiently and to fix nitrogen (to improve the fertility of soil), and an enormous effort is currently being expended to create organisms that can convert biological waste to alcohol. A lethal gene from *Bacillus thuringensis*, a bacterium pathogenic to the black cutworm, has been engineered into *Pseudomonas fluorescens*, which lives in association with maize and soybean roots. The black cutworm causes extensive crop damage and is usually combatted with noxious insecticides. Inoculation of soil with the engineered *P. fluorescens* resulted in death of the cutworm. This type of genetic engineering of bacteria will surely have a great impact on world economy and environmental quality in the future.

Genetic Engineering of Plants

Altering the genotypes of plants is an important application of recombinant DNA technology. In Chapter 11, we discussed the bacterium *Agrobacterium tumefaciens* and the Ti plasmid, which produces crown gall tumors in plants. These tumors result from integration of the plasmid DNA into the plant chromosome.

It is possible by genetic engineering to introduce genes from one plant into this plasmid and then, by infecting a second plant with the bacterium, transfer the genes of the first plant to the second plant. For example, this approach has been used to transfer genes that allow plants to grow under high osmotic stress in an attempt to allow plants to grow in arid, high-salinity soils.

The first engineered recombinant plant of commercial value was developed in 1985. Glyphosate is an economically important herbicide (weed killer), which inhibits a particular essential enzyme in many plants. Most herbicides, however, cannot be applied to fields growing crops because both the crop and the weeds would be killed. (The chlorinated acids that selectively kill dicotyledonous plants but not the grasses, such as maize and the cereals, are out of favor because of their persistence in soil, toxicity to animals, and possible carcinogenicity in humans.) The target gene of glyphosate is also present in *Salmonella*. A resistant form of the gene was obtained by mutagenesis and growth of *Salmonella* in the presence of glyphosate; the gene was cloned in *E. coli* and then recloned in *Agrobacterium*. Infection of plants with purified Ti plasmid containing the glyphosate-resistance gene has yielded varieties of maize, cotton, and tobacco that are resistant to glyphosate. Thus, fields of these crops can be sprayed with glyphosate at any stage of growth of the crop.

Production of Drugs

Genetic engineering is having an impact on clinical medicine. The initial focus was on developing organisms that would overproduce antibiotics, thereby reducing production costs, and this has been accomplished for several antibiotics. In addition, it is now possible to isolate large quantities of biologically active compounds that previously could be isolated only in small quantities from enormous amounts of tissue. Somatostatin is one example discussed earlier. Another example is the antiviral agent α-interferon, which could not even be tested clinically before genetic engineering provided a source of pure material. This substance reduces the duration of viral infection, is effective against herpesvirus infection of the eye, reduces the incidence of attacks of multiple sclerosis, suppresses atherosclerosis in rats on high cholesterol diets, and is being examined for its antitumor activity. In 1985, tumor necrosis factor, a powerful anticancer substance in rats, was cloned, and animal studies began shortly afterward. Interleukin-2, a substance that stimulates multiplication of certain cells in the immune system, has also been cloned and is overproduced. It shows promise in shrinking certain cancers in humans.

Synthetic Vaccines

A major breakthrough in disease prevention has been the development of synthetic vaccines. Production of certain vaccines, such as anti-hepatitis B, has been difficult because of the extreme hazards of working with large quantities of the virus. The danger would be avoided if the viral antigen could be cloned and purified in *E. coli* because the pure antigen could be given as a vaccine. Several viral antigens have been cloned, but because of either thermal instability or poor antigenicity when pure, most attempts to make vaccines in this way have been unsuccessful. The use of vaccinia virus as a carrier, however, has been fruitful. The procedure makes use of the fact that viral antigens are on the surface of virus particles and that some of these antigens can be engineered into the coat of vaccinia. The use of vaccinia virus is not straightforward because the virus is huge, containing a single DNA molecule with a molecular weight of 160×10^6, far too large for controlled cutting by restriction enzymes and subsequent splicing. Therefore, the viral antigen genes are first cloned in an *E. coli* plasmid. Animal cell lines that support the growth of vaccinia are then infected with both normal vaccinia DNA and the plasmid DNA containing the viral-antigen gene. Genetic recombination occurs in the infected cells, and some progeny vaccinia particles are formed that contain the cloned gene and hence the foreign viral antigen in the vaccinia coat. Various procedures are used to isolate these vaccinia hybrids. By 1985, vaccinia hybrids with surface antigens of hepatitis B, influenza virus, and vesicular stomatitis virus (which kills cattle, horses, and pigs) had been prepared and shown to be useful vaccines in animal tests.

SUMMARY

Since the early 1980s, genetic engineering has become commonplace in all areas of biology. Striking evidence of this is that more than 200 biotechnology companies have formed since that time, and many new scientific journals and newspapers have been created that are devoted exclusively to genetic engineering. With the legal consequences of developing new products, patent lawyers are even being required to take courses in molecular biology and microbial genetics. It is likely that this growing billion-dollar industry will someday employ many readers of this book.

KEY TERMS

double digest	polynucleotide kinase
expression vector	restriction map
oligonucleotide primers	site-directed
partial digest	mutagenesis
polymerase chain	*Taq* DNA
reaction (PCR)	polymerase

QUESTIONS AND PROBLEMS

1. A prokaryotic gene is cloned in a vector next to a promoter and transformed into another prokaryote. Hybridization with a radioactive mRNA probe indicates that the gene is present in the cell, but the gene product is not made. What are potential reasons for the lack of synthesis of the gene product?

2. Give several reasons why a cloned eukaryotic gene may not yield functional protein in a bacterial host.

3. The DNA of temperate phage P4 is linear, is double-stranded, is 11.5 kb long, and has cohesive ends. Digestion with the *Bam*HI restriction endonuclease yields fragments 6.4, 4.1, and 1.0 kb in length. Partial digestion with the *Bam*HI enzyme yields fragments 10.5, 7.4, 6.4, 4.1, and 1.0 kb in length. Circular P4 DNA made with DNA ligase can be digested with the *Bam*HI enzyme to yield fragments 6.4 and 5.1 kb in length.
   a. What is the order of fragments in the DNA?
   b. DNA is extracted from cells lysogenic for P4 and digested completely with the *Bam*HI enzyme. Each of the fragments is tested for hybridization with radioactive denatured P4 phage DNA. Fragments whose lengths are 15.0, 12.5, and 6.4 kb are labeled with radioactivity. What conclusions can you draw from this experiment about prophage structure?

4. Wild-type phage X4 DNA is treated with a restriction enzyme, and five fragments are separated and purified by gel electrophoresis. These fragments were mapped by

comparison to the genetic map of X4. This was done by performing 25 transfection experiments using the five mutations, a^-, b^-, c^-, d^-, and e^-, which are equally spaced from along the DNA. The experiment consists of transformation of a Rec$^+$ cell with intact DNA from a single mutant and a purified fragment. The data are shown below:

	I	II	III	IV	V
a^-	21	14	21,842	3	8
b^-	6	0	4	4856	2
c^-	32,681	16	8	11	13
d^-	1	3	0	1	1825
e^-	22	17,813	27	18	30

The values indicate the number of plaques per microgram of intact DNA molecules. What is the order of the fragments starting at the left end of the map?

5. A DNA fragment was isolated that contains the coding sequences of the A and B genes of *E. coli*. In an effort to make large quantities of the gene products, the fragment was joined to a plasmid vector, and the DNA was used to transform an $A^- B^-$ bacterium. Colonies containing the plasmid were isolated easily, but none were $A^+ B^+$. However, $A^+ B^-$ and $A^- B^+$ colonies were found. What are possible reasons for this observation?

6. The restriction enzyme *Eco*RI cleaves a particular phage DNA molecule into five fragments of sizes 2.9, 4.5, 6.2, 7.4, and 8.0 kb. If the DNA has previously been radioactively labeled at the 5' ends, the 6.2-kb and 8.0-kb fragments are found to be labeled. *Bam*HI cleaves the same molecule into three fragments, whose sizes are 6.0, 10.1, and 12.9 kb, with the 6.0-kb and 10.0-kb fragments labeled with ^{32}P. When both enzymes are used together, the fragments have sizes 1.0, 2.0, 2.9, 3.5, 6.0, 6.2, and 7.4 kb. What are the restriction maps for the two enzymes?

7. In analyzing DNA by treatment with a restriction endonuclease followed by gel electrophoresis, in a particular experiment it is observed that in addition to the expected bands, DNA is present in all regions of the gel starting from the position of the most slowly moving fragment and past that of the most rapidly moving fragment. If the DNA is analyzed before treatment with the restriction enzyme, no such material is seen. Suggest an explanation.

8. Can the cDNA technique be used with any eukaryotic gene? (This is a practical and not a theoretical question.)

9. Plasmid pBR607 DNA is circular and double-stranded with a total length of 2.6 kb. This plasmid carries two genes whose protein products confer resistance to tetracycline (Tet^r) and ampicillin (Amp^r) in host bacteria. The DNA has a single site for each of the following restriction enzymes: *Eco*RI, *Bam*HI, *Hin*dIII, *Pst*I, and *Sal*I. Cloning DNA into the *Eco*RI site does not affect resistance to either drug. Cloning DNA into the *Bam*HI, *Hin*dIII, and *Sal*I sites abolishes Tet^r. Cloning into the *Pst*I site abolishes Amp^r. Digestion with the following mixtures of restriction enzymes yields fragments with the sizes (in kilobases) indicated in the column at the right:

*Eco*RI, *Pst*I	0.46, 2.14
*Eco*RI, *Bam*HI	0.2, 2.4
*Eco*RI, *Hin*dIII	0.05, 2.55
*Eco*RI, *Sal*I	0.55, 2.05
*Eco*RI, *Bam*HI, *Pst*I	0.2, 0.46, 1.94

Position the *Pst*I, *Bam*HI, *Hin*dIII, and *Sal*I cleavage sites on a restriction map, relative to the *Eco*RI cleavage site.

10. Vector pHUB1 is a 5.7-kb circular, double-stranded DNA. This plasmid carries a gene whose protein product confers resistance to tetracycline (Tet^r) on the host bacterium. The DNA has one site each for the following restriction enzymes: *Eco*RI, *Hpa*I, *Bam*HI, *Pst*I, *Sal*I, and *Bgl*II. Cloning into the *Bam*HI and *Sal*I sites abolishes Tet^r; cloning into the other sites does not. Digestion with various combinations of restriction enzymes yields the DNA fragments shown (with sizes in kilobases):

*Eco*RI	5.7
*Eco*RI, *Bam*HI	0.4, 5.3
*Eco*RI, *Hpa*I	0.5, 5.2
*Eco*RI, *Sal*I	0.7, 5.0
*Eco*RI, *Bgl*II	1.1, 4.6
*Eco*RI, *Pst*I	2.4, 3.3
*Pst*I, *Bgl*II	1.3, 4.4
*Bgl*II, *Hpa*I	0.6, 5.1

Position the cleavage sites for all these enzymes on a map of the pHUB1 DNA. Draw the map as a circle with 5.7 kb marked off on the circumference.

REFERENCES

*Anderson, W. F. 1984. Prospects for human gene therapy. *Science*, 226, 401.

Ausubel, F., R. Brent, R. E. Kingston, D. D. Moore, J. G. Seidman, and J. A. Smith. 1987. *Current Protocols in Molecular Biology*. John Wiley and Sons.

Berger, S., and A. Kimmel. 1987. *Guide to Molecular Cloning Techniques. Meth. Enzymol.* vol. 152, Academic Press, New York.

Casadaban, M. J., and S. N. Cohen. 1980. Analysis of gene control signals by DNA fusion and cloning in *E. coli. J. Mol. Biol.*, 138, 179.

*Gilbert, W., and L. Villa-Komaroff. 1980. Useful proteins from recombinant bacteria. *Scientific Am.*, April, p. 74.

*Fersht, A., and G. Winter. 1992. Protein engineering. *Trends Biochem. Sci.*, 17, 292.

*Freifelder, D. 1987. *Problems for Molecular Biology*. Jones and Bartlett, Boston.

Kunkel, J. Roberts, and R. Zakour. 1985. Rapid and efficient site-specific mutagenesis without phenotypic selection. *Proc. Natl. Acad. Sci. USA*, 82, 488.

McPherson, M. 1991. *Directed Mutagenesis: A Practical Approach*. IRL Press, New York.

*Sambrook, J., E. F. Fritsch, and T. Maniatis. 1989. *Molecular Cloning: A Laboratory Manual*. Cold Spring Harbor Laboratory.

Seeberg, P. H., et al. 1978. Synthesis of growth hormone by bacteria. *Nature*, 276, 795.

Shortle, D., and D. Nathans. 1978. Local mutagenesis: a method for generating viral mutants with base substitutions in preselected regions of the viral genome. *Proc. Natl. Acad. Sci. USA*, 72, 2170.

Smith, H. O., and M. L. Bernstiel. 1976. A simple method for DNA restriction site mapping. *Nucl. Acid Res.*, 3, 2387.

*Verma, I. 1990. Gene therapy. *Scientific Am.*, 262, 68.

*Watson, J., M. Gilman, J. Witkowski, and M. Zoller. 1992. *Recombinant DNA*, Second edition. Scientific American Books.

Zoller, M. J., and M. Smith. 1982. Oligonucleotide-directed mutagenesis using M13-derived vectors: an efficient and general procedure for the production of point mutations in any fragment of DNA. *Nucl. Acid. Res.*, 10, 6487.

*Resources for additional information.

Answers to Problems

Chapter 1

1. The genotype is the genetic characteristics of an organism, and phenotype is the observable properties of an organism. Not all genetic characteristics produce an observable effect, so the genotype is not identical to phenotype.

2. A null mutation completely eliminates activity of the gene product. A conditional mutation eliminates activity of the gene product under certain nonpermissive conditions but retains at least partial activity of the gene product under permissive conditions. Nonpermissive and permissive conditions may be determined by the growth conditions or other genetic properties of the strain of the organism used.

3. Several examples of conditional mutations are: temperature-sensitive (Ts) mutations, cold-sensitive (Cs) mutations, and suppressible mutations.

4. A genetic selection allows growth only of cells with the desired genetic properties (e.g., antibiotic resistant mutants)—other cell types will not form colonies under the conditions used. A suitable genetic selection is essential to isolate very rare mutations. A genetic screen allows all cell types to grow, but colonies of cells with specific genetic properties can be distinguished from other colonies (e.g., on color indicator plates).

5. Complementation is due to a diffusible gene product that can provide a missing function. Complementation is usually tested in strains with two copies of the same DNA segment but each with a different allele of the gene (e.g., mutant vs. wild type or one mutant vs. a different mutant). Complementation does not involve a physical exchange between the DNA molecules. Recombination requires the physical breakage and covalent joining of two DNA molecules, which can result in replacement of one allele of a gene with a different allele.

6. $(2 \times 10^{-6})(8 \times 10^{-5}) = 1.6 \times 10^{-10}$.

7. The frequency at which true reverents are obtained is highest for a point mutation. The frequency at which true reverents are obtained for a deletion is zero.

8. R is a precursor to Z, perhaps occurring immediately before Z in the biosynthetic pathway. The mutation inactivates an enzyme that causes conversion of R to the next step in the pathway.

9. x^+y^+ (wild-type) and x^-y^- (double mutant).

10. *a d b*. If a and d were on opposite sides of *b*, *a-d* would be 4%. Thus, the order must be *a d b* or *d a b*. For the first order, *a-d* would be 1.2%; for the second order, *b-d* would have to be larger than *a-b*, which is not the case.

11. The results suggest that at least four proteins are required.

12. The mutations complement. At least two genes are required.

13. (a) The gene order is *a c b d*. (b) $0.15 \times 0.04 = 0.006$.

14. *e g f* is the gene order. The spacing between *e* and *g* is three times that between *f* and *g*.

15. (a) *kyuQ*. (b) A regulatory mutant that prevents synthesis of each gene product.

449

16. Four genes. The groups are: (1,7), (2,4), (5), and (3,6).

17. The new *putP* mutation must map within deletion interval 8 because it cannot recombine with *del(put-550), del(put-715),* or *del(put-563),* which all remove region 8, but it can recombine with each of the other deletions which remove different regions of the *putP* gene. If the mutation mapped within intervals 1–7 then *del(put-515) would not* have been able to repair it. If the mutation mapped in intervals 9-10 then *del(put-550) would* have been able to repair it.

18. The predicted map is shown below:

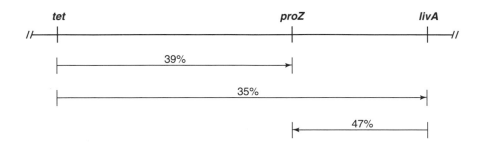

19. When Tet^r was selected, the rarest class of recombinants was ProZ$^+$ LivA$^+$. Because the rarest class of recombinants requires four crossovers, these results indicate that *proZ* is the middle gene.

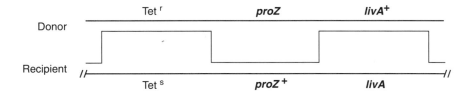

20. All the chromosomal genes are functional except the *hisD* gene and the *hisA* gene. The *hisA* gene will be complemented by the *hisA$^+$* gene from the second copy of the gene, but no functional copy of the *hisD* gene is present. The cell is missing a gene product needed to make histidine and thus will be a histidine auxotroph.

21. The following linkage map can be drawn from the two-factor cross data. Note that the observed linkage between two genes may be different depending on which is the selected marker.

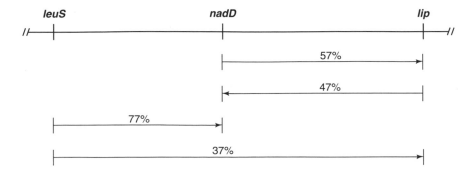

22. The order predicted from the three-factor crosses is shown in the following figures. This order agrees with the order predicted from the two-factor crosses shown in question 21.

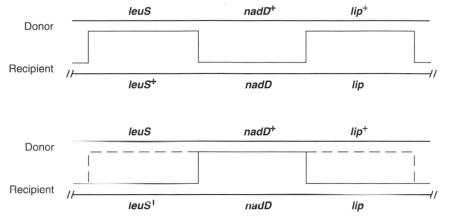

Select *lip*⁺
Rare class of recombinants is *nadD leuS* thus *nadD* is the middle gene

Select *nadD*⁺
No rare class of recombinants, thus the selected marker *nadD*⁺ is the middle gene

23. The following linkage map can be drawn from the two-factor cross data. Note that the observed linkage between two genes may be different depending on which is the selected marker.

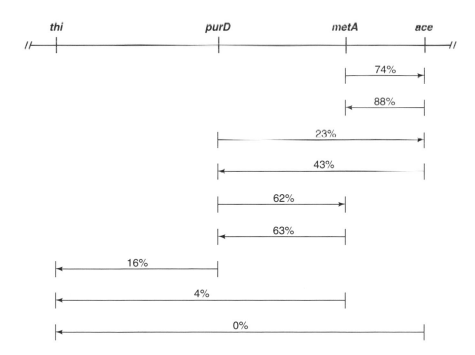

24. The order predicted from the three-factor crosses is shown in the following figures. This order agrees with the order predicted from the two-factor crosses shown in question 23.

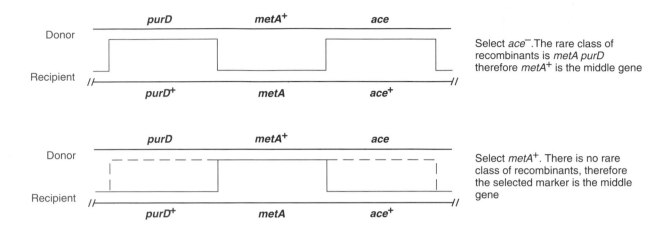

Select *ace⁻*.The rare class of recombinants is *metA purD* therefore *metA⁺* is the middle gene

Select *metA⁺*. There is no rare class of recombinants, therefore the selected marker is the middle gene

25. The merodiploid is Leu⁺. At low temperatures both copies of the *leu* genes are functional. At high temperatures the *leu⁺* copy is functional. Thus, the merodiploid would be Leu⁺ independent of temperature.

Chapter 2

1. Thymine, DNA; uracil, RNA

2. Ribose has an OH group on the 2 carbon, and 2-deoxyribose has a H atom at the same position.

3. An N-glycosidic bond.

4. The 3' and 5' carbon atoms.

5. One 3'-OH and one 5'-P group.

6. One.

7. AT, 2; GC, 3.

8. A 5'-phosphate and a 3'-hydroxyl group.

9. 45/10 = 4.5.

10. The sugars have opposite orientations in the two strands; thus, at one end of a double-stranded DNA molecule, one strand terminates with a 3-OH group and the other strand with a 5-P group.

11. A nuclease is an enzyme capable of breaking a phosphodiester bond. An endonuclease can break any phosphodiester bond, whereas an exonuclease can only remove a terminal nucleotide or a short terminal oligonucleotide.

12. Yes. The base-pairing rule would be satisfied perfectly and the two components of the double-stranded portion would be antiparallel.

13. Molecule 1, because of the long tract of GC pairs in molecule 2.

14. Pairing will occur randomly. Thus, there are three bands, in the ratios 1 $^{15}N^{15}N$: 2 $^{14}N^{15}N$: 1 $^{14}N^{14}N$.

15. (a) 4 × 10 = 40 base pairs. (b) Four nodes.

16. Naturally occurring DNA is negatively supercoiled. Ethidium bromide adds positive supercoiling. Thus, the compact supercoil progressively loses its supercoiling until a slower moving nontwisted circle has formed. As more ethidium bromide is bound, the DNA becomes progressively positively supercoiled and hence more compact, so the value of *s* increases.

17. The charge per nucleotide is the same for molecules of all sizes, so the net force on

all molecules is in a single direction. The pore size of the gel determines the rate at which a molecule of a particular size can move through the gel. Larger molecules cannot penetrate the pores as easily as smaller molecules and hence move more slowly through the gel.

18. The 5-terminal bases are obtained from the 3 termini of the complementary strand. Thus, the two strands are 5'-TGATACGACGAAGTACTGG and 3'-ACTATGCTGCTTCATGACC.

19. The lanes of the autoradiogram correspond to dideoxy sequencing reactions in the following order: ddATP, ddCTP, ddGTP, ddTTP. The mutant sequence can be easily noted by looking for the band that is present in one of the sequencing lanes for the mutant gene (5-8) but absent from the sequencing lanes for the wild-type gene (1-4).

Chapter 3

1. Set 3, in a hydrophobic cluster.

2. As the size of an object increases, the ratio of its surface area to its volume decreases. Since polar amino acids are more often located on the surface than within a folded polypeptide chain and the reverse is true of nonpolar amino acids, the ratio of polar to nonpolar amino acids also decreases.

3. The side chain of glycine (a hydrogen atom) is so small that there is no steric interference with tight folding.

4. Both will tend to be near an amino acid of opposite charge.

Chapter 4

1. 2 hours = 6 doubling times. Thus, $2^6 = 64$.

2. The total dilution is $(100)(100) = 10^4$. Thus, the cell density is $72/[(0.1)](10^4) = 7.2 \times 10^6$ per ml.

3. The microscopic count shows that the cell density is $(453)(1000)/0.001 = 4.53 \times 10^8$ per ml. Since plating yields a much smaller number, it is clear that many of the cells are dead.

4. Draw a curve relating absorbance and cell density; it will yield a straight line. You will find that 0.7 corresponds to 7×10^8 cells/ml.

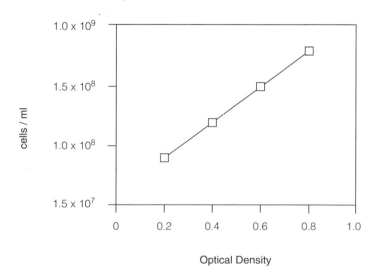

5. A shows that Leu is required. B implies that Arg is needed. C indicates that Trp is not needed. Thus, the genotype is *arg⁻ leu⁻*.

6. $75/0.3 = 250$ minutes.

7. The most common reason for such an observation is that the concentration of the required nutrient is not high enough.

8. B is either a precursor to A, or at least B can be converted to A.

9. One single-strand break causes loss of supercoiling of one loop rather than of the entire DNA molecule.

10. *aux–2001* = requires a vitamin or alanine; *aux–2002* = requires histidine; *aux–2003* = requires both adenine and thiamine.

Chapter 5

1. Just one.

2. One. If the bacterium has not lysed, all progeny will be in the same position.

3. Use the Poisson equation. With a 1:1 ratio of phage to bacteria, the fraction uninfected is e^{-1}.

4. Phage-host specificity refers to the fact that a particular phage can grow on only one or a small number of bacterial species or strains. The most frequent cause is the inability of a phage to adsorb, except to certain bacteria.

5. No. Phage can develop only in a metabolizing bacterium.

6. A phage can multiply only in a growing bacterium, and the nutrients in the growth medium become depleted.

7. Host restriction.

8. 50 to a few hundred.

9. Clearly, the receptors never evolved in the first place in order that the bacterium could be infected. Thus, the receptors must be important to the bacterium. For example, the receptor for phage T1 is a transport system for the Fe^{2+} ion. Receptor sites might gradually change in time to reduce the accessibility of certain phages but the change will be limited by the number of amino acids that can be altered without loss of essential activity.

10. The bacteria had stopped growing when the phage were added, so phage growth is not possible.

11. T4 will be turbid, because half of the bacteria will never be infected; that is B/4 can continue to grow in the plaque. T4h will be clear, because all bacteria will be infected.

12. The cell culture lyses visibly at 60 minutes. This time is 20 minutes after the MOI becomes greater than 1 phage per cell. The calculation is the following: After 20 min there are $200 \times 5000 = 10^6$ phage/ml. At t = 30 min the bacteria have doubled to ca. 4×10^7 per ml. At t = 40 the number of phage per ml is $200 \times 10^6 = 2 \times 10^8$, which is greater than the number of bacteria. Thus, lysis will occur one phage life cycle, or 20 minutes, later at t = 60.

13. From the Poisson distribution, $e^{-2} = 0.135$ of the bacteria are uninfected. Therefore the number of plaques is $200(1 - 0.135) = 173$.

14. (a) These are bacteria to which no phage adsorb. Since the total MOI = 5, the fraction of cells that are uninfected is $e^{-5} = 0.007$. The number of uninfected cells is 7×10^5. (b) This number includes bacteria to which P4, but not P2, adsorb, as well as the infected cells. The MOI of P2 is 3; the fraction of cells getting no P2 is a $e^{-3} = 0.05$. Since 0.007 get neither P2 nor P4, then $0.05 - 0.007 = 0.043$ get one or more P4 but no P2. Hence, $0.043 \times 10^8 = 4.3 \times 10^6$ bacteria get P4 but no P2. (c) These are the bacteria to which P2 adsorb but no P4 adsorb. Using the above reasoning, $e^{-2} = 0.135$ get no P4, and 0.007 get neither; hence $0.135 - 0.007 = 0.128$, and 1.28×10^7 cells adsorb P2 but no P4. (d) These are the bacteria infected by at least one P2 and one P4. The number of cells getting at least one P4 and one P2 is $10^8 - (1.28 \times 10^7) - (4.3 \times 10^6)(7 \times 10^5) = 8.22 \times 10^7$.

15. A temperate phage may choose between a lytic and lysogenic cycle; a virulent phage can engage only in a lytic cycle.

Chapter 6

1. From the 5' end to the 3' end.

2. UAA, UAG, UGA.

3. AUG; methionine.

4. UAG.

5. $(3 \times 124) + 3$ (termination) + 3 (initiation) = 378.

6. There is no way of knowing because both codons could be altered to yield the stop codons UAA and UAG.

7. (1), (3), (4).

8. The amino end is made first.

9. The 70S ribosome forms when a 50S subunit joins the 30S preinitiation complex.

10. NH_2-Arg-Leu-COOH, when reading starts exactly at the 5 terminus. Reading can begin at other positions also yielding NH_2-Gly-Tyr-Arg- Lys-COOH and NH_2-Val-Ile-Gly-Lys-COOH and fragments thereof.

11. Synthesis would have to await the completion of a molecule of mRNA. In the existing system, protein synthesis can occur while the mRNA is being copied from the DNA. Thus, protein synthesis can start earlier than would be possible with reverse polarity, and the mRNA is relatively resistant to nuclease attack.

Chapter 7

1. Turning on and off a set of genes as a unit, simultaneously or by a single signal.

2. The ratio of the amount of each component is maintained constant without having each one separately regulated.

3. Negative regulation.

4. Not always, especially if needed in large quantities. If it were regulated, it may be autoregulated.

5. It typically binds to a repressor and changes the affinity of the repressor for DNA.

6. The promoter is inaccessible to RNA polymerase.

7. β-Galactosidase (*lacZ* product), lactose permease (*lacY* product), and the transacetylase (*lacA* product).

8. Inducible, because the *lacI* product is diffusible.

9. Constitutive.

10. The major reason is that all promoters do not have the same strength, so the rate of initiation of transcription varies from one gene to the next.

11. No, because the repressor is a diffusible protein.

12. Yes, since binding of the repressor to the operator interferes with binding of RNA polymerase to the promoter.

13. Low.

14. No. In the first case, it could not make cAMP. In the second case, it could not make CRP.

15. No.

16. No. There is often a polar effect in that the relative amounts of each protein decrease toward the 3' end of the mRNA. Also, the translational start sites in front of each gene will affect the rate of translation of each gene individually.

17. No. Nothing is specifically bound to an attenuator.

18. In attenuation the structure of the termination sequence is altered such that RNA polymerase either does or does not see it. In antitermination RNA polymerase is altered in order that it can ignore certain termination sites.

19. At very low levels all inducible operons are transcribed—for example, a repressor might come off the operator for an instant and RNA polymerase will initiate transcription. Hence each cell will contain a few β-galactosidase and permease molecules. The permease will bring in a few lactose molecules, so derepression can occur.

20. (a) Yes, I, yes; (b) yes, I, yes; (c) yes, C, yes; (d) yes, I, yes; (e) yes, C, yes; (f) yes, I, yes; (g) no, neither, no; (h) yes, I, yes; (i) no, neither, no; (j) yes, constitutive, no.

21. If neither glucose nor lactose is present, two proteins (the repressor and CRP-cAMP) are bound to the DNA. If glucose alone is present, only one protein (the repressor) is bound to the DNA.

22. The cell cannot be derepressed. The simplest explanation is that the repressor cannot bind the inducer. The partial diploid given at the end of the problem would not make enzyme, because the constitutive component makes no enzyme and the other component cannot be induced.

23. (a) The strain is constitutive, that is, $lacI^-$ or $lacO^c$. (b) The second mutant must be $lacZ^-$, since no β-galactosidase is made; also, it must have a normal repressor-operator system, since permease synthesis responds to lactose. That is, its genotype is $lacI^+lacO^+lacZ^-lacY^+$. Since the partial diploid is regulated, the operator in the operon with a functional $lacZ$ gene must be wild-type. Hence, the first mutant must be $lacI^-$.

24. Glucose is irrelevant to the functioning of the trp operon. When tryptophan is present (a,c), the repressor is active and is bound to the DNA.

25. It clearly lacks a general system that regulates sugar metabolism. The only thing we know about is the cAMP-dependent system, namely, that several mutations prevent activity of this system. For example, the mutant could make either (1) a defective CRP protein, and either fail to bind to the promoters or be unresponsive to cAMP, or (2) an inactive adenyl cyclase, making no cAMP.

26. The operator is probably next to gene E. Deletions that remove the operator make transcription constitutive. The promoter cannot be between the operator and gene E, because in that case all deletions that remove the operator would also remove the promoter, and no constitutive deletions would be found. The order is E o p, which is consistent with the fact that some deletions are not constitutive; these would be larger deletions that include both the operator and the promoter. Thus, the gene order (transcription from left to right) is p o E D C B A.

27. Since addition of Q causes Qase to be made, the system is regulated. Deletion of kyu prevents induction, suggesting that kyu encodes a positive regulatory element. A partial diploid with both wild-type kyu and the deletion is regulated, indicating that the Kyu product is present and that it is a positive regulator activated by an inducer. Examination of mutants confirms this view. The $kyu-1$ mutant never makes Qase but a partial diploid with wild-type is inducible, indicating that the $kyu-1$ mutation is recessive. It seems likely that the Kyu product binds Q and is activated. The $kyu-2$ mutation is constitutive and dominant, suggesting that it is able to turn on transcription of Qase without Q. The Kyu-1 product probably fails to bind Q and the active positive regulator is a Q-Kyu complex. The Kyu-2 product is probably able to bind to the promoter without first binding Q.

28. (a) Three kinds of repressor mutations might occur: (1) a repressor that cannot bind X—the operon is on; (2) a repressor that cannot bind to operator even when X is present—the operon is on; (3) a repressor that binds to the operator without X—the operon is off. (b) Phenotypes of partial diploid with wild-type and each of the mutants are: first mutant, inducible; second mutant, inducible; third mutant, unable to synthesize X under all conditions.

Chapter 8

1. All, 1/2, 1/4 respectively.

2. No. After one round of replication it has hybrid density but a 15N strand always remains in the circle.

3. Polymerizing activity, 5'-3' exonuclease, 3'-5' exonuclease.

4. From the 3' end to the 5' end.

5. Deoxynucleoside triphosphates.

6. To a 3'-OH group.

7. The 5' → 3' activity removes ribonucleotides from the 5' termini of precursor fragments, and the 3' → 5' exonuclease removes a base that has been incorrectly added to the growing end of a DNA strand.

8. Pol III is responsible for the addition of all nucleotides in the growing fork; Pol I removes RNA from the 5' end of precursor fragments and replaces it with deoxynucleotides.

9. DNA polymerases join a 5'-triphosphate to a 3'-OH group and in so doing remove two phosphates, so that the phosphodiester bond contains one phosphate; a ligase joins a 5'-monophosphate to a 3'-OH group.

10. Rolling circle replication must be initiated by a single-strand break. θ Replication does not need such a break.

11. DNA polymerase requires a primer and RNA polymerase does not.

Chapter 9

1. In the same strand.

2. Increased.

3. It will be unaffected since phage particles do not contain any repair enzymes.

4. Photoreactivation.

5. The UvrABC endonuclease, polymerase I, and DNA ligase.

6. Incision refers to the production of one single-strand break, usually near a nucleotide that is to be removed. Excision refers to the second break and subsequent removal of a nucleotide or an oligonucleotide.

7. It is inducible and it is error-prone.

8. Like all light-absorbing phenomena, there is a photoreceptor that absorbs in a particular range of wavelengths. Thus, we should expect a wavelength response curve.

9. Statement 2.

10. (a) Either the damage to T4 DNA is, for some reason, barely repairable by the Uvr system, or T4 possesses its own repair system. (b) The mutant probably lacks a T4-encoded repair system.

11. (a) X is not inducible, because the enzymes were present before protein synthesis was blocked by chloramphenicol. (b) X would probably be considered inducible. The residual 5% could be due either to a second noninducible system, which can excise thymine dimers, or to a small amount of synthesis of X proteins when chloramphenicol is present.

12. The phage apparently makes its own excision enzyme.

13. (a) The larger the DNA molecule, the greater is the probability that the DNA will receive a lethal alteration. (b) The phage may possess its own repair system.

Chapter 10

1. Yes, because, strictly speaking, that is the definition of a mutation.

2. No. The change does not always cause an amino acid replacement, and an amino acid change does not always alter protein function.

3. Transition, b; transversion, a, c.

4. Transitions.

5. A particular base pair, such as AT, can undergo two transversions, namely, AT → TA and AT → CG, but only one transition, AT → GC. This pattern is true for all base pairs. Thus, the theoretical ratio of transversions to transitions is 2:1. Actually, transversions are quite rare because all base changes are not equally frequent; in fact, Pu → Pu and Py → Py changes are fairly rare.

6. A missense mutation results in an amino acid substitution. A nonsense mutation results in premature termination of a protein.

7. An amino acid substitution in the protein product causes weakening of the structure such that a shape change occurs in the protein above or below a critical temperature.

8. A tautomer can cause mutations if it has two forms in equilibrium with another and each form has different base-pairing properties. A tautomeric molecule can be incorporated in an aberrant form and then switch to a form that cannot pair with the base in the DNA. Alternatively, when in the template strand, it can temporarily switch its base-pairing properties and allow incorporation of a base that would not have been added to the end of the growing chain if the switch had not occurred.

9. It tautomerizes, which allows base substitutions, and it affects nucleotide pools, which increases incorporation errors.

10. Yes, because the entire frame of reading of the bases is changed, which results in alterations of most amino acids downstream from the mutation. Only two types of base addition or deletion would not cause a phenotypic change: one that occurs in the space between genes and one that occurs very near the end of a gene.

11. Three. Otherwise, there would have been a frameshift and all downstream bases would have been changed.

12. A deletion.

13. Chain termination mutations, since the reversion event would have to eliminate the chain termination site. Of course, since a chain termination codon contains three bases, it is possible that a second-site reversion could occur in one of the two positions of the chain termination codon other than that which had originally been altered. However, this would certainly produce an amino acid change at the mutated site, so reversion would be unlikely.

14. Many substances are not mutagenic by themselves but are converted to mutagens by enzymes in the liver.

15. This mutant could not be isolated because there is no temperature at which it could grow.

16. The ratio of polymerizing activity to exonuclease function will decrease in the antimutator because errors will be removed more often.

17. They clearly interact. Since a change in the sign of the charge yields a mutant and a second sign-change in another amino acid yields a revertant, amino acids 28 and 76 are probably held together by an ionic bond.

18. $2000 \times (1 \times 10^{-5}) = 0.02$ mutation per cell per generation.

19. $\mu = -\ln (5/12)/(5 \times 10^8) = 1.8 \times 10^{-9}$.

20. A: $-\ln (22/40)/(5.6 \times 10^8) = 1.1 \times 10^{-9}$. B: $-\ln (15/37)/(5 \times 10^8) = 1.8 \times 10^{-9}$.

21. Ease is, of course, often a matter of opinion, but in this case there is little doubt that the reverse mutation (pro^- to pro^+) is the more easily detected. Simply by placing bacteria or other microorganisms on two types of solid medium, one with proline and the other lacking proline, rare pro^+ cells can be detected by their ability to grow on both media.

22. Proline determines the way the polypeptide backbone folds, so any change from proline (change a) usually yields a mutant phenotype. Arginine to lysine (change b) is probably ineffective because both have the same charge and nearly the same size. Substitution of threonine, which is weakly charged, by isoleucine, which is quite hydrophobic (change c), would very likely cause mutation. Valine to isoleucine (change d), both of which are nonpolar and have nearly the same size, would probably have little or no effect. Glycine to alanine (change e), both of which are nonpolar, would have little effect unless there were insufficient space for the slightly larger alanine side chain. Histidine to tyrosine (change f) probably would cause a phenotypic change because of the differences in charge and lack of flexibility, though in many proteins the change might be ineffective. Glycine to phenylalanine (change g) would probably have an effect because phenylalanine is so bulky that it is likely to cause a change in shape of the protein. For any of the changes mentioned, it is always possible that the shape change is in a part of the molecule such as an exposed loop on the surface that is fairly tolerant of changes.

23. Amino acid substitutions at many positions in a protein have little or no effect on its activity; only certain substitutions occurring at critical positions have a detectable effect.

24. It will be incorporated into DNA opposite a T. In a later round of replication it will pair mostly with T but occasionally with C, leading to a GC pair, at the site of an AT pair. Therefore, the change is an AT to GC transition.

25. The site of the removed guanine serves as a template for an A; thus, the change is GC to TA, a transversion.

26. Yes, because nitrous acid causes both AT to GC and GC to AT transitions.

27. (a) A double point mutation. (b) No. Frequencies give no information about the type of base change.

28. Although a second-site mutation can restore activity, the new interaction that replaces the original intramolecular interaction destroyed by the mutation may not be as strong as the wild-type interaction and hence may be more easily disrupted by thermal motion.

29. The mutation frequency in the Lac$^-$ culture is 9×10^{-9}. The probability of production of a suppressor is independent of the presence of a suppressor-sensitive mutation (which the *lac*$^-$ mutation is), so the mutation rate in the Lac$^+$ cell is also 9×10^{-9}.

Chapter 11

1. Not really. The largest plasmids have a low copy number but the small plasmids can exist in any number.

2. θ replication.

3. Yes, during transfer.

4. In both. Transfer replication occurs in the donor and conversion of the transferred single strand to double-stranded DNA occurs in the recipient.

5. A nick at the transfer origin to generate a 3'-OH group.

6. The donor

7. The genes reside in the plasmid. Thus, F pili are made by pili genes in F, and R pili are made by genes in an R factor.

8. (1) They are involved in recognition of donor and recipient; and (2) the donor pilus attaches to the recipient cell, retracts and brings the two cells in close contact.

9. The genes required for transfer—for example, pili genes and replication genes.

10. A conjugative plasmid is able to form a donor-recipient pair. A mobilizable plasmid carries genes needed for DNA transfer. A self-transmissible plasmid is both conjugative and mobilizable.

11. All three terms describe F. ColE1 is only conjugative.

12. It cannot mediate effective contact (formation of donor-recipient pairs).

13. The origin from the high-copy-number plasmid. Hence, the copy number will be high.

14. They must have the same, or at least similar, inhibitors of replication initiation.

15. F contains DNA sequences that are homologous to sequences in the chromosome. Recombination between these sequences gives rise to insertion.

16. No. ColE1 is nonconjugative, so the cell could not transfer its DNA unless F were also present.

17. RTF, which carries the replication and transfer genes, and the r determinant, which carries the gene for antibiotic resistance.

18. Yes, The simplest plasmid would be a tiny DNA fragment carrying no information other than a copy of the replication origin of the host cell.

19. The plasmid has a temperature-sensitive mutation in some replication function.

20. Since an RNA primer is not needed to initiate rolling circle replication, a reasonable guess would be that transcription of an inducible gene is needed for transfer.

21. Synthesis in the donor provides the single strand that is copied. Synthesis in the recipient converts the transferred strand to double-stranded DNA.

22. (a) Lac$^+$. (b) Yes. (c) *Flac* has integrated into the chromosome. The chromosome is now replicating from the origin of F and hence does not need the DnaA protein. (d) No.

23. (a) Lac⁻. (b) F(Ts) *lac*⁺ has integrated into the bacterial chromosome. (c) Integration has occurred within a *gal* gene.

24. 5.

25. Early in the development of the colony a cell has divided to yield a daughter cell lacking F' *lac*.

26. No. The donor cannot make the enzyme because of the streptomycin. The females also cannot, because they are *lac*⁻. Cells that have received the plasmid also are unable to make the enzyme, because without functional primase, the transferred circle (transferred by looped rolling circle replication) will not be converted to double-stranded DNA. Hence, the *lac* gene of the transferred plasmid will not be transcribed, and enzyme will not be made.

Chapter 12

1. Terminal inverted repeats, and a gene encoding a transposase.

2. A unique sequence flanked by a sequence in direct or inverted repeat (the termini of the transposable element), the entire three-component unit flanked by a short sequence in direct repeat (the target sequence).

3. No. Each target sequence will be different.

4. The number of nucleotide pairs.

5. 5 to 10 nucleotide pairs.

6. A transposable element consisting of a unique sequence flanked by two copies of a simple transposon. The simple transposons can be in an inverted or direct-repeat array.

7. Although the termini of simple transposons can be in direct repeat, these transposons have inverted-repeat termini, so there will always be inverted terminal sequences.

8. No.

9. Many of the IS elements contain transcription termination sites. Hence if they are downstream from a promoter, whether within a gene or between genes of a polycistronic region, they will cause premature termination and hence loss of expression of downstream genes.

10. When transposition occurs, a copy of the transposable element appears at a new position without loss of the original element.

11. The transposon itself and the target sequence.

12. Yes, since within a given DNA segment, for example, the *E. coli lac* operon, certain sites are occupied more often than others. It is likely that transposition can occur to any target sequence, yet certain sequences are more likely than others.

13. The fused plasmid is a cointegrate; it contains two copies of the transposon.

14. The transposase.

15. The enzyme is a resolvase. There are two copies of the transposable element, as is always the case with transposition in bacteria.

16. Replicon, fusion, deletion, inversion, transposition, and mutation.

17. The total amount of DNA always increases when transposition occurs, because of the duplication of the target sequence. With most transposons the transposon itself is also duplicated.

18. An effect of sequence on the probability of the first breakage event seems most likely.

19. (1) If it produces within a gene an insertion that is not a multiple of three base pairs, it can eliminate a natural stop codon. (2) If it is quite long and inserts between two cistrons in a polycistronic mRNA, the probability of initiation of translation of downstream cistrons can be reduced. (3) It can contain an in-phase stop codon quite far from the start codon of a subsequent cistron.

20. Heat denaturation followed by quick cooling and electron microscopic viewing might show a stem-and-loop structure. This would be suggestive but by no means would it be proof. Only base sequencing can answer the question, because if a transposon is present in the sequence, a target sequence must be duplicated. If a short duplication is seen, this is again only suggestive. If a duplication is not contained in the sequence, one can say with certainty that no transposon (of the types known) is present.

21. (a) Integration of the plasmid into the chromosome by homologous recombination; reversion to temperature-insensitivity; transposition of the Amp^r gene from the plasmid to the chromosome. (b) Integration (if homology-dependent).

22. You could start by forming heteroduplexes with a λ DNA molecule lacking the insertion. Each insertion will produce a single-stranded loop. If the loops are not the same size, the transposons must be different. If the loops are the same size, the two transposons are probably the same. A better test would be to move one of the transposons (A) to a nonhomologous DNA molecule. Then the DNA containing transposon A can be renatured with DNA containing transposon B. Only if the transposons are the same will heteroduplexes form. Radioactive RNA could also be obtained by in vitro transcription of the DNA containing A, and this RNA could be used in a hybridization experiment with the DNA containing B. Successful hybridization would indicate that the transposons are the same.

23. (a) The IS elements are inserted in two different orientations, and there is a transcription stop sequence in only one orientation. (b) Heteroduplexes between the DNA of the two phages would indicate that the transposons are the same or different.

24. (a) A high frequency of adjacent mutations suggests that a deletion including these genes has occurred. The deletion frequency is sufficiently high that one might reasonably suspect that a transposon is adjacent to the genes and that the phenomenon is an example of transposon-mediated deletion. (b) Since many mutations result from SOS repair, which is absent in $recA^-$ cells, and since the transposition is independent of RecA function, study of the phenomenon in a $recA^-$ cell would be useful. If the deletion frequency were the same in Rec$^+$ and Rec$^-$ cells, a transposon would be implicated. (c) Most mutagens would increase the frequency of chk^- mutations with little or no effect on deletion frequency, if transposition were involved. There might be an effect if the site represented a mutational hot spot and the phenotype were a result of a pair of point mutations.

25. (1) Precise excision of the transposon; (2) a deletion that either just removes part or all of the Tet gene or the transposon plus adjacent DNA; (3) a mutation in the Tet gene.

26. No. The phage DNA sequences are inserted but the terminal bacterial DNA sequences are not.

27. No, because transposition occurs throughout the life cycle.

28. Certainly some should exist, but at exceedingly low frequency because the bacterial sequences at the termini are different. For example, suppose the termini, which include a few thousand base pairs, never contain overlapping bacterial sequences; then, the number of different termini would be about 100. Hence, only 1 molecule in $(100)^2$, or 1 in 10^4, would have unfrayed ends.

29. No. First of all, most bacterial genes are normally turned off, and most bacterial gene products are not needed by Mu. In addition, if a gene product were needed by Mu, it is unlikely that it would have to be induced by the infection. Hence, the gene product would exist in the cell at the time of infection.

Chapter 13

1. Progeny of a transformed cell all had the new character; in genetic terms, the cells bred true.

2. The ability of a cell extract to transform was not reduced by treatment of the extract with enzymes that depolymerize proteins or RNA.

3. (a) The solution containing 10^{-7} mg/ml DNA would contain 0.02% of 10^{-7} mg = 2×10^{-14} g protein. The molecular weight of a protein containing 300 amino acids is 300 times the average molecular weight of an amino acid, or approximately $300 \times 100 = 3 \times 10^4$. Thus, the number of protein molecules is $(6 \times 10^{23})(2 \times 10^{-14})/(3 \times 10^4) = 4 \times 10^5$. (b) The numbers do not exclude the possibility that the transformation is protein-mediated, since the maximum estimate of the number of protein molecules (4×10^5) exceeds the number of transformants 400-fold.

4. Isolate DNA and centrifuge it in a sucrose gradient. Fractionate the centrifuge tube and test each sample for transformability using a very low DNA concentration. Try to find two genetic markers that cotransform, that is, two for which the recipient bacterium can acquire both markers even when the DNA is very dilute. You will observe that the genetic element that carries linked markers always sediments more rapidly than the bulk of the elements that carry only one marker. If this can be shown for several pairs of markers, you will have shown that transformability sediments with DNA and that cotransformability requires DNA of higher molecular weight. This should be pretty convincing. You could then also break the DNA molecules and show that as the average molecular weight of the DNA decreases, there is no loss of transformability of individual markers but that linkage is lost.

5. Show that transformability as a function of pH follows curves for DNA denaturation rather than protein denaturation. Show that transformation occurs at an ionic strength (e.g., 0.3 M) at which DNA-protein interactions are very weak. Separate protein and DNA by gel electrophoresis or gel chromatography, and test purified fractions for transformability.

6. Isolate the DNA and centrifuge to equilibrium in buoyant CsCl. Since DNA and protein have different densities, they will come to equilibrium at different positions in the gradient. Fractionate the centrifuge tube. The fractions containing DNA only or protein only can be tested separately for transforming activity.

7. White (nontransformant, or a cell in which the lac^+ DNA was integrated and in which there was a conversion from + to −), red (one lac^+ strand integrated and a conversion from − to +), and sectored (one lac^+ strand integrated and no conversion).

8. Mix the bacteria in the presence of a large amount of DNase. If recombination resulted from transformation, all transformation would be eliminated, because the DNase would destroy all of the free DNA in the culture. Conjugation would be completely resistant to the enzyme.

9. The first selection says that B is nearer to A than C is to A. Thus, the order is either C A B or A B C. The second selection says that B and C can cotransform. If C A B were the correct order, then cotransformation of B and C would mean that A and C should also cotransform. The latter is not the case, so the order must be A B C.

10. A 10-fold dilution of the DNA decreased the number of A^+ transformants by a factor of 10, as expected. However, the cotransformants decreased considerably more. Thus, A and B are not linked.

11. In the $lacZ1/lac^+$ heterozygous region either no mismatch repair occurs or correction tends toward the + genotype. In contrast, mismatch repair usually corrects a $lacZ2/lac^+$ mismatch to the $lacZ2$ sequence.

12. Since one strand of every DNA fragment is digested by nucleotides, only half of the DNA (5000 cpm) will be associated with the cells.

13. From the first part of the experiment one can obtain the following cotransformation frequencies: A^+B^-, (12 + 3)/250 = 0.06; A^+B^+, (12 + 100)/250 = 0.45. The second part of the experiment yields the frequencies B^-C^+, (17 + 13)/250 = 0.12; A^+C^+, (17 + 85)/250 = 0.41. Therefore, A and C are near one another, and B is nearer to C than B is to A. Furthermore, B cannot be between A and C, because A is nearer to C than B is to C. Thus, the gene order is A C B.

Chapter 14

1. In an Hfr cell F is integrated into the chromosome.

2. It prevents the Hfr from producing a colony.

3. An Hfr cell can transfer more DNA, but usually only a fragment of the chromosome is transferred. An F^+ or F cell generally transfers the intact plasmid.

4. The order is $p\,f\,q$ and the map distances are origin-7-p-4-f-8-q.

5. Aberrant excision of F from an Hfr cell, such that adjacent bacterial genes are contained in the circular plasmid.

6. (a) Met$^-$ is the counterselective marker, and prevents growth of the Hfr cells. (b) his, leu, trp, in that order. (c) An interrupted mating, that is, a time-of-entry experiment.

7. (a) True, because a enters before c. (b) False, because b enters before c. (c) True,

because some exchanges will be between a and b. (d) False, see (c). (e) True, because b is between a and c. (f) True, because the a^- and b^- markers are near, and on, the same DNA molecule. (g) True, because b enters after a.

8. The larger the fraction the nearer the particular *gal* mutation is to the *bio* locus. Thus, the gene order is *galA galB galC bio.*

9. Between c and d, because x is closely linked to both genes.

10. a, 10; b, 15; c, 20; d, 30. Gene d is probably very near the *str* locus.

11. The arg^- mutations have different map locations and must be in different genes; one is transferred before *met*; the other is transferred after *met*.

12. *pro* is a terminal marker. F is closely linked to *pro* (that is, all *pro* recombinants are donors) and hence F $^+$.

13. DNA-binding, proteolysis.

14. Presynaptic binding and conjunction. Drifting could occur, but homologous alignment would never be achieved.

15. No. Base-pairing occurs in the next stage.

16. DNA-binding is necessary for conjunction. Helicase activity is necessary for strand assimilation.

17. No.

18. F'f, F'ef, ... , by the technique of looking for early transfer of a later marker. F'g could be obtained if a *recA* recipient were used.

19. Genes h and *tet* must be very near one another, so selection for the h marker of the Hfr tends to select also for the *tet* marker of the Hfr.

20. All receive the marker, but the positions of the exchange determine whether the marker is present in the recombinant.

21. V13 contains an XP1 prophage between e and f. When the prophage is transferred to the recipient, no repressor is present and phage induction occurs. Thus, recombinants can form only if mating is interrupted before gene f enters the recipient. Strain S139 probably also contains an XP1 prophage, so the repressor contained in the recipient strain prevents induction.

Chapter 15

1. In a simultaneous infection the T4 nuclease destroys the T7 DNA.

2. To prevent incorporation of cytosine into T4 DNA.

3. The T4-induced nucleases will destroy all newly synthesized phage DNA molecules since, in cases a and b, glucosylation cannot occur.

4. Modify *E. coli* RNA polymerase (1) to prevent recognition of promoters that have been used and are no longer needed and (2) to permit recognition of the next promoter that is needed.

5. Yes. If the headful rule is followed, the deletion will produce terminal redundancy.

6. The early transcripts always contain genes for DNA replication and genes that (in varied ways) regulate transcription. Late mRNA generally encodes the structural protein genes and the lysis system. T7 is an exception in that the replication and structural protein genes are initially on a single transcript.

7. The r^+ phage will lyse both cells and produce a clear plaque. The *rII* phage will lyse only the B cells and hence will produce a turbid plaque.

8. Recall that only r^+ recombinants plate on K12(λ). Thus, since recombination is reciprocal in mass lysates, the number of recombinants is twice the number of r^+ phage, or 52. Thus, the recombination frequency is 52/482, or 10.8%.

9. (a) Cross 4 shows that A and D overlap, and 6 says that B and D overlap. Cross 1 indicates that A and B do not overlap. Thus, these three deletions have the order A D B. Cross 5 demonstrates that B and C overlap, and 3 shows that C and D do not. Thus, B, C, and D have the order D B C. The complete order is A D B C, with overlap between each pair of adjacent deletions. (b) Between the termini of A and B.

10. The structural proteins from the late transcripts are needed in stoichiometric amounts. The early proteins are mostly enzymes needed in tiny amounts. Since both early and late mRNA molecules have nearly the same half-lives, more copies of late mRNA are needed than early mRNA.

11. *B A D*.

12. Between *B* and *D* and covered by *A*.

13. It is either a double mutant or a deletion.

14. Since there are no plaques in the background, the *rIIA* mutation is a deletion. Mutant 1 complements and hence is in the rIIB cistron. Mutant 2 recombines and hence contains an *rIIA* mutation that is not covered by the tester *rIIA* deletion.

15. The deletion spans the junction of the rIIA and rIIB cistrons, removing a portion of each cistron.

16. In the wild-type DNA the terminal redundancy is two genes long. Two genes are deleted, so the terminal redundancy should increase to four genes in the deletion— that is, set 2.

17. Use the Poisson distribution. The fraction of cells infected by two or more phage = $1 - P(0) - P(1) = 0.8$. Therefore 80% of 10^8 bacteria (8×10^7 bacteria) are infected by two or more phage.

18. There are several approaches, each of which involves isolating phage mutants that eliminate the activity of X. It is also necessary to have an in vitro assay for X; this might be enzymatic or immunological or might require isolation of the protein. First, consider a heat-sensitive mutant. With this, it can be shown that at high temperature, enzymatic activity or perhaps immunological reactivity (not as good a test) is substantially reduced. Second, consider a chain-termination mutation. Here it can be shown by sedimentation or by gel electrophoresis that the protein has a lower molecular weight than the wild-type protein. Note the greater power of the second approach. With the heat-sensitive mutant, one merely correlates a property in vitro with that in vivo, but with a chain-termination mutant one observes not only the absence of the wild-type protein but the appearance of a new and smaller molecule.

19. The best way is to use a temperature-sensitive mutant and raise the temperature of the infected cell at various times. If the gene product is needed only in the interval between T_1 and T_2, then raising the temperature later than T_2 will not affect phage development, and a period of heating at any time between zero time and $T = T_1$, followed by cooling, will usually not affect phage development.

Chapter 16

1. T4, cut to fill a head; λ, cut at *cos* sites.

2. L1, R1, R2, and R4.

3. No late mRNA is made. Therefore, there are no structural proteins, DNA concatemers are not cleaved, and the cells do not lyse.

4. (a) Messenger RNA terminates at the first leftward terminator tL1 of λ, and N protein is needed to prevent termination. This conclusion is correct as long as the *trp* operon lacks a promoter. (b) Yes. If part of gene *N* is deleted, tL1 must also be deleted and the mRNA from the main leftward promoter will extend through the *trp* operon.

5. For the dimer, after the first DNA molecule is cut out and packaged, the remaining DNA has no cos site and hence cannot be packaged. Thus, only one particle can be formed from a dimeric circle. By the same reasoning, two particles are packaged from a trimeric circle.

6. Recombination produces a circular dimer in some of the infected cells.

7. No, because T4 can be transcribed and the T4-encoded nucleases will destroy the λ DNA.

8. Probably a mutation in the termination sequence tR2. This mutation allows Q protein to be made without N protein. The plaques are small because in the absence of an active N protein, the *O* and *P* genes are not transcribed at a normal level.

9. λ cannot be plated on the *dnaA* bacteria at a nonpermissive temperature because the gene is essential for bacterial multiplication; hence, at high temperature there will

be no lawn. The method is to adsorb λ to the *dnaA* mutant at 42°C at an MOI < 1 and plate the infected cells on a wild-type indicator at 42°C. If DnaA protein is not needed, each infected cell will produce a plaque. If it is needed, no plaques will result. In fact, DnaA protein is not needed.

10. The mutant probably has a very leaky mutation in gene *N* in the region of the N protein that interacts with RNA polymerase.

11. The mutant has a chain-termination mutation. The amino acid put in by either supE or supD cells does not create a functional protein, but the amino acid inserted at the mutant site by supF cells does yield an active protein.

12. The branch grows in length until the Ter system becomes active. At this point, λ units are cleaved from the branch at roughly the same rate as the growth of the branch. This accounts for the fairly constant length of the branch. Cleavage cannot occur unless a head is assembled. Since E protein is the major head protein, the Ter system is inactive with E⁻ mutants. Thus, the branch will continue to increase in length.

13. In the absence of N protein, the bacteria-killing functions (such as lysozyme production) are not activated. Thus, the cells survive. However, the fact that progeny cells receive a λ means that λ DNA must be able to replicate somewhat in the absence of N. We know that N stimulates transcription of the *OP* region by antiterminating at tR1. If replication occurs without N, the terminator tR1 must be a weak one, allowing a small amount of transcription of the *O* and *P* genes. A small amount of transcription may lead to synthesis of just enough O and P protein for one round of DNA replication per cell generation, which would enable the λ plasmid to be maintained from one cell generation to the next. One would expect fairly frequent segregation of λ-free cells, which is indeed the case.

Chapter 17

1. λ*imm434* and λ*imm21*.

2. λ*immλ*, λ*imm21*.

3. Generally their repressors and repressor-binding sites (operators) must be the same, though it is sufficient if the repressor of each phage can recognize the operator of the other.

4. A phage that can plate on a homoimmune lysogen.

5. Both make clear plaques.

6. No difference.

7. At the normal bacterial attachment site for λ, since the hybrid phage has λ att.

8. Bacterial, BOB; phage, POP; *gal*, BOP; *bio*, POB.

9. *D E F A B C.*

10. Integration—integrase and the bacterial integration host factor; excision—the phage Int and Xis proteins.

11. pRE is activated by the cII protein; pRM is activated by the cI protein.

12. λ*gal* site = attL (BOP), λ*bio* site = attR (POB).

13. Mutations in the *cI, cII, cIII,* and *int* genes would prevent lysogenization. Since an active N product is needed to produce the cII and cIII products in appreciable quantities, N mutants will lysogenize quite poorly. Mutations in *att* will also prevent lysogenization. *cro* mutants also have a greatly reduced lysogenization frequency, but the reasons are fairly complex.

14. No. Operator mutants are *cis*-dominant so they cannot be complemented.

15. The left. If it inserted in the right, the order would be *gal* BOP *R A* POP *R A gal* BOB *bio*.

16. (a) Repression or establishment of lysogeny. (b) Complementation between two proteins was needed to establish repression.

17. Only *N⁻*, because it fails to make the cII protein.

18. No, because the *bio* substitution eliminates the *int* gene. Int-promoted recombination is the only possible recombination mechanism that could carry out the exchange if the other recombination systems were absent.

19. Phage production is in each case a result of prophage induction. Induction by ultraviolet light cannot occur in a recA⁻ cell, because the RecA product facilitates for cleavage of the cI repressor. Induction occurs with a temperature-sensitive repressor in both mutant and wild-type since it is the high temperature that is inactivating the repressor.

20. If enough λ DNA molecules enter a bacterium, the intracellular repressor will be titrated.

21. (a) No, because in the absence of active repressor the pRM promoter is inactive. (b) Yes, because the renatured repressor activates pRM. (c) After 10 minutes enough DNA replication has occurred without repressor synthesis that there is sufficient repressor to bind all of the operators. After 5 minutes repressor is still sufficient.

22. The genotype will be *cI857 P⁺ A⁺*, because the only mode of phage production is Ter-mediated excision at the two cos sites.

23. The phage has a chain-termination mutation in one of the genes *cI*, *cII*, or *cIII*, and strain B (but not A) has a nonsense suppressor.

24. (a) The frequency with which a lysogen for a wild-type phage is spontaneously induced is relatively low, and the phage released form turbid plaques. However, if a mutation has occurred in the gene making the *cI* repressor, a lysogen containing this mutant (which will form a clear plaque) will always produce phage. (b) The gene cI makes repressor. Genes *cII* and *cIII* are needed in the wild-type state for the establishment of lysogeny, but not for its maintenance. Thus, a *cII⁻* or *cIII⁻* mutation in a prophage does not result in escape from repression.

25. Using the integrase of λ*b2* (λ*bio* has no *int* gene), an Int-promoted recombinational event has occurred yielding a dimeric circle containing DOB and POP, namely, ΔOP × POP → ΔOB + POP, and the dimer has integrated using the newly generated POP site.

26. (a) Prophage DNA can replicate, even though excision does not occur. (b) The replication forks can leave the prophage and enter the bacterial DNA region. (c) Excision must be fairly slow, because often the replication forks leave the prophage before excision occurs.

27. Integration into the right attachment site would require an exchange between BOP and POB, which is the reaction that requires the Xis protein. In an infection, the Xis protein is not made to any great extent.

28. A phage-resistant mutant is usually formed by plating about 10⁸ phage on a lawn of about the same number of bacteria and picking surviving colonies. This will not work with a λ lysogen because all cells will survive. However, if λ*imm*434 were used, only adsorption-defective bacteria would form colonies.

29. No λ*ind⁻* is dominant preventing UV induction.

Chapter 18

1. None, because specialized transducing particles are produced only by lysogens.

2. Specialized only: integrates into the chromosome and either does not cause host DNA fragmentation during phage development or cannot package host DNA fragments. Generalized only: fragments host DNA, can package host DNA fragments, and probably do not integrate. Both: integrates, fragments host chromosome, and can package host fragments.

3. They are formed only by aberrant excision of a prophage. In a lytic infection λ DNA is not inserted in the *E. coli* chromosome.

4. (1) Their excision system is perfect (unlikely). (2) When aberrant excision occurs the fragment size is too large or too small to be packaged. (3) No known host genes are near enough to the prophage to be picked up.

5. At each end of the prophage map essential genes are present that are absent in both types of transducing particles. Thus, lytic growth is only possible when a helper phage providing the missing functions is present. In single infection, transduction occurs by substitution because neither the right nor left prophage attachment sites can recombine with the bacterial attachment site. In mixed infection, either the transducing DNA and the helper phage DNA recombine or the attachment sites of the transducing DNA can recombine with a newly generated prophage attachment site.

6. If obtained by infecting a λ lysogen; the phage DNA would then be no different from any other bacterial DNA.

7. The DNA in the transducing particles was replicated before infection. Also, either there is no replication of bacterial DNA after infection or replicated DNA never gets into transducing particles.

8. Three orders are possible in such a cross: *A B C* (I), *B A C* (II), and *A C B* (III). Drawing the two crosses for each of the orders shows that order II can be eliminated because a quadruple exchange would be needed in the first cross and a double exchange in the reciprocal cross, which would not yield the observed equal frequencies. Double exchanges yield wild-type transductants with both orders I and III in both crosses. Thus, orders I and III cannot be distinguished by the data. Selection of another recombinant class would yield the order.

9. Crosses 1 and 3 are reciprocal crosses. With order *ant trpA34 trpA213* cross 3 would have more cotransductants of Ant$^+$ with Trp$^+$ than would the order *ant trpA223 trpA34*. The data then yield the order *ant trpA34 trpA233*. Crosses 2 and 4 are also reciprocal crosses. The same reasoning yields the order *ant trpA223 trpA46*. Thus, the overall order is *ant trpA34 trpA223 trpA46*.

10. (a) Drawing out the three orders shows that only the order *pyrF trpA trpC* yields the rarest recombinant, *trpC$^+$ pyrF$^-$ trpA$^+$*, as the product of a quadruple exchange. (b) The cotransduction frequencies are: *pyrF-trpC*, (548 + 3)/1220 – 0.452; *pyrF-trpA*, 548/1220 = 0.443; *trpA-trpC*, (548 + 579)/1220 = 0.92. (c) Use the Wu formula. The values are: *pyrF-trpA*, 0.466 minutes; *trpA-trpC*, 0.055 minutes; pyrF-trpC, 0.476 minutes. (d) The map appears to be *pyrF*-0.466-*trpA*-0.055-*trpC*, with *pyrF-trpC* also being 0.476 minutes. Note that 0.466 + 0.055 does not equal 0.476. The value of 0.055 is accurate since it is determined from a large number of colonies. However, the difference between 0.466 and 0.476 represents only three colonies, which is not statistically significant.

Chapter 19

1. Mate Hfr gal lac$^+$ × F$^-$trp$^-$lacStr^r for 25 minutes (longer may allow the trp$^+$ allele to be transferred. Select Lac$^+$Str^r colonies on lactose-EMB agar and test these for the Gal$^-$ phenotype by replica-plating onto galactose-EMB (or tetrazolium) plates. When completed, be sure to check that the strain is trp$^-$ by plating on minimal plates without tryptophan.

2. First, mate an Hfr lac$^+$ Str^s strain with the lac$^-$ Str^s strain and select a Lac$^+$Str^r strain on lactose-EMB plates containing streptomycin. Then, mate the Hfr Δlac with this new recipient. Since the deletion is transferred early, no special selection is needed. Plating the cells on lactose-EMB plates with streptomycin will be sufficient in that between 1% and 10% of the colonies will be Lac$^-$ and each will carry the deletion.

3. Lysogenize the *gal$^+$* strain with λcI857. Grow the lysogens at 42°C on galactose-tetrazolium plates. Pick survivors (which must have lost at least the repressor gene) that are Gal$^-$. These will be deletions that extend from the gal locus through the prophage at least past the *cI* gene. Test the Gal$^-$ survivors for the bio gene by plating on medium lacking biotin. Gal$^-$Bio$^-$colonies will have a deletion that extends from the *gal* locus to the *bio* locus and includes *att*λ.

4. As in Problem 3, lysogenize the *gal$^+$* strain with λcI857. Grow the lysogens at 42°C on galactose-tetrazolium plates. Pick survivors (which must have lost at least the repressor gene) that are Gal$^-$. These will be deletions that extend from the *gal* locus through the prophage at least past the *cI* gene. Test the Gal$^-$ survivors for UV sensitivity. These will be deletions that at least enter the *uvrB* gene. An alternative procedure, which does not use λ, is the following: Plate the original strain on plates containing potassium chlorate. Colonies that grow will be chlorate resistant. There are several chlorate-resistance genes, but screening for Gal$^-$ on galactose-tetrazolium plates ensures that one of the *chl* genes near *uvrB* will be deleted. Test these for UV sensitivity. This second procedure is the fastest, because two days are saved, namely, the days required to lysogenize and test the lysogen. On the other hand, the chlorate test is inconvenient, as chlorate resistance is manifested only in the strict absence of molecular O$_2$, requiring facilities for anaerobic growth.

5. Plate the recipient with an excess of T6 and pick Tsx^r colonies. Recall that some of these contain deletions that extend into *pur*. Thus, select Pur$^-$ colonies by replica

plating. Mate the Hfr *lac⁻* strain with the new *pur⁻* recipient and plate on minimal plates lacking a purine. The *pur* and *lac* loci are sufficiently near that the Pur⁺ recombinants can be tested for the linked *lac⁻* mutation by replica-plating onto lactose-EMB plates.

6. First, cross T4 *ac41* and T4 *rII*, plate on strain B on plates containing an acridine, and select large plaques. These are *ac41 rII* recombinants. Test by plating on K12(λ). Then, prepare *E. coli* B/4 (T4-resistant cells) by plating B cells on a plate spread with T4, and picking survivors. Then, cross T4 *ac41 rII* and T4*h* and plate on B/4 on plates containing an acridine. Pick large plaques, which should be *ac41 rII h* recombinants. It is worthwhile to make a final check for the rII mutation by plating on K12(λ) to ensure that the combinations *ac41 h* or *ac r⁺ h* do not produce a large plaque.

7. Grow the *leu⁻* strain on medium containing chlorate and select Chl^r cells. Test these on galactose-tetrazolium plates to isolate Gal-Chl^r double mutants. Most of these will be *gal-chlD* deletions. Check these for UV sensitivity to find a deletion that extends from *gal* to *uvrB*. Plate this strain on a plate spread with 10^8 λ*cI⁻* mutants (*cI⁻* to avoid lysogenization) and select survivors, which are λ-resistant. Test these on maltose-tetrazolium plates to be sure that they are *lamB⁻*. Grow P1 on the uvrA- mutant, which is also *mal⁺*, and transduce the *leu⁻uvrB malB⁻* strain to Mal⁺ by selection on maltose-minimal plates containing leucine. Check Mal⁺ transductants for UV sensitivity. It is not obvious that the double mutant will be more sensitive than the single mutant. However, it is worth comparing the UV sensitivity of the Mal⁺ transductants to that of either *uvrA⁻* or *uvrB⁻* cells. If the UV sensitivity were greater than either of the single mutants, you could feel confident that you have the double mutant. The best test would be to cross a *uvrB⁺* allele into several Mal⁺ transductants (either by transduction or Hfr crosses, selecting for Gal⁺ or Bio⁺) and see whether the recombinants retain the sensitivity of a *uvrA⁻* cell. If so, the Mal⁺ transductant is also *uvrA⁻*.

8. First, make it *thy* Str^r using the trimethoprim selection. Then, make it Tsx^r by plating with an excess of phage T6 (*tsx* is on the opposite side of the map from both *thyA* and *str*, which are quite near one another). Take an Hfr that transfers thy fairly early in either of the orders thy-str or str-thy and mate it with the F⁻ cells. If the thy-str Hfr is used, keep the mating shorter than the time of entry of *leu*. Select against the Hfr with T6 and for the Thy⁺ trait by plating on leucine-minimal plates lacking thymine. Test Thy⁺ recombinants for streptomycin sensitivity.

9. Continued protein synthesis without cell division will cause the cells to elongate. In fact, one expects the cells to become twice as long in the first generation time, three times as long in the second generation time, and so forth. This does occur, and such filamentous bacteria are often called "snakes." In a culture containing a few *dna*(Ts) mutants only the mutants will form snakes at elevated temperature; the wild-type cells will grow and divide normally. The culture is then passed through a nitrocellulose filter with pores about 2 μm in diameter. Normal cells will pass through the filter, but mutants will be retained on the filter. The filter is then washed and the cells removed from it are plated at 30°C. Colonies are replica-plated onto fresh plates and tested for the ability to grow at 42°C. Some of the colonies are *dna*(Ts) mutants and some are Ts cell-division mutants. These are distinguished by measuring the ability of a liquid culture to incorporate radioactive thymidine at 42°C. Those that cannot are *dna*(Ts) mutants. This was the screening technique used in the 1960s in the original isolation of *dna*(Ts) mutants.

10. Prepare a Leu⁻ lysogen with λ*cI857*. Superinfect a culture of the lysogen with λ::Tn*10* and plate on medium containing ampicillin. Since λ::Tn10 cannot multiply in the lysogen, these cells will contain Tn*10* in the chromosome. Test these for Leu⁻ by replica-plating to a pair of minimal plates, one of which contains leucine. Isolate a Leu⁻ colony, prepare a liquid culture, and cure it of the prophage by growth at 42°C for 5 minutes, and several hours at 30°C. Plate at 42°C and pick survivors. These are Leu⁻ nonlysogens.

11. Infect an Ara⁺ strain containing a *lac* deletion with Mu(Ts) and plate at 30°C. Cross-streak colonies against Mu to select for Mu(Ts) lysogens. Replica-plate Mu(Ts) lysogens to arabinose-tetrazolium plates to screen for Ara⁻ cells (red colonies). These cells presumably contain a Mu(Ts) prophage in an ara gene. Lysogenize these cells with λp(*lacP⁻ lacZ⁺*, Mu) and test for λ lysogeny by the λ cross-streak test. Plate a λ lysogen at 42°C on lactose-EMB plates containing arabinose (an inducer of the arabinose operon) and select purple (Lac⁺) colonies. Replica-plate them to lactose-EMB plates lacking arabinose; all should be Lac⁻. Colonies that are Lac⁺ on (lactose-

arabinose)-EMB plates and Lac⁻ on lactose-EMB plates have the *ara* promoter fused to the *lacZ* gene.

12. Cross *bio-11 c⁺* with *cI857 Q⁻* and plate on a P2 lysogen at 42°C. Pick clear plaques and test for *Q⁻* by stabbing into *sup⁺* and *sup⁰* bacteria (*Q⁻*phage grow on *sup⁺* but not *sup⁰*). You now have *bio11 cI857 Q⁻*. Cross with *red⁻ c⁺* and plate on *recA sup⁺* cells at 42°C. Select clear plaques (*λcI857 red⁻*) and test for the *Q⁻* marker by stabbing into *sup⁺* and *sup⁰* bacteria.

13. Cross the two phages and plate on *recA sup⁺* cells at 42°C. Select clear plaques. Test for *cI857 A⁺ R⁺* double exchanges by stabbing into *sup⁰* and *sup⁺* bacteria. Discard those that grow on *sup⁰* cells. A single exchange between *A* and *bio-11* will give *cI857 A⁻ R⁻*. However, *A* and *R* are so far apart that multiple exchanges may occur. Plate *A⁻* and *R⁻* phage on *sup⁰* bacteria to determine the reversion frequency. Do the same with the clear mutant that failed to grow on *sup⁰* cells. Those that produce no revertants (i.e., frequency $< 10^{-10}$) are the double mutants.

14. Cross the phage and plate on *recA sup⁺* cells at 42°C, and pick clear plaques. The parent *λcI857 R⁻* and the recombinant *λcI857 A⁻ R⁻* will both grow but the *bio-11 c⁺ A⁻* parent will not. There will also be *λcI857 A⁺ R⁺* recombinants. Test all clear plaques by stabbing them into sup⁺ and sup⁰ bacteria and discard those that grow on *sup⁰* cells (the *A⁺ R⁺* recombinants). Since the recombination frequency is 10%, about 10% of the clear-plaque phage that fail to grow on *sup⁰* cells will be *A⁻ R⁻*. These can be identified by measuring the reversion frequency; if about 20 of the plaques are tested, a nonreverting double mutant should be found.

Chapter 20

1. Blunt ends, cohesive ends with 3' extensions, and cohesive ends with 5' extensions.
2. They are palindromes.
3. No.
4. Yes, note that the sequences have the form of Pu GCGC Py, which is the *Hae*II site.
5. They are found in most, if not all, bacterial species.
6. No. The recognition sites must share common sequences.
7. They must both recognize the same base sequence.
8. It can add nucleotides to an extended single-stranded 3' terminus of a DNA molecule without the need of a template.
9. Blunt-end ligation, addition of linkers, homopolymer tail-joining.
10. Prepare fragments by cleavage with two different restriction enzymes such that their termini are different; put different linkers on a blunt-ended molecule.
11. These are circularized fragments.
12. One gene can be used to detect the plasmid in a transformation experiment; and if there is a restriction site in the other gene, lack of resistance to that antibiotic can be used to show that insertion has occurred.
13. A particular fragment is isolated from a gel after electrophoretic separation, or a particular cDNA is used.
14. The transducing phage produced by the joining techniques must contain inserted DNA.
15. Both enzymes recognize the same base sequence. They might even be the same enzyme.
16. A mutation may alter one base in a restriction site and thereby cause two potential fragments to remain uncleaved.
17. The *A* gene probably contains an *Eco*RI site, so the gene is destroyed. It does not contain an *Hae*III site.
18. In principle, either could be used. If you used *Eco*RI, the plasmid could be detected by growth on medium using lactose as the sole carbon source. Colonies that grow could then be tested for antibiotic sensitivity to find those containing DNA inserted in the amp^r gene. Note that this process would take two steps, and, in fact, two days of growth. If you used *Sal*I, transformants with inserted DNA could be isolated in a

single step by plating transformed bacteria on a lactose color-indicator medium containing ampicillin.

19. (a) The tet ^{r} gene has not been cleaved, so addition of tetracycline to the medium will require that the colonies be Tetr and hence carry the plasmid. (b) Tet^r Kan^r and Tet^r Kan^s. (c) Tet^r Kan^s, because insertion will occur in the cleaved kan gene.

20. A frameshift of two bases is generated, so all colonies will be Lac$^-$.

21. Since annealing occurs at random, some clones will contain more than one copy of the *lacZ* gene and one copy of *lacY*, and others will contain more than one copy of *lacY* and one copy of *lacZ*. Furthermore, some will have the gene order *lacP lacZ lacY,* and others the gene order *lacP lacY lacZ*. Because of the polarity of synthesis of the gene products of a polycistronic mRNA, the former will make more *lacZ* and the latter will make more *lacY*.

22. Select an enzyme whose restriction site is further from the lac promoter than the *Bam*HI site. Let us assume it is *Hae*I. If the gene of interest does not contain a *Hae*I site, do the following. Cleave the plasmid with both *Bam*HI and *Hae*I and retain the fragment (there will be two) that contains the replication origin and other essential genes. One end of the fragment will have a *Bam*HI terminus, and the other end will have a *Hae*I terminus. Cleave the donor DNA with both enzymes and join the fragments to the isolated plasmid fragment. Since each fragment has one *Bam*HI terminus and one *Hae*I terminus, joining can occur only in a particular orientation.

23. f = $(2 \times 10^4)/(3 \times 10^9)$ = 6.67×10^{-6}. Therefore, N = ln (0.01)/ln $(1 - 6.67 \times 10^{-6})$ = 6.91×10^6.

24. The target sequence of enzyme 1 is part of the target sequence for enzyme 2. For example, enzyme 2 might cut in the sequence GATATC and enzyme 1 might act at ATAT, which is part of the target sequence for enzyme 2.

25. Most likely, the enzymatic digestion was not carried to completion and the extra fragment is a result of incomplete cutting at a particular site. The fragment moves slowly because its size is equal to the sum of two adjacent fragments.

26. There are three possibilities, of which the first two are quite unlikely: (1) A single cut is made precisely in the center of the molecule; (2) several cuts are made in positions such that all fragments have sizes that are indistinguishable by gel electrophoresis; (3) the molecule is a circle and a single cut is made.

27. Note that the 3.6-kb and 5.3-kb fragments are now joined to form the 8.9-kb fragment. This suggests that they are terminal fragments of a linear molecule and that the intracellular DNA is circular.

28. The plasmid is circular (that is, no free end), so all termini generated by the enzyme will have the same cohesive end.

Chapter 21

1. It has been inserted in the incorrect orientation. The gene may have been from its ribosome binding site.

2. The bacterium does not recognize the eukaryotic promoter. The primary transcript contains introns that are not properly removed in the bacterium. The protein may be degraded by proteases.

3. (a) The order is 1.0—6.4—4.1. The reasoning is the following. First, in the linear DNA the 6.4-kb fragment can be linked to either of the other fragments, when digestion with BamHI is partial; therefore, the 6.4-kb fragment must be in the middle. Second, in the circular DNA, where the cohesive ends are ligated together, the 1.0-kb and 4.1-kb fragments are joined, showing that they are at the ends. (b) The P4 phage attachment site in the phage DNA lies in either the 1.0-kb or 4.1-kb BamHI fragment and not in the 6.4-kb fragment, which is preserved after integration of prophage. The BamHI sites in the host DNA that are nearest the prophage are placed to give host-phage fragments of 15.0 and 12.5 kb.

4. III-IV-I-V-II.

5. The fragments have eliminated either the promoters or the ribosome binding sites of both genes. The plasmid contains both of these, but they are on the same strand of the plasmid DNA. If genes A and B are transcribed in opposite directions, only one can be expressed from the plasmid.

6. The labeled fragments indicate the termini, which are 6.2 and 8.0 for *Eco*RI, and 6.0 and 10.1 for *Bam*HI. Therefore, the *Bam*HI map is 6.0, 12.9, 10.1, in which we arbitrarily place the 6.0 fragment at the left. If the 6.2-kb terminus of the *Eco*RI map were at the left, a 0.2-kb fragment would be in the double digest, but no such fragment is present. If the 8.0-kb terminus were at the left, a 2.0-kb fragment would be present, as it is. Thus, the 8.0-kb fragment is at the left, and the 6.2-kb fragment is at the right. Now, consider the 4.5-kb fragment. If it were next to the 6.2-kb fragment, the double digest would have a 0.6-kb fragment, which is not present. Also, if it were next to the 8.0-kb fragment, the double digest would have a 6.5-kb fragment, which it does not. Therefore, the 4.5-kb fragment cannot be next to either the 6.2-kb or the 8.0-kb fragment and must be in the center of the molecule. Now, consider the 7.4-kb fragment. If it were next to the 8.0-kb fragment, there would be a 2.5-kb fragment, which there is. If it were instead next to the 6.2-kb fragment, there would be a 3.5-kb fragment, which is not present. Thus, the 7.4-kb and 6.2-kb fragments are adjacent. Analysis of the position of the 2.9-kb fragment shows that it is next to the 8.0-kb fragment, which agrees with the position of the 7.4-kb fragment. Therefore, if the *Bam*HI map has the order 6.0-12.9-10.1, the *Eco*RI map has the order 8.0-7.4-4.5-2.9-6.2, with the same orientation.

7. The restriction enzyme is impure and contains at least one nuclease that is not site specific.

8. No, because a source of fairly pure mRNA is needed. This is really only practical with particular mRNA molecules made in very large quantities by particular cells, such as the cells making hemoglobin. If the mRNA is a minor species, there might be no way to isolate or identify it. Normally, one would do this by hybridization to a DNA molecule. However, often there is no way to obtain the DNA molecule except by cloning, and the c-DNA method may be the only way to do the cloning.

9.

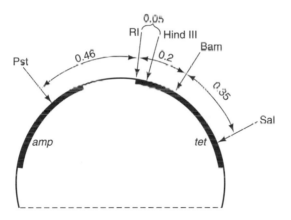

10.

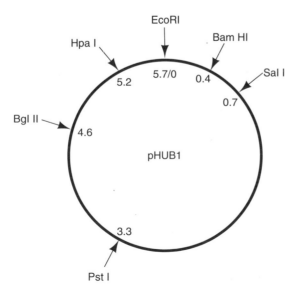

Index

473